Records of the Indian Museum

(A journal of Indian zoology)

(Volume XI) 1915

Unknown

Alpha Editions

This edition published in 2020

ISBN : 9789354019395

Design and Setting By
Alpha Editions
email - alphaedis@gmail.com

RECORDS

of the

INDIAN MUSEUM

(A JOURNAL OF INDIAN ZOOLOGY)

Vol. XI, 1915.

EDITED BY

THE SUPERINTENDENT

OF THE

INDIAN MUSEUM.

Calcutta:

PUBLISHED BY ORDER OF THE TRUSTEES OF THE INDIAN MUSEUM,

BAPTIST MISSION PRESS.

1915.

CONTENTS.

ERRATA.

Page 567, line 5. *For* "ployps" *read* "polyps."
,, ,, ,, 13. *For* "gemmatta" *read* "gemmata."
,, 568, ,, 1. *For* "Hillard" *read* "Billard."

CONTENTS.

—◇—

LIST OF PLATES.

LIST OF AUTHORS.

INDEX.

—:—

N.B.—An asterisk (*) preceding a line denotes a new variety or subspecies; a dagger (†) indicates a new species; a double dagger (‡), a new genus or sub-genus: synonyms are printed in italics.

A

B

C

D

E

I

J

K

L

M

N

O

P

R

S

T

U

V

W

X

Z

I. INDIAN BORING SPONGES OF THE FAMILY CLIONIDAE.

By N. ANNANDALE, *D.Sc., F.A.S.B., Superintendent, Indian Museum.*

(Plate i.)

Among the sponges found in excavations in shells and corals by far the best known are those of the family Clionidae. Having recently had occasion to inquire, in connection with other work, into the species that occur in Indian seas (that is to say, the Bay of Bengal with its subordinate gulfs and straits and the Arabian Sea, with which it is convenient to include the Persian Gulf and those parts of the Indian Ocean that lie immediately south and south-west of the Indian Peninsula), I found in our collection so large a proportion of the species known from Oriental waters—as well as several hitherto undescribed—that it seems worth while to bring together in a single paper references to all the former, with such notes as my material suggests, with keys to species and genera and descriptions of new forms.

The specimens examined have included a large part of the collection made by the late Dr. John Anderson in the Mergui Archipelago off the coast of Tenasserim, and described by the late Dr. H. J. Carter in Vol. XXI of the *Journal of the Linnean Society* (Zool). in 1887[1]; as well as examples of sponges extricated from shells and corals from various sources in the general collection of the Indian Museum and specimens specially collected in the Gulf of Manaar and Palk Straits and in lagoons on the east coast of India by Mr. S. W. Kemp, Mr. J. Hornell and myself. I have to thank Messrs. Kemp and Hornell for valuable assistance in this direction.

Fam. CLIONIDAE.

The taxonomy and systematic position of the Clionidae have been considered most fully by Topsent in his papers on the family in vols. V[2] and IX of the *Archives de Zoologie expérimental et général* (1887 and 1891) and I have little to add to the general conclusions there set forth. References to more recent literature are given below in connection with the different species discussed.

Six genera are now recognized by Topsent as constituting the family, namely *Cliona*, Grant, *Clionopsis*, Thiele, *Alectona*,

[1] This paper, with many others originally published in the same Journal, was re-issued by Anderson in 1889 in vol. i of his *Fauna of the Mergui Archipelago*.

Carter, *Thoosa*, Hancock, *Dotona*, Carter and *Cliothosa*, Topsent: but the last seems to me of doubtful validity.

Of these six genera all but *Clionopsis* are known to occur in Oriental waters. *Clionopsis*[1] is at present recorded only from the Pacific Coast of S. America and from an unknown locality probably in the Tropics. *Alectona*[2] and *Dotona*[3] both occur in the Gulf of Manaar, although I have not been so fortunate as to find examples of either. *Cliona* and *Thoosa* are well represented in the Indian marine fauna, while a specimen that would be assigned by Topsent to his genus *Cliothosa* has been found in a shell from the Andamans. I am not satisfied that this last "genus" represents more than a phase of certain species of *Thoosa* (see p. 22, *postea*).

KEY TO THE GENERA OF CLIONIDAE.

I. Microscleres essentially spirasters.
  - A. Macroscleres both amphioxi and styli (usually tylostyli), or either alone: if both present the amphioxi never the larger. Microscleres often variable and sometimes divisible into two groups but never of two quite distinct kinds .. .. *Cliona*.
  - B. Macroscleres amphioxi and tylostyli, of which the former are the larger. Microscleres slender, elongate, zig-zag spirasters and short, stout, irregularly contorted ones .. .. *Clionopsis*.
  - C. Macroscleres reduced to minute simple styli or amphioxi and confined to the external papillae. Microscleres relatively large spiral spirasters and minute straight ones of amphiaster-like form .. .. *Dotona*.

II. Microscleres essentially amphiasters.
  - A. Macroscleres, if present straight or regularly curved amphioxi or tylostyli, occurring in the internal galleries. Typical amphiasters consisting of a cylindrical stem bearing at or near both ends a ring of relatively large bosses and terminating in similar bosses.[4] Other forms

[1] Thiele, *Zool. Jahrb.*, suppl. VI, Vol. III, p. 412 (1905): Topsent, *Bull. Mus. Océanog. Monaco*, No. 120 (1908).

[2] The fullest description, illustrated by numerous figures, is that given by Topsent in his "Étude monographique des Spongiaires de France" (*Arch. Zool. expérim.* VIII, p. 24: 1900). The original description, by Carter, is in *Journ. Roy. Micro. Soc.* II, p. 493 (1879).

[3] Carter, *Ann. Mag. Nat. Hist.* (5) VI, p. 57 (1880): Topsent, "Spongiaires des Açores," *Res. Camp. Sci. Monaco*, XXV, p. 108 (1904).

[4] In *Thoosa laeviaster*, described on p. 22, both lateral and terminal bosses are reduced to smooth conical projections.

of microscleres present also, but never spiny diactinial spicules of relatively large size and of polyactinial origin .. *Thoosa.*

B. [Macroscleres normal tylostyli, occurring as in *Thoosa.* The only microscleres amphiasters consisting of a cylindrical stem bearing at the ends a circle of relatively long horizontal branches which are inflated at the tip or terminate in several minute hooks: the whole spicule smooth and slender .. .. .. *Cliothosa.*]

C. Macroscleres entirely absent; their place taken in the external papillae, but not in the galleries, by relatively large spiny or nodular diactinial spicules some of which reveal their polyactinial origin by being definitely bent or geniculate in the middle, or even by bearing extra rays, complete or rudimentary, in this position. Amphiasters like the typical ones of *Thoosa* but with the lateral bosses far removed from the extremities, which are not always capitate .. .. *Alectona.*

Genus Cliona, Grant.

1826. *Cliona*, Grant, *Edinb. Phil. Journ.* I, p. 78.
1849. ,, Hancock, *Ann. Mag. Sci. Nat.* (2) III, p. 305.
1888. ,, Topsent, *Arch. Zool. expérim.* (2) V², p. 76.
1891. ,, *Id.*, *ibid.*, IX, p. 556.
1900. ,, *Id.*, *ibid.*, (3) VIII, p. 32.
1900. *Dyscliona*, Kirkpatrick, *Ann. Mag. Nat. Hist.* (7) VI, p. 353.
1907. *Cliona*, Topsent, *Arch. Zool. expérim.* (4) VII, p. xvii.

Further references will be found in Topsent's papers, which are essential for a study of the Clionid genera and particularly for that of *Cliona*. In 1891 he arranged the species in six groups as follows:—

Group I. Spicules including tylostyles, diactinial macroscleres and spirasters (microscleres).

Group II. Spicules consisting of tylostyles and diactinial macroscleres only.

Group III. Spicules consisting of tylostyles and microscleres only.

Group IV. Spicules consisting of amphioxi and microscleres only.

Group V. Spicules consisting of tylostyles only.

Group VI. Spicules consisting of amphioxi only.

This grouping is convenient for the purposes of a provisional classification, which is all that is possible until the life-histories of the species are known: but it must be remembered that in at least one species (*Cliona celata*, Grant) phases occur in the life of an individual sponge that would fall respectively into groups I, II and V. The sponge in its younger stage possesses tylostyles, diactinial spicules and microscleres, but as it grows it loses first the diactinial spicules and then, sometimes, the microscleres, so that in its mature form it has only tylostyles. It is possible, and indeed probable, that other species resemble it in this respect, so that groups V and VI may actually consist of species whose earlier stages are unknown and if known would fall into other groups, or even in some cases of species known by other names and assigned to other groups at different phases of development.

Taking the groups as they stand, we find that among the species known from Indian seas all but group VI are represented. Group V, so far as hitherto described species are concerned, has not withstood recent criticism and research [1], but a new species belonging to it is described here on p. 14. In the following key to the species found in the Indian Ocean (including the Red Sea, the Bay of Bengal with its appurtenances and the western part of the Malay Archipelago) I have found it more convenient to make the primary division between species that possess and those that do not possess microscleres. Even so, it is necessary to include *C. celata* under three separate headings in accordance with its three phases of development.

Of the sixteen species now known from the Indian Ocean at least twelve have been found in the Bay of Bengal or the Gulf of Manaar. Of these, four are of very wide distribution (*C. celata*, *C. vastifica*, *C. carpenteri*, *C. viridis*): *C. carpenteri* is essentially a circumtropical sponge, but the other three are cosmopolitan. Three species have a wide distribution in the Indo-Pacific Region, namely *C. margaritiferae*, *C. mucronata* and *C. orientalis*; while five (*C. annulifera*, *C. indica*, *C. ensifera*, *C. acustella* and *C. warreni*) have been definitely recorded only from the Bay of Bengal and Ceylon. Of the four species not known from these seas, two were originally described, or are only known definitely, from the "Indian Ocean," namely *C. michelini* and *C. millepunctata*, but the original specimen of the latter was doubtfully ascribed to the N. Atlantic. One species (*C. mussae*) has been found only in the Red Sea, and one (*C. patera*) in the western part of the Malay Archipelago.

I have not included *C. gracilis*, Hancock, among the species known from the Indian Ocean, although Topsent (1887, p. 77) has done so; because the latter author's brief description of his specimen from that area ("Spicules en épingle=150μ de long, spic. en zigzag=15-20μ") is totally at variance with Hancock's original diagnosis, which is supported by good figures, and some mistake

[1] See Topsent, *Arch. Zool. expérim.* (3) VIII, p. 78 (1900).

in the identification must have occurred. The *Cliona ? sceptrellifera*, of Carter[1], if he rightly associated the isolated spicules on which it was based, is probably a *Thoosa* or an *Alectona*, but I have been unable to find these spicules in that part of his original material at my disposal.

The names of species on which notes are given are distinguished by an asterisk in the key. I have not seen the following forms :—

C. indica, Topsent, *Arch. Zool. expérim.* (2) IX, p. 574 (1891).
C. michelini, *id*, *ibid.*, vol. V², p. 79 (1887).
C. mussae (Keller), *Zeitschr. wiss. Zool.* III, p. 321 (1891).
C. warreni[2] Carter, *Ann. Mag. Nat. Hist.* (5) VII, p. 370 (1881).
C. millepunctata Hancock, *Ann. Mag. Nat. Hist.* (2) III, p. 341 (1849) ; Topsent, *op. cit.*, 1887, p. 78.

Cliona has a wide bathymetric range. In the Bay of Bengal one species has been found at a depth of over 700 fathoms[3] (*C. annulifera*, p. 9) and another (*C. vastifica*, p. 8) in lagoons of brackish water actually above sea-level. The genus is, however, best represented in comparatively shallow water below low tide. On beds of gregarious sedentary molluscs such as *Ostrea* or *Margaritifera* a single species usually predominates and becomes very abundant, but in the less vigorous parts of coral-reefs several are sometimes found together in a flourishing condition. More than one may also occur in a single shell, either Gastropod or Lamellibranch, that is of suitable size, thickness, etc., but does not belong to a markedly gregarious species.

KEY TO THE SPECIES OF *Cliona* KNOWN FROM THE INDIAN OCEAN.

I. **Species with microscleres.**
  A. *Macroscleres both diactinial and tylostyle.*
    1. Diactinial spicules smooth, hair-like, fasciculated .. .. *C. celata* (A) *
    2. Diactinial spicules granular, spindle-shaped, moderately stout, not fasciculated.
      a. Microscleres sinuate, truncate. *C. vastifica.**
      b. Microscleres straight, spindle-shaped .. .. *C. carpenteri.**
    3. Diactinial spicules cylindrical, irregularly spiny .. *C. margaritiferae.**

[1] *Fauna of Mergui I* (*Journ. Linn. Soc.* (Zool.) XXI : 1887), p. 76. "Spongiaires des Açores," *Rés. Camp. Sci. Monaco*, XXV, p. 108 (1904).
[2] Topsent (*Arch. Zool. expérim.* (3) VIII, p. 54) regards this species as identical with *C. quadrata*, Hancock.
[3] *C. abyssorum*, Carter was taken at the mouth of the English Channel in 500 fathoms (*Ann. Mag. Nat. Hist.* (4) XIV, p. 249, 1874). This is apparently the only other species as yet recorded from depths of like magnitude.

B. *All the macroscleres tylostyles.*

1. Shaft of macroscleres bearing a single convex ring a short distance below the head .. *C. annulifera.**
2. Shaft of tylostyles normally smooth.
   - a. Tylostyles definitely of two sorts; one sort normal, the other very short and bearing a sharp subsidiary spine at its point .. .. .. *C. mucronata.**
   - b. No "mucronate" spicules of this type.
     - i. Spines on all the microscleres very small and set close together; two groups of zigzag microscleres, one very slender .. *C. indica.*
     - ii. Spines on microscleres stout, very irregular, often blunt but never widely separated; microscleres not divisible into two groups .. *C. michelini.*
     - iii. Spines of microscleres relatively long, sharply pointed, normally arranged in a spiral band winding round the spicule.
       - α. Some of the macroscleres conspicuously but gradually expanded before narrowing to the point; hair-like tylostyles not present. *C. ensifera.**
       - β. None of the macroscleres of expanded form; hair-like tylostyles, sometimes with spiny heads, often present. *C. viridis.**
     - iv. Spines of microscleres as in iii, but arranged in a sinuous band outlining one side of the spicule .. .. *C. orientalis.**

C. *All the macroscleres amphioxi.*

(Microscleres short, straight, approaching the amphiaster type in different degrees) .. .. *C. acustella.**

II. Species without microscleres.

Macroscleres both diactinial and tylostyle.

1. Diactinial spicules hair-like, fasciculated .. .. .. *C. celata.* (B).*

2. Diactinial spicules moderately stout, (smooth), spindle-shaped. *C. mussae.*

B. *All the spicules tylostyles.*
1. Sponge forming a gigantic free cup; spicules relatively stout .. *C. patera.**
2. Sponge confined to its excavations or forming a small rounded mass; spicules relatively slender.
a. Head of spicules spherical .. *C. warreni.*
b. Head of spicule elliptical .. *C. millepunctata.*
c. Head of spicule usually trilobed. *C. celata* (C).*

Cliona celata, Grant.

1900. Topsent, *Arch. Zool. expérim.* (3) VIII, p. 32, pl. i, figs. 5, 6–9, pl. ii. fig. 1.
1909. Hentschel, "Tetraxonida" in Michaelsen and Hartmeyer's *Faun Südwest. Australiens*, p. 386.
1911. Row, *Journ. Linn. Soc.* (Zool.) XXXI, p. 305.

Topsent, in the paper cited after his name (1900), has discussed the structure and synonomy of this species in detail. As he had shown in previous papers, the spiculation undergoes great changes in the life of the individual sponge. At first three kinds of spicules are present—tylostyles, diactinial macroscleres and microscleres of the zigzag spiraster type The last disappear first, and then, in some cases, the diactinial microscleres, which, even in the young sponge, are much reduced and have the form of hair-like bodies adhering in bundles. There are three specimens from the Bay of Bengal in the collection of the Indian Museum which illustrate three different phases of growth in an interesting manner. One of them is clearly young and retains the full spiculation. It consists of a few galleries, with about half a dozen apertures, in a nodule of calcareous alga dredged by the "Investigator" in 28 fathoms off the coast of Burma.

The other two specimens are both in chank-shells (*Pyrula rappa*, L.) from the east coast of India. One was taken at the town of Madras in shallow water by Prof. K. Ramunni Menon, who has kindly given it to me. The shell was apparently vacant when collected but still retained its horny epidermis. The whole of its subtance is permeated by the sponge, in which only tylostyle spicules remain. The external apertures are, however, small (about 1 mm. in diameter) and the sponge is wholly confined in the thickness of the shell.

The third specimen was dredged by Mr. J. Hornell of the Madras Fisheries, whom I have to thank for it, in the Gulf of Manaar near Tuticorin in 6½ fathoms. The shell in this case had evidently been "dead" for some time and its epidermis had wholly disappeared. The apertures made by the sponge are much larger (2 to 3·25 mm. in diameter) and it has begun to grow out over the inner surface of the shell in the form of a uniform crust, much as in a specimen figured by Topsent (1887, pl. i, fig. 3).

Cliona celata probably occurs in all seas. It was originally described from the British coasts and has since been found at several places on the Atlantic side of North America, in the Red Sea and the adjacent parts of the Indian Ocean, off the south and south-west coasts of Australia, off New Guinea, Ceylon, Singapore, etc. I have examined specimens from several of these localities.

Cliona vastifica, Hancock.

1900. Topsent, *Arch. Zool. expérim.* (3) VIII, p. 56, pl. ii, figs. 3–9.

1909. Hentschel, "Tetraxonida" in Michaelsen and Hartmeyer's *Faun. Südwest Australiens*, p. 387.

This is another cosmopolitan species described at length by Topsent in his "Étude Monographique des Spongiaires de France" (*op. cit. supra*) as well as in his previous papers on the family (1887 and 1891) in the same journal. *Cliona velans*, Hentschel (*op. cit.*, p. 388, fig. 19) from S. W. Australia is evidently very closely related to *C. vastifica*, but is apparently distinguished by its method of growth and by having the heads of the tylostyles imperfectly differentiated.

In the littoral zone of Indian seas *C. vastifica* appears to be by far the commonest species and, as already stated, makes its way well into brackish water. I have found it in that medium in the Chilka Lake in Orissa and the Ganjam district of the Madras Presidency (in shells of *Ostrea* and *Purpura*), in the Adyar River at Madras and in the Ennur Backwater in the same district, in both places in shells of *Ostrea*. In the Persian Gulf it is common in, and apparently destructive to, pearl-shells (*Avicula* and *Margaritifera*); I have seen it in a *Placuna*-shell from Palk Straits (5½ fathoms), in shells of *Oliva* and *Malleus* from the Andamans, of *Voluta* and *Ostrea* from New South Wales. In Indian seas it occurs most frequently in the shells of gregarious sedentary bivalves, to which it probably causes great damage, but only in very shallow water. In European seas it is common; it has been recorded by Topsent and others from many widely separated regions.

Cliona carpenteri, Hancock.

1887. *Cliona carpenteri*, Topsent, *Arch. Zool. expérim.* V² (suppl.), p. 77, pl. vii, fig. 4.

1887 (1889). *Cliona bacillifera*, Carter, *Faun. Mergui Arch.* I: *Journ. Linn. Soc.* (*Zool.*) XXI, p. 76.

This species, as Topsent has pointed out, is easily distinguished from its allies, and in particular from *C. vastifica*, by its straight, spindle-shaped microscleres. Carter's *Cliona bacillifera* from Mergui, of which the type (or a schizotype) is in the Indian

Museum, falls well within the limits of the species as defined by the former author.

C. carpenteri is a tropical sponge distributed all round the globe. Topsent found it more frequently than any other in shells he examined from the Gulf of Mexico, the Pacific coast of Central America, the Gaboon, the Indian Ocean, etc. It does not appear, however, to be common in the Bay of Bengal. In addition to the type-specimen of Carter's species, which is in a dead oyster-shell, I have examined specimens in a shell of *Malleus* from Singapore and in one of *Voluta* from New South Wales.

Cliona margaritiferae, Dendy.

1905. Dendy, "Porifera" in Herdman's *Rep. Ceylon Pearl Oyster Fish.* V, p. 128, pl. v. fig. 9.

1909. Hentschel, "Tetraxonida" in *Faun. Südwest Australiens*, p. 386.

I have included this species (p. 5) among those that possess macroscleres of two kinds, but Dendy evidently regards the larger amphioxi as modified spirasters and points out that there are transitionary forms of spicules between them and the small microscleres. This is true; but there seems to me to be a slight but definite break in the series and it is at any rate more convenient to regard the large spiny amphioxi for the present as the equivalents of the granular amphioxi of such species as *C. vastifica*.

C. margaritiferae was originally described from the shell of the pearl-oyster of the Ceylon banks (*Margaritifera vulgaris*). I have found it in the same shell from the type-locality (*T. Southwell*) and also in a piece of Madreporarian coral from the Palk Straits (off Tondi, 5½ fathom: *J. Hornell*). Hentschel examined specimens in a shell of *Chama*, sp., from Michaelsen and Hartmeyer's Australian collection.

Cliona annulifera, sp. nov.

(Plate i, figs. 1-4.)

A *Cliona* with tylostyle macroscleres and spirasters of the normal type, the former bearing a single convex ring round the shaft; some of the latter unusually large. The gemmules are provided with spirasters of a specialized form.

The only known specimen is in a dead Gastropod shell (*Xenophora pallidula*, Rve.).

General structure. The sponge consists of a series of subspherical or ovoid chambers connected by short horizontal tubules and bearing the papillae on short vertical ones. The chambers form a single horizontal layer. The greatest longitudinal diameter of the larger chambers is about 1·3 mm. and their greatest depth about 0·9 mm. The average length of the connecting tubules (which, of course, represent the thickness of the wall of shell left between the chambers) is about 0·425 mm. and the diameter 0·119 mm. The papillae as a rule are borne only on the surface nearest the outer surface of the shell. The tubules connecting them with the chambers are longer than the horizontal tubules, but always much shorter than any diameter of the chambers. The chambers are by no means solid, their internal structure being coarsely reticulate. Delicate cellular diaphragms can sometimes be detected at or near one extremity of the connecting tubules.

Papillae. I have been able to find only two kinds of papillae, corresponding to those styled "poriferous" and "mixed" by Topsent (1887). The largest poriferous papillae have a diameter of about 0·225 mm They are readily distinguished by the

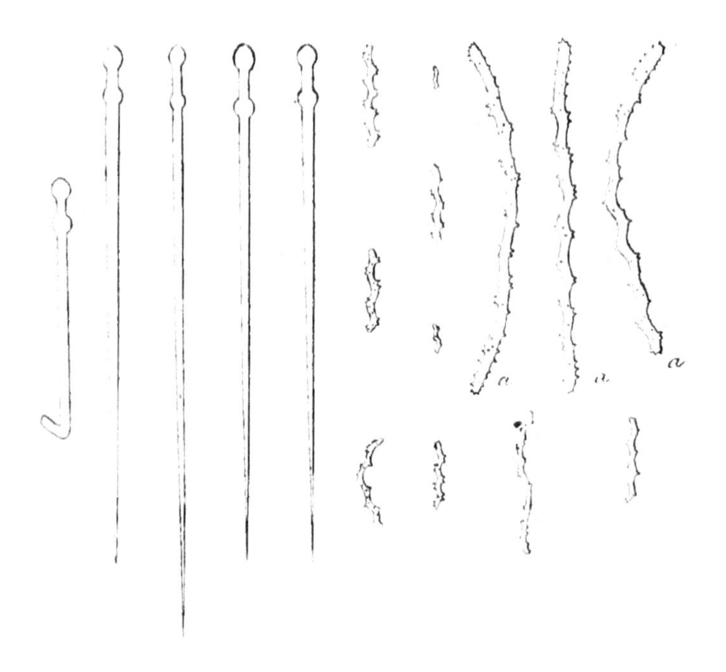

FIG. 1.—Spicules of *cliona annulifera*.
a. Gemmule-spicule.

absence of a central orifice and by the absence or paucity (at any rate when they are contracted) of projecting spicules upon their surface, which is flat and horizontal. It is closed by a minutely perforate membrane on and in which the calcareous particles derived from the shell and carried out through the oscula lie in considerable numbers, being too large to enter the pores. In profile these papillae are flat and table-like, extending beyond their supporting tubules, which are cylindrical, for a considerable distance on either side; the free surface forms an acute angle with the projecting lateral margin. The mixed papillae are about the same size but have a central star-shaped or oval orifice of relatively large diameter. This is surrounded by a number of poriferous lobes through which tylostyle spicules project upwards and outwards (pl. i, figs. 2, 3, 4). In profile these papillae, with their supporting tubule, are trumpet-shaped. Their outer walls (pl. i, fig. 4) are coated with minute calcareous particles considerably smaller than those which lie scattered in the interior of the sponge and on the poriferous papillae. They are covered by a delicate cortex, which protects the calcareous particles against strong acid unless the surface is subjected to its action for a considerable period. The mixed and the poriferous tubules are about equally abundant.

Skeleton.—In the chambers the macroscleres lie scattered, irregularly and somewhat sparsely, parallel to the outer walls. As a rule they are more abundant in the upper than in the lower parts. Occasionally they seem to radiate from the chambers into the connecting tubules, but this arrangement is never of a very regular nature and no trace of it can often be detected. In the vertical tubules the macroscleres form supporting columns, their heads resting in a more or less complete, and more or less regular, ring at the base of the tubule and their points directed upwards. In the case of the mixed papillae the heads are rarely on anything like a uniform level and the points project outwards as well as upwards. The ordinary (*i.e.* the smaller) microscleres lie scattered, somewhat sparsely and almost uniformly, throughout the sponge, but their main axis is always approximately parallel to the outer surface. The gemmules have a special skeleton, which is described below.

Spicules.—The macroscleres are small, slender tylostyles, as a rule quite straight, sharply and gradually pointed at one extremity and bearing a well-differentiated head at the other. The head is most frequently somewhat heart-shaped, but in many cases almost spherical and occasionally with a tendency to be trilobed. It contains as a rule a single minute expansion of the axial tubule of the spicule. There is no distinct contraction of the shaft below the head but, at about 1/10 the distance between it and the point, the shaft is surrounded by a single convex ring. The extent to which this ring is developed varies somewhat, but its presence and position seem to be practically constant features of the species. The average length of the macroscleres is 0·2 mm., the extremes being 0·148 and 0·234 mm.

The microscleres are all slender spirasters of the normal zigzag type, but they differ greatly in size and two groups may be distinguished amongst them in accordance with this character. Those of the small type are, when well developed, from 0·008 mm. to 0·042 mm. in length and have as a rule from 4 to 8 bends, but are sometimes irregularly sinuous. Their spines are arranged in a regular spiral. These spicules lie scattered throughout the sponge.

The larger microscleres (fig. 1A). are as much as 0·126 mm. long, or even longer. They have more numerous and as a rule less well-defined whorls. The spicules of this type are found only on the gemmules.

Gemmules.—Gemmules are abundant in the only specimen examined, most of the chambers containing from one to three (pl. i, fig. 1). They lie at the periphery of the lower part of the chamber and are as a rule somewhat lenticular in shape. The external surface is frequently flattened by pressure against the wall of the excavation. The greatest diameter rarely exceeds 0·56 mm. The internal structure is that of a typical sponge-gemmule, that is to say, each gemmule consists of a mass of cells closely packed together and filled with granules of food-substance. There is a thin horny external coat. The most remarkable feature, however, lies in the spicular coat that occurs on the surface of the gemmule in contact with the sponge, for the spicules of which it consists differ considerable from those of the general choanesome. The spicules have already been described. They lie embedded horizontally in the horny coat on one side of the gemmule only, being completely absent on the side that is in contact with the wall of the excavation.

Locality.—Off the coast of Ceylon: 703 fathoms (*R.I.M.S.* "*Investigator*").

Type. No. Z.E.V. 6424/7, *Ind. Mus.*: in spirit.

C. annulifera. is related to *C. viridis* (Schmidt), from which it differs in the form of its megascleres. It is remarkable for the regularity and distinctness of its chambers and especially for the peculiar spiculation of its gemmules, a feature in which it apparently differs from all other known marine sponges. That a deep-sea sponge should possess gemmules at all is a remarkable fact, and one to the signification of which I hope to devote attention in a later paper.

Cliona mucronata, Sollas.

1878. Sollas, *Ann. Mag. Nat. Hist.* (5) I, p. 54, pl. i, figs. 1, 2-7, 9, 10, 15, 17, pl. ii, figs. 1-9.
1887. Topsent, *Arch. Zool. expérim.* (2) V[2], p. 37.
1897. *Id.*, *Rév. Suisse Zool.* IV, p. 440.

The peculiar short, stout, mucronate tylostyles that form a considerable element in the spiculation of this species are quite characteristic. In the only specimen I have examined, they seem to be grouped together at certain points in the interior of the sponge, but this specimen is very imperfect, having been overwhelmed in its excavations by other sponges. Many of the tylostyles are of the normal type, but very slender.

C. mucronata was originally described from a coral (*Isis*, sp.) of unknown *provénance*. Topsent found it common in corals from the Bay of Amboina, and the only specimen in our collection is in a fragment of dead Madreporarian from the Indian shore of the Gulf of Manaar (Kilakarai: *S. W. Kemp*).

Cliona ensifera, Sollas.

1878. Sollas, *Ann. Mag. Nat. Hist.* (5) I, 61, pl. i, figs. 1, 18; pl. ii, figs. 10-5.

1887 (1889). Carter, *Faun. Mergui*, I, p. 75.
1891. Topsent, *Arch. Zool. expérim.* (2) IX, p. 570.

This species is closely allied to *C. mucronata*, with which it has been found on more than one occasion, including that on which the type-specimens of both species were discovered. Its tylostyles are, like those of *C. mucronata*, of two types, one of which is remarkable for the great expansion of the lower part of the shaft. The tapering of the point is, however, regular and the spicules is never mucronate. The other type of tylostyles is slender and in no way remarkable. The species is apparently more robust in its growth than *C. mucronata*.

C. ensifera, which was originally described as occurring in the coral *Isis* from an unknown locality, is abundant in dead reef-corals from the Mergui and Andaman archipelagoes.

Cliona viridis (Schmidt).

1887 (1889). *Cliona ? stellifera* (in part), Carter, *Faun. Mergui* I, p. 75.
1900. *Cliona viridis*, Topsent, *Arch. Zool. expérim.* (3) VIII, p. 84, pl. ii, figs. 11-14; pl. iii, figs. 2, 3; pl. iv, fig. 2.

Topsent has discussed this species and its synonomy in great detail and further references are unnecessary. It may be noted, however, that Carter's provisional species *Cliona stellifera* was founded on the macroscleres of this *Cliona* and the microscleres of a parasitic *Chondrilla*. I have found the two sponges in close association in his original specimen of dead coral from Mergui.

C. viridis is a cosmopolitan species evidently common in dead coral in the Mergui Archipelago and off the coast of the mainland of Burma. It was originally described from the Mediterranean and is known from the Gulf of Mexico, the Red Sea and many other widely separated localities.

Cliona orientalis, Thiele.

1887 (1889). *Suberites coronarius*, Carter (*nec. id.*, 1882) *Faun. Mergui* I, p. 74, pl. vii, figs. 4, 5.
1900. *Cliona orientalis*, Theile, *Abh. senckhenb. Natur. Gesellsch.* XXV, p. 71, pl. iii, fig. 24.

Thiele pointed out in 1900 (*op. cit.*) that the sponge described by Carter from Mergui under the name of *Suberites coronarius* was not identical with the species the latter had previously described under the same name from the West Indies, but actually a species of *Cliona*. He redescribed it with fresh figures of the spicules and named it *Cliona orientalis*. A re-examination of a part of Carter's Burmese material shows that Thiele was right in both contentions.

C. orientalis is closely allied to *C. viridis*, from which it may be distinguished by the arrangement of the spines on the microscleres. These, instead of running in a spiral round the spicule, are confined to its outline on one side. Carter's figures, although they illustrate this point clearly, are poor and misleading in other respects. The free form of the sponge closely resembles that of *C. viridis*.

C. orientalis has been found only in the Mergui Archipelago (in dead coral) and off Ternate in the Malay Archipelago.

Cliona acustella, sp. nov.

This is a species belonging to Topsent's fourth group, having microscleres and amphioxous macroscleres only. The latter, al-

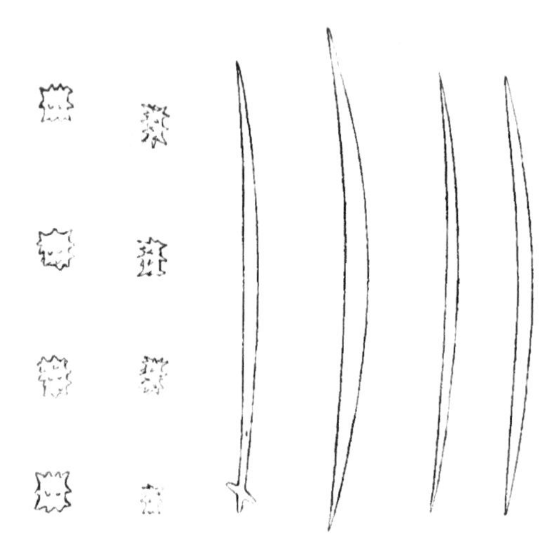

FIG. 2.—Spicules of *Cliona acustella*.

though many of them can be referred to the spiraster type, exhibit a marked tendency to assume a simple amphiaster-like form.

General structure.—Only dried specimens are available for examination, and of these I have been able to extract only minute fragments lacking the external papillae. Chambers excavated apparently by this sponge are, however, abundant in oyster-shells from several adjacent localities. The apertures on the surface of the shell are small and sparsely scattered: their diameter does not exceed 0·4 m. These apertures are connected with the chambers by very short vertical tubules. The chambers are subcircular or polygonal, not more than 3 mm. in diameter and separated only by very narrow partitions of shell. They are arranged in several horizontal layers. The tubules connecting them horizontally and vertically are very slender as well as short.

Spicules.—The macroscleres are smooth, slender, sharply-pointed, somewhat spindle-shaped amphioxi on an average 0·1447 mm. long by 0·008 mm. broad. They are never strongly arched or geniculate. Spicules of this type are fairly abundant.

The microscleres are minute, straight, truncate, cylindrical bodies bearing relatively large spines which often show a tendency to group themselves in three rings (two terminal and one median), but sometimes cover the spicule quite irregularly. Their average length is 0·012 mm. and breadth, with the spines, 0·008 mm.

Distribution.—Apparently common in shells of *Ostrea imbricata* and *O. cuculata* in from 15 to 30 fathoms of water off the coast of Orissa and the Ganjam district of Madras in the Bay of Bengal (*S.S. 'Golden Crown'*).

Type.—No. Z.E.V. 6415/17, *Ind. Mus.*

The microscleres of this species appear at first sight to be in many cases amphiasters rather than spirasters, but actually exhibit (fig. 2) a fairly regular transition between the two types. Some of them are not unlike the small spicules of *Dotona.* The species at present stands alone in the genus so far as its spiculation is concerned, but I have little doubt as to its validity, although the circumstances in which it was found seem at first sight a little suspicious. Large numbers of oyster-shells, all of which were unfortunately cleaned and dried before being examined, were obtained by the trawler 'Golden Crown' off the east coast of India in 1909. The majority of them were found, on recent examination, to be perforated and in many cases partially disintegrated by the burrows of a *Cliona*, of which minute fragments were extracted from broken shells. Spicule-preparations made from other pieces of the same shells contained in some cases only spicules identical with those which occurred in the fragments of sponge extracted, *viz.* smooth amphioxi and microscleres of the type described and figured above. No tylostyles could be found either in the spicule-preparation or in the fragments of sponge. Other fragments of sponge extracted from shells were clearly no part of a Clionid but represented two species of *Eurypon.* Many spicule-preparations contained a mixture of the spicules of the *Cliona* with those of one or other representative of the latter genus. No actual confusion is possible, however, between the two very different genera thus associated.

Cliona patera (Hardwicke).

1822. *Spongia patera*, Hardwicke, *Asiat. Researches* XIV, p. 180, pl. i.

1858. *Poterion neptuni*, Schlegel, *Handleid. Dierkunde* II, p. 542.

1880. *Poterion patera*, Sollas, *Ann. Mag. Nat. Hist.* (5) VI, p. 441.

1908. *Poterion patera*, Vosmaer, *Versl Gew. Verg. Wis-en-Naturk. Afd.* XVII (1), p. 16.

1909. *Cliona patera*, Topsent, *Arch. Zool. expérim.* (4) IX, p. lxix.

Although this large and conspicuous sponge has been known, so far as its external form is concerned, for nearly a century, its true systematic position has only been discovered, by Vosmaer and Topsent, in the last few years. There is a fine series of dried specimens from Singapore, the original locality, in the Indian Museum: but they do not include the type. The species seems to me to be very closely related to *Cliona celata*, from which it differs in its stouter spicules but which it resembles in its general structure and in particular in that of the papillae. So far as these are concerned it agrees more closely with *C. celata* than with *C. viridis*, of which Vosmaer was apparently prepared to regard it as a variety.

Some of our specimens contain at the base both Lamellibranch and Gastropod shells, as well as many small pebbles. The latter, being of hard stone, are intact, as are also some of the shells. Others, however, both of bivalves and of Gastropods, have had ramifying grooves excavated on their surface by the sponge. In one Lamellibranch shell that was partially embedded in it the grooves are entirely confined to the embedded position. At least one Gastropod shell, which was extracted from the centre of the basal portion of a large specimen, is wholly permeated and nearly destroyed by excavations filled with sponge substance. I am convinced by these facts that the excavations in shells found in large specimens of *C. patera* are of a secondary nature, and it seems improbable, in any case, that so large a sponge, if it commenced life in the thickness of any Molluscan shell, should not have completely destroyed that shell before reaching its full size.

So far as I am aware, *C. patera* has as yet been found only in the neighbourhood of Singapore and Java, where it is abundant. If it occurred in the Gulf of Manaar, where several large collections of sponges have been made, so conspicuous an object could hardly have escaped notice. Indeed, its place seems to be taken in the seas round Ceylon and India by the Halichondrine sponge *Petrosia testudinaria* (Lamarck), which bears a considerable superficial resemblance to it, although the "cup" and the "stalk" are not so clearly differentiated.

Genus **Thoosa**, Hancock.

1849. *Thoosa*, Hancock, *Ann. Mag. Nat. Hist.* (2) III, p. 345.
1887. ,, Topsent, *Arch. Zool. expérim.* (2) V^2, p. 88.
1891. ,, *id.*, *ibid.*, (2) IX, p. 577.
1905. ? *Cliothosa*, *id.*, *Bul. Mus. Hist. Nat. Paris*, XI, p. 95.

This genus is much less well known than *Cliona*. Most of the species, being of tropical origin and having a very inconspicuous appearance externally, have been described from dried specimens extracted from shells or corals, and many of these have been imperfect. Possibly it will ultimately be proved that several quite distinct genera are included under the name. The genus as at present constituted is remarkable for the great diversity of its microscleres, which always include some form of amphiaster, as well as, in many cases, degenerate forms of euasters. True spirasters seem to be invariably absent.

The typical spicule is characteristic. It consists of a relatively stout cylindrical stem, as a rule quite straight, and of two circles of horizontal branches, which surround the stem at or near its extremities. The stem is quite smooth. In most cases the lateral branches are very short and greatly inflated at their tips, so that they have actually the form of subspherical bosses or prominences. They are never numerous, four to six being the normal number in each ring. In the more highly developed forms the prominences are covered with short spines, and the extremities of the stem are inflated and spiny also. Both the terminal and the lateral prominences may, however, be greatly reduced and take the form of smooth rounded or conical projections.

Another form of amphiaster that is often, though not invariably, present also consists of a smooth cylindrical stem surrounded at the ends by a ring of horizontal branches. Both the stem (as a rule) and the branches (always) are, however, more slender and the latter are much produced. The extremities of neither are regularly spiny, but each branch terminates either in a minute inflation or in several small hook-like spines.

A third form of microscleres that often occurs has been shown to be a degenerate oxyaster, although in its common form it has little resemblance to that type of spicule. As a rule it consists merely of two slender, more or less strongly curved spines attached to a minute centrum and having the appearance of the horns of some Ruminant attached to a fragment of the skull, or that of a sea-gull in flight as seen from a distance, or rather as conventionally represented in pictures. Occasionally more than two spines are present, and the spicule may assume a star-like form. Other microscleres, which resembles toxa but probably have the same origin, also occur in some species.

Yet another type of aster is often found. It has the form of a flat, spiny plate or a spiny cylinder and is referred to by Topsent as a pseudosteraster. I have not come across this form of spicule myself in the specimens I have examined.

The macroscleres, if present, are either amphioxi or tylostyles, but they are often absent.

The distribution of *Thoosa* is essentially tropical, but several of the species are as yet recorded only from specimens of unknown history. They appear to occur mainly in shells of solid structure or reef-corals from shallow water, but one species described here (*T. investigatoris*, p. 18), was found in a thin Gastropod shell from a depth of over 700 fathoms.

The following species have been recorded, or are here recorded for the first time, from the seas of British India and Ceylon:—

Thoosa radiata, Topsent.
T. socialis, Carter.
T. armata, Topsent.
T. hancocci, Topsent.
Thoosa, investigatoris, nov.
T. fischeri, Topsent.
T. laeviaster, nov.

Of these I have not seen *T. socialis* [1] and *T. fischeri*,[2] both of which are only known from Ceylon.

It does not seem advisable at present to attempt to draw up a key to the Indian species. One to all those known in 1891 is given by Topsent on pp. 585-586 of his paper cited after that date on p. 16, and no new species have been published since. Two are described in this paper.

Thoosa investigatoris, sp. nov.

(Plate i, figs. 5, 6).

This is a species with megascleres in the form of pin-like tylostyles and with three types of amphiasters as microscleres, *viz.* (1)

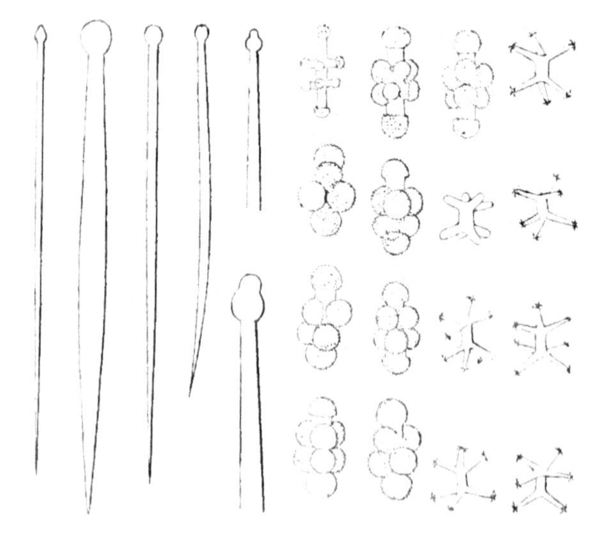

FIG. 3.—Spicules of *Thoosa investigatoris*.

nodular amphiasters typical of the genus, (2) smooth amphiasters with horizontal branches ending in a circle of hooklets, and (3) much stouter smooth amphiasters without hooks or spines of any kind.

General structure.—The sponge consists of a number of tubules which anastomose in one plane and swell out at intervals into not very clearly differentiated chambers of a flattened form and of irregullar outline. The whole structure is fragile and delicate, offering in this respect a strong contrast to *Cliona annulifera*,

[1] Carter, 1880, p. 56 (*v.* p. 2, footnote 3).
[2] Topsent, 1891, p. 582 (*v.* p. 16).

which was taken at the same station. The papillae are borne on very short pedicels, as a rule only on what may be called the upper surface of the sponge; occasionally they are also found on the lower surface. I have not been able to detect cellular diaphragms.

Papillae.—Two kinds of papillae have been observed, one of which is apparently inhalent, while the other is probably of a mixed nature. The latter is considerably larger than the former; its diameter is on an average, in normal circumstances, about 1 mm., whereas that of the smaller papillae is only about 0·4 mm. In both cases the vertical walls are straight and the actual papilla does not expand much beyond them. The exhalent apertures are circular and are protected, as is explained below under the heading "Skeleton", by a peculiar arrangement of spicules.

Spicules.—The macroscleres are tylostyles with well differentiated heads usually spherical in form and frequently containing a single large vacuole. The stem is usually curved and spindle-shaped, tapering considerably towards both extremities and considerably swollen in the middle. More slender tylostyles occur in which the stem is much less swollen, but there are also intermediate forms. The shape of the head is not constant, for, especially in the more slender macroscleres, it is sometimes trilobed and sometimes flattened above; occasionally it is even acorn-shaped or quite asymmetrical. In the stouter tylostyles the total length is on an average about 0·34 mm., the greatest thickness of the shaft about 0·02 mm. and that of the head slightly less.

The nodular amphiasters have both the lateral and the terminal nodules or bosses relatively large, nearly spherical and covered densely with minute straight spines. They are joined to the stem, which they often conceal almost completely, by very short smooth pedicels. The average length of the spicule of this type is about 0·0369 mm. and the greatest breadth across the lateral nodules 0·0164 mm.

The smooth amphiasters with terminal hooks on the lateral branches are of the habitual form. Their stem is rather stout and the lateral branches taper straightly towards the tip, which usually bears about six hooklets. The greatest length of the spicule is on an average about 0·0164 mm. and the greatest breadth from tip to tip of the branches 0·0246 mm.

The third type of amphiaster, which is very scarce, is about the same size as the second, which it resembles considerably, but the branches are stouter and bear no terminal hooklets, nor are they inflated at the tips.

Skeleton.—The spicules are arranged to form a skeletal structure in a somewhat more regular manner than is the case in most species of Clionidae.

In the horizontal tubules the macroscleres lie parallel to the surface and in a large proportion of cases point in the same direction. They exhibit, in quite a definite manner, evidence of fasciculation, although in this part of the sponge they do not appear to be bound together by any horny substance. At certain points, probably where the aperture for a new papilla is about to be excavated, a stout chitinoid covering is secreted over the sponge and the macroscleres adopt a convergent arrangement and are densely massed together. At such places the nodular microscleres are sometimes present in large numbers and form a layer several spicules thick over the protecting mass. The papillae are protected by a dense ring of vertical macroscleres fortified with chitinoid substance and arranged concentrically in several or many circles with the heads resting at the base of the very short vertical tubule. Within this ring, in the case of exhalent orifices, there is an arrangement of convergent macroscleres with their tips meeting almost horizontally and their heads set in a broad spiral of about 1½ turns. Presumably the tips can be separated in the living sponge by rotation of the heads. The whole arrangement is strikingly reminiscent of the diaphragm in the stage of a compound microscope. The smooth amphiasters are scattered in the flesh of the tubules and chambers. Neither they nor the nodular amphiasters play any part in the protection of the external papillae.

Gemmules.—I have found several gemmules in the specimen examined. They are spherical masses of cells of the usual type, but have no horny protective membrane. Each is about 0·374 mm. in diameter. Each gemmule occupies a separate chamber which it fills completely. There is a slender strand of cells connecting it with the active part of the sponge.

Type.—No. Z.E.V. 6430/7, *Ind. Mus*, in spirit: in a dead Gastropod shell.

Locality.—Off Ceylon: 703 fathoms (*R.I.M.S.* '*Investigator*').

The form of the nodular microscleres is characteristic, in particular in the large size of the lateral and terminal bosses; otherwise they resemble those of *T. socialis*, Carter. The species is evidently related to *T. armata*, which, however, has the spicules of this type with the bosses perfectly smooth as well as relatively smaller.

A noteworthy feature of *T. investigatoris* is its power of secreting a horny covering for its growing-points when they come in contact with foreign bodies. I hope to show in a subsequent paper that it protects itself in this manner against aggression on the part of a sponge of the genus *Coppatias* that is parasitic in its burrows. At most of the points at which new galleries are being formed in the shell no such covering can be detected, but at some, probably where the sponge is in contact with the outer layers of the shell, and is about to form a new exhalent or inhalent papilla, there is a thick one. It is only where such a covering occurs that the nodular amphioxi are found, and if the covering is very thick, a number of these spicules can usually be discovered in which the spines on the nodules seem to be completely worn away and the nodules themselves even to some extent destroyed. Such spicules lie in or on the outer or distal part of the covering. These facts would suggest that spicules of this peculiar type play an important part in the perforation of the compact outer layers of the shells in which the sponge constructs its burrows.

Thoosa armata, Topsent.

1887. Topsent, *Arch. Zool. expérim.* (2) V², p. 81, pl. vii, fig. 9.
1891. *Id.*, *ibid.* IX, p. 579.
1904. *Id.*, "Spongiaires des Açores" *Rés. Camp. Sci. Monaco*, fasc. XXI, p. 106, pl. xi, fig. 5.

In preparations of *Cliona vastifica* from a shell of *Malleus* from the Andaman Is., I find, mingled with the spicules of that species, others of three types that agree well with those of *Thoosa armata* as described and figured by Topsent. They are nodular amphiasters, reduced oxyasters consisting of a pair of long horn-like spines arising from a minute centrum, and smooth, sharply pointed amphioxi. The spicules of other types figured by Topsent I have not found in this very imperfect specimen.

As to the smooth amphioxi, they certainly do not belong to the *Cliona* and no trace of any other sponge but the *Cliona* and the *Thoosa* is present in some of my preparations. Topsent in his original description of *T. armata* described amphioxi of the kind as an essential element in the spiculation of the species, but did not find them in the specimen from the Azores he described in 1904. In my specimen, in parts of which they seem to be definitely associated to form a skeletal structure, they are on an average 0·09 mm. long and 0·002 mm. broad at the thickest part. They are thus rather larger than in Topsent's original example.

Thoosa armata was described from a dried sponge in an oyster-shell from the Gaboon (West Africa), and has also been found in a dead coral in the Azores. It has not hitherto been known from the Indian Ocean. The extraordinary larva was described and figured by Topsent (*op. cit.*) in 1904.

Thoosa hancocci, Topsent.

1887. Topsent, *Arch. Zool. expérim.* (2) V², p. 80, pl. vii, fig. 12.
1891. *Id.*, *ibid.*, IX, pp. 577, 580.
1898. Lindgren, *Zool. Jahrb.* (Syst. Abth.) XI, p. 320.
1905. Topsent, *Bull. Mus. Hist. Nat. Paris*, XI, p. 94.

Topsent and Lindgren have described this species as having spicules of three types, (*a*) tylostyles, (*b*) nodular amphiasters, and (*c*) slender amphiasters—Lindgren calls them spirasters—with lateral branches terminating in minute hooks. Topsent (*op. cit.*, 1905) has also described a closely similar species without spicules of the last type (*c*), and founded for its reception the new genus *Cliothosa*. The only known species of this supposed genus (*C. seurati*, Topsent) only differs from *T. hancocci*, apart from the supposed generic character, in having the head of the tylostyle oval (instead of usually spherical) and with a group of minute vacuoles in its centre.

In the collection of the Indian Museum there are two shells from the Andamans, one of a *Tridacna* and one of a *Malleus*,[1] that contain the burrows of a Clionid which agrees well with Topsent's description of *T. hancocci* so far as the general structure and the colour are concerned. In the *Tridacna*-shell the papillae of the sponge have been destroyed, but they are well preserved in that of the *Malleus*. In neither specimen have I been able to find a single nodular amphiaster, although there has been no difficulty in removing the papillae for microscopic examination from one of them. The slender amphiasters are abundant in both specimens, scattered in the galleries of the sponge, and the majority of the tylostyles in the galleries have spherical heads, but those in the papillae are variable in shape. In no single spicule can I detect a group of vacuoles in this part.

The question naturally arises, Is *Cliothosa* a distinct genus or merely a phase of *Thoosa*? In considering this question the facts known in reference to other species of the family must be noted. In the first place, it is known that *Cliona celata*[2] may lose two types of spicules in the course of its latter development and that *Thoosa armata*[3] does the same at an earlier stage. Secondly, we know that the nodular amphiasters are sometimes scarce in *T. hancocci* itself and, apparently, may be either confined to the papillae[4] or scattered throughout the sponge.[5] Thirdly, in the type-specimens of *T. investigatoris* (*antea*, p. 18) and *T. laeviaster* (p. 23, *postea*) these spicules were not found in the fully formed papillae but in what were apparently papillae in the process of formation. Furthermore, in the case of the former species, they sometimes exhibited distinct traces of wear in that position. All these facts seem to me to point to the possibility of there being a stage, perhaps but seldom attained, in the life-cycle of *Thoosa* at which the characteristic spicules of the genus disappear and the sponge gains nominal generic distinction under the title *Cliothosa*. If I am right, there can, I think, be no doubt that at least one of my specimens from the Andamans has reached this stage.

Thoosa hancocci was originally described from a *Tridacna*-shell from the Indian Ocean. It is apparently common in coral from shallow water in the neighbourhood of Java and was taken by Prof. Stanley Gardiner, also in coral, in the Maldives (*fide* Topsent, 1905, p. 94).

Thoosa laeviaster, sp. nov.

Spicules and fragments of the sponge of this species were found in the piece of dead coral referred to by Carter, whose

[1] One valve of the individual in the other valve of which *Thoosa armata* was found intermingled with *Cliona vastifica*.
[2] Topsent, **1900**, p. 42, etc.
[3] Topsent, **1904**, p. 111: see synonomy of *T. armata*, p. 21.
[4] Topsent, **1905**, p. 94.
[5] Lindgren, **1898**, p. 321.

notice they apparently escaped, in his account of the sponges of the Mergui Archipelago: *Fauna of the Mergui Archipelago* I, p. 75. It is remarkable in the form of its nodular amphiasters, the "nodules" of which are reduced to short, slender, blunt or pointed branches totally devoid of spines. Reduced spirasters of the type common in the genus are also present, while the macroscleres are smooth amphioxi.

General structure.—Nothing is known of the general structure except that the sponge consists, in part at any rate, of slender apparently cylindrical branches ramifying in dead coral.

Papillae.—The papillae, of which several imperfect examples were extracted, are evidently very small, probably not more than 0·3 mm. in diameter. They are protected by dense masses of upright macroscleres.

Spicules.—The macroscleres are small, slender, smooth, sharply pointed, spindle-shaped amphioxi; a large proportion of them are definitely geniculate in the middle. The average length is 0·08 mm. and the average breadth in the middle 0·003 mm.

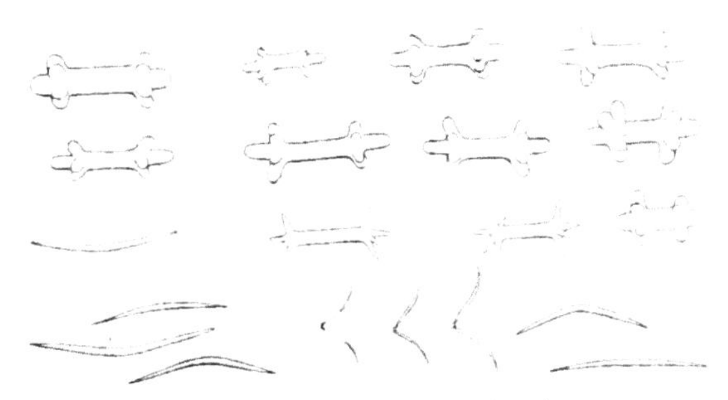

FIG. 4.—Spicules of *Thoosa laeviaster.*

Only two types of microscleres can be distinguished; (*a*) smooth, rather slender amphiasters surrounded at some little distance from each extremity by a circle of several (normally 4) horizontal branches, which are also smooth and relatively slender. These are usually blunt but sometimes pointed; they are always stouter at the base than at the tips. The length of each branch is usually equal to the distance of its base from the nearest extremity of the shaft, which terminates in the same manner as the branches, and the distance apart of the two circles is considerably greater. The average length of the spicule of this type is from about 0·041 to 0·08 mm.; the average thickness of the shaft from about 0·0065 to 0·013 mm. and the breadth from tip to tip across the branches from 0·0246 to 0·0328 mm., but all these measurements are variable. (*b*) The second type of microsclere is a reduced oxyaster consisting of a pair of relatively long and slender curved horn-like spines attached close together to a minute centrum.

Skeleton.—From the fragment of sponge extracted from the coral it is evident that the macroscleres are arranged much as in

T. investigatoris. The reduced oxyasters are scattered in the main body of the sponge, while the amphiasters are collected in small groups and associated with films of horny substance in the interior. They also appear, therefore, to have been arranged in the same manner as their homologues in *T. investigatoris.*

Type.—A microscopic preparation mounted in Canada balsam. No. Z.E.V. 6639/7. *Ind. Mus.*

Locality.—King Id., Mergui Archipelago; in dead coral (*J. Anderson*).

The form of the amphiaster is unlike that of any other species in the genus, for the " nodules " of these spicules, even when they are smooth as in *Thoosa armata*, are usually short, stout and rounded. It is clear, however, that their form in *T. laeviaster* does not depart very widely from the generic type and is really nearer that of the normal amphiaster of *T. radiata* (Topsent, 1887, pl. vii, fig. 11) than that of the homologous spicule of *T. armata.*

EXPLANATION OF PLATE I.

FIGS. 1, 2, 3, 4. *Cliona annulifera*, sp. nov.

1.—Part of type specimen extricated from the shell in which the sponge had burrowed, viewed from below as an opaque object, × *ca.* 30. *g.* = gemmule: *s.* = space from which shell-substance has been removed.

2.—A single chamber mounted, without staining, in Canada balsam and viewed from one side by transmitted light, × *ca.* 57. *i.p.* = inhalent papilla: *m. p.* = mixed papilla: *G.*=gemmule.

3.—Another chamber stained with borax carmine and similarly mounted, seen from above by transmitted light, × *ca.* 57. Lettering as in figs. 1 and 2.

4.—A mixed papilla, viewed obliquely from one side as a solid object, more highly magnified. *c.* = calcareous granules.

,, 5, 6. *Thoosa investigatoris*, sp. nov.

5.—Part of type-specimen extricated from the shell in which the sponge had burrowed, viewed from above by transmitted light, × *ca.* 35. *e.p.* = exhalent papilla: *h.r.* = horny ring surrounding papilla: *n.g.* = commencement of a new gallery: *s.* = space from which shell-substance has been removed.

6. Another part of the same sponge only partially extricated, viewed obliquely by transmitted light, and more highly magnified. *i.p.* = inhalent papilla: S.= fragment of shell.

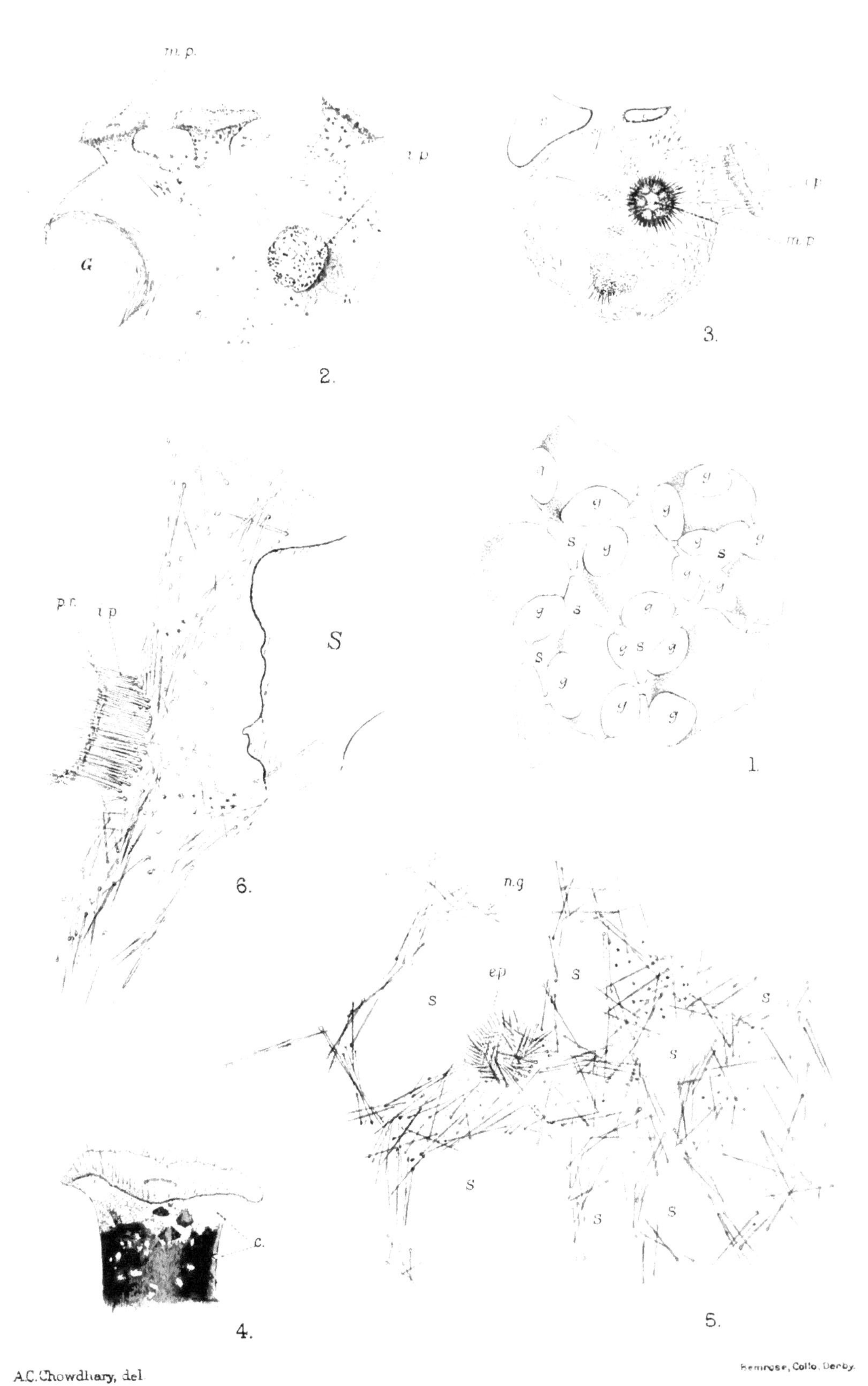

A.C. Chowdhary, del. Bemrose, Collo, Derby.

Figs. 1-4. Cliona annulifera. Figs. 5, 6. Thoosa investigatoris.

II. HERMIT-CRABS FROM THE CHILKA LAKE.

By J. R. HENDERSON, *M.B., C.M., F.L.S., Superintendent, Madras Government Museum.*

The small collection of Paguridæ which forms the subject of this paper was obtained by Dr. N. Annandale and Mr. S. W. Kemp, during their survey of the Chilka Lake, on the Orissa Coast in the Bay of Bengal. Of the five species taken it has only been found necessary to describe one as new. In each of the previously known species reference is made to Col. A. Alcock's "Catalogue of the Indian Decapod Crustacea in the collection of the Indian Museum," part II, Anomura, Fasc. I, Pagurides (1905), where a full bibliography will be found.

Clibanarius padavensis, de Man.

Alcock, p. 44, pl. iv, fig. 2.

Station 22[1], 8698/10. Five specimens of moderate size, including a female with ova.

Station 75, 8700/10. A young specimen with the carapace measuring only 7 mm. in length, yet possessing all the characteristic colour markings.

Station 82, 8705/10 A male, and a female with ova, of moderate size, the carapace of the former measuring 16 mm. in length; also a very young specimen.

Station 83, 8696/10. Six individuals of moderate size, the carapace of the largest (a female) measuring 18 mm. in length.

This species, which shows a special predilection for brackish water, occurs in suitable localities round the Indian coast from Burma to Bombay. It has also been recorded from Singapore, Queensland, New Guinea and New Caledonia.

Clibanarius longitarsis, de Haan.[2]

Alcock, p. 158.

Station 142, 8968/10. In *Purpura* shells. Two specimens, male and female, the latter which is slightly larger with the carapace 11 mm. long.

[1] An explanation of the station numbers will be given in a subsequent paper dealing in a general manner with the results of our survey of the Lake. (N. A.)

[2] I take this species as characterized by de Man in his account of the Crustacea collected by Dr. Brock in the Malay Archipelago (Archiv. f. Nat. LIII, p. 441, 1887).

This record is interesting as it tends to show that the species, which was previously known only from the southern part of the east coast so far as India is concerned, probably occurs in suitable localities all along this coast. It is the commonest brackish-water pagurid on the Coromandel coast.

C. longitarsis has been found in various localities from East Africa to Japan.

Clibanarius olivaceus, n. sp.

Station 22, 8698/10. Two males and two females. Also 8710/10, a male, and 8708/10 a very young specimen in a fragmentary condition which probably belongs to the present species.

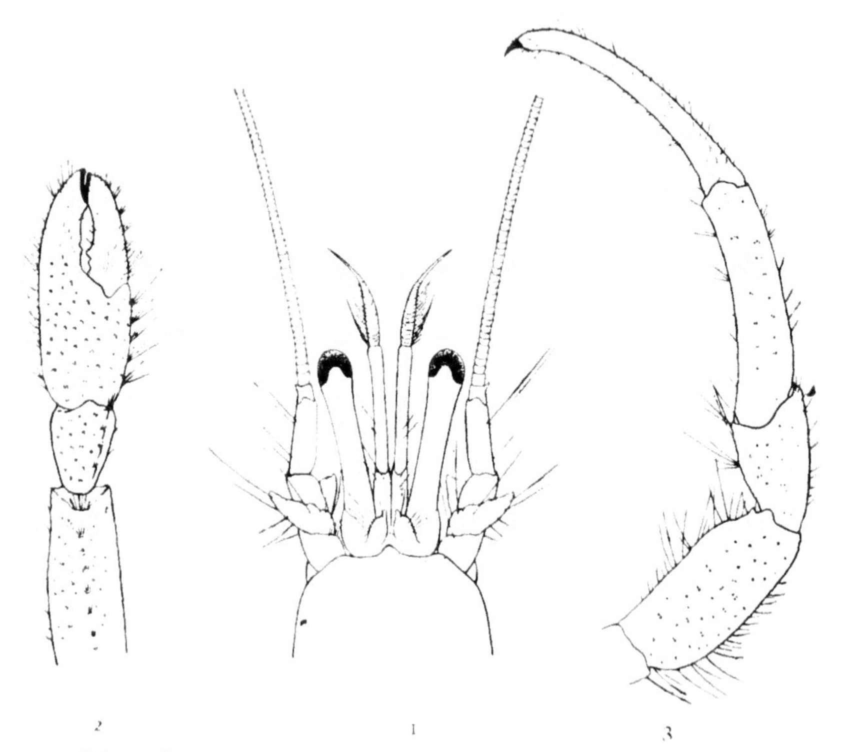

Clibanarius olivaceus, n. sp.: 1. anterior part of carapace, eyestalks, etc. from above; 2. left cheliped from above; 3. last four segments of second leg from above.

Station 142, 8967/10. A female with the carapace 9 mm. long.

Some small individuals of this species were taken by Dr. Annandale near the mouth of the Adyar River, Madras, in October 1913.

Carapace with the usual tufts of setae, which are most numerous towards the sides and immediately behind the cervical groove. Rostrum moderately prominent, reaching a little beyond the antennal angles of the carapace.

Eyestalks distinctly longer than the anterior border of the carapace and almost as long as the antennular peduncles; the eye occupies about one-tenth of the length of the stalk. Ophthalmic scales with the outer edge setose and faintly spinose.

Antennal acicle setose and slightly spinose, the proximal spinule being most prominent, scarcely reaching the terminal joint of the peduncle; flagellum about one and a half times as long as the carapace.

Chelipeds subequal and similar: merus with the upper margin obscurely serrulate and two spinules at the distal end of the outer lower margin; carpus with a distinct spinule at the distal end of the upper margin and two or three smaller spinules further back on the margin; hand only slightly roughened and comparatively free of setae, with no spines anywhere on its palmar surface; fingers rougher and more setose than the palm, the projections almost becoming spinose towards the finger-tips, when closed exhibiting an intervening hiatus. The length of the hand, including the fingers, is about twice its breadth.

The second and third legs exceed the chelipeds on both sides by the length of the dactyli and nearly half the propodi; a spinule is present at the lower distal end of the merus and another at the upper distal end of the carpus in both pairs of legs. The dactylus of the third leg is about one-fourth longer than the corresponding propodus; the dactylus of the second leg while shorter is still distinctly longer than its propodus.

The colour of spirit specimens is yellowish, with the chelipeds and second and third pairs of legs olive green. The only distinct bands of colour are three pale red lines, on the inner upper and outer surfaces of the eyestalks, though in the specimen from station 142 the upper line is obsolescent. The meral and carpal joints of the second and third legs, particularly the former, show a faint bluish green colour near the upper posterior surface and a reddish brown tinge near the lower margin, but these colours are not sufficiently circumscribed to constitute bands. The dactyli of the second and third legs are buff-coloured.

Length of carapace (in a male) 12 mm., breadth of anterior border of the carapace 4 mm., length of eyestalk 5·5 mm., length of dactylus of second left leg 9·5 mm., length of propodus of second left leg 7·5 mm. The left eyestalk of the type specimen is shorter than the right and is evidently in process of regeneration.

This species is closely related to *C. padavensis*, de Man, and *C. longitarsis* (de Haan), both of which commonly occur in brackish water in India. It agrees with them in the long dactyli of the second and third legs, and in the long eyestalks, which are longer than the anterior border of the carapace, but is readily distinguished from both by its colouration.

In *C. padavensis*[1], there are very distinct deep red or

[1] My remarks on *C. padavensis*, in the Journ. As. Soc. Bengal LXV, pt. ii, p. 520, 1896, were made in error and really apply to *C. longitarsis*.

crimson lines on the second and third legs, eyestalks and chelipeds. In *C. longitarsis*, a pale blue band bordered above and below by red brown occurs on the joints of the second and third legs, being best marked on the propodi, and there are no colour lines on the eyestalks or chelipeds These distinctive marks are present at all stages of growth, and I am of opinion that colour is a fairly reliable character in separating the species of *Clibanarius*.

In addition to the colour differences, the hand of *C. olivaceus* is much smoother and less hirsute on the upper surface, while the inner margin is devoid of the spinules which occur in the two other species.

Clibanarius sp.

Station 82, 8709/10. Five very small individuals which are too young to identify satisfactorily. They perhaps belong to the last species.

Diogenes miles (Herbst).

Alcock, p. 67, pl. vi, fig. 5.

Ganjam Coast outside the southern part of the Chilka Lake, in a *Voluta* shell, 8706/10. A female with ova in which the carapace measures 15 mm. in length.

This species, which is common on the east coast of India, but so far as I know does not affect brackish water, has a characteristically flattened carapace, and the hand of the left or larger cheliped can be bent almost at right angles to the wrist, the result of living in shells with a long narrow aperture.

Diogenes avarus, Heller.

Alcock, p. 68, pl. vi, fig. 6.

Station 71, 8703/10. Three small specimens.

Station 91, 8764/10. Eleven small specimens, including several females with ova in which the carapace measures less than 5 mm. in length Also 8707/10; two small specimens.

Station 93, 8701/10. Seven small specimens, one of which, a female with ova, has the carapace 5 mm. long.

Station 94, 8702/10. Two minute specimens.

This small species is common in the South Indian backwaters, but also occurs in the sea, both between tidemarks in places such as the shores of the Gulf of Manar where the surf is not excessive, and in shallow water. As was first pointed out by de Man the lengthening of the carpus and band of the larger cheliped, which is so characteristic a feature of adult males, is much less marked in females and young males. Some individuals appear to attain maturity while of small size; in the present collection there are several females with ova in which the carapace measures 5 mm. in length or even less, while individuals from other localities are found at least double this size.

Cœnobita rugosus, Milne-Edwards.

Alcock, p. 143, pl. xiv, fig. 3.

Station 107, 8966/10. A small specimen in a *Natica* shell.

Station 123, 8965/10. A small specimen.

These specimens, which are evidently immature, as the carapace of the larger one measures only 10 mm. in length, appear to belong to this common Indo-Pacific species. They possess a stridulating organ composed of an oblique row of elongated parallel teeth, on the outer surface of the left palm, and the outer surface of the propodite of the third left leg is separated from the anterior surface by a distinct ridge.

Cœnobita cavipes, Stimpson.

Alcock, p. 146, pl. xiv, fig. 1.
= *C. violascens*, Heller, and *C. compressus*, Ortmann.

Station 79, 8699/10. Three young specimens.

Station 82, 8695/10. A number of young specimens, the largest of which is a female with the carapace 14 mm. long.

Station 95, 8697/10. A half-grown male.

The lower part of the outer surface of the left palm is smooth, and as pointed out by Stimpson is mahogany coloured.

This species is very common in the neighbourhood of the backwaters along the east coast of India, and is frequently found at some distance from the water. It is most active at night.

III. NOTES ON SOME SOUTH INDIAN BATRACHIA.

By C. R. NARAYAN RAO, *Central College, Bangalore.*

I. The Larvae of **Microhyla rubra** and **Rana breviceps.**

These tadpoles have been described by Mr. H. S. Ferguson, F.L.S., late Director of the Trevandrum Museum, in his paper on "A List of Travancore Batrachians," published in the *Journal of the Bombay Natural History Society* (Vol. XV, p. 499). I am of opinion that Mr. Ferguson has mixed up the larvae of *M. rubra* with those of the allied species *M. ornata*, and there is considerable difference between his account of the tadpoles of *R. breviceps* and the specimens I have collected. These facts sufficiently justify the publication of the following notes, in which I purpose to describe the specimens in full and at the end indicate the chief points wherein I differ. I might add here that examples of all these larvae have been sent to Dr. N. Annandale whom I have to thank for examining them.

Larva of **M. rubra.**

H. S. Ferguson, *J.B.N.H.S.*, Vol. XV, 1904, p.506: Boulenger, *Fauna*, p. 491.

Towards the middle of July, a few specimens of this tadpole were obtained at Bangalore from a pond in which rain water had collected. Other tadpoles found in their company were those of *M. ornata*, *R. breviceps* and *Rhacophorus maculatus*. The tadpoles were allowed to complete their metamorphosis in the college aquarium. They may be described as follows :—

The head and body.—Head depressed and almost flat, snout broadly rounded but not squarish. Both dorsal and ventral surfaces of trunk flat. In horizontal section, the body is nearly elliptical. Skin smooth.

Eye and nostril.—Nostrils nearer to the snout than the eyes, and are dorsal. The inter-orbital space nearly six times the internasal. Eyes lateral, visible from below and by no means prominent. Pupil round. (It is vertical in the adult).

Mouth.—Very small, nearly terminal or dorsal: broadly triangular or nearly oval. Upper lip better developed, with a horny edge. Beaks, horny teeth and papillae absent.

Sensory glands and pits.—A conspicuous white glandular area, somewhat dome-shaped, just behind the mouth or between the nostrils. A number of sensory pits round the mouth, especially about the corners. A fine white glandular streak from nostrils to the outer angle of the eyes, extending along the sides of the body. In a few specimens a similar dorso-median streak is occasionally present.

Spiracle.—Situated in the midventral line, large and broadly "Λ"-shaped, opening directed backwards and is far from the snout. Behind, another pore involved in the lower caudal crest is present, marked abdominal pore in figure B. There is reason to suppose this to represent a secondary spiracle. Water comes out in two streams as may be experimented with carmine solution.

Vent.—Slightly sinistral, inconspicuous, covered over by the lower tail lobe.

Tail.—Muscular portion thick at the root and ends in a very

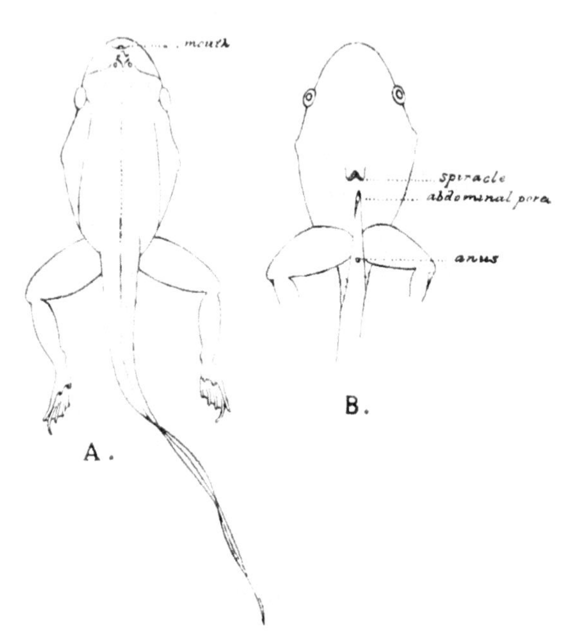

FIG. 1.—Tadpoles of *Microhyla rubra*.
A. Dorsal view. B. Ventral view.

fine flagellum. At the greatest width, i.e. between the thighs, the ventral crest is more than four times the upper membrane. The former begins behind the spiracle and surrounds the second pore. The lobes are delicate and transparent.

Colour.—Live specimens are olive above, beautifully marbled. Spirit specimens do not, however, show this delicate scheme of colour. Following the glandular line of the head and body, is a dark band which throws it in relief. Limbs barred. A brown band across the thighs. Abdomen immaculate, occasionally the throat is bronzed.

Dimensions.—The following are the average measurements of four tadpoles with well-developed hind limbs:—

Total length	..	40 mm.
Length of head and body	..	15 mm.
Maximum width of body	..	9 mm.
Do. depth do.	..	7 mm.
Do. do. tail	..	9 mm.

Biological.—The period occupied by development in the aquarium is roughly twenty days, and it must also represent the time taken in nature. *Microhyla* like the other genera of the family Engystomatidae spawns in localities which dry up very soon, and the tadpoles are also otherwise exposed to attacks by ducks and geese. Rapid metamorphosis is apparently a provision, in the case of these thoroughly terrestrial forms, for the preservation of the species.

The larvae float on the surface and the highly contractile mouth is a character which they share in common with the other species, *M. ornata*, noticed by Capt. S. S. Flower and Mr. Ferguson. The food of the tadpoles consists of small micro-organisms such as water fleas, Infusoria and Rotifers. The fine flagellum at the end of the tail is kept lashing the water. As soon as the fore-limbs develop, the larvae leave the water and squat on the stones in the aquarium, and if these are removed they easily perish. The metatarsal tubercles are well-developed and the baby frogs with short stumpy tails use them in burrowing. The web which completely invests the toes in the larval stage atrophies when the tadpoles leave the water.

Points of Difference.

I shall next proceed to enumerate briefly the points in which I differ from Mr. Ferguson.

(1) He remarks that the nostrils are nearer the eye than the end of snout.

> I make out in my specimens that the converse is true; the distance between the eye and the nostril is at least $1\frac{1}{2}$ mm. greater than that between the nostrils and snout.

(2) The spiraculum is described by him as being directed *downwards* and backwards.

> I notice that the spiracle is directed *downwards* and backwards in the larvae of *M. ornata* in which the abdomen is laterally compressed; while in *M. rubra*, the body being dorso-ventrally depressed, the spiraculum opens backwards as a rule.

(3) Again Mr. Ferguson observes that the spiraculum is close to the anus which also opens in middle line in the lower edge of the subcaudal crest.

> It is obvious that he mistakes the abdominal pore for the anus which for anatomical reasons cannot occupy that position. The anus, however, is normal in position between the hind legs and is slightly sinistral.

(4) In describing the colour, Mr. Ferguson observes that in life the body is almost transparent.

> I am perfectly certain that the tadpoles of *M. rubra* are opaque while the transparent character of the larvae of *M. ornata* is noticed by Capt. S. S. Flower and Mr. A. L. Butler.

(5) Further down Mr. Ferguson notices that the dark marks form a more or less diamond-shaped figure on the back.

> This is again a feature characteristic of the tadpoles of *M. ornata* and not met with in the allied form, viz. *M. rubra*. In the former species, if we follow the progress of metamorphosis, we may notice the diamond-shaped figure developing in the adult into "a large dark marking on the back, beginning between the eyes and widening as it extends to the hind part of the body." Boulenger (*Fauna*, p. 412).

Habits of the Adult.

This frog does not appear to extend into the Malay Peninsula as may be judged from Mr. A. L. Butler's account of the batrachians of that region (*J.B.N.H.S.*, Vol. XV, p. 387), nor does it occur in such abundance as the other little frog *M. ornata*. It is a deep digger as is evidenced by the presence of two powerfully developed metatarsal tubercles, and I have myself obtained specimens nearly two feet from the surface. The frog does not come out of the burrow during the hot weather and only a very heavy shower of rain, an inch and a half or two, can induce it to leave its hiding place. During the breeding season which in Madras comes off between November and January, and in Bangalore between June and September, the batrachian generally remains on the surface hiding by day under stones, flower pots or in hedges and coming out to feed or spawn by night. The frogs are very good jumpers, but if kept long in water show signs of distress. They feed voraciously on young termites and can stand captivity well. The call notes resemble the shrill chirping of a tree cricket from which they however differ in being an interrupted cry. It is by no means difficult to distinguish the cry of this species in the general babel of amphibian voices that ensue a heavy shower of rain in the night.

Larva of **R. breviceps**.

H. S. Ferguson, *J.B.N.H.S.*, Vol. XV, 1904, p. 502; Boulenger, *Fauna*, p. 451.

These tadpoles were taken in conjunction with the larvae of *M. rubra* and were reared in the college aquarium. They differ from Mr. Ferguson's account in so many particulars that I have no doubt that he is describing some other species. My specimens may be described thus:—

The head and body.—Body short and oval. Dorsal and ventral surfaces moderately flat or slightly arched. Snout obtuse or rounded. Length of body about one and a half times the breadth. Mouth ventral.

Nostril and eye.—Small, not prominent, nostrils dorsal, nearer to eyes than to mouth. The inter-orbital space is slightly more than twice the distance between the nostrils. Eyes dorso-lateral.

Mouth.—Ventral, small. Lower lip better developed and directed forwards. Both lips are bare of papillae, which, however, are large and are aggregated in two or three rows in the corners of the mouth. Occasionally in a few cases a small ovoid gland may be present in the same region. Beaks horny and not powerful; both finely serrated; lower jaw broadly **V**-shaped and the upper crescentic. Dental formula, 1 : 1/3.

Sensory pits and glands.—Occur generally scattered on the head. A fine row of whitish glands from the eyes to the tympanum. A dorso-median streak sometimes found.

Spiracle.—Tubular, sinistral, pointing upwards and slightly backwards; a fairly circular opening; somewhat low on the side, nearer to eye than to root of tail.

Vent.—Also sinistral, a fairly prominent tube.

Tail.—Tip not pointed; dorsal lobe beginning much behind the root of legs, is strongly arched. The ventral poorly developed, the outer margin of which is almost parallel to the long axis of the muscular portion. The greatest depth of tail is [illegible] of the total length, and at this part the lower membrane is only $\frac{1}{3}$ of the upper. Muscular portion strongly developed.

Skin and colouration.—Skin either granular or warty with strongly developed tubercles. Dorsal surface deep grey with broadly **V**-shaped dark mark between the eyes, and **M**-shaped, sometimes broken, marks on the back. Ventral surface whitish and sides finely dotted in a few specimens. The muscular portion and lobes and tail deeply blotched. Limbs barred.

Limbs.—Short, toes poorly webbed at the base. The metatarsal tubercle well-developed, about the size of the first toe. Subarticular tubercles well formed.

Dimensions.—A fully grown tadpole measures as follows:—

Total length	50 mm.
Length of body	20 mm.
,, ,, tail	30 mm.
Maximum breadth of body	13 mm.
,, depth ,, ,, ...	9 mm.
,, ,, ,, tail . ..	8 mm.

Biological.—The time occupied by the development of this frog is almost the same as that taken by the other burrowing types, viz. 18 to 20 days. I have noticed that these larvae remain at the bottom of the aquarium, occasionally coming to the surface to breathe air. When disturbed, they would move on their legs, rather than swim. They were fed on weeds and also on dead tadpoles. Foul water is death to them. Like the larvae of other

Engystomatidae, they leave the water as soon as the front limbs sprout.

Points of Difference.

The particulars in which the above description differs from Mr. Ferguson's may be briefly indicated below.

(1) He states at page 502 of the journal cited above, that the length of the body is one and three quarters its breadth

> I have measured ten full-grown specimens and I find the average ratio of length to breadth is as 15 : 10 mm , in other words the length is one and a half times the breadth.

(2) Further he describes that the distance between the eyes is one and a quarter that between the nostrils and is equal to the width of the mouth.

> In measuring the same ten specimens, I find that the inter-orbital space is more than twice the internasal, and is one and two-thirds of the width of the mouth.

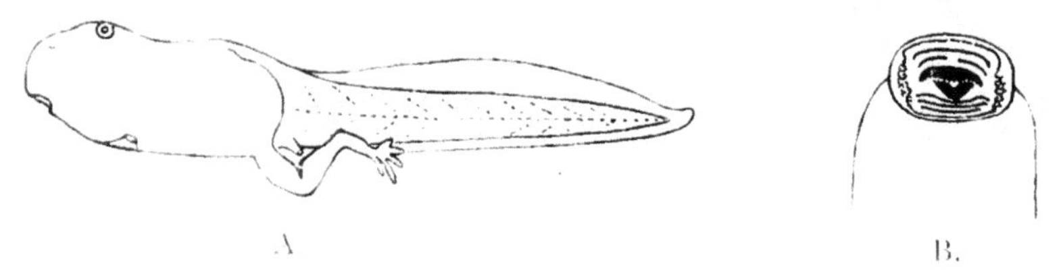

FIG. 2. Tadpoles of *Rana breviceps*.
A. Lateral view. B. mouth

(3) Mr. Ferguson states in regard to the spiraculum that it is visible above and below.

> In all my specimens the spiracle is so low on the side that it is visible from neither view.

(4) He makes out that the anal opening is on the middle line.

> All the adult tadpoles in my collection possess a sinistral vent.

(5) The tail is described in the Travancore specimens as being acutely pointed.

> Almost all the specimens in my collection show a rounded tip.

(6) In the description of mouth parts, Mr. Ferguson states that the upper mandible has a blunt tooth-like prominence and that the outermost row of teeth on the lower lip is less than half the length of the middle row which again is shorter than the upper.

> The prominence spoken of, perhaps such as is met with in the larva of *R. tigrina*, is not discoverable and as regards

the rows of horny teeth, the first two rows are nearly equal, while the third is only slightly shorter than either the first or the second.

(7) The total length of Mr. Ferguson's specimens is 41 mm. The maximum length of my specimens exceeds this by at least 9 mm.

Habits of the Adult.

The adult frog is thoroughly terrestrial and the burrowing habits have produced an external appearance not unlike that of *Cacopus systoma*: a rounded snout, small mouth, a stout body, short hind limbs, a powerful metatarsal tubercle and very slight web. It leads a solitary life and congregates only during the pairing season. A light vertebral line is present in most specimens and its occurrence is purely arbitrary. It is not one of the concert-giving frogs like *R. cyanophlyctis* and the call notes may be expressed by the short syllables "Rut-Rut-Rut," uttered in quick succession. The batrachian is entirely nocturnal in its habits and young frogs stand captivity much better than adult examples.

II. The Distribution and Habits of **Bufo fergusoni**.

This little toad has been described by Dr. G. A. Boulenger, (*J.B.N.H.S.*, Vol. VII, p. 317).

In the article quoted above (*viz.* "A list of Travancore Batrachians"), Mr. Ferguson makes the following remark in the opening lines: "There have been so far thirty-four species of Batrachians described as occurring in Travancore, three of which have not been found elsewhere as yet. They are *Rana aurantiaca*, *Ixalus travancoricus* and *Bufo fergusoni*" (*J.B.N.H.S.*, Vol. XV, 1904, p. 499).

I have no doubt that this species of *Bufo* enjoys a much wider distribution. In 1903, two specimens were taken in the compound of the then residence of Dr. William Miller in Nungambakam, Madras: one of which was sent to Dr. G. A. Boulenger, who in acknowledging receipt of the toad, mentions that it is also known from Ceylon. Since then specimens have been obtained from S. Malabar and the outlying districts of Mysore. It is possible that the little toad may be found in North India, though, however, its occurrence is not yet reported.[1]

The following is a short account of the observations made on the habits of this animal. It is entirely nocturnal and does not appear to occur in any large numbers and is certainly one of the rare toads. When given loose earth, it burrows with great ease. It feeds almost exclusively on termites. It does not touch black ants, smaller beetles and earthworms which form the staple

[1] I think that this toad is replaced in Northern India by *B.* [illegible] Lütken.—[illegible]

food of the bigger toads like *B. melanostictus*. Walking is the normal mode of progression and it can also run, especially if quarry is sighted at a distance. When the animal walks, the body is lifted from the ground, but is still underhung from the limbs, and the movement has all the awkwardness of a *Calotes*, which arises from the inequalities of the limbs. In trying to take a wider range of view of the surroundings, the body is supported on the four legs and the animal may move in that condition somewhat mammal-wise. In running the head is kept low. When left on the table it gently crawls round the edge (body almost touching the surface) measuring the height, and prefers to remain quiet in the centre to performing the heroic feat of jumping off. Even if pressed under the arm pit, it does not utter the plaintive metallic cry characteristic of the common toad. When held, it does not struggle to escape, but will remain quiet and even pick up white ants from off one's hand. If thrown in water, especially if it is deep, it darts here and there and then is easily drowned if not rescued in time.

IV. SOME ORIENTAL SAWFLIES IN THE INDIAN MUSEUM.

By S. A. ROHWER, *Bureau of Entomology, United States Department of Agriculture, Washington, D.C.*

In the fall of 1912 the writer received, on his request, the unnamed sawflies of the Indian Museum for study. This collection contained some new species and genera and certain species previously described. A report of the named species and descriptions of the new species and genera will be found on the following pages. With permission of the authorities of the Indian Museum certain duplicates were retained; these have been placed in the collections of the United States National Museum. Thanks are due the authorities of the Indian Museum for the privilege of studying this collection, for their generosity in giving duplicates, and for the extending of the original time limit.

Superfamily *SIRICOIDEA*.

Genus **Sirex**, Linnaeus.

Sirex imperialis, Kirby.

One male from Shillong, Assam (*La Touche*).

This male has the apical margins of the tergites rufous and the wings more yellowish than the description of the female indicates.

Genus **Xiphydria**, Latreille.

Xiphydria orientalis, Westwood.

One female from Kurseong, East Himalayas, collected May 21-29, 1906, at an altitude of 5,000 feet (*N. A.*).

This specimen differs from the original description in the antefurcal second recurrent vein, but is no doubt Westwood's species. The mandibles are quadridentate; the head around the ocelli is striato-punctate.

Superfamily *TENTHREDINOIDEA*.

Family CIMBICIDAE.

Genus **Abia**, Leach.

Abia melanoceros, Cameron.

One male from Khasi Hills, Assam.

Family ARGIDAE.

Genus **Cibdela**, Konow.

Cibdela janthina (Klug)

One male from Sadiya, Assam.

Genus **Athermantus**, Kirby.

In 1882 (List Hym. Brit. Mus., vol. I, p. 54) Kirby characterized his genus *Athermantus* for *Hylotoma imperialis*, Smith. Kirby's description is brief and the figure is in part incorrect, so this genus has not been well understood. Konow in his table in the Genera Insectorum (fasc. 29, 1905, p.13) separates Kirby's genus from the other Argini without a superapical spur on the hind tibiae, by the compressed hind tibiae. In his table of the genera of Argidae in Zeit. Hym. Dipt., vol. 7, 1907, p. 185, Konow abandons the character of compressed hind tibiae and separates *Athermantus* from the other genera on venational characters. Due to the inaccuracy of the artist and the probability that Konow had never seen a specimen of *Athermantus*, this separation cannot be used. Mr. Meade-Waldo has kindly examined the type of *Athermantus imperialis* (Smith) and from his notes and the descriptions there can be no doubt that the specimen before me is correctly determined. The following descriptive notes are given to more properly establish the identity of this genus:—

Closely allied to *Cibdela*, Konow, but may be separated from that genus by the following comparison:

ATHERMANTUS, Kirby.	CIBDELA, Konow.
Facial quadrangle much broader than the length of the eye.	Facial quadrangle with its width subequal or but little greater than the length of the eye.
Posterior orbits much broader than the cephalo-caudad diameter of the eye.	Posterior orbits much narrower than the cephalo-caudad diameter of the eye.
Malar space longer than the length of pedicellum.	Malar space narrower than the length of pedicellum.
Postocellar area well defined.	Postocellar area obsolete.
Lateral ocelli behind the supra-orbital line.	Lateral ocelli with their anterior margins on the supra-orbital line.
Propodeum without a median furrow.	Propodeum with a faint median furrow.
Posterior tibiae compressed.	Posterior tibiae not compressed.

In Konow's table in Zeit. Hym. Dipt., vol. 7, 1907, p. 185, this should fall next to *Cibdela*, but would be separated by the above comparison.

Athermantus imperialis (Smith).

One female from Kurseong, East Himalayas, collected August 6, 1909, at an altitude of 6,000 feet (*E. D'Abreu*). Indian Museum No. $\frac{4630}{19}$.

Genus Arge, Schrank.

Arge fumipennis (Smith).

Two females from Almora, Kumaon, collected September 3-12, 1911, at an altitude of 5,500 feet (*C. Paiva*).

Arge luteiventris (Cameron).

Fourteen specimens, males and females, from Shillong, Assam (*La Touche*).

Arge xanthogastra (Cameron).

Two specimens from Almora, Kumaon, collected June 27, 1911, at an altitude of 5,500 feet (*C. Paiva*).

Arge albocincta (Cameron).

Hylotoma albocincta, Cameron, Trans. Ent. Soc. London, 1876, p. 459.

It may be advisable to make a new genus for this characteristic species which has the large eyes almost touching the base of the mandibles, but until more material has been studied the author feels loath to propose such a genus. The following characters apply to the specimen at hand: Emargination of the clypeus sub-V-ed; supraclypeal foveæ deep, elongate; frontal basin well defined, two and one-third times as long as its dorsal width; a shallow depression in front of the anterior ocellus; postocellar furrow angulate anteriorly; postocellar area not defined laterally; head and thorax shining, with sparse, fine punctures; basal vein almost the length of the intracostal vein basad of cubitus; second cubital cell parallel-sided, about three times as long as apical width; apical abdominal segment with dense white hair.

One female, Shillong, Assam (*La Touche*) in the Indian Museum Collection No. $\frac{1706}{15}$.

Genus Pampsilota, Konow.

Pampsilota sinensis (Kirby).

Hylotoma microcephala, Cameron, Trans. Ent. Soc. London, 1876, p. 460. (*nec* Vollenhoven).

Hylotoma sinensis, Kirby, List Hym. Brit. Mus., vol. I, 1882, p. 72, pl. 5, fig. 2.

Eight specimens, males and females, from Assam (Sadiya, 5 specimens, "Sibs" (Sibsagar) N.E. Assam 1 specimen, and 2 without definite locality) forwarded by the Indian Museum, agree well with Cameron's and Kirby's accounts. They belong to the genus *Pampsilota* which easily explains Kirby's inability to detect superapical spurs

Pampsilota nigriceps, sp. n.

This species is probably more closely allied to *sinensis* (Kirby) than any other described species of this genus, but it may be differentiated from Kirby's species by the black legs and black tergites.

Female.—Length 11 mm. Anterior margin of the clypeus very slightly incurved ; supraclypeal area black ; frontal fovea open below, extending parallel until it reaches the anterior margin of the anterior ocellus, with an accentuated triangular-shaped depression opposite the upper margin of the antennae: postocellar line distinctly shorter than the ocellocular line; postocellar furrow well defined, about the width of the posterior ocellus behind the posterior ocelli ; postocellar area strongly convex, slightly parted by a median furrow; antennae typical for the genus, extending to the posterior margin of the scutellum. Rufo-ferrugineous ; head except the palpi black ; scutellum, metanotum, tergites except the lateral margin and the apical portion of the posterior ones, the sheath above, mesosternum and legs *black* ; the anterior legs beneath are piceous ; wings dusky hyaline, venation black.

Male.—Length 8 mm. What appears to be the male of this species has the frontal fovea closed below and differs in colour in having the scutum and the basal portion of the prescutum black, and in the piceous stigma and paler wings.

India. Described from two females, one type, and one male, allotype: the type female from Mungphu, Sikkim : the paratype female from Sikkim, May 1912 ; the allotype from Sadon, Upper Burma, collected at an altitude of 7,000 feet, April, 1911 (*E. Colenso*).

Type and *allotype* in Indian Museum, type No. $\frac{7921}{8}$, allotype [illegible].

Paratype—(Female) Type Cat. No. 18530, U.S.N.M.

Family TENTHREDINIDAE.

Genus Xenapates, Kirby.

Xenapates incerta (Cameron).

Two females from Sikkim, East Himalayas, collected May, 1912. One female from Ghumti, Darjiling district, East Himalayas, collected July, 1911, at a calculated altitude of 4,000 feet, by F. H. Gravely. One male from Kurseong, East Himalayas, collected June 29, 1910, at an altitude between 4,000 and 8,000 feet (*N. Annandale*). One female from Sadiya, Assam.

Monostegidea, gen. n.

This genus belongs to the tribe Allantini where it is related to *Monostegia*, Costa and *Monsoma*, MacGillivray. It may be separated from both of these by having the antennae long and slender and the third and fourth joints subequal.

Malar space distinct ; clypeus arcuately emarginate (the depth of the emargination varies considerably) ; antennal furrows complete but not strong ; orbital carina obsolete ; posterior orbits narrower than or subequal with cephalo-caudad diameter of the eye ; antennae long and slender, pedicellum wider than long, third and fourth joints subequal ; head and thorax shining, almost impunctate ; tarsal claws cleft, inner tooth shorter ; first transverse cubitus present ; nervulus its length from the basal ; hind wings with one discal cell, nervellus at right angles with the anal vein ; apical joints of the hind tarsi shorter than the two preceding.

Type.—*Poecilosoma nigriceps*, Cameron.

Monostegidea nigriceps (Cameron) Rohwer.

Poecilosoma nigriceps, Cameron. Jn. Bombay Nat. Hist. Soc., Vol. XIV, 1902, p. 442.

This species is represented in material received from the Indian Museum from the following localities :—

Darjiling, altitude 7,000 feet. Two males collected by C. Paiva. Kurseong, altitude 5,000 feet. Two females and three males : collector not given.

Siliguri, base of the East Himalayas. Three females collected by Museum collector.

Ghumti, altitude 4,000 feet. One female collected by F. H. Gravely.

Monostegidea leucomelaena, sp. n.

This resembles to some extent *Taxonus fulvipes* as described by Cameron, but it is not in agreement with Cameron's description in many characters. It can readily be separated by the pale spot on the scutellum.

Female.—Length 6 mm. Anterior margin of the clypeus deeply, narrowly, arcuately emarginate, the lobes broad, obtusely rounded apically : supraclypeal area rectangular in outline, strongly convex : supraclypeal foveae greatly reduced, below the antennal fovea ; middle fovea crescent-shaped ; antennal furrows uninterrupted, not complete dorsally : no depression in front of the anterior ocellus : postocellar furrows distinctly defined, angulate anteriorly ; postocellar line less than one half as long as the ocellocular line, shorter than the ocellocipital line ; third, fourth and fifth antennal joints subequal ; stigma gently rounded on the lower margin ; transverse radius joining the radius slightly beyond the middle of the third cubital cell ; sheath straight above, obtuse at apex and obliquely rounded below. Black with white markings ; head black ; mouth parts except the apices of the mandibles, face below the antennae, inner orbits to the top of the eye, posterior orbits to above the upper margin of the eye, white : pronotum black ; tegulae white : mesoscutum black : scutellum black with a small white spot anteriorly; mesepisternum white except along the dorsal margin ; entire ventral part of the insect white ; abdomen black above, except the narrow lateral margin of the segments ; venter white : legs yellowish ; white posterior tarsi dusky : antennae black ; wings hyaline, iridescent ; venation dark brown, stigma pale brown, paler at base.

Male.—Length 5 mm. Agrees with female except in the sexual characters.

Darjiling, East Himalayas. Described from one female and six males collected at an altitude of 7,000 feet, May 25-29 (*E. Brunetti*).

Type, *allotype* and *paratype* in the Indian Museum. Type No. $\frac{4673}{19}$; and allotype, No. $\frac{4663}{19}$.

Paratype in U.S.N.M.

Genus **Tenthredella**, Rohwer.

Tenthredella assamensis (Konow).

One female from the Khasi Hills, Assam, which agrees very well with the original description. The basin is shallow, bounded by low rounded walls; clypeal lobes obtusely triangular; postocellar area slightly wider than long.

Tenthredella carinifrons (Cameron).

One female from Darjiling, collected May 28, 1910, at an altitude of 7,000 feet (*E. Brunetti*).

Tenthredella segrega (Konow).

Two females from Sikkim, Eastern Himalayas.

Tenthredella turneri, Rohwer.

One female from Shillong, Assam.

Tenthredella xanthoptera (Cameron).

One female from Kurseong, East Himalayas, collected September 7, 1909, at an altitude of 5,000 feet.

This specimen, which agrees well with the original description, has the scutellum pyramidal; anterior ocellus in a well defined, shining basin which is deeper than the rest of the frontal basin ; labrum longer than wide, rounded apically; sheath subparallel-sided, with apex regularly rounded: mesepisternum strongly angulate below.

Tenthredella annandalei, sp. n.

This species may be readily separated from the other species of this group by the black posterior tibiae, the pale posterior femora, the pyramidal scutellum, and in having the postocellar area broader than long.

Female.—Length 14 mm.; length of the antennae 10·5 mm. Labrum broadly rounded, surface subopaque with a few large setigerous punctures; clypeus shining with a deep, narrow, arcuate emargination, the lobes broad, truncate; supraclypeal foveae deep, punctiform; supraclypeal area flat, slightly carinate dorsally; head concave in front; a deep longitudinal depression from the base of the antenna to the anterior ocellus, but below the anterior ocellus broadens into a diamond-shaped area which meets immediately behind the anterior ocellus; antennal furrows poorly defined; ocelli little less than an equilateral triangle; postocellar area sharply defined laterally, and defined anteriorly by shallow postocellar furrow, one-fifth longer than the latrad width; head shining, practically impunctate; antennae slender, the third and fourth joints subequal; thorax shining; scutellum pyramidal; stigma long and narrow, tapering apically; transverse radius strongly curved, received in the third cubital cell distinctly beyond the middle but not in the apical third; third cubital cell on the radius longer than the first and second; second recurrent vein strongly bullated, joining the cubitus a little less than the length of the curved second transverse cubitus beyond the base of the third cubital cell; spurs of the posterior tibiae of equal length, about half as long as the posterior basitarsus; sheath straight above, apex rounded, the lower margin convex. Rufo-ferrugineous with distinct, erect black hairs; clypeus, labrum, mandibles except apices, inner orbits of the eyes and tegulae rufo-stramineous; antennae black; scape stramineous beneath; apical three abdominal segments black; legs colour of the body except the posterior tibiae beyond the basal third and the posterior tarsi which are black; wings yellow hyaline, strongly dusky beyond the apex of the stigma; most of the venation colour of the wing but the median and basal veins black; stigma yellow.

Kurseong, East Himalayas. Described from one female collected at an altitude between 4,500 and 5,000 feet, on June 22, 1910 by Dr. N. Annandale, after whom the species is named.

Type.—Indian Museum No. $\frac{1698}{19}$.

Genus **Parastatis**, Kirby.

This is hardly more than a species group of *Tenthredo*, Linnaeus.

Parastatis indica, Kirby.

Four specimens from Sikkim.

Genus **Pëus**, Konow.

Pëus privus, Konow.

One female from Kurseong, East Himalayas, collected August 14, 1909, at an altitude of 6,000 feet (*D'Abreu*).

Genus **Fethalia**, Cameron.

Konow in the Genera Insectorum, 1905, Fasc. 29, p. 132, places the genus *Fethaila*, Cameron, as a synonym of *Tenthredo* (*Allantus*), but a careful examination of Cameron's description indicates that this genus is good and is more closely allied to the genera *Pëus*, Konow, and *Jermakia*, Jakovlev. Cameron says that there is no "blotch" on the abdomen. This is taken to mean that the first tergite is without a longitudinal furrow. The three genera of the Tenthredinini which do not have longitudinal furrows are *Jermakia*, *Fethalia* and *Pëus*. *Jermakia* can be readily separated from *Fethalia* and *Pëus* by the obsolete malar space. The only character in Cameron's description which will separate *Fethalia* from *Pëus* is the short antennae. Until examination of Cameron's type has been made these two genera had best be regarded as distinct. It may be, however, that they are not separable as the relative length of the antennae when taken alone can hardly be considered as a generic character in this group.

Genus **Pachyprotasis**, Hartig.

Pachyprotasis versicolor, Cameron.

Three females and four males from Darjiling, East Himalayas, collected May 25, 1910, at an altitude of 7,000 feet (*E. Brunetti*).

Genus **Athalia**, Leach.

Athalia infumata (Marlatt).

One male from Bijrani, Naini-Tal District, base of West Himalayas, collected March 10, 1910.

Athalia proxima (Klug).

Six specimens from Calcutta; two from Shillong, Assam (*La Touche*); one from Mangaldai, Assam; two from Bangalore, South India; and one from Sonali, Purneah District, Behar.

Genus **Anapeptamena**, Konow.

Anapeptamena viridipes (Cameron).

One female from Siliguri, base of East Himalayas, collected July 18-20, 1907.

In the original description, fifth line from the bottom of page "second" should be changed to "third", and in the fourth line from bottom of page "third" should be changed to "second".

Busarbidea, gen. n.

Type.—*Busarbidea himalaiensis*, new species.

Clypeus arcuately emarginate; malar space wanting; inner margins of the eyes parallel; pentagonal area large, well defined and with a transverse carina from its lateral margin to near inner margin of eye; posterior orbits rather narrow, with a strong, well defined carina; antennae slender, the third joint distinctly longer than the fourth; pedicellum subequal in length with the scape, much longer than wide; basal vein curved, joining the subcosta well basad to the origin of the cubitus, somewhat divergent with the first recurrent; nervulus at about middle of cell; costa enlarged apically; lanceolate cell with a nearly straight cross-vein; hind wings with two discal cells and a petiolate anal cell; claws cleft, inner tooth shorter; hind basitarsus subequal with following joints.

This genus, which belongs to the Selandriinae, is very closely related to *Anapeptamena*, Konow, but may be distinguished by the presence of a cross-vein in the lanceolate cell.

Busarbidea himalaiensis, sp. n.

Female.—Length 5 mm. Anterior margin of the clypeus depressed, rather deeply arcuately emarginate, basal portion convex; supraclypeal area uniformly convex; supraclypeal foveae deep, punctiform; middle fovea nearly quadrate in outline, not sharply separated from the pentagonal area; pentagonal area broader on its ventral margin than the dorsad-ventrad length; from its dorsal margin is a short rather poorly defined carina which extends posteriorly one-third the length of the postocellar area; postocellar area sharply defined laterally narrowing anteriorly; posterior margin subequal with the median cephla-caudad length; postocellar line one-fifth shorter than the ocellocular line; antennae distinctly compressed beyond the fifth joint; fourth and fifth joints subequal; the third slightly longer than the fourth; head and thorax shining, impunctate; stigma evenly rounded below; second cubital cell shorter on both radius and cubitus than the third; transverse radius strongly curved, received at the apical third; third transverse cubitus twice as long as the second; sheath robust, straight above, obtuse apically, oblique beneath. Black; clypeus, labrum, palpi, first two joints of the flagellum, tegulae, posterior angles of the pronotum, legs yellowish; wings hyaline, faintly dusky; venation dark brown; head and thorax with short black hair.

Male.—Length 4 mm. Differs from the description of the female in having the body markings piceous; hypopygidium truncate apically, the angles rounded.

East Himalayas. Described from one female (type) collected at Siliguri, April 18-20, 1907, by a Museum collector, and from one male (allotype), and one female (paratype) from Kurseong, at an altitude of 5,000 feet.

Type and *allotype* in the Indian Museum. Type No. [illegible], allotype $\frac{2601}{12}$.

Paratype in U.S.N.M.

Genus **Aneugmenus**, Hartig.

Aneugmenus annandalei, sp. n.

This species is related to the European *morio* but may be separated from it by head characters, judging from specimens of *morio* determined by Konow, now in the National Museum.

Female.—Length 5 mm. Labrum truncate, convex; clypeus convex; anterior margin shallowly arcuately emarginate, its surface with scattered punctures; supraclypeal area low, flat; supraclypeal elongate, not connected with the antennal foveae; middle fovea well defined, quadrate in outline; frontal foveae punctiform, deep, lower margin slightly below the line drawn tangent to the dorsal margin of the middle fovea, pentagonal area indicated by a **U**-shaped raised area the dorsal margin of which is inside of the lateral ocelli; posterior orbits and genae without carinae; postocellar area convex, poorly defined laterally by short furrows, not defined anteriorly; postocellar line subequal with the ocellocular line; pedicellum wider apically, longer than its apical width; third antennal joint little shorter than the fourth and fifth; head and thorax shining; nervulus less than its length from the end of the cell; first transverse cubitus obsolete; stigma broad, uniformly rounded beneath; transverse radius oblique, joining the radius at about the apical third or a little beyond that; third transverse cubitus about three times as long as second; lanceolate cell of the hind wings petiolate, petiole half as long as the nervellus; posterior basitarsus somewhat shorter than the following joints; claws cleft. Black; palpi and legs, except the infuscate apical joints of the tarsi, yellow: head and thorax with short gray hair; wings dusky hyaline; venation black.

Male.—Length 3·5 mm. Differs from the female in having the bases of the coxae black.

Bangalore, South India. Described from one female (type) and one male (allotype) collected at a calculated altitude of 3,000 feet, October 15, 1910 by Dr. N. Annandale, after whom the species is named.

Type.—In the Indian Museum No. $\frac{1759}{19}$.

Allotype in U.S.N.M

Genus **Nesoselandria**, Rohwer.

Nesoselandria rufiventris, sp. n.

Readily separated by the fulvous abdomen.

Female.—Length 4 mm. Anterior margin of the clypeus truncate; supraclypeal area convex; supraclypeal foveae obsolete; middle fovea transverse, well defined; lateral foveae punctiform with their lower margin tangent with the upper margin of the middle fovea; head above the antennae without transverse carinae; pentagonal area obsolete; postocellar area indicated anteriorly by punctiform foveae; ocellocipital line distinctly longer than the intraocellar line; postocellar line a trifle longer than the ocellocular line; fourth antennal joint slightly longer than the fifth; stigma gently rounded below, broader at the basal third; sheath subacuminate, narrow. Black; apical margin of the clypeus, legs and abdomen fulvous; wings dusky hyaline; venation dark brown; head and thorax with thin gray hairs.

Male.—Length 3·75 mm. Differs from the above description of the female in having the apical three abdominal segments black and having the clypeus entirely black; hypopygidium nearly rounded apically.

India and Assam. Described from two females, one type, from Calcutta, collected November 22, 1907, and from one male collected at Mazbat, Mangaldai, Assam, January 8, 1911 (*S. W. Kemp*). One male from Margherita, Assam.

Type and *allotype* in Indian Museum; type No. $\frac{4596}{19}$; allotype No. $\frac{4536}{19}$.

Paratype.—Cat. No. 18910, U.S.N.M.

Genus Neostromboceros, Rohwer.

Neostromboceros coeruleiceps (Cameron).

One female from the Assam-Bhutan Frontier, Mangaldai District, N.E., collected December 26, 1910 (*S. W. Kemp*); two males from the same locality collected by the same collector January 1-2, 1911; and two males from Sadiya, Assam.

Neostromboceros similaris, sp. n.

From *trifoveatus*, Cameron, to which this species runs in Enslin's table, this species may be separated by the white labrum.

Female.—Length 7 mm. Anterior margin of the labrum broadly rounded; anterior margin of the clypeus truncate; supraclypeal area rectangular in outline, flat; supraclypeal foveae connected with the antennal foveae; median fovea a U-shaped depression around a median tubercle, the ends of the U deeper; frontal foveae punctiform, their lower margins tangent to a line drawn through the median fovea; antennal furrow distinct above crest; postocellar area wider posteriorly, the anterior lateral part sharply defined, the posterior lateral part poorly defined; flagellum somewhat flattened; first transverse cubitus wanting; sheath truncate apically; inner tooth of claws smaller than outer. Blue-black; labrum, posterior margin of pronotum, tegulae and perapteron white; palpi whitish, infuscate; legs yellowish white, bases of coxae and femora, and the posterior tarsi more or less infuscate; wings hyaline, slightly dusky; venation dark brown; head and thorax with short gray hair.

Male.—Length 6 mm. Hypopygidium broadly rounded apically. Very like female.

Mazbat, Mangaldai, Assam. Described from three females (one type) and two males (one allotype) collected January 8, 1911 (*S. W. Kemp*).

Type, *allotype* and *paratype* in Indian Museum; type No. $\frac{1741}{19}$, allotype No. $\frac{1736}{19}$, paratype $\frac{1731}{19}$.

Paratypes (male and female) in U.S.N.M.

Genus **Stromboceros**, Konow.

Stromboceros tarsalis (Rohwer).

Three females from Sadiya, Assam; and two females from Margherita, Assam.

This is a good species.

Stromboceros phaleratus, Konow.

One male from Sikkim collected by Knyvett; and one male from Margherita, Assam.

Stromboceros ruficornis, sp. n.

This species is readily separated from all other species of *Stromboceros* occurring in the oriental region by having the basal joints of the flagellum rufous.

Female.—Length 8 mm. Anterior margin of the clypeus depressed, the middle very slightly emarginate, the basal portion subconvex; supraclypeal area flat; supraclypeal foveae oblique, deep, not connected with the antennal foveae; middle fovea represented by an inverted U-shaped furrow around a flattened tubercle; frontal foveae deep, rounded, the lower margins tangent to a line drawn across the top of the median tubercle; postocellar area well defined laterally, not defined anteriorly; postocellar line but little more than half the length of the ocellocular line; head shining, impunctate; antennae slightly tapering apically; the third joint slightly longer than the fourth; thorax shining; stigma uniformly rounded below; first transverse cubitus obsolete; nervulus slightly basad of the middle; transverse radius oblique, received at the apical third; third transverse cubitus oblique at about the same angle as the transverse radius, slightly more than twice as long as the second transverse cubitus; posterior basitarsus subequal in length with the following joints. Black; three basal joints of the flagellum rufous; anterior knees, bases of the posterior coxae, posterior trochanters and the band on the posterior tibiae white; head and thorax with dense gray hair; wings hyaline; venation black.

Darjiling, Eastern Himalayas. Described from one female collected May 27, 1910, at an altitude of 7,000 feet (*E. Brunetti*).

Type.—In the Indian Museum No. $\frac{4682}{19}$.

Genus Canonias, Konow.

Canonias assamensis, sp. n.

This species differs in minor colour characters from *inopinus*, Konow.

Female.—Length 8 mm. Anterior margin of the clypeus slightly arcuately emarginate, the angles sharp; supraclypeal area flat; supraclypeal foveae elongate, deep; antennal foveae obsolete, middle fovea quadrate in outline, open above; pentagonal area well defined, from its broadest portion there is a transverse carina which touches the inner margin of the eye; postocellar area depressed, sharply defined laterally by deep foveae, not defined anteriorly, about four times as wide as the cephal-caudad length; postocellar line about one-fourth longer than the ocellocular line; antennae long and slender, tapering apically; third joint distinctly shorter than the fourth; head and thorax shining, impunctate; stigma broadest at middle, tapering each way; transverse radius strongly oblique joining the radius slightly before the apical third; second cubital cell longer on both radius and cubitus than cubital third. Black; scape, pedicellum, sixth, seventh and eighth antennal joints, tergite, posterior margin of the angles of the pronotum, tegulae, perapteron, palpi, anterior legs, intermediate legs (except a fuscous band on the tibiae and the basitarsus), the posterior legs (except the tibiae and basitarsus) *white* or *yellowish white*; venter and the sides of the tergites and the apical two tergites fulvous; wings hyaline, iridescent; venation dark brown; head and thorax without pubescence.

Margherita, Assam. Described from three females (one type).

Type and *paratype* in Indian Museum; type No. $\frac{260}{9}$, paratype No $\frac{261}{9}$.

Paratype in U.S.N.M.

Genus Beleses, Cameron.

Beleses nigriceps, sp. n.

This species is readily separated from the other species of this genus by the black head. As far as the males and females have been associated it is the only species in which colour antigeny occurs. Except for the colour of the legs and having only one discal cell in the hind wings the female agrees with Cameron's *Sunoxa purpureifrons*.

Female.—Length 6 mm. Anterior margins with the clypeus truncate, surface coarsely irregularly punctured; supraclypeal area flat, narrow; supraclypeal foveae deep, confluent with the antennal foveae; middle fovea represented by the shallow transverse impression; front and posterior orbits shining, impunctate; postocellar area sharply defined laterally but not defined anteriorly; postocellar line subequal with the ocellocular; flagellum gradually thickened until it reaches the apex of the second joint; the second joint one-fourth longer than the third; the fourth and following joints compressed; thorax shining, impunctate; stigma gently rounded below; second cubital cell longer on both radius and cubitus than third, which is twice as wide apically as basally; the transverse radius curved, joining the radius about the same distance from the second transverse cubitus as the second recurrent is from the same vein; claws cleft with the inner teeth exceeding the outer; sheath straight above, truncate apically, oblique below. Head and posterior femora beyond middle, four posterior tarsi and the antennae black, the rest of the insect rufous; head and throax covered with short gray hairs: wings distinctly hyaline, venation dark brown.

Male.—Length 5 mm. Differs from the above description of the female in having the abdomen, thorax and legs, except the anterior tibiae and the trochanter which are fulvous, dark piceous and the hairs on the thorax and the head blackish; hypopygidium rounded apically.

South India. Described from one female from Marikuppam, collected October 19, 1910, at an altitude of 3,500 feet; and from two males, one allotype, from Bangalore, collected September 12, 1910, at an altitude of 3,000 feet.

Type and *allotype* in the Indian Museum; type No. $\frac{1756}{19}$, allotype No. $\frac{1753}{19}$.

Paratype.—Cat. No. 18909, U.S.N.M.

Genus **Entomostethus**, Enslin.

Entomostethus assamensis (Rohwer).

One male and two females from Ghumti, Darjiling District, East Himalayas, collected July 1911, at an altitude of 4000 feet (*F. H. Gravely*); five females and four males from Kurseong, East Himalayas, collected July 1908, at an altitude of 5,000 feet; five females and eleven males from Darjiling, East Himalayas, collected September 29 1908, at an altitude of 6,000 feet (*E. Brunetti*); nine females and ten males from the same locality, collected May 29, 1910, at an altitude of 7,000 feet (*E. Brunetti*); one female collected at the same locality and elevation, August 11, 1909 (*J. T. Jenkins*); one male from the same locality and elevation, collected August 9, 1909 (*C. Paiva*); and one male from Gangtok, Sikkim, collected September 8, 1909, at an altitude of 6,750 feet.

Two males from Kurseong have the legs slightly darker than typical; and all the specimens indicate that the basitarsi are usually black. The female has the sheath straight above and broadly rounded from the tip.

This species differs from *laticarinatus*, Cameron, which may belong to the same genus, by the colour of the legs, as Cameron's species is said to have the femora pale beneath.

Entomostethus hirticornis (Rohwer).

One female from Kurseong, East Himalayas, collected September 7, 1909, at an altitude of 5,000 feet; four females from Ghumti,

Darjiling District, East Himalayas, July 1911, at a calculated altitude of 4,000 feet (*F. H. Gravely*); one female from Gangtok, Sikkim, collected September 9, 1909, at an altitude of 6,150 feet; and one female from Kurseong, East Himalayas, collected July 3, 1908, at an altitude of 5,000 feet.

Genus **Cladius**, Rossi.

Cladius orientalis, Cameron.

One female from Simla, collected July 20, 1911, at an altitude of 7,000 feet (*N. Annandale*).

Genus **Hemichroa**, Stephens.

Hemichroa major, sp. n.

This species is readily separated from all the other species of *Hemichroa* by its larger size, general colour and yellow wings.

Female.—Length 11·5 mm.; length of the anterior wings 12 mm. Labrum obtusely pointed; clypeus obtusely, arcuately emarginate, the arcuation conforming in outline with the obtusely triangular lobes, surface shining; supraclypeal area more convex dorsally; median fovea rectangular in outline, open above, being confluent with the ocellar basin, the middle with a small punctiform pit; ocellar basin pentagonal in outline, well defined, meeting on the postocellar line, in front of the anterior ocellus it has a shallow, poorly defined depression; head shining with only setigerous punctures; postocellar area well defined laterally but not well differentiated anteriorly as the postocellar furrow is subobsolete and angulate anteriorly; postocellar line subequal with ocellocular line; antennae strongly tapering, the third joint distinctly longer than the fourth; thorax shining; inner tooth of the claws longer than the outer; stigma broader at base, gradually tapering to the apex; third cubital cell nearly parallel-sided, one-fourth longer than its apical width; nervellus in the middle of the cell; sheath stout, straight above; truncate apically, tapering below, the upper angle sharp, the lower angle rounded. Rufo-ferrugineous; head and antennae piceous; thorax beneath and on the sides black to piceous; legs black; wings bright yellow, venation ferrugineous.

Darjiling, Eastern Himalayas. Described from one female.

Type.—Indian Museum No. 5666/[illegible].

V. INDIAN TETRIGINAE (ACRYDIINAE).

By J. L. HANCOCK

(Plate xiv.)

Several months ago the extensive collection of these small Orthoptera in the Indian Museum was placed in my hands for study by Dr. N. Annandale, Superintendent of the Museum. At the time I received the collection Dr. Annandale stated in a letter that: "a large proportion of the collection was named by the late Mr. Kirby just before his death, and I am sending these specimens also." A considerable number of the remaining specimens, not seen by Mr Kirby, were named by Saussure and others. I find after going over the collection that the part examined by Mr. Kirby bears evidence that he had not passed final judgment on many of the specimens. This is shown from a number of cases where a hastily written label, with a specific name, is attached to one insect among a series containing from one to several species, so that the remaining ones were left undetermined.

The Indian Museum collection contains such a large percentage of the described species of the Indian Empire, besides so many new ones, that I concluded to give a review of the recorded Indian species of this Orthopteran family. I have given a synopsis of the subfamilies, and the genera; and in most cases in the large genera I have given tables for the separation of species. The literature of all the species has been added but in conjunction with this part Kirby's remarkable Catalogue of Orthoptera, Volume III, will be found invaluable for reference. The latter, however, includes the literature only to the end of 1898. Since then, a number of Indian species have been described, which are recorded in the present paper. Included at the end are some species in the Indian Museum outside the Indian Empire. Those from Ceylon are for the sake of convenience incorporated in the text with the Indian species.[1]

Synoptical Table of Subfamilies and Genera of India.

1. Antennae with all the articles excepting the basal and the small atrophied apical

[1] As the proofs of this article come to hand I find that Kirby's volume on Orthoptera has just been published in the "Fauna of British India" 1914. It refers to a number of Tettigid species described in the present paper, and in order to clarify the confusion that may arise from the difference in determinations the names given by Kirby, and those that I have applied, are placed in parallel columns on p. 132 of this article.

articles deplanate triquitrous, composed of eight to ten joints; vertex transverse, very broad, or strongly acuminate produced .. .. TRIPETALOCERINAE.
1. Gen. *Birmana*, Brunn.

1. 1. Antennae filiform.
2. Anterior femora more or less compressed carinate above.
3. Frontal costa widely forked, the rami forming a frontal scutellum .. CLADONOTINAE.
4. Pronotum extremely compressed, above wholly foliaceous .. 2. Gen. *Oxyphyllum*, Hanc.
4. 4. Pronotum acute tectiform, anteriorly angulate .. .. 3. Gen. *Deltonotus*, Hanc.
5. Dorsum of pronotum bearing a ramose process; body and legs ornate with spiniform tubercles .. 4. Gen. *Cladonotus*, Sauss.
5. 5. Dorsum of pronotum not at all ornate with a ramose process.
5. 5*a*. Elytra and wings wanting.
6. Posterior angles of the lateral lobes of pronotum widely laminate expanded, erose, and produced in spiniform tubercles; pronotum truncate anteriorly, dorsum fossulate-reticulate, between the shoulders cristulate .. 5. Gen. *Tettilobus*, Hanc.
6. 6. Posterior angles of the lateral lobes of pronotum laminate obliquely truncate behind, setose: dorsum compressed gibbose between the shoulders, posteriorly abbreviated, the apex truncate-emarginate .. .. 6. Gen. *Gignotettix*, Hanc.
5. 5*b*. Elytra minute: wings wanting; median carina of pronotum cristulate. 7. Gen. *Potua*, Bolivar.
3. 3. Frontal costa furcillate, but the rami diverge only moderately, or parallel, very frequently separated by a sulcus.
7. Pronotum truncate anteriorly, posterior angles of the lateral lobes of pronotum laminate produced outwards, acute, or posteriorly obliquely truncate, rarely turned down.
8. Posterior angles of the lateral lobes of the pronotum acute produced outwards often spined; the first joint of the posterior tarsi longer than the third; posterior tibiae strongly ampliate, or margins dilated toward the apices; very frequently not or sparingly spinose .. SCELIMENINAE.

9. Antennae inserted slightly or distinctly below the eyes.
10. Margins of hind tibiae strongly expanded, the first joint of the posterior tarsi dilated and much wider than the third.
11. Paired ocelli placed nearly between the lower third of the eyes; anterior and middle femora very narrow elongate; posterior angles of the lateral lobes of the pronotum turned down but the margin on each side armed with a straight more or less cylindrical sharp spine .. .. 8. Gen. *Amphibotettix*, Hanc.

11. 11. Paired ocelli placed nearly on a line with the lower border of the eyes; humeral angles of pronotum unarmed; posterior angles of the lateral lobes laminate and on each side produced in a curved spine; hind tibial margins distinctly membraneous expanded .. 9. Gen. *Scelimena*, Serv.

10. 10. Margins of posterior tibiae little expanded; first joint of hind tarsi not widely expanded; but little wider than the third, rarely narrow.
12. Stature large, body prolongate.
13. Pronotum with the dorsum often lightly gibbulose and fossulate; humeral angles often armed with denticles or tubercles; vertex unarmed; margins of hind tibiae bearing minute denticles 10. Gen. *Eugavialidium*, Hanc.

13. 13. Pronotum with the dorsum distinctly deplanate, more or less reticulate punctate; vertex armed on each side with an elevated tubercle; margins of hind tibiae serrulate .. 11. Gen *Gavialidium*, Sauss.

12. 12. Stature moderate not so prolongate.
14. Vertex very narrow, often one half the width of one of the eyes or even narrower; head exserted; posterior angles of the lateral lobes of the pronotum either unarmed, angulate and prominent or produced in a spine on each side. 12. Gen. *Bolotettix*, Hanc.

14. 14. Vertex equal to or wider than one of the eyes; pronotum above rugose tuberculose; lateral carinae behind the humeral angles compressed sinuate; median carina gibbose .. .. 13. Gen. *Thoradonta*, Hanc.

9. 9. Antennae inserted between the lower part of the eyes.

15. Vertex narrower or subequal to one of the eyes.
16. Head lightly exserted; posterior angles of the lateral lobes of the pronotum prominent, acute, or produced in a spine more or less transverse, or directed obliquely forward .. 14. Gen. *Criotettix*, Bol.

16. 16. Head not at all exserted; frontal costa roundly produced in advance of the eyes; posterior angles of the lateral subspiniform produced, or oblique and obtuse .. 15. Gen. *Loxilobus*, Hanc.

15. 15. Vertex as wide or wider than one of the eyes.
17. Dorsum of pronotum little rugose, or rugulose; head not at all exserted; eyes not elevated; antennae short; spine on each side of the lateral lobes of pronotum obliquely directed backward. 16. Gen. *Acanthalobus*, Hanc.

17. 17. Dorsum of pronotum subsmooth, punctate, spine on each side of the lateral lobes directed obliquely forward and curvate .. .. 17. Gen. *Tettitellum*, Hanc.

8. 8. Posterior angles of the lateral lobes of the pronotum little produced outwards, obliquely truncate behind, very rarely acute spinose; first and third joints of the hind tarsi nearly equal in length .. METRODORINAE.
18. Head distinctly exserted; vertex very narrow, the eyes strongly approximated and elevated .. .. 18. Gen. *Systolederus*, Bol.

18. 18. Head little exserted; lateral lobes with the inferior margins widely and roundly dilated; posterior angles behind transversely widely truncate; vertex very narrow; stature small, apterous. 19. *Eurymorphopus*, Hanc.
19. Vertex strongly produced, median carina prominently projecting from the front border; in profile acute angulate produced; face strongly oblique; frontal costa sinuate between the eyes. 20. Gen. *Spadotettix*, Hanc.

19. 19. Vertex little produced, in profile obtuse angulate; body apterous. 21. Gen. *Apterotettix*, Hanc.
20. Body very small, apterous; vertex very wide and not advanced as far as the front border of the eyes; pronotum subtectiform forward, deplanate posterior-

ly, hind process abbreviated, apex truncate; first joint of hind tarsi much longer than the third .. 22. Gen. *Amphinotus*, Hanc.

20. 20. Body larger, moderately crassate; vertex wider than one of the eyes; first and third articles of the hind tarsi equal in length; paired ocelli placed little below the middle of the eyes; posterior angles of the lateral lobes of the pronotum straight or obtuse, behind widely truncate .. .. 23. *Mazarredia*, Bol.

21. Body narrow elongate, posterior angles of the lateral lobes of pronotum subrounded; vertex cuspidate on each side or elevated styliform .. Gen. 24. *Xistra*, Bol.

22. Posterior angles of the lateral lobes of pronotum not at all produced outwards, narrowed, toward the apex rounded; pronotum between the shoulders strongly elevated in an obtuse gibbosity.
25. Gen. *Xistrella*, Bol.

22 22. Posterior angles of the lateral lobes of pronotum laminate, prominent, acute produced or obliquely truncate; median carina of the pronotum and the disc on either side bearing gibbosities.
26. Gen. *Lamellitettix*, Hanc.

21. 21. Body strongly crassate; vertex wider than one of the eyes, imperfectly marginate; eyes more or less conoidal in form, antennae inserted between the lower part of the eyes; lateral lobes of the pronotum with the posterior angles laminate dilated, widely truncate behind, prominent subacute, or rarely not reflexed outward, obliquely truncate behind .. .. 27. Gen. *Hyboella*, Hanc.

7. 7. Pronotum anteriorly truncate, or rarely angulate produced; posterior angles of the lateral lobes of pronotum turned down, more or less rounded; first articles of the hind tarsi longer than the third .. .. .. TETTIGINAE.

23. Vertex viewed in profile not at all produced.

24. Vertex very narrow, strongly narrowed forward drawing the eyes near together in front; eyes not elevated; pronotum above smooth, carinae not at all elevated .. .. 28. Gen *Teredorus*, Hanc.

24. 24. Vertex narrower than one of the eyes, anteriorly truncate; frontal costa often sinuate between the eyes; pronotum granulate or little rugose, the carinae little compressed elevated: first article of the posterior tarsi longer than the third .. .. 29. Gen. *Paratettix*, Bol.

23. 23. Vertex viewed in profile produced before the eyes, angulate, viewed from above wider than one of the eyes, not at all narrowed forward .. 30. Gen. *Acrydium*, Goeffr

25. Antennae inserted between the eyes, slender filiform.

26. Pronotum above little rugose, often bearing round or abbreviated lineate tubercles; frontal costa arcuate or roundly produced between the eyes; body moderately slender; vertex narrowed forward, fossulate on each side 31. Gen. *Coptotettix*, Bol.

26. 26. Pronotum above granulate, or barely punctate, rarely rugose; median carina of pronotum percurrent, not at all interrupted .. .. 32. Gen. *Hedotettix*, Bol.

25. 25. Antennae inserted between the lower border or angles of the eyes or below the eyes.

27. Head more or less exserted; frontal costa arcuate produced between the middle of the eyes; vertex narrower than one of the eyes and truncate; paired ocelli placed nearly on a line with the middle of the eyes. 33. Gen. *Euparatettix*, Hanc.

27. 27. Head distinctly exserted; paired ocelli placed between the lower third of the eyes; antennae inserted below the eyes; frontal costa little arcuate elevated between the antennae, but not above between the middle of the eyes; median carina of pronotum often undulate or sinuate; hind process with the lateral carinae toward the apex entire or frequently minutely crenulate, or bearing very small dilated lobes. 34. Gen. *Indatettix*, Hanc.

2. 2. Anterior femora above distinctly sulcate; pronotum anteriorly produced over the head in a cornute process; antennae having sixteen to twenty-two articles. BATRACHIDINAE.

35. Gen. *Saussurella*. Bol.

TRIPETALOCERINAE.

Genus **Birmana**, Brunner.

Brunner, Ann. Mus. Genova, xxxiii, p. 113, 1893.
Hancock, Gen. Ins. Orth. Acrid. Tetr., p. 4, 1906.

Birmana gracilis, Brunner.

Brunner, Ann. Mus. Genova, xxxiii, p. 114, pl. 5, fig. 47, 1893.

Habitat.—Burma. Not represented in the material under consideration.

CLADONOTINAE.

Genus **Oxyphyllum**, Hancock.

Hancock, Trans. Ent. Soc. London, p. 393, 1908.

Oxyphyllum pennatum, Hancock.

Hancock, Trans. Ent. Soc. Lond., pp. 393, 394, pl. xxii, fig. 3, 1908.

Habitat.—Darjiling, India. Not in the present collection.

Genus **Deltonotus**, Hancock.

Deltonotus subcullatus, Walker.

Tettix subcullatus, Walk., Cat. Derm. Salt. Brit. Mus., V, p. 830, 1871.

Deltonotus tectiformis, Hancock, Spol. Zeylan., ii, p. 154, pl. i, fig. 2, 1904.

Hanc., Gen. Ins. Orth. Acrid. Tetr., p. 14, pl. i, fig. 1, 1906.

Habitat.—Kandy, Ceylon, June 12, 1900; one example, Ind. Mus. coll.

Deltonotus gibbiceps, Bolivar.

Poecilotettix gibbiceps, Bol., Ann. Soc. Ent. France, lxx, p. 580, 1902.

Deltonotus gibbiceps, Hanc., Gen. Ins. Orth. Acrid. Tetr., p. 14, 1906.

Habitat.—Madura. Not in the present collection.

Genus **Cladonotus**, Saussure.

The three described species of this Ceylonese genus may be distinguished by the following table:—

1. Pronotal cornu curved forward at the middle, and furcate at the apex ... *humbertianus*, Saussure.

1. 1. Pronotal cornu not distinctly curved forward.

2. Cornu obliquely ascendant, truncate, and dentate in front and behind .. *turrifer*, Walker.

2. 2. Cornu nearly vertically ascendant, dentate in front and distinctly broadened toward the apical half .. *latiramus*, Hancock.

Cladonotus humbertianus, Saussure.

Sauss., Ann. Soc. Ent. France, i, p. 478, 1861; Bolivar, Ann. Soc. Ent. Belg., xxxi, p. 209, pl. 4, fig. 10 1887; Hanc., Spol. Zeylan., ii, p. 113, 1904.
Habitat.—Ceylon

Cladonotus turrifer, Walker.

Walker, Cat. Derm. Salt. Brit. Mus., p. 843, 1871.
Habitat.—Ceylon.

Cladonotus latiramus, Hancock.

Hanc., Spol. Zeylan., ii, p. 114, pl. i, fig. 1, 1904; Hanc., Gen. Ins. Orth. Acrid. Tetr., p. 16, pl. i, fig. 3, 1906.
Habitat.—Ceylon. Author's coll.

Genus **Tettilobus**, Hancock.

Tettilobus spinifrons, Hancock.

Hanc., Trans. Ent. Soc. London, pp. 396, 397, pl. xxii, fig. 4, 1908.
Habitat.—Ceylon. Not in the present collection.

Tettilobus pelops, Walker.

Cladonotus pelops, Walk., Cat. Derm. Salt. Brit. Mus., p. 843, 1871.
Habitat.—Ceylon. Not in the present coll.

Genus **Gignotettix**, Hancock.

Gignotettix burri, Hancock.

Hancock. Trans. Ent. Soc. Lond., p. 398, pl. xxii, fig. 5, 1908.
Habitat.—Ceylon. Not in the present coll.

Genus **Potua**, Bolivar.

Potua sabulosa, sp. nov.

Body very small, rugose scabrous; ferrugineous. Head not at all exserted: face large; eyes moderately small, not prominent, subconoidal in profile. Vertex rugose, distinctly wider than one of the eyes, the frontal carinulae laterally rounded and little elevated on each side, subhigher than the eyes, fossulate on each side, the short median carina little produced. Pronotum anteriorly truncate; median carina of the pronotum strongly elevated forward in a compressed gibbosity, reaching from the front border to the humeral angles, rounded forward and abruptly sloping backward: behind the humeral angles elevated in a lower second gibbosity, posteriorly tuberculate; disc little elevated at the middle and bearing a short oblique gibbulate carina on each side, behind the shoulders depressed: posterior process abbreviated, extended to the knees of the hind femora, and above strongly rugose-tuberculate; elytra minute, elongate; wings wanting; posterior femora stout, externally scabrous and obtuse tuberculate; margin below curvate and subtuberculate-erose; the three pulvilli of the first joint of the hind tarsi subequal in length.

Entire length of male 6·5 mm; pronotum 5 mm.; posterior femora 3·8 mm.

Habitat.—Yenna Valley, Satara Dist., Bombay Pres., 2500–3500 ft., Apr. 17, 1912 (*F. H. Gravely*).

SCELIMENINAE.

Genus **Amphibotettix**, Hancock.

Hancock, Ent. News, xviii, p. 86, 1906; Hanc., Gen. Ins. Orth. Acrid. Tetr., p. 22, 1906.

Amphibotettix rosaceus, sp. nov.

Allied to *A. longipes*, Hancock, but larger in stature, the spines of the pronotum being little stouter and not quite so long produced. Body coloured fuscous or black, the sides of the pronotum and dorsum obscurely suffused with rose, the lateral carinae forward, the tubercles and the spines of the lateral lobes bright rose colour. Head scarcely at all exserted; eyes slightly elevated and strongly globose. Vertex narrower than one of the eyes, narrowed forward, not advanced as far as the eyes; frontal costa protuberant between the antennae, produced little beyond the eyes; antennae inserted below the eyes; lower part of face strongly obliquely retreating. Pronotum deplanate above, strongly elongate, irregularly depressed forward before the shoulders at the sulci, transversely fossulate behind the shoulders, lengthily produced backward beyond the apices of the hind femora; dorsum rather smooth, minutely granulate, between the shoulders little elevated, behind the shoulders bearing a pair of obtuse subcarinated nodes, and further backwards on base of process presenting another pair of very obtuse rounded nodes; humeral angles bicarinate, hind process above subrounded; median carina of pronotum very low, following the inequalities, and forward near the convex margin turned upward but not produced in a tubercle, yet very slightly subtuberculate; lateral carinae on the shoulders obscurely subtuberculate and marked with rose colour in the type, the lateral carinae before the shoulders on each side forward near the sulci terminating in a rose-coloured spot; sides of pronotum at the front of lateral lobes produced on each side in a tubercle; posterior angles of the lateral lobes turned down not at all laminate, the lateral margins just before the angles ou wardly produced in an acute strong spine on each side, directed obliquely forward and little curvate toward the apex. Elytra moderately wide at the base and distinctly narrowed acuminate backward to the apices, externally strongly impresso-punctate; wings extended nearly to the apex of the process. Anterior and middle femora strongly elongate and narrow, margins of the anterior above subbicrenulate toward the base; middle femoral margins above subundulate; hind femora slender elongate; carinae of posterior tibiae strongly dilated; the first joints of the hind tarsi strongly membraneous expanded; the first and second pulvilli small subacute, and widely separated, dividing the article into thirds, the third pulvilli strongly obtuse and planate below.

Entire length of body, male, 25 mm.; pronotum 24 mm.; post. femora 8 mm.

Habitat.—Thingannyinaung to Sukli, Dawna Hills, Tenasserim, 900–2100 ft. elevation, Nov. 23, 1911 (*F. H. Gravely*). One example, Ind. Mus. coll.

This species differs from *Scelimena sanguinulenta*, which also has rose-coloured spines, in being longer, and the lateral spines are not so long produced. The new species is devoid of the median produced tubercle at the front of the pronotum, which is styliform and strongly produced in *longipes*. From the latter species it also differs in the legs being less attenuate, though very slender, and in the lateral spines being less cylindrical.

Genus **Scelimena**, Serville.

Table for distinguishing the Indian species.

1. Vertex narrower than one of the eyes, narrowed forward.
2. Humeral angles not at all provided with evident denticles.
3. Lateral lobes of the pronotum bearing one spine only on each side.
4. Spine on each side of the lateral lobes triangular acute and straight, the apex sharp, not at all curvate, coloured yellow; vertex very narrow; dorsum convex between the shoulders; body more or less fuscous or greyish-fuscous marked with yellow .. .. *harpago*, Serville.

4. 4. Spine on each side of the lateral lobes of pronotum produced, slender, and hooked forward, often bright rose-coloured or coral red .. .. *gavialis*, Saussure.

3. 3. Lateral lobes of pronotum provided with one spine and a tubercle on each side, the latter placed just before the spine of the posterior angles; body often greyish-fuscous marked with yellow; inferior margins of femora very strongly dentate .. .. .. *logani*, Hancock.

1. 1. Vertex not quite so narrowed forward.

2. 2. Humeral angles barely behind the apices and also the lateral carinae forward before the shoulders on each side slightly compressed obtuse denticulate; pronotal process strongly produced backward beyond the hind tibial apices; spine on each side of the lateral lobes posteriorly slender, sharp and curvate forward . *spinata*, sp. nov.

Body six millimetres (species probably described from larva or pupa?) .. *uncinata*, Serville.

The above representatives fall into the series of species having the hind tibial margins distinctly membraneous expanded, and the first joint of the hind tarsi similarly strongly dilated; the pronotum between the shoulders convex, not so distinctly deplanate as in *Eugavialidium*, Hanc.; the front margin of the pronotum entire, or provided only with very small front tubercles, not at all produced, placed on either side of the lateral front margin of the lobes; the paired ocelli placed low between the eyes yet somewhat higher than in the latter genus; the vertex narrower than one of the eyes or at most subequal; the apex of the pronotum bifid and the hind femoral margins below strongly dentate.

Scelimena harpago, Serville.

Tetrix harpago, Serv., Ins. Orth., p. 763, 1839; DeHaan Temminck. Verhandel., Orth., p. 161, 1842; Bol., Ann. Ent. Belg., p. 217, pl. 4, fig. 13, 1887.

Habitat.—Igatpuri, W. Ghats, Bombay Pres., Nov. 21, 1909: Medha, Satara Dist., Oct. 22, 1912 (*N. Annandale*): Kasara, W. base of West Ghats, Bombay Pres., Nov. 23, 1899; Datar Hill nr. Junagadh, Kathiawar, "in or near a stream" (*S. P. A.*), Nechal, W. Ghats, Satara Dist., 2000 ft. (*F. H. Gravely*): Medha, Yenna Valley, Satara Dist., 2100 ft., Apr. 17, 1912 (*F. H. Gravely*): Tambi, Koyna Valley, Satara Dist., 2100 ft., Mar. 24, 1912 (*F. H. Gravely*).

Most of the specimens are dark or fuscous on the dorsum; some have a greyish-fuscous cast, while several are suffused with reddish-ochre on the pronotum. There is one male of the latter colour from Medha which is very much smaller in stature, the entire length being 16 mm., the pronotum 14·7 mm. It has all the characters of the normal-sized individuals. The average male and female in the present series measures: entire length 19-22·5 mm.; the pronotum 18-21 mm. Serville (*l. c.*) gives the entire length of male and female as 21 mm.

Scelimena gavialis, Saussure.

Scelymena gavialis, Sauss., Ann. Soc. Ent. France, iv, p. 845, 1861; *Scelymena nodosa*, Walker, Cat. Derm. Salt. Brit. Mus., p. 840, 1871; *Scelimena gavialis*, Hancock, Spol. Zeylan., ii, p. 154, pl. i, fig. 4, 1904.

Habitat.—Labugama, W. P. Ceylon (Hancock coll.); Madulsima, Ceylon (*T. B. Fletcher*, Hancock coll.). Ind. Mus. coll.

This black species has the pronotum in front, the tip of the hind process, and the lateral spine on each side bright coral-red. In the Ceylonese species *logani*, Hanc., the coral-red is replaced with yellow, and the posterior angles of the lateral lobes of the pronotum are armed on each side with a spine and a denticle as indicated in the table.

Scelimena ? producta, Brunner.

Brunner reports this species from Carin Cheba in Rev. Syst. Orth., p. 103, 1893, and the previously reported habitat of this species is Java. The female representative which Brunner referred to, is much larger than the typical *S. producta*, Serville, and I think it is the species which I have described as *Eugavialidium discalis*, Hancock.

Scelimena spinata, sp. nov.

Near *Scelimena harpago*, Serv. Body above on the pronotum ferruginеous and greyish-fuscous, often more reddish on the disc, marked with ochre or yellow, the femoral denticles below light yellow. Vertex toward the front narrower than one of the eyes, not so distinctly narrowed forward as in *gavialis*, widened backward between the eyes; paired ocelli placed between the lower part of the eyes; antennae inserted distinctly below the eyes; frontal costa protuberant between the antennae. Pronotum truncate anteriorly, the front margin devoid of produced tubercles, but the lateral margins below the eyes bearing a minute tubercle, subelevated, on each side; dorsum rather smooth, not at all deeply fossulate, between the shoulders convex, the disc before the shoulders bearing two short supernumerary carinulae and minute side offshoots forward; behind the disc depressed, and on the dorsum above the hind femora bearing two pairs of very low nodes more or less carinated, the hind pair longer and nearer together; median carina of pronotum very low, thin, and percurrent; lateral carinae extended forward but less distinct on the shoulders, and barely behind the apices of the humeral angles the margin little compressed obtuse dentate; the lateral carinae at the terminus forward subdentate; hind process very long produced backward beyond the apices of the extended hind tibiae, the apex bifid; posterior angles of the lateral lobes turned outwards and produced in a moderately strong spine distinctly curved forward. Elytra moderately wide, elongate, the distal fourth narrowed to the apices and angulate, externally impresso-punctate; wings fully explicate, extended backward to the apex of the pronotal process. Anterior femora little compressed elongate, margin above little compressed at the basal half, and bearing a small subacute tubercle, below bidentate; middle femora above undulate, below acute bidentate; posterior femora moderately stout, the superior carina crenulate and bearing an acute antegenicular denticle, below strongly dentate, often quadridentate, the three denticles at the middle strongly produced spinose; hind tibial margins and first joint of posterior tarsi widely expanded, the first two pulvilli acute and placed backward leaving a wide basal space, the three pulvilli subequal in length.

Entire length male and female 21·7 mm.; pronotum 21–25 mm.; post. femora 9 mm.

Habitat.—Trevandrum, Travancore, Aug. 1890; Kellar, Travancore; Trevandrum State. Ind. Mus. coll.

Scelimena uncinata, Serville.

Tetrix uncinata, Serv., Hist. Nat. Ins. Orth., pp. 763, 764, 1893; *Scelimena uncinata*, Bol., Ann. Soc. Ent. Belg., xxxi, p. 218, 1887.

Judging from the description of this questionable species which has a length of only six millimetres, it appears that the type was a larva or pupa. This leaves the identity of the species in doubt. The type came from Bombay. In the Indian Museum collection are two specimens from Sibsagar, N. E. Assam, which are labelled *Scelimena uncinata*, Serville, by Saussure. These are pupa of some species near *Eugavialidium india*, Hanc.

Genus Eugavialidium, Hancock.

The members of this genus have the paired ocelli placed between the extreme lower part of the eyes; the dorsum of pronotum deplanate, the front margin of the pronotum more or less ornate with denticles or tubercles, often armed with a produced tubercle at the middle above the occiput, or when absent there, they appear in front on either side of the lateral lobes; the femoral margins more or less tuberculate; the lateral carinae on either side of the shoulders often tuberculate or dentate, and the median carina of the pronotum sometimes tuberculate; the hind process strongly prolonged backward beyond the hind femoral apices; the margins of the hind tibiae and first joint of the posterior tarsi moderately expanded, but not so strongly membraneous dilated as in *Scelimena*; the lateral lobes have the posterior angles outwardly produced, acute, triangular, or on either side bearing a spine often curved forward.

Table separating the Indian species and one from China.

1. Posterior angles of the lateral lobes of pronotum produced in a sharp spine more or less distinctly curvate forward.
2. Lateral carinae on each side of the disc of dorsum provided with more than one distinct denticle.
3. Lateral carinae on each side of dorsum including hind process ornate with many small denticles throughout; dorsum bearing distinctly elevated nodes; pronotum of female 21 mm .. *multidentatum*, sp. nov.
4. Dorsum of pronotum somewhat smoother.

3. 3. Lateral carinae on each side of disc of dorsum provided with denticles, and often tuberculate backward as far as the base of process only; hind process very long produced; median carina of pronotum not at all tuberculate over occiput at the front margin; pronotum of female 29·5 mm. .. .. *discalis*, sp. nov.

2. 2. Lateral carinae on each side of disc of dorsum entire, not at all dentate, but pale bimaculate, or very indistinctly subtuberculate; body and legs often adorned with yellow spots: front margin of pronotum bearing a tubercle on each side below the eyes, and one at the middle above the occiput: pronotum of female 19·5 mm. .. *indicum*, Hancock.

4. 4. Dorsum above rugose, reticulate; front margin of pronotum at the middle over the occiput and on each side below the eyes ornate with a tubercle; humeral angle on each side bearing a small pale denticle; pronotum strongly prolongate: pronotum of female 29 mm. *chinensis*, sp. nov.[1]

1. 1. Posterior angles of the lateral lobes of pronotum more or less triangular or acute produced, straight, not at all curvate.

5. Lateral lobes of pronotum with the posterior angles slender spiniform produced; anterior border of pronotum on each side bearing three tubercles, but the middle of dorsum over the occiput

[1] For description of this Chinese species, see report of species outside of India at the end of this paper.

not tuberculate; lateral carinae on each side of disc forward bidentate, the denticles at the humeral angles obtuse crenulate; pronotum of female 22·8 mm. .. .. .. *kempi*, sp. nov.

5. 5. Lateral lobes of pronotum with the posterior angles little prominent, triangular, not aculeate.

6. Anterior border of the lateral lobes of pronotum dentate; humeral angles very obtuse crenulate, provided with a pale tubercle on each side; longitudinal carinae ornate with pale crenules; pronotum of female 17 mm. . *birmanicum*, Brunner.

6. 6. Anterior border of the lateral lobes of the pronotum entire, not at all dentate but rounded; humeral angles obtuse and acute marginate; posterior angles of the lateral lobes of pronotum produced in an acute triangular lobe, profoundly abruptly sinuate behind; pronotum of male 17 mm. .. .. *feae*, Bolivar.

7. Pronotum bearing large tubercles on the disc; humeral angles obtuse, provided with abbreviated external carinulae: dorsum with two tubercles behind the humeral angles, and four at the middle of the process; posterior angles of the lateral lobes produced, but not at all spinose, margin behind with a profound subrectangular sinus; pronotum of female 19·5 mm. .. *flavopictus*, Bolivar.

7. 7. Pronotum not bearing elevated tubercles on the disc.

8. Dorsum above smooth punctate, little subreticulate; humeral angles very obtuse; median carina little elevated, percurrent, tuberculate in front at the middle over occiput; lateral and median carina not at all appreciably tuberculate; margins of first joint of posterior tarsi somewhat dilated: pronotum of female 21·7 mm. .. *angulatum*, sp. nov.

8. 8. Dorsum very lightly rugose, coarsely granulose; median carina bearing three small tubercles forward before the humeral angles and tuberculate backward in a subindistinct series to the base of process; lateral carinae subindistinctly bituberculate on each side

of shoulders and barely subtuberculate backward to base of process; margins of first joint of hind tarsi not at all expanded; pronotum of male 19 mm. *saussurei*, sp. nov.

Eugavialidium multidentatum, sp. nov.

Near *E. birmanicum*, Brunner. Body and legs marked with fuscous and yellow, and yellow tubercles dot the course of the lateral carinae throughout each side of the dorsum, the three tubercles in front yellow, the spines of the lateral lobes pinkish-yellow, and underneath the body darker pinkish-yellow. Head not at all exserted, eyes globose; vertex subwider than one of the eyes; paired ocelli placed between the extreme lower part of the eyes; antennae long, very slender filiform, inserted distinctly below the eyes; frontal costa protuberant between the antennae. Pronotum deplanate, strongly inequal; median carina of pronotum irregularly undulate, and at the front margin terminating in a produced elevated tubercle; dorsum minutely punctate or bearing pale granulations, between the shoulders on the disc provided with short elevated subcarinate tubercles, behind the shoulders depressed, and above the apices of the elytra on each side of the dorsum provided with an elevated subcarinate boss, further backward on base of process bearing two elongate elevated nodes; hind process acuminate, subrounded above toward the apex, but flattened forward toward the base; lateral carinae throughout on each side dotted with a series of small and rather widely separated obtuse denticles, the one at the humeral angle scarcely more distinct and obtuse, and those backward toward the apex becoming smaller and less distinct, but marked by minute pale maculae; lateral lobes at the front margin produced in a denticle on each side; the posterior angles produced in a strong spine slightly curvate forward and moderately broad at the base; elytra acuminate toward the apices: wings extended to or barely beyond the apex of the hind process. Anterior femoral carinae above subtuberculate, two toward the base often subacute, margins below bituberculate; middle femora above subtrilobate, below bidentate; posterior femoral carinae above indistinctly minutely trilobate, below provided with small obtuse denticles very slightly produced; margins of hind tibiae sparingly minutely dentate on outer margins, the margins moderately expanded; the first articles of the posterior tarsi moderately expanded, the three pulvilli nearly equal in length.

Entire length of female 20·5–22·5 mm.: pronotum 19·5–21·5 mm.; post. femora 8·5 mm.

Habitat.—Sukli, east side of Dawna Hills, 2100 ft., Nov. 22, 1911 (*F. H. Gravely*); Dawna Hills, 2000–3000 ft., Tenasserim, Mar., 1909 (*N. Annandale*).

This species may be distinguished from *E. birmanicum*, Brunner, by the curvate spines of the lateral lobes; the longer pronotum, and the shorter hind femora, the tuberculate margins of the femora, and the dentate lateral carinae of the pronotum. In *birmanicum* the longitudinal carinae are ornate with pale crenules, and the humeral angles are provided with a pale obtuse tubercle on each side, the lateral spines straight and triangular, not at all aculeate.

Eugavialidium discalis, sp. nov.

Near *Scelimena producta*, Serville. Colour greyish-ferrugineous. Vertex wider than one of the eyes, scarcely at all narrowed forward; eyes little prominent, subsessile, viewed from above reniform, from the side globose; paired ocelli placed between the extreme lower angles of the eyes: antennae inserted far below the eyes; frontal costa protuberant between the antennae. Pronotum truncate anteriorly; the front margin on either side of the lateral lobes bearing a produced tubercle, not tuberculate at the middle above the occiput; dorsum deplanate, between the shoulders on the disc bearing low, short, lineate tubercles, depressed behind the shoulders, and backward provided with four very low obtuse nodes; median carina very low; lateral carinae on each side bearing a series of denticles often placed as far back as the base of the hind process but here very minute; humeral angles ornate with a distinct obtusely elevated tubercle on each side; the denticles or tubercles each side of the forward disc larger than those backward; hind process lengthily produced backward beyond the apices of the extended hind tibiae; posterior angles of the lateral lobes produced in a curved spine on each side, strongly produced in the male, rather stout in the female. Elytra elongate, acuminate, externally impresso-punctate; wings extended backward almost to the apex of process. Anterior femora elongate, margins crenulate, above undulate and compressed-sublobate toward the base, below subtri-tuberculate; middle femora above subtrilobate, below subbituberculate; hind femora elongate, the superior carinae crenulate and more or less quadricompressed, with two acute denticles toward the knees, margins below often quadridentate but little produced; hind tibial margins and first joint of the hind tarsi moderately expanded, but not so widely dilated as in typical *Scelimena*; inner margins of hind tibiae entire; the first and second pulvilli of the first joint of posterior tarsi small, acute, more widely separated than the second and third.

Entire length of male and female 25–30·5 mm.; pronotum 24–29·5 mm.; hind femora 7–9 mm.

Habitat.—Sibsagar, N.-E. Assam (*S. E. Peal*); Upper Assam (*Doherty*).

This species differs from *Scelimena producta*, Serville, in the larger stature, and in being wider between the shoulders, in the dentate humeral angles and lateral carinae of pronotum, and in the tuberculate femora. It is readily distinguished when it is compared with a series of *S. producta*, Serv., from Java.

Eugavialidium indicum, Hancock.

Scelimena idia, Hanc., Trans. Ent. Soc. London, p. 219, 1907; Hanc., Rec. Indian Museum, viii, p. 311, 1913.

Habitat.—Assam (*H. H. G. Austen*).

The type in the author's collection is from Cherrapungi, Assam. The type is conspicuously ornate with small yellow spots forward on the greyish-fuscous dorsum, and also on the legs; the median carina of the pronotum at the front bears a tubercle; the lateral carinae of the dorsum of pronotum are not furnished with denticles or tubercles; the lateral lobes at the front under the eyes bear a tubercle on each side; the spines of the posterior angles of the lateral lobes are little curvate forward at the apices and more slender in the male; disc of the pronotum above provided on each side with a more or less distinct short branching carinula. In the specimens in the Indian Museum collection the bright yellow maculae are more or less obscured; the hind femora are missing, but in the type the hind tibiae and first joint of the posterior tarsi show moderate expansion of the margins but not nearly so strongly dilated as in typical *Scelimena*; the outer margins of the hind tibiae bear small acute denticles. In the two females in the Indian Museum the pronotum measures 20 and 22 mm. in length.

Eugavialidium kempi, sp. nov.

Near *E. birmanicum*, Brunner. Greyish-fuscous, yellow maculate, pronotal process suffused with, and spotted with chrome yellow. Vertex little wider than one of the eyes. Pronotum at the front border trituberculate, but the middle of the dorsum behind the occiput not at all tuberculate; median carina of pronotum irregularly compressed, spotted with yellow; just behind the sulci, and behind the shoulders gibbulate; dorsum fossulate behind the shoulders, on either side above the base of the hind femora obtuse nodulose, and further backward bearing a pair of elevated nodes; lateral carinae subcompressed and on each side of disc at the humeral angles and before the angles bearing a pale yellow denticle, backward yellow maculate; anterior and middle femora slender elongate, margins of anterior femora above subentire, bearing one very small tubercle, below subentire or bearing two minute tubercles; middle femora yellow maculate, and subtrilobate above, below subbituberculate; hind femora elongate, superior margin crenulate and above the middle and at the distal fourth subacute dentate, below subentire.

Entire length of female 23·5 mm.; pronotum 23 mm.; post. femora 8·7 mm.

Habitat.—Above Panji, 4000 ft., "Rebang stream under stones" (*Kemp*).

This species was mistaken for females of *E. indicum*, Hancock, in my former paper in the Records of the Indian Museum, VIII, p. 311, 1913. I have since examined two females of *E. indicum*, which are mentioned under the preceding heading. In *E. kempi* the posterior angles of the lateral lobes of the pronotum are produced outwards on each side in a straight narrow spine, whereas in *E. indicum* the spine on each side of the lateral lobes is curvate forward. In *E. birmanicum* the posterior angles are triangularly produced, not aculeate, and the body is shorter.

Eugavialidium birmanicum, Brunner.

Gavialidium birmanicum, Brunn., Ann. Mus. Genova xxxiii, p. 104, pl. 5, fig. 37, 1893; Hanc., Gen. Ins. Orth. Acrid. Tetr. *E birmani* (cum), p. 25, 1906.

Habitat.—Burma (*Brunner*).

Eugavialidium feae, Bolivar.

Bol., Real. Soc. Espan. Nat. Hist., ix, p. 396, 1909.

Habitat.—Carin Cheba (*Bolivar*).

Eugavialidium flavopictum, Bolivar.

Bol., Real. Soc. Espan. Nat. Hist., ix, pp. 394, 395, 1909.

Habitat.—Calcutta, India (*Bolivar*).

Eugavialidium angulatum, sp. nov.

Colour ochreous. Vertex subequal in width to one of the eyes, on either side dentate but not elevated above the eyes; dorsum of pronotum plain, punctate, and minutely subreticulate; between the shoulders provided with a short elevated line or ruga on each side; humeral angles very obtuse-convex; the lateral and median carinae unarmed, not at all tuberculate; median carina little elevated subgibbose at the sulci forward and terminating in front over the occiput in a tubercle; prozonal carinae forward behind the front margin compressed parallel; the front lateral margin on either side armed with a tubercle; posterior angles of the lateral lobes of the pronotum turned outwards and little triangulate protuberant, not spined; hind process of pronotum produced beyond the apices of the hind femora about the length of the femora; wings fully explicate, reaching to the apex of the process; posterior femora rather stout, margins above crenulate with indistinct antegenicular lobe, margin below entire; posterior tibial margins dentate, a little expanded toward the apices, the first joints of the posterior tarsi with the margins little expanded; the three pulvilli equal in length and planate below.

Entire length of female 23 mm.; pronotum 21·5 mm.; post. femora 8·5 mm.

Habitat.—Calcutta.

This species is labelled: "*Gavialidium philippinum*, Bol.," evidently in Saussure's handwriting. It is much smaller than that species and it differs in the shape of the posterior angles of the lateral lobes.

Eugavialidium saussurei, sp. nov.

Similar to the preceding. Ferruginous. Vertex subequal in width to one of the eyes, on either side dentate, barely elevated. Dorsum of pronotum above barely rugose, coarsely granulate; median carina of pronotum before the humeral angles provided with three small tubercles and also tuberculate backward as far as the base of the hind process: humeral angles very obtuse; posterior angles of the lateral lobes of the pronotum little acute triangulate produced outwards, not spined; front margin of pronotum on either side of the lobes bearing a small tubercle; wings fully explicate reaching nearly to the apex of process; anterior and middle femora little compressed, margins undulate; posterior femora stout, marigns crenulate otherwise unarmed; hind tibial margins dentate, very moderately expanded toward the apices; first articles of the hind tarsi not at all expanded, narrow, the first pulvillus very small, the second and third longer and equal in length.

Entire length of male 20 mm.; pronotum 19 mm.; posterior femora 7·5 mm.

Habitat.—Calcutta, India.

This species like the preceding is labelled, "*Gavialidium philippinum*, Boi.," by Saussure. It is allied to *E. angulatum*, Hanc., but differs in the tuberculate median carina of the pronotum and as shown in the table of species. This species as well as the preceding somewhat resemble members of the genus *Gavialidium* in the dentate character of the vertex, but the lateral carinulae on each side are not elevated above the eyes.

Genus **Gavialidium**, Saussure.

Gavialidium crocodilus, Saussure.

Sauss., Ann. Soc. Ent. France, iv, p. 481, 1861; Bol., Ann. Soc. Ent. Belg., xxxi, p. 219, 1887; Hanc., Spol. Zeylanica, ii, pp. 122, 123, pl. 2, fig. 11, 1904; Hanc., Gen. Ins. Orth. Acrid. Tetr., pp. 22, 25, pl. 2, fig 16, 1906.

Habitat.—Peradeniya, Ceylon; Pundaluoya, Ceylon.

One of these examples is labelled and the species determined by Saussure, which helps to authenticate this species.

In regard to the species *Gavialidium alligator*, Saussure (*Scelymena alligator*), its status is in doubt. It appears to me that the pupa of *crocodilus* served as a type for Saussure's *alligator*. I have a large series of the former species collected in Ceylon by Fletcher and Green, which show diverse variations, some being smaller than the normal size. From an examination of these specimens some of the immature pupa agree with the description of *alligator* given by Saussure. I am not sure, but the latter seems to be the pupa of *crocodilus* and therefore is synonymous.

Genus Bolotettix, Hancock.

This genus occupies a position midway between *Criotettix*, Bolivar, on the one hand, and *Systolederus* on the other. The representatives are small in stature, the eyes more or less exserted and near together, the vertex viewed from above very narrow, often one half or even one-third the width of one of the eyes, whereas, in *Systolederus*, the eyes are still closer together, separated only by the very narrow vertex as viewed in front and from above. In *Criotettix* the vertex is wider, and the antennae inserted between the eyes. The lateral lobes in *Bolotettix* have the posterior angles either turned down, or laterly reflexed outwards, little prominent, or produced in a spine on each side. Represented by a number of species in India, and other sections in the oriental region.

Table for the separation of Bolotettix *of India.*

1. Pronotal process shortened, not extended beyond the hind femoral apices; dorsum little rugose above; posterior angles of the lateral lobes obliquely sublaminate produced, apex of angle subacute: pronotum of female 9 mm. *anomalus*, Hancock.

1. 1. Pronotal process subulate, posteriorly extended beyond the knees of the hind femora.

2. Lateral lobes of the pronotum with the posterior angles acute spinate.

3. Legs distinctly fusco-annulate; posterior femora bearing oblique grey fascia; pronotum of male and female 11–14 mm. .. .. .. *oculatus*, Bolivar.

3. 3. Legs not distinctly fusco-annulate.

4. Sides of the pronotum and the four anterior femora testaceous-yellow; posterior femora below bearing a deep black longitudinal fascia; pronotum of male and female 11·5–13·5 mm. .. *armatus*, sp. nov.

4. 4. Sides of pronotum yellowish, legs yellowish obscurely marked with fuscous: pronotum of female 12 mm. .. *pictipes*, sp. nov.

2. 2. Lateral lobes of pronotum with the posterior angles either turned down, or obliquely reflexed outward but not acute spined.

5. Posterior angles of the lateral lobes of the pronotum turned down, not at all reflexed outward, rounded below; pronotum of female 12·5 mm. .. *inermis*, sp. nov.

5. 5. Posterior angles of the lateral lobes reflexed outwards.

6. Posterior angles of the lateral lobes subquadrate, little reflexed outwards, obliquely truncate behind: pronotum of male and female 12·5–13·5 mm. .. *lobatus*, Hancock.
   Pronotum of female 10 mm. .. *exsertus*, Bolivar.
   Pronotum of female 14 mm. *quadratus*, sp. nov.

6. 6. Posterior angles of the lateral lobes triangulate produced, margin behind subsinuate-truncate, pronotum of male and female 12–15 mm. *triangularis*, sp. nov.

Bolotettix anomalus, Hancock.

Systolederus anomalus, Hanc., Spolia Zeylanica, vi, p. 146, 1910.

Habitat.—Madulsima, Ceylon (*Hancock*).

Bolotettix oculatus, Bolivar.

Criotettix oculatus, Bol., Ann. Mus. Civ. di Genova, xxxix, p. 71, 1898.

Habitat.—Kodaikanal, S. India (*Bolivar*).

The type came from Sumatra, and this species is reported from Java.

Bolotettix armatus, sp. nov.

Near *lobatus*, Hancock. A graceful-bodied species. Head fuscous, pronotum above dark reddish-ochre forward, and backward on the hind process becoming very pale toward the apex, the four anterior legs, sides of the pronotum, and the lower part of the lateral lobes and spines light testaceous-yellow, the hind femora pale yellow with a deep black longitudinal fascia below the lower external carina. Vertex strongly narrower than one of the eyes, narrowed forward toward the front tricarinate, the median carina very little projecting; head and eyes distinctly exserted; eyes strongly elevated above the dorsum of pronotum and globose; antennae inserted below the eyes; frontal costa rather widely arcuate-elevated between the antennae and depressed between the eyes. Pronotum above plain, the dorsum little turned up in front; between the shoulders barely rugulose, behind the shoulders depressed subfossulate; median carina of pronotum percurrent substraight but little compressed forward between the sulci; lateral carinae little compressed subbicarinate on the shoulders; hind process long acute subulate, surpassing the hind femoral apices; posterior angles of the lateral lobes of the pronotum laminate and produced in a straight acute spine on each side, spine subcarinate and transverse, behind the spine the margin sinuate. Elytra small, elongate; wings extended barely beyond the pronotal apex, coloured black-infumate. Anterior and middle femoral margins entire; posterior femora elongate, externally bearing distinct oblique rugae, and above with a series of rounded tubercles; the three pulvilli of the first joint of the hind tarsi subequal in length, or the first barely longer than the second.

Entire length of male and female 12·5–14·5 mm.; pronotum 11·5–13·5 mm.; posterior femora 5·2–6·3 mm.

Habitat.—Sukli, Dawna Hills, 2100 ft., Nov. 22, 1911 (*F. H. Gravely*).

This species has the head and eyes more exserted than in *lobatus*, Hanc.; the vetex narrower, and it, moreover, has strongly produced spines arming the lateral lobes directed at a right angle to the body. This species approaches *Systolederus* in the exserted and very narrow vertex.

Bolotettix pictipes, sp. nov.

Near *armatus*, Hancock. Above obscure yellowish-fuscous, the lower sides of the body and legs light yellow, obscure fusco-variegated, the hind femora pale fasciate at the middle and dark below. Head exserted; eyes globose; frontal costa distinctly produced between the eyes; vertex strongly narrower than one of the eyes in front, on either side of the mid-carina the space very little wider than in *armatus*, in the latter species the space on either side very narrow sulcate and the vertex tricarinate. Pronotum above with the dorsum planate, granulose, between the shoulders bearing two short carinulae, behind the shoulders subfossulate, posteriorly planate and obscurely subtuberculate; median carina of pronotum percurrent, very thin and low, subundulate posteriorly, forward behind the front margin depressed; hind process long acute subulate, surpassing the hind femoral apices; posterior angles of the lateral lobes of pronotum outwardly reflexed on each side and produced in a very narrow acute spine subtransverse, very slightly directed backward, the margin behind the base of spine rectangulate sinuate. Elytra short, ovate; wings extended to the apex of the pronotal process; femoral margins entire; the three pulvilli of the first joint of the hind tarsi equal in length.

Entire length of female 13·2 mm.; pronotum 12·5 mm.; posterior femora 5·8 mm.

Habitat.—Madras, Shevaroys, 4000 ft., Aug. 1907 (*C. W. M., T. B. Fletcher*) in author's collection.

Bolotettix inermis, sp. nov.

Body above fuscous, sides and legs paler and variegated with fuscous, the hind femora mottled with yellow and fuscous, dark toward the apices and light basally, below the lower external carina black. Head very little exserted; eyes globose; vertex scarcely narrowed toward the front, subnarrower than one of the eyes, middle carinate, on each side fossulate, frontal carinulae laterally little elevated-acute, front subrounded truncate; antennae very long filiform, inserted far below the eyes; frontal costa compresso-elevated between the antennae, depressed above between the eyes and distinctly sinuate at the median ocellus as viewed in profile. Pronotum above plain; median carina percurrent, elevated little arcuate between the sulci forward; between the shoulders provided on each side with a very distinct oblique rugula or line; lateral carinae percurrent on the shoulders, and before the shoulders distinctly compressed elevated; prozonal carinae behind the front border forward distinctly compressed, subparallel; hind process subulate, extended beyond the hind femoral apices; posterior angles of the lateral lobes of pronotum turned down, lower margin rounded. Elytra short, ovate, in the type black, with pale yellow apices; wings extended little beyond the pronotal apex; the four anterior femora elongate, compressed, margins entire; posterior femora stout, superior margins minutely serrulate-granose; the three pulvilli of the first joint of hind tarsi subequal in length.

Entire length of female 13·5 mm.; pronotum 12·5 mm.; posterior femora 6·5 mm.

Habitat.—Ghumti, Darjiling Dist., E. Himalayas, 4000 ft., July, 1911 (*F. H. Gravely*).

In one of the specimens the hind process of pronotum is slightly less produced beyond the apices of the hind femora.

Bolotettix lobatus, Hancock.

Systolederus lobatus, Hanc., Mem. Dept. Agric. India, iv, pp. 143, 144, 1912.

This species has the body above infuscate, sides and legs paler, variegated with light ochre, hind femora with a pale median stripe, and below the lower external carina longitudinally black fasciate. Vertex strongly narrower than one of the eyes, distinctly narrowed forward between the curvate frontal carinulae; middle carinate between the forward half of the eyes and very slightly projecting, on either side of the mid-carina elongate fossulate; head exserted; eyes prominent and globose, higher than the dorsum of pronotum; antennae inserted far below the eyes; paired ocelli placed between the lower fourth of the eyes; frontal costa compressed arcuate between the antennae, little depressed between the eyes; strongly sinuate at the median ocellus in profile. Pronotum above plain subcylindrical forward, shining granulose, between the shoulders the dorsum convex, disc on either side presenting somewhat distinct oblique lines, behind the shoulders depressed subfossulate; lateral carina very slightly compressed, often reddish in colour; median carina percurrent, little elevated; prozonal carina forward behind the front margin very thin and parallel; posterior process extended beyond the hind femoral apices; posterior angles of the lateral lobes of the pronotum little obliquely laminate, reflexed outward, and obliquely truncate behind, the angles subacute. Elytra small, elongate, margin above substraight, below widely rounded and both extremities rounded, externally punctate, the upper third part light ochreous, lower part infuscate; wings nearly reaching to the apex of the pronotal process, infumate. Femora elongate; anterior femoral margins subundulate; posterior femoral margins granose; the three pulvilli of the first joint of the hind tarsi acute, subequal, or the third barely longer than the second.

Entire length of male and female 13·5–14·5 mm.; pronotum 12·5–13·5 mm.; posterior femora 6·2–7 mm.

Habitat.—Ghumti, Darjiling Dist., E. Himalayas, 4000 ft., July, 1911 (*F. H. Gravely*); Kurseong, E. Himalayas, 5000 ft., July 5, 1908 (*N. Annandale*).

The type in the author's collection is from Lebong, Darjiling Dist., 5000 ft. As it is an imperfect specimen, I have drawn the above description from fresh examples.[1]

Bolotettix exsertus, Bolivar.

Criotettix exsertus, Bol., Ann. Soc. Ent. France, lxx, p. lxx, 583, 1902.

Habitat.—Kodaikanal, S. India (*Bolivar*).

Bolotettix quadratus, sp. nov.

Body pale grey or yellow, more or less marked with fuscous or black, on each side of the lower part of the pronotal lobes pale, posterior femora with a median longitudinal light fascia, and externally above marked with fuscous and with a longitudinal black fascia below but the fascia interrupted with yellow at the distal third; hind tibiae annulate with light and dark; underneath the body black and pale variegated. Head and eyes exserted, eyes higher than the dorsum of pronotum; vertex strongly narrower than one of the eyes, equal to about one-third the width, tricarinate forward; frontal costa between the antennae roundly compressed-elevated, depressed above between the eyes, sinuate at the median ocellus. Pronotum with the dorsum plain, little rugose between the shoulders; prozonal carina behind the anterior margin parallel; median carina percurrent, little compressed-elevated before the shoulders, behind the anterior margin concave, the margin anteriorly slightly elevated; lateral carinae in front of the shoulders barely compressed; humeral angles subbicarinate; hind process long subulate, extended beyond the hind femoral apices; posterior angles of the lateral lobes subquadrate, obliquely reflexed, and truncate behind, the apices little prominent. Elytra elongate, the apices rounded; wings black or infumate, extended beyond the pronotal apex. Anterior and middle femora elongate, margins entire; hind femora slender, margins minutely serrulate, hind

[1] Kirby had examined these specimens in the Indian Museum, and on one of the specimens he placed a label bearing the name: "*Systolederus cinereus*, Bol.," while on the second specimen he had affixed a label with the determination: "*Mazarredia lugubris*, sp. nov."

tibiae little curvate toward the base; the three pulvilli of the first joint of the hind tarsi subequal in length.

Entire length of female 15 mm.; pronotum 14 mm.; posterior femora 6 mm.

Habitat.—Singla, Darjiling District, 1500 ft., June, 1913 (*Lord Carmichael coll.*); Sikkim, Darjiling Dist. (*L. Mandelli*).

This species resembles *triangularis* in the narrow vertex, but differs in the lateral lobes which are subquadrate and little prominent as compared to the triangulate produced angles in *triangularis*. The exserted head suggests its approach to *Systolederus*, but the vertex is wider than in typical representatives of that genus.

Bolotettix triangularis, sp. nov.

Allied to *armatus*, Hancock, slightly larger in stature. Body reddish-ochreous, front of head and the sides of the lateral lobes mottled with black, underneath the body black, femora pale yellow, the hind femora marked with a longitudinal black fascia below the lower external carina, and faint traces of fuscous bars on the upper part. Vertex strongly narrower than one of the eyes, tricarinate, the median carina very little projecting; head and eyes exserted; frontal costa arcuate-elevated between the antennae, sinuate at the median ocellus. Pronotum plain above, dorsum rugulose between the shoulders; humeral angles bicarinate; median carina of pronotum substraight, percurrent, and subobsolete near the front border; posterior process subulate, long surpassing the hind femoral apices; posterior angles of the lateral lobes of the pronotum laminate dilated laterally and distinctly triangulate produced, subacute, the margin behind subsinuate truncate. Elytra small, elongate-ovate; wings extended little beyond the pronotal apex. Anterior and middle femoral margins entire; posterior femoral margins granose or entire; the third pulvilli of the first joint of the hind tarsi little longer than the second.

Entire length of male and female 13·5-16 mm.; pronotum 12-15 mm.; posterior femora of the male 6 mm.

Habitat.—Sibsagar, N.-E. Assam (*S. E. Peal*).

The two specimens in the Indian Museum collection were determined as "*Systolederus angusticeps*, Stal," presumably by Saussure. It is hardly necessary to state that the latter, a Philippine species, is of much larger stature and has acute spines arming the lateral lobes of the pronotum.

Genus **Thoradonta**, Hancock.

Hancock, Trans. Ent. Soc. London, p. 407, 1907.

Thoradonta spiculoba, Hancock.

Hanc., Mem. Dept. Agric. India, iv, p. 138, 1912; Hanc., Records Ind. Mus., viii, pp. 312, 313, 1913.

Habitat.—Calcutta (*N. Annandale*; *Brunetti*; and *F. H. Gravely*); Rangoon, 1905 (*Brunetti*); Kandy, Ceylon (*Hancock coll.*). I have previously reported this species from: Bihar, Pupri, Muzaffarpur; Pusa; Durbhanga; Dibrugarh, N. E. Assam.

Thoradonta sinuata, sp. nov.

Colour ferrugineous. Head not at all exserted; vertex wider than one of the eyes; frontal costa compresso-elevated between the antennae. Pronotum above rugose, tuberculose; median carina strongly sinuate, tuberculate, in front of the shoulders and behind the humeral angles gibbulate; prozonal carinae forward behind the anterior margin convergent backward; lateral carinae behind the humeral angles sinuate and compressed; dorsum on the disc between the shoulders widened, tuberculose, and bearing a short carinula on each side, behind the shoulders bifossulate; humeral angles with the carinae compressed: posterior process subulate, extended little beyond the hind femora apices; posterior angles of the lateral lobes of the pronotum laminate expanded and abruptly constricted and produced in a narrow transverse sharp spine on each side, sinuate in front and behind the angle; margins of the lateral lobes minutely serrulate. Elytra wide at the middle, narrowed forward and strongly narrowed toward the apices; wings extended nearly to the apex of the pronotal process. Anterior femoral margins undulate; middle femoral margins subtrilobate above and below; posterior femora externally rugose; margins serrulate; the first two pulvilli of the first article of the hind tarsi spinose, the third acute.

Entire length of female 9 mm.; pronotum 8 mm.; posterior femora 5 mm.

Habitat.—Moleshwar, W. of Yenna Valley, Satara Dist. 3200 ft., April 23, 1912 (*F. H. Gravely*).

This species differs from *spiculoba*, Hancock, in having the forward gibbosity in front of the shoulders on the pronotum lower and smaller, and in the less dilated posterior angles of the lateral lobes of the pronotum, and in the abruptly contracted and transversely produced narrow sharp spine, in contrast with the less contracted suboblique spine on each side in *spiculoba*, the spines in the latter having the bases wider.

Thoradonta apiculata, sp. nov.

Near *spiculoba* and *sinuata*, Hancock. Colour greyish-rufescent, sometimes infuscate, the hind tibiae pale annulate. Head not at all exserted; vertex wider than one of the eyes; frontal costa distinctly protuberant between the antennae. Pronotum above rugose-granulose; median carina of the pronotum compressed-gibbulate before the shoulders, but not so elevated as in *spiculoba*, and posteriorly sinuate; hind process subulate and extended much beyond the hind femoral apices; posterior angles of the lateral lobes of the pronotum little dilated-laminate, and produced on each side in a small sharp spine with a wide base, the margin in front oblique and not at all sinuate, but behind right-angle sinuate; wings extended to the apex of the pronotal process.

Entire length of male and female 10–11 mm.; pronotum 9·5–10·5 mm.

Habitat.—Upper Assam; Sukli, Dawna Hills, 900-2100 ft., Oct. 23, 1911 (*F. H. Gravely*); Tenasserim Valley, Lower Burma (*Doherty*); Darjiling Dist. Singla, 1500 ft., May, 1913 (*Lord Carmichael coll.*); Sibsagar, N. E. Assam (*S. E. Peal*).

Thoradonta nodulosa, Stal.

Tettix nodulosa, Stal, Eugenies Resa, Orth., p. 348, 1860; *Criotettix nodulosus*, Bol., Ann. Soc. Ent. Belg., xxxi, p. 230. 1887; Brunn., Rev. Syst. Orth., p. 105, Genova, 1893.

Habitat.—Carin Cheba (*Brunner*).

The habitat of this species is Java and Malacca, and Brunner's record may refer to one of the species I have just described. Specimens of *Thoradonta nodulosa*, Stal, are in my collection taken in Java by Jacobson, and they differ from any of the Indian species, in the posterior angles of the lateral lobes of the pronotum. In *nodulosus*, Stal, the angles are shortly acuminate, while in the above mentioned species they are spined, though in *apiculata* and *spiculoba* the base of the spines is widened.

Genus Criotettix, Bolivar.

Plate xiv.

Table for separating the Indian species of Criotettix.

1. Vertex not at all produced.
2. Hind process of pronotum not, or very little surpassing the hind femoral apices: wings shorter than the process: frontal costa arcuate produced before the eyes: dorsum narrow between the shoulders; pronotum of male 9 mm. .. *rugosus*, Bolivar.

2. 2. Hind process of pronotum lengthily surpassing the hind femoral apices; wings extended to the apex of the pronotal process.

3. Stature moderately small; pronotum of female not exceeding 13 or 14 mm. in length.
4. Vertex very narrow, strongly narrower than one of the eyes; spine on each side of the posterior angles of the lateral lobes of pronotum long, straight, and sharp: pronotum of male and female 10·5–12·5 mm. .. *tricarinatus*, Bolivar.

4. 4. Vertex subequal in width to one of the eyes: dorsum between the shoulders convex; spine on each side of the lateral lobes transverse slender elongate and sharp.

5. Lateral lobes of pronotum with the posterior angles spined.

6. Spine on each side of the lateral lobes straight; pronotum of male 11 mm. *indicus*, Bolivar; *orientalis*, Hancock.

6. 6. Spine on each side of the lateral lobes curvate; base of pronotal process above tuberculose; pronotum of male and female 9·8–12 mm. .. *spinilobus*, Hancock.

5. 5. Lateral lobes of pronotum with the posterior angles laminate, subquadrate, the angle but little prominent; hind process above subnodulose; vertex narrower than one of the eyes; pronotum of male and female 10·8–11 mm. .. *pallidus*, sp. nov.

3. 3. Stature larger, above little rugose subtuberculose; posterior angles of the lateral lobes subquadrate, obliquely truncate behind, the angle little prominent, not at all spined; colour ferrugineous; vertex narrower than one of the eyes; pronotum of male and female 13–15·5 mm. .. .. .. *dohertyi*, sp. nov

7. Lateral lobes of the pronotum with the posterior angles barely produced.

8. Pronotum slender subulate posteriorly, strongly produced backward; dorsum above somewhat smooth, behind the shoulders lightly fossulate; posterior angles of the lateral lobes little prominent, sinuate behind; pronotum 17 mm. sex? .. .. *aequalis*, Hancock.

8. 8. Pronotum wider between the shoulders, behind the shoulders strongly fossulate; hind process produced beyond the hind femora 4 mm.; apices of the posterior angles of the lateral lobes little prominent, not at all produced or spined; pronotum of male 15 mm. *montanus*, Hancock.

7. 7. Lateral lobes at the posterior angles produced, acute.

9. Spines of the posterior angles of the lateral lobes of the pronotum more or less obliquely directed backward; vertex subnarrower than one of the eyes.

10. Stature of moderate size; above dull rugose-granulate; dorsum behind the shoulders fossulate, tuberculose on the base of process; colour fuscous, or sometimes marked with ochre, the hind femora obscurely marked, or with bars of ochre; pronotum of male and female 13·5–17 mm. .. .. *annandalei*, sp. nov.

10. 10. Stature large, above shiny; rather broadly depressed behind the shoulders and on either side between the shoulders: colour often ochreous-brown above; pronotum of male and female 16·7–20·5 mm. .. .. *gravelyi*, sp. nov.

9. 9. Spine on each side of the posterior angles of the lateral lobes of pronotum directed at a right angle, transverse, not oblique.

11. Posterior femoral margins above tri- or quadridentate; dorsum moderately wide between the shoulders; abdomen yellow maculate; pronotum of male 17 mm. .. .. *flavopictus*, Bolivar.

11. 11. Posterior femoral margins above subcrenulate, or barely lobate; dorsum wider between the shoulders; abdomen often white maculate.

12. Body above greyish-ochreous; sides of the body and hind femora covered with pale granulations; hind process often pale maculate toward the apex; body below fuscous and light, palpi white; pronotum of male and female 17·5–21 mm. .. .. .. *grandis*, Hancock.

12. 12. Body above fuscous or greyish-fuscous; hind femora obscurely mottled or with bars of ochre; pronotum of male and female 17·5–20·5 mm. .. .. *maximus*, Hanc.

Pronotum of female 18 mm.

race or var. *extremus*, Hancock.

1. 1. Vertex little produced; dorsum of pronotum rugulose; posterior angles of the lateral lobes of pronotum depressed, acute, barely produced; pronotum of male and female 14 mm. . .. *vidali*, Bolivar.

Criotettix rugosus, Bolivar.

(Fig. 2, Plate xiv.)

Bol., Ann. Soc. Ent. Belg., xxxi, p. 228, 1887; Brunn., Ann. Mus. Genova, xxxiii, p. 105, 1893.

Habitat.—Rangoon, Burma (*N. Annandale*). Brunner records this species from Lower Burma.

Criotettix tricarinatus, Bolivar.

(Fig. 14, Plate xiv.)

Bol., Ann. Soc. Ent. Belg., xxxi, p. 224, 1887; Hanc., Spolia Zeylanica, ii, p. 128, pl. 3, fig. 15, 1904.

Habitat.—Sigiriya, Ceylon, Sept. 1909 (*E. E. Green*); Kandy, Ceylon, Apr. 1907; Peradeniya, Ceylon, July 1913 (*A. R.*). Bolivar records this species from Kodaikanal, S. India, in Ann. Soc. Ent. France, lxx, p. 583, 1902.

Criotettix spinilobus, Hancock.

(Fig. 13. Plate xiv.)

Hanc., Spol. Zeylan., ii, p. 130, pl. 3, fig. 12, 1904; Hanc., Gen. Ins. Orth. Acrid. Tetriginae, p. 28, fig. 13, 1906.

Habitat.—Pundaluoya, Ceylon; Vurkalay, Travancore coast, S. India (*N. Annandale*).

This small species has the vertex subequal in width to one of the eyes, and the posterior angles of the lateral lobes on each side have a strongly produced sharp spine, curvate forward. The specimen from Travancore in the Indian Museum bears Kirby's label on which is written "*Criotettix obscurus* Kb. type." The specimen is identical with the type of *spinilobus* in the author's collection. This species resembles *indicus*, Bol., but may be distinguished by the curved spines.

Criotettix indicus, Bolivar.

Bol., Ann. Soc. Ent. France, lxx, p. 581, 1902.

Habitat.—? S. India (*Bolivar*). Not represented in the Indian Museum coll.

Criotettix orientalis, Hancock.

(Fig. 12, Plate xiv.)

Hanc., Records Ind. Mus., vii, p. 312, pl. xv, fig. 4, 1913.

Habitat.—Dibrugarh, N. E. Assam.

Criotettix pallidus, sp. nov.

(Fig. 5, Plate xiv.)

Near *indicus*, Bolivar. Stature small; vertex distinctly narrower than one of the eyes, little narrowed forward; eyes globose: head not at all exserted; antennae inserted between the lower part of the eyes; frontal costa compresso-elevated between the antennae. Pronotum little rugose-granulose, between the shoulders little convex, disc on each side bearing a short carinula; prozonal carinae behind the anterior border convergent backward; dorsum behind the shoulders, depressed, fossulate, and backward on the process bearing subelevated nodules; hind process subulate, extended beyond the hind femoral apices; median carina of the pronotum very low, thin, and sinuous; posterior angles of the lateral lobes laminate expanded, little triangulate and prominent, behind truncate, but the hind margin concave. Elytra ovate, apices rounded; wings fully explicate extended backward to the end of the pronotal process. Anterior femoral margins entire; middle femoral margins above and below undulate; hind femora somewhat stout, margins entire, minutely crenulate, externally above with a series of obtuse tubercles, and at the middle bearing distinct oblique rugulae; the pulvilli of the first joint of hind tarsi acute spiculate.

Entire length of male and female 11·5 mm.; pronotum 10·8 mm.; hind femora 5 mm.

Habitat.—Tenasserim Valley, Lower Burma (*Doherty*).

Criotettix dohertyi, sp. nov.

(Fig. 4, Plate xiv.)

Colour ferrugineous. Vertex strongly narrower than one of the eyes, little narrowed forward; eyes globose; head not at all exserted; antennae inserted between the lower angles of the eyes, not wholly between the eyes, and lower than in *tricarinatus*. Frontal costa compresso-elevated between the antennae, a little depressed above between the eyes, and distinctly sinuate at the median ocellus. Pronotum with the dorsum rugose-granulate, behind the shoulders depressed, between the shoulders convex and rather wide; on each side bearing a very thin carinula; on the process rugose-granulate, and bearing more or less irregular subelevated obtuse tubercles; posterior process subulate, long produced beyond the hind femora apices; median carina of pronotum very low, thin, and sinuous; prozonal carinae forward behind the front margin convergent backward, indistinctly expressed; posterior angles of the lateral lobes of the pronotum laminate expanded, little acute, prominent, margin behind the angle truncate, but very slightly concave. Elytra ovate, apices rounded-truncate; wings extended to the apex of the pronotal process. Anterior femoral margins entire; middle femoral carinae above entire, below subundulate; middle femora of male stouter, less elongate; hind femoral margins granulose, entire, the external face below subinfuscate; hind tibiae plain dark brown; the third pulvillus of the first joint of the hind tarsi little longer than the second, the first two pulvilli acute spinose.

Entire length of male and female 13–16·5 mm. pronotum 12–15·5 mm.; posterior femora 5·5-7·3 mm.

Habitat.—Upper Assam (*Doherty*).

Two specimens in the Indian Museum have labels apparently in Saussure's handwriting, bearing the name "*Paratettix varia-*

balis, Bol." This species recalls the genus *Paratettix* in the characters of the vertex, yet the lateral lobes at the posterior angles are laminate, the angle truncate behind and the apex prominent. It materially differs from *Paratettix variabilis*, Bol., in many respects.

Criotettix aequalis, Hancock.

(Fig. 3, Plate xiv.)

Hanc., Mem. Dept. Agric. India, iv, p. 136, 1912.
Habitat.—Bengal, Probsering, Lebong, 5000 ft. (Author's coll.)

Criotettix montanus, Hancock.

(Fig. 1, Plate xiv.)

Hanc., Mem. Dept. Acric. India, iv, pp. 133, 134, 1912.
Habitat.—Punjab, Simla, 7000 ft. (Author's coll.)

Criotettix annandalei, sp. nov.

(Fig. 6, Plate xiv.)

Near *gravelyi*, Hancock. Stature smaller: coloured fuscous, obscurely variegated with lighter brown, hind femora obscurely pale mottled. Vertex narrower than one of the eyes, subgranulose, fossulate on each side of the median carina, widened backward; head not exserted; antennae inserted between the lower angles of the eyes, not wholly between the eyes; frontal costa compresso-elevated arcuate between the antennae, not so roundly produced as in *gravelyi*. Pronotum deplanate on the dorsum, little convex between the shoulders, dull rugose-granulose, behind the shoulders bifossulate and the surface on the base of the process more or less pitted and tuberculose as in *Eugavialidium*, prozonal carinae forward behind the front border lightly expressed, subparallel; posterior process subulate and acute produced much beyond the hind femoral apices; median carina of pronotum very low and thin, following the inequalities, obsolete forward behind the anterior margin; elytra elongate-ovate, apices rounded; wings extended to the apex of the pronotal process; posterior angles of the lateral lobes laminate and produced in an oblique acute spine on each side, anterior femoral margins subentire; middle femoral margins above minutely crenulate, subundulate, below undulate, very indistinctly bilobate; hind femora rather stout, margins above crenulate and often bearing very indistinct pale crenulate lobes, or absent, below margins subentire; pulvilli of the first joint of the hind tarsi subacute, not at all spinose, the third pulvillus longer than the second and planate below.

Entire length of male and female 14·5 17 mm.; pronotum 13·5 mm.; posterior femora 6–7 mm.

Habitat.—Paresnath, W. Bengal (Chota Nagpur), 4300 ft., April 15, 1909 (*N. Annandale*).

Two of the specimens in the Indian Museum collection were labelled by Kirby "*Criotettix exsertus*, Bol." The latter species has the head more exserted, and is otherwise very different from *Criotettix annandalei*, sp. nov., described above, and as noted under the genus *Bolotettix*, Hancock.

Criotettix gravelyi, sp. nov.

(Fig. 9, Plate xiv.)

Stature large, dorsum ochreous or dark ochreous-brown, and subglabrous, the sides of the body, the under parts, legs, and wings, more or less black; posterior femoral carinae pale maculate, externally the hind femora marked with ochre, vertex smooth, in front subnarrower or subequal in width to one of the eyes, very slightly subnarrowed forward, median carina little expressed at the front; frontal costa roundly compressed between the antennae; eyes not exserted; antennae inserted between the lower angles of the eyes. Pronotum with the dorsum deplanate, rather smooth, little convex between the shoulders, broadly depressed behind the shoulders, posteriorly long subulate; prozonal carinae behind the front margin subparallel or indistinctly convergent backward; median carina of pronotum substraight, very low, obliterated forward behind the front border, little rounded subnodulose forward between the sulci; hind process produced much beyond the apices of the hind femora; posterior angles of the lateral lobes laminate, the apices produced in suboblique spines; elytra oblong-ovate, apices rounded-truncate; wings reaching to the apex of the pronotal process; margins of four anterior femora subentire; the superior carina of the middle femora granulate subundulate, below biundulate, or entire; posterior femoral margins above minutely crenulate, with very indistinct elevated pale lobes; externally above bearing a series of rounded tubercles, and at the middle bearing oblique rugulae; the three pulvilli of the first joint of hind tarsi subequal in length, the third subplanate below.

Entire length of male and female 17·5—20·5 mm.; pronotum 17–20 mm ; hind femora 7–9 mm.

Habitat.—Ghumti, Darjiling Dist., E. Himalayas, 4000 ft., July 1911 (*F. H. Gravely*); Sikkim, Darjiling Dist. (*L. Mandelli*).

Criotettix flavopictus, Bolivar.

(Fig. 10, Plate xiv.)

Bol., Ann. Soc. Ent. France, lxx, p. 582, 1902.

Habitat.—Thingannyinaung to Sukli, Dawna Hills, 900–2100 ft., Nov. 23, 1911 (*F. H. Gravely*); Misty Hollow, W. side of Dawna Hills, 2200 ft., Nov. 22, 1911 (*F. H. Gravely*): Ind. Mus. coll. Anamalais, about 2500 ft., Jan. 21, 1912 (*T. B. Fletcher author's coll.*). Kodaikanal (Castets, Decoly; *Bolivar*).

Criotettix maximus, Hancock.

(Fig. 7, Plate xiv.)

Hanc., Records Indian Museum, viii, pp. 311, 312, pl. xv, fig. 1, 1913.

Habitat.—Ghumti, Darjiling Dist., E. Himalayas, 4000 ft., July, 1911 (*F. H. Gravely*). Yembung, 1100 ft.; Janakmukh, 600 ft. (*author's coll.*).

Note: In the figure given in the Rec. Ind. Mus. the elytra are drawn too large by the artist.

Criotettix extremus, Hancock.

(Fig. 11, Plate xiv.)

Hanc., Mem. Dept. Agric. India, iv, pp. 132, 133, 1912.

Habitat.—Madras, Shevaroys, 4000 ft. (*author's coll.*).

It is quite probable that this is a variety or race of *maximus* Hanc., and I have so regarded it in the table separating the species of *Criotettix*.

Criotettix vidali, Bolivar.

Bol., Ann. Soc. Ent. Belg., xxxi, p. 227, 1887; Brunn. Ann. Mus. Genova, xxxiii, p. 105, 1853.

Habitat.—Carin Cheba (*Brunner*). This species was described by Bolivar from Philippine specimens, and I am including it here on the authority of Brunner.

Criotettix grandis, Hancock.

(Fig. 8, Plate xiv.)

Hancock, Mem. Dept. Agric. India, iv, pp. 134, 135, 1912.

Habitat.—Darjiling Dist., Singla, 1500 ft., Mar., 1913 (*Lord Carmichael coll.*).

The type in the author's collection is from Assam, Cherapunji, Khasi Hills. The specimens in the Indian Museum do not differ from the type, except in the colour of the body. In these specimens the colour is grey above, the sides of the body and hind femora covered with pale granulations, the pronotal process toward the apex darker and often minutely pale spotted; the anterior and middle tibiae annulate with white; body below fuscous and light; palpi white; hind femora with obscure pale bars; spines of the lateral lobes of pronotum acute, but not long produced. In the males the spines narrower. In one specimen from Upper Burma, Shan Hills (*J. C. Brown*), the spines of the lateral lobes are more acutely produced. The type specimen is somewhat faded with age, and the colour of the fresh specimens add materially to the original description.

Genus **Loxilobus**, Hancock.

Hancock, Spol. Zeylan., ii, p. 134, 1904.

The members of this genus have the vertex subequal or wider than one of the eyes, and narrowed forward; the frontal costa roundly produced before the eyes; the antennae inserted between the eyes, and the head not at all exserted. The pronotum above little rugose, often tuberculate, and with elongate lines; the posterior angles of the lateral lobes laminate, triangulate, and acute, the margin of lobes behind truncate or obtuse sinuate; wings often extended backward little beyond the pronotal apex. This genus seems to occupy a place between *Criotettix* on the one hand and *Coptotettix* on the other.

Loxilobus acutus, Hancock.

Hanc., Spol. Zeylan., ii, p. 134, figs. 3 and 16, 1904; Hanc., Gen. Ins. Orth. Acrid. Tetr., p. 29, pl. ii, fig. 17. 1906; Hanc., Mem. Dept. Agric. India, iv, p. 137, 1912.

Habitat.—Ceylon. Author's coll.

Loxilobus subulatus, Bolivar.

Criotettix subulatus, Bol., Ann Soc. Ent. Belg., xxxi, p. 227, 1887.

Habitat.—" Indes Orientalis" (*Bolivar*).

Loxilobus hancocki, Kirby.

Loxilobus rugosus, Hanc., Spol. Zeylan , ii, p. 135, 1908; Kirby, Syn. Cat. Orth. Brit. Mus., iii, p. 18, 1910.

Habitat.—Bombay, India; Ceylon. Author's collection.

Loxilobus assamus, Hancock.

Hanc., Trans. Ent. Soc. Lond., p. 223, 1901; Hanc. Mem. Dept Agric. India, v, p. 136, 1912.

Habitat.—Assam; Bengal, Lebong. Author's coll.

Loxilobus parvispinus, sp. nov.

Resembling *acutus*, Hancock, but having the dorsum of pronotum little rugose; median carina of the pronotum thin, very low, irregularly compressed backwards, the rugae not so distinctly elevated; the posterior angles of the lateral lobes of the pronotum laminate, the apex on each side bearing a very small minute acute spine, the margin behind sinuate; the lateral carinae and the median carina of the pronotum much thinner and not so compressed as in *acutus*; colour dark ferrugineous, the sides of the abdomen and wings black, the hind femora with an obscure longitudinal black fascia below on the outer face: wings extended little beyond the pronotal apex.

Entire length of male 13 mm. ; pronotum 12 mm. ; posterior femora 5 mm.

Habitat.—Pusa, Bihar, July 9, 1910 (*T. B. Fletcher*). Author's coll.

This species was taken for *L. acutus*, and the male was described as that species in my report in the Mem. Dept. India, IV, p. 137, 1912. The present description is supplemental to that account, and refers to *parvispinus* instead of *acutus*.

Loxilobus striatus, sp. nov.

Near *acutus*, Hancock. Greyish-fuscous above, sides of body, and legs, paler reddish-ochre; hind femora below the lower external carina marked with a longitudinal black fascia; wings black or infumate. Vertex wider than one of the eyes, slightly narrowed forward, granulose; frontal costa arcuate produced between the eyes; antennae inserted distinctly between the lower fourth of the eyes. Pronotum granose, interspersed with very small tubercles; dorsum between the shoulders little convex, moderately wide, behind the shoulders deplanate, subulate posteriorly, surpassing the hind femoral apices; median carina of pronotum very low, thin, and irregularly interrupted and compressed; posterior angles of the lateral lobes of the pronotum laminate and obliquely truncate behind, the apices subacute, little prominent. Elytra elongate, the apices rounded-truncate; wings extended little beyond the pronotal apex. Anterior and middle femoral margins above entire, the inferior margins subundulate; hind femora moderately stout, margins minutely crenulate, the antegenicular spine acute, the third pulvilli of the first joint of the hind tarsi nearly as long as the first and second united, the first and second pulvilli spinose.

Entire length of male 11·5 mm. ; pronotum 10·5 mm. ; posterior femora 5·5 mm.

Habitat.—Calcutta, Aug. 26, 1904 (*Brunetti*); Thingannyinaung to Sukli, Dawna Hills, Lower Burma, 900-2100 ft., Nov. 23, 1911 (*F. H. Gravely*).

This species differs from *parvispinus* in the posterior angles of the lateral lobes being obliquely truncate behind, instead of sinuate and spined; while in *acutus* the angle of the lobes has the margins behind subtransverse.

Genus Acanthalobus, Hancock.

Table for the separation of the Indian species.

1. Vertex little wider than one of the eyes, lateral lobules in front marginate; spine on each side of the posterior angles of the lateral lobes of pronotum acute and obliquely produced, the margin behind widely concave-sinuate .. .. *miliarius*, Bolivar.
cuneatus, Hancock

1. 1. Vertex distinctly wider than one of the eyes, narrowed toward the front, imperfectly marginate, or subtruncate.
   2. Hind femora with the superior margin above distinctly lobate-serrulate, and toward the base costate; inferior margin subtuberculate; spine on each side of the lateral lobes of the pronotum moderately produced, acute, and subtransverse, the margin behind strongly sinuate .. .. *curticornis*, sp. nov.
2. 2. Hind femoral margins above not at all or scarcely dentate; spine on each side of the lateral lobes of the pronotum acute and oblique, and strongly produced.
   3. Median carina of the pronotum behind the shoulders depressed, dorsum between the humeral angles convex; posterior femora above sparingly lobate .. .. *saginatus*, Bolivar.
3. 3. Median carina of the pronotum behind the shoulders not or barely depressed; dorsum between the humeral angles subdeplanate; posterior femora above entire, not at all dentate .. .. *bispinosus*, Dalm.

Acanthalobus miliarius, Bolivar.

Criotettix miliarius, Bol., Ann. Soc. Ent. Belg., xxxi, p. 226, 1887.

Acanthalobus miliarius, Hanc. Spol. Zeylan., ii p. 155, pl. 2, fig. 8, 1904.

Hanc. Gen. Ins. Orth. Acrid. Tetr., p. 29, pl. 2, fig. 19, 1906.

Habitat.—Trincomalee, Nov. 1906; Colombo; Ceylon. Ind. Mus. coll.

Acanthalobus cuneatus, Hancock, *l. c.* 1904, is apparently the short-wing form of *miliarius*.

Acanthalobus curticornis, sp. nov.

Yellowish-rufescent, darker on the dorsum of pronotum forward and on the front of the head. Vertex strongly wider than one of the eyes, narrowed forward toward the front, subfossulate on each side, front imperfectly marginate, subtruncate, median carina very small, little elevated and subproduced; antennae very short, inserted between the lower part of the eyes; frontal costa arcuate-elevated between the antennae and depressed at the vertex. Pronotum with the dorsum deplanate, between the shoulders convex and rather wide, bearing rounded tubercles irregularly distributed; median carina low, interrupted and sparingly tuberculate, forward between the sulci elevated-crassate, subnodulose: prozonal carinae behind the anterior border convergent backward, and distinctly expressed; hind process of pronotum strongly crassate and rounded, strongly produced backward beyond the hind femoral apices; posterior angles of the lateral lobes of the pronotum laminate expanded, the angle contracted and acute produced in a spine on each side, subtransversely directed, the margin behind distinctly right-angulate sinuate. Elytra oblong, narrowed toward the apices and rounded; wings largely covered by the hind process, and extended to the apex of the pronotum, the part showing narrow; anterior femoral carinae entire; middle femoral margins above subundulate, below indistinctly bituberculate; posterior femoral margins above with the forward half costate, backward dentate, and minutely serrulate; in the type two of the denticles very distinct, and two less distinct, margin below provided with a series of barely elevated tubercles; hind tibiae sinuate-curvate; pulvilli of the first joint of the hind tarsi elongate, acute

Entire length of female 19 mm.; pronotum 18·5 mm.; posterior femora 7 mm.

Habitat.—Medha, Yenna Valley, Satara Dist., Bombay Pres., 2200 ft., Apr. 17, 1912 (*F. H. Gravely*).

The spines of the lateral lobes in this species are shorter than in *saginatus*, or in *miliarius*, and they are more transverse; moreover, there are none of the series of short lines on the dorsum of pronotum, though tubercles are distinctly evident.

Acanthalobus saginatus, Bolivar.

Criotettix saginatus, Bol., Ann. Soc. Ent. Belg., xxxi, pp. 225, 226, 1887; Brunner, Ann. Mus. Genova, xxxiii, p. 104, pl. 5, fig. 38, 1893.

Habitat.—S. India (*Bolivar*); Rangoon, L. Burma (*Brunner*). Not in the present collection.

The name *Acanthalobus saginatus* is given as a synonym for *Tettix inornata*, Walker, in Kirby's Cat. Orth., III, p. 17, 1910. This does not seem justified from Walker's description in which he states in referring to the pronotum: "three spines on each side, the hind spine longer than the two others, and inclined obliquely backward." As there is only one spine on each side of the pronotum in *saginatus*, Walker's species cannot be interpreted as this species, but it may belong to the genus *Hexocera*.

Acanthalobus bispinosus, Dalm.

Acrydium bispinosum, Dalm., Vet.-Akad. Handl., p. 77, 1818; Kirby, Cat. Orth., iii, p. 18, 1910; Syn. *Tettix pallitarsus*, Walk.; *Tettix armiger*, Walk.; *Tettix latispinus*, Walk.

Habitat. Sibsagar, N. E. Assam (*S. E. Peal*); Assam. Ind. Mus. coll.

Genus **Tettitelum**, nov.

Head not at all exserted; vertex very wide, strongly wider than one of the eyes, on each side fossulate, and provided laterally with a rounded compressed carinula; middle subcarinate forward barely subproduced; face broad, in profile little oblique; eyes viewed from above little prominent, but viewed from the side moderately small; paired ocelli placed between the middle of the eyes; antennae inserted between the lower part of the eyes, or barely lower; frontal costa widely sulcate between the paired ocelli, the rami subparallel, in profile scarcely elevated between the antennae; apical articles of palpi narrow. Pronotum above plain, subcylindrical forward, anteriorly truncate, posteriorly long acuminate, the apex acute subspinate: humeral angles wanting, the dorsum forward roundly sloping at the sides; prozonal carinae behind the anterior margin obsolete; posterior angles of the lateral lobes of the pronotum laminate-expanded, with the inferior margin before the angle, or the angle produced in an acute spine, curvate forward. Elytra elongate, apices rounded; wings fully explicate and reaching to the pronotal apex; anterior and middle femora elongate, margins entire, minutely serrulate. Blades of the female ovipositor rather short and straight, not at all curvate at the apices, hirsute, but not at all dentate, the upper blade rather broad. Near *Acanthalobus*, Hancock.

Tettitelum hastatum, sp. nov.

Colour yellowish, variegated with fuscous, the base of the pronotal process above suffused with black, lower margins of the lateral lobes, the elytra, and underneath the body pale yellow, abdomen greyish-fuscous, backward densely white maculate toward the extremity. Face robust; vertex not advanced as far as the eyes, strongly wider than one of the eyes, marginate in front, on either side fossulate, subampliate on the occiput; frontal costa compressed but scarcely elevated between the antennae. Pronotum above plain, minutely rugulose-granose and punctate; dorsum between the shoulders very obtuse tectiform, little subdepressed behind; median carina of pronotum very low, percurrent, and barely elevated forward before the shoulders; lateral carinae before the shoulders and also the prozonal carinae behind the anterior border wanting: the forward part of the pronotum cylindrical; posterior process long subulate, the apex sharply pointed; posterior angles of the lateral lobes of pronotum laminate expanded, the lower margin produced laterally in a curvate acute spine on each side, directed forward; anterior and middle femora elongate, margins minutely serrulate.

Entire length of female 17·5 mm.; pronotum 16 mm.: (hind femora missing).

Habitat.—Kawkareik, Amherst Dist., Lower Burma, Mar. 5, 1908 (*N. Annandale*).

The type bears a label on which Kirby has written: "*Eugavialidium hastatum*, Kb., Type."

METRODORINAE.

Genus **Systolederus**, Bolivar.

Systolederus greeni, Bolivar.

Bol., Ann. Soc. Ent. France, lxx, p. 584, 1892; Hanc., Spol. Zeylan., ii, p. 155, pl. 2, fig. 9, 1904; Hanc., Gen. Ins. Orth. Acrid. Tetr., p. 34, pl. 2, fig. 14, 1906.

Habitat.—Nilgiris, S. India; Maskeliya, Ceylon; Pundaluoya, Ceylon; Madulsima, Ceylon, July 13, 1908 (*T. B. Fletcher*; *Hancock*). Ind. Mus. coll.

Systolederus cinereus, Brunner.

Brunn., Ann. Mus. Genova, xxxiii, p. 105, 1893.
Habitat.—Carin Cheba (*Brunner*).

Genus **Eurymorphopus**, Hancock.

Eurymorphopus latilobus, Hancock.

Hanc., Spolia Zeylanica, v, pp. 113, 114, fig. 1, 1908.

Habitat.—Kandy, Ceylon, May 29, 1910; Undugoda, Ceylon, Sept 1909. Ind. Mus. coll.

This small species is easily recognized by the narrow vertex, and the strongly dilated margins of the lateral lobes of the pronotum.

Genus **Spadotettix**, Hancock.

Hanc., Spol. Zeylan., vi, pp. 146, 147, 1910; Hanc., Mem. Dept. Agric. India, iv, pp. 141, 142, 1912.

Spadotettix fletcheri, Hancock.

Spol. Zeylan., vi, pp. 147, 148, figs. 1, 2, 1910.
Habitat.—Ceylon. Author's coll.

Spadotettix provertex, Hancock.

Mem. Dept. Agric. India, iv, pp. 142, 143, 1912.
Habitat.—Madras, India. Author's coll.

Genus **Apterotettix**, Hancock.

Hanc., Spol. Zeylan., ii, pp. 108, 140, 1904; Hanc., Gen. Ins., pp. 30, 35, 1906.

Apterotettix obtusus, Hancock.

Hanc., Spol. Zeylan., ii, p. 155, pl. 3, fig. 13, 1904; Gen. Ins., p. 35, fig. 16, 1906.

Habitat.—Peradeniya, Ceylon. Ind. Mus. coll.

Genus **Amphinotus,** nov.

Stature small and apterous, with the head moderately exserted. Vertex wide, subwidened forward, strongly wider than one of the eyes, bifossulate forward, submammilate between the posterior part of the eyes, the frontal carinulae laterally little compressed, abruptly terminated but not cuspidate, open each side of the mid-carina, middle carinate forward, compressed and little produced; face little oblique; eyes prominent, rather small, viewed from above subpedunculate and reniform; frontal costa strongly sinuate between the eyes, rather widely furcillate, not forked above the paired ocelli, the rami compresso-elevated between the antennae; paired ocelli placed nearly on a line with the lower border of the eyes: antennae rather short, filiform and inserted below the eyes. Pronotum truncate anteriorly, subtectiform forward, deplanate posteriorly, hind process abbreviated and truncate at the apex; median carina compressed, strongly elevated between the shoulders; humeral angles wanting; lateral carinae of the posterior process compressed; posterior angles of the lateral lobes of the pronotum oblique, the angle little prominent outwards and obliquely truncate behind. Elytra and wings wanting; legs elongate; the anterior and middle femoral margins undulate; posterior femora armed with denticles on the external longitudinal carina, the middle denticle compressed obtuse triangularly elevated; hind tibiae armed with spines; the first joint of the posterior tarsi strongly longer than the first, the three pulvilli often planate below and equal in length.

The type is *Amphinotus pygmaeus*, sp. nov.

This genus recalls the Cladonotinae in some respects, yet it is near *Mazarredia*, Bol. It differs in the subsessile eyes, wider vertex, lower position of the paired ocelli, absence of the humeral angles, less laminate posterior angles of the lateral lobes of the pronotum, the absence of elytra and wings, and long first joint of the hind tarsi.

Amphinotus pygmaeus, sp. nov.

Body very small, one of the smallest known Tettigids; coloured greyish, with two black bars marking the sides at the base of the hind process, divided by an oblique lighter line. Head little exserted; vertex very wide, front not advanced as far as the eyes, widened forward between the anterior carinulae, about twice the width of one of the eyes, bifossulate forward, and bearing small submammilate ridges just behind the fossae, frontal carinulae rounded, but abruptly terminated at the inner sides of the eyes and angulate, but not elevated above the eyes; median carina of the vertex compressed-rounded, little produced; frontal costa in profile strongly sinuate between the eyes, little protuberant between the antennae. Pronotum truncate anteriorly, or barely obtuse angulate; dorsum little compressed at the sides; the short prozonal carinulae behind the anterior margin distinctly compressed, elevated; median carina of the pronotum thinly compressed subcristate, highest forward between the sulci, sloping forward and little sloping backward; hind process strongly abbreviated, truncate, with the lateral angles obtuse, on each side of process bearing a compressed convex carina: humeral angles wanting, but instead the dorsum bearing an expressed lineate tubercle on each side, and two between the shoulders; posterior angles of the lateral lobes oblique, little prominent laterally and obliquely truncate behind. Anterior and middle femoral margins undulate; posterior femoral carinae bidentate on the outer face, the middle denticle larger, compressed, and produced, the superior and inferior margins entire; hind tibiae armed with spines and narrow; the first joint of the hind tarsi nearly twice the length of the third.

Entire length of male and female, 5–7·5 mm.; pronotum 3–4 mm.; posterior femora 3·7–4 mm.

Habitat.—Hakgala, Ceylon, Mar. 1907 (*E. E. Green*); Punduluoya, Ceylon, Feb. 1899. Two adults and a larva in the author's collection.

Genus **Mazarredia**, Bolivar.

Table for separating the Indian species.

1. Disc of pronotum above strongly unequal, somewhat elevated between the shoulders and bearing high subcarinate tubercles.
2. Dorsum of pronotum behind the shoulders profoundly fossulate.
3. Frontal costa not at all sinuate between the eyes.
4. Posterior angles of the lateral lobes of the pronotum little produced, apex subrounded; elytra oblong ovate; length of male and female pronotum 17·2–17·8 mm. .. .. *inequalis*, Brunner.

4. 4. Posterior angles of the lateral lobes of the pronotum widely dilated, subobtuse-truncate; elytra elliptical acuminate toward the apices; median carina of pronotum forward between the sulci strongly gibbose-crenulate; length of female pronotum 13·5 mm. *singlaensis*, sp. nov.

1. 1. Disc of pronotum between the shoulders lower.

2. 2. Dorsum of pronotum behind the disc distinctly though moderately fossulate.

3. 3. Frontal costa sinuate between the eyes.

5. Median carina of the pronotum lightly cristate before the shoulders, but depressed in front; posterior angles of the lateral lobes of the pronotum little expanded, obtuse; length of male and female pronotum 13·5–15·5 mm ... *sikkimensis*, Bolivar.

5. 5. Median carina of the pronotum subcristate before the shoulders, but not depressed in front; posterior angles of the lateral lobes of pronotum little produced, obtuse; length of female pronotum 17 mm. .. .. .. *sculpta*, Bolivar.

6. Disc of the pronotum plain, convex between the shoulders; median carina lightly and subequally elevated.

7. Vertex nearly twice the width of one of the eyes; (8 and 8·8).

8. Pronotal process subulate, surpassing the hind femoral apices; length of female pronotum 14·7 mm. *lativertex*, Brunner.

8. 8. Pronotal process not reaching to the knees of the hind femora: length of female pronotum 10·5 mm. *ghumtiana*, sp. nov.

9. Head little exserted, lateral carinulae not angulate or cuspidate.

10. Stature very small, body apterous but with small elytra: vertex subnarrower than one of the eyes, not at all produced; pronotum of male 5 mm., of female 7·3 mm. .. .. *perplexa*, sp. nov.

10. 10. Stature larger.

7. 7. Vertex subwider than one of the eyes.

11. Pronotum above granose, subrugose; eyes strongly exserted: length of pronotum 12 mm. .. .. *laticeps*, Bolivar.

11. 11. Pronotum above granose, and bearing irregular lineate tubercles on the process; eyes moderately exserted; frontal costa strongly protuberant between the antennae; length of pronotum male and female 14–16 mm. .. .. *dubia*, Hancock.

Length of pronotum male and female 13–15 mm. .. .. *convergens*. Brunner.

6. 6. Disc of the pronotum deplanate.

12. Median carina of the pronotum gibbose only between the sulci forward; dorsum rugulose: vertex strongly wider

than one of the eyes; length of female pronotum 13·5 mm. .. *latifrons*, Hancock.

12. 12. Median carina of the pronotum behind the anterior margin forward compresso-elevated, cristate and entire, vertex very little wider than one of the eyes; length of female pronotum 13 mm. *cristulata*, Bolivar.

9. 9. Head distinctly exserted.

13. Vertex on each side bearing a minute angulate lobe, narrowed toward the front; antennae inserted far below the eyes; elytral apices yellow; lateral lobes of pronotum with the posterior angles outwardly produced, obtuse; length of female pronotum 11·8 mm. *ophthalmica*, Bolivar.

13. 13. Vertex on each side cuspidate; head little exserted; pronotum above rugulose, carinae subacute; dorsum depressed behind the shoulders; length of male and female pronotum 10–11·8 mm. .. .. *insularis*, Bolivar.

Mazarredia inequalis, Brunner.

Brunn., Ann. Mus. Genova, xxxiii, p. 106, pl. 5, fig. 39, 1893.

Habitat.— Burma (*Brunner*).

Mazarredia singlaensis, sp. nov.

Near *inequalis*, Brunner. Yellowish-grey sparingly marked with fuscous. Body rugose; vertex very little wider than one of the eyes, subnarrowed forward, not at all produced at the middle, imperfectly marginate, elongate fossulate on each side of the mid-carinula; frontal costa not sinuate between the eyes but depressed, between antennae distinctly protuberant, viewed in front rather widely sulcate, branching above the paired ocelli, in profile sinuate at the median ocellus; antennae inserted little below the eyes; palpi compressed and white. Pronotum deplanate, above coarsely granulose and sparingly tuberculate; dorsum between the shoulders moderately dilated; median carina of the pronotum very unequal, forward between the sulci compressed-elevated in a rough gibbosity which gradually slopes forward, uneven, and subdepressed just behind the front border, strongly and abruptly sinuate and crenulate behind, the median carina posteriorly sinuate and subtuberculate; the dorsum between the shoulders little elevated, and bearing on each side of the middle an elongate elevated carinula; behind the shoulders strongly fossulate: posterior angles of the lateral lobes expanded laminate, the angle obliquely truncate behind, the apex somewhat prominent; elytra subeliptical: wings reaching to the pronotal apex: the four anterior femora compressed, margins of anterior above and below undulate: middle femoral margins above and below subtrilobate, minutely crenulate: hind femoral margins above serrulate, and sublobate, inferior carina very narrow, thin, and minutely undulate-crenulate; the third pulvilli of the first joint of the hind tarsi equal in length to the first and second combined, the first and second speculate, the first minute.

Entire length of female 14 mm.; pronotum 13 mm.: post. fem. 6 mm.

Habitat. Singla, Darjiling Dist., 1500 ft., June 1913 (*Lord Carmichael's coll.*), Ind Mus. coll.

Mazarredia sikkimensis, Bolivar.

Bol., Bol. Soc. Espan., ix, pp. 398, 399, 1909.
Habitat.—Sikkim, E. Himalayas. Ind. Mus. coll.

Mazarredia sculpta, Bolivar.

Bol., Ann. Soc. Ent. Belg., xxxi, p. 238, 1887; Brunn., Ann. Mus. Genova, xxxiii, p. 107, 1893; Hanc., Trans. Ent. Soc. London, p. 405, 1908.

Habitat.—Sukna, Darjiling Dist., 1000 ft., May 1913 (*Lord Carmichael's coll.*), Ind. Mus. coll. Also reported from Tenasserim; Pegu; Assam: Oriental India.

Mazarredia lativertex, Brunner.

Brunn., Ann. Mus. Genova, xxxiii, p. 108, pl. 5, fig. 41, 1893.
Habitat.—Burma (*Brunner*).

Mazarredia ghumtiana, sp. nov.

Near *lativertex*, Brunner. Ferrugineous-fuscous. Head large; body somewhat smooth granulose, devoid of a gibbosity and elevated tubercles. Vertex very wide, nearly twice the width of one of the eyes, horizontal, on each side fossulate, the lateral carinulae rounded, the front margin barely advanced beyond the eyes; frontal costa scarcely sinuate between the eyes; antennae inserted below the eyes. Pronotum truncate anteriorly, transversely rounded before the shoulders; hind process cuneate, not reaching or barely extended to the hind femoral knees; dorsum above deplanate, little convex between the shoulders, subfossulate behind the shoulders; median carina very low undulate; prozonal carinae behind the front border merely granulose lines convergent backward, not at all expressed; antehumeral carinae on each side low and not at all compressed, behind the shoulders laterally bicarinate; the hind process above toward the apex tricarinate and planate, and above little rugulose; posterior angles of the lateral lobes oblique, very little prominent, subnarrowed, obliquely truncate behind: elytra very small, elongate, apices rounded; wings present, but not reaching to the pronotal apex; anterior and middle femora subentire, granose; hind femora large, margins granose. minutely serrulate, antegenicular lobe acute; the three pulvilli of the first joint of the hind tarsi subequal in length, subobtuse.

Entire length of female 11·5 mm. (to end of pronotum): pronotum 10·5 mm.; hind femora 8 mm.

Habitat.—Ghumti, Darjiling Dist., E. Himalayas, Mar. 27, 1910 (*F. H. Gravely*); a larval specimen which may be this, or an allied species, is from Kurseong, 5000 ft., E. Himalayas (*N. Annandale*). The latter is labelled apparently by Kirby: "*Coptotettix acuteterminatus*, Brunn."

Mazarredia perplexa, sp. nov.

Stature small; coloured grey obscurely variegated with fuscous; head little exserted; eyes prominent somewhat elevated; vertex subnarrower than one of the eyes, narrowed forward, not at all produced, elongate fossulate on each side, laterally the carinulae little roundly elevated; antennae inserted distinctly below the eyes; paired ocelli placed between the lower third of the eyes: frontal costa depressed above between the eyes, rounded and protuberant between the antennae. Pronotum truncate anteriorly, posteriorly abbreviated, distinctly flattened backward and cuneate produced only to the knees of the hind femora; median carina little compresso-elevated forward between the sulci, depressed-undulate between and behind the shoulders; prozonal carinae forward behind the front margin parallel or subdivergent backward; on either side of the disc bicarinate; the lateral carinae becoming obsolete toward the apex; hind process little rugulose on the base; posterior angles of the lateral lobes subrounded truncate, not at all laminate outward nor prominent; elytra small and elliptical; wings wanting or undeveloped; anterior and middle femora elongate, margins entire; the third pulvilli of the first joint of the hind tarsi as long as the first and second combined, the first and second spinose.

Entire length of male 6·5 mm.; female 8·5 mm.; pronotum 5·7 mm.; hind femora female 5·2 mm.

Habitat.—Sikkim (*Knyvett*). Ind. Mus. coll.

Mazarredia dubia, Hancock.

Hanc., Mem. Dept. Agric. India, iv, p. 139 1912.

Habitat.—Bengal, Proobsering, Lebong, Darjiling, 5000 ft. Author's coll.

Mazarredia convergens, Brunner.

Brunn., Ann. Mus. Genova, xxxiii, p. 107, pl. 5, fig. 40, 1893.

Habitat.—Burma (*Brunner*).

Mazarredia laticeps, Bolivar.

Bol., Bol. Soc. Espan., ix, p. 399, 1909.
Habitat.—Upper Assam (*Bolivar*).

Mazarredia latifrons, Hancock.

Hanc., Mem. Dept. Agric. India, iv, p. 139, 1912.
Habitat.—Bengal, Proobsering, Lebong. Darjiling Dist., 5000 ft. Author's coll.

Mazarredia cristulata, Bolivar.

Bol., Ann. Soc. Ent. France, lxx, p. 584, 1902.
Habitat.—S. India (*Bolivar*)

Mazarredia ophthalmica, Bolivar.

Bol., Bol. Soc. Espan., ix, p. 399, 1909.
Habitat.—" Sibsagar, N. E. Assam (*S. E. P.*)" (*Bolivar*).

Mazarredia insularis, Bolivar

Bol., Ann. Soc. Ent. Belg., xxxi, p. 239, 1887 : Hanc., Spol. Zeylan., ii, p. 155, pl. 2, fig 7, 1904.

Habitat.—Kandy, Ceylon, July 1910. Ind. Mus. coll.

Genus **Xistra,** Bolivar.

Xistra stylata, Hancock.

Hanc., Trans. Ent. Soc. London, p. 231. 1907.
Habitat.—Ceylon. Author's coll.

Xistra dubia, Brunner.

Brunn., Ann. Mus. Genova, xxxiii, p. 108, pl. 5, fig. 42, 1893.

Xistra sikkimensis, sp. nov.

Near *sagittata*, Bolivar. Colour pale ochreous above, with two narrow black markings behind the shoulders. Head little exserted; eyes elevated higher than the dorsum of pronotum; vertex subequal in width to one of the eyes, little narrowed forward, middle carinate, subtruncate in front, the anterior carinulae laterally reflexed, angulate-subcuspidate, but not elevated higher than the eyes; antennae inserted below the eyes; paired ocelli placed between the middle of the eyes: frontal costa moderately produced between the eyes, arcuate, slightly sinuate above between the eyes, and distinctly sinuate below at the median ocellus. Pronotum lengthily subulate; dorsum tectiform, punctate and granose; median carina of the pronotum percurrent, little compressed, sinuate behind the anterior border, and distinctly compressed forward on either side at the median sulcus, and the dorsum strongly depressed; elytra ovate, apices rounded; wings caudate; middle femora elongate, margins entire; posterior angles of the lateral lobes of the pronotum hebitate.

Entire length of female 14 mm.; pronotum 10·8 mm.

Habitat.—Sikkim. Ind. Mus. coll.

This species differs from *sagittata* in the subcuspidate frontal carinulae of the vertex; in the tectiform dorsum, and slightly elevated undulate median carina forward. It is not at all typical of the genus *Xistra*, and approaches *Paratettix* in some respects. This species, like *saggittata* and *tricristata* belong to a series by themselves which really form a subgenus.

Genus **Xistrella**, Bolivar.

Bol., Real. Soc. Espan. Nat. Hist., ix, pp. 400, 401, 1909.

Xistrella dromadaria, Bolivar.

Bol., Real. Soc. Espan. Nat. Hist., ix, p. 401, 1909.

Habitat.—Sikkim (*L. M.*). Indian Museum coll.

The type came from the same locality. The female in the Ind. Mus. coll. has the pronotum lengthily subulate, and the median carina between the shoulders strongly elevated in an obtuse gibbosity. The length of the pronotum 17 mm.

Genus **Lamellitettix**, Hancock.

Lamellitettix fletcheri, sp. nov.

Near *acutus*, Hancock. Differing in the smaller stature: the vertex on either side more acute cuspidate; the frontal costa less protuberant between the antennae; median carina of the pronotum bigibbulate, the first gibbosity very small and rounded placed between the sulci forward, the second gibbosity joined on either side with a strongly compressed subtransverse ruga; median carina on the hind process sinuate. The strongly elevated carinate tubercles on each side between the shoulders in *acutus* reduced in *fletcheri* to short compressed carinulae; the posterior angles of the lateral lobes of the pronotum little acute produced in this species, whereas, in *acutus*, they are strongly transversely acute produced, the triangulate spine on each side having a broader base.

Entire length of male 9·5 mm; pronotum 8·7 mm: hind femora 4·5 mm.

Habitat.—Anamalais, Castlecroft Estate, India, 400 ft., Jan. 23, 1912 (*T. B. Fletcher*). Author's coll.

Lamellitettix pluricarinatus, Hancock.

Hancock, Trans. Ent. Soc. Lond., p. 404, 1908.

Habitat.—Deltota, Ceylon. Author's coll.

Genus **Hyboella**, nov. (*Coptotettix* in part).

Body more or less crassate; head not at all exserted; vertex wider than one of the eyes, narrowed forward, granose, imperfectly marginate, on either side bearing a very small looped carinula, more distinct next to the eyes, very lightly fossulate, middle minutely carinate; frontal costa produced, often depressed between the eyes, compressed-arcuate between the antennae; eyes more or less conoidal in form; paired ocelli placed little below the middle, or between the middle of the eyes; antennae filiform, of moderate length, and inserted between the extreme lower part of the eyes, or barely below. Pronotum anteriorly truncate, or subconvex, or barely produced at the middle: hind process posteriorly cuneate, not extended beyond the knees of the hind femora; or subulate and extended beyond the apices of the hind femora; median carina of the pronotum compressed elevated forward between the sulci, the dorsum tectiform forward, behind the shoulders depressed or planate, the median carina often bearing tubercles: middle of front border not at all or little produced over the occiput; posterior angles of the lateral lobes of the pronotum laminate dilated, subtransversely truncate behind, the angles subacute produced, or the posterior angles more or less reflexed outwards and obliquely truncate behind, rarely not reflexed outwards. Elytra and wings wanting, or when present the elytra elongate lanceolate, narrowed toward the apices, and the wings abbreviated, or extended beyond the pronotal apex. Anterior femora elongate, margins entire; middle femoral margins entire or undulate; posterior femora distinctly crassate, strongly widened toward the base, external face above bearing large tubercles, the superior carina minutely serrulate, entire, or bearing denticles toward the apices; the first joint of the posterior tarsi distinctly longer than the third.

This genus differs from *Copotettix*, Bolivar, in the stouter stature, the often depressed frontal costa between the eyes; the position of the antennae; the dilated or reflexed posterior angles of the lateral lobes of the pronotum, which are truncate behind; the compressed tectiform pronotum forward; and in the frequent imperfect development or absence of the wings. In the character of the posterior angles of the lateral lobes, this genus resembles *Mazarredia* somewhat. It includes a number of species in India, and some species in the Oriental Region outside of India. The type is *Hyboella tentata*, sp. nov.

Table for separating the Indian species.

1. Elytra and wings wanting.
2. Antennae inserted below the eyes; disc of pronotum elevated; lateral lobes of pronotum strongly arcuate backwards; pronotum of male 9·3 mm. *latifrons*, Brunner.

Stature smaller, pronotum of male 6·5 mm. .. .. *acuteterminata*, Brunner.

2. 2. Antennae inserted between the lower part of the eyes; lateral lobes of pronotum moderately expanded, the posterior angles obliquely truncate; disc of pronotum very little elevated; pronotum of female 10 mm. .. *nullipennis*, Hancock.

1. 1. Elytra present, of moderate size.

3. Pronotum viewed in front not quadrate.

4. Dorsum of pronotum somewhat smooth, tumid; median carina of the pronotum not produced at the front border; body crassate; pronotum of male and female 10–11 mm. .. .. .. *tumida*, Hancock.

4. 4. Dorsum of pronotum tuberculate, strongly tectiform forward; front border subangulate; the median carina of pronotum little produced; posterior angles of the lateral lobes of the pronotum widely dilated, triangulate produced; pronotum of male and female 11–13·8 mm. .. .. .. *tentata*, sp. nov.

Front border of pronotum truncate; pronotum of female 14 mm. .. *dilatata*, de Haan.

3. 3. Pronotum viewed in front subquadrate, above tectiform, sides concave; body strongly crassate; hind femora wide; frontal costa barely elevated; hind process of pronotum cuneate; wings reaching to the apex of pronotal process or little beyond; pronotum of male and female 12–15 mm. .. .. *obesa*, sp. nov.

5. Vertex viewed from above planate, little produced, in profile angulate produced; eyes strongly conoidal; posterior process of pronotum subulate distinctly surpassing the hind femoral apices; stature large; pronotum of female 17·5 mm. .. .. *angulifrons*, sp. nov.

5. 5. Vertex viewed from above fossulate on each side; stature smaller; frontal costa arcuate produced before the eyes.

6. Posterior angles of the lateral lobes of pronotum oblique, apices obtuse, pronotum little wide between the shoulders, behind the shoulders subfossulate; median carina arcuate forward and backward sinuate-tuberculate; pro-

notum of male and female 10·5-12·5 mm. . . . *conioptica*, sp. nov.

6. 6. Posterior angles of the lateral lobes of the pronotum subexpanded outwardly, behind widely truncate; pronotum of male and female 11·5 and 9 mm. *problematica*, Bolivar.

Hyboella latifrons, Brunner.

Coptotettix latifrons, Brunn., Ann Mus. Genova xxxiii, p. 112, pl. 5, fig. 44, 1893.

Habitat.—Burma (*Brunner*).

Hyboella acuteterminata, Brunner.

Coptotettix acuteterminatus, Brunn., Ann. Mus. Genova, xxxiii, p. 112, 1893.

Habitat.—Pegu (*Brunner*).

Hyboella nullipennis, Hancock.

Coptotettix nullipennis, Hanc., Records Ind. Mus., viii, p. 314, pl. xv, fig. 2, 1913.

Habitat.—Janakmukh, India (*Kemp*). Ind. Mus. coll.

Hyboella tumida, Hancock.

Coptotettix tumidus, Hanc., Records Ind. Mus., viii, pp. 313, 314, pl. xv, fig. 3, 1913.

Habitat.—Dibrugarh, N.-E. Assam (*Kemp*). Ind. Mus. coll.

Hyboella tentata, sp. nov.

Body crassate; face large: head not at all exserted, little retracted under the pronotum. Colour yellowish-ferrugineous, or subinfuscate. Vertex wider than one of the eyes, narrowed forward, subampliate backward, in front imperfectly marginate, open either side of the mid-carina, bearing very small subcompressed flexed carinulae laterally next to the eyes, and little subfossulate on each side: antennae inserted between the lower margin of the eyes; paired ocelli conspicuously placed before the middle of the eyes; frontal costa compressed arcuate between the antennae, depressed above, and sinuate at the median ocellus. Pronotum tuberculose granose above, the dorsum tectiform forward, the disc somewhat tumose, posteriorly planate; hind process acute cuneate abbreviated and not reaching to, or extended to the hind femoral knees; median carina of pronotum forward crassate and compressed-elevated arcuate, from the front to a point backward above the articulation of the hind femora, backward very low and bearing compressed tubercles; at the front margin the median carina crassate and little obtuse produced over the occiput; prozonal carinae behind the anterior border small, imperfectly developed;

humeral angles strongly depressed; posterior angles of the lateral lobes of the pronotum strongly laminate dilated laterally, apex subacute, widely transversely truncate behind: elytral sinus shallow: elytra narrow, elongate, lanceolate; wings undeveloped, but not entirely wanting: anterior femora elongate, margins entire, subundulate; middle femora little compressed, margins undulate; hind femora strongly crassate, very wide toward the base, superior margin granose and minutely serrulate, subbidentate toward the knees, the antegenicular spine acute; hind tibiae with both margins armed with strong spines; first article of the posterior tarsi strongly longer than the third, the third pulvillus as long as the first and second combined, planate below, the first and second spiculate.

Entire length of male and female 12–14·5 mm. (to end of process); pronotum 11–13·8 mm.; posterior femora 8–9 mm.

Habitat.—Sibsagar, N.-E. Assam (*S. E. Peal*); Upper Assam (*Doherty*).

Two examples are labelled with the name "*Tettix dilatatus.*" This species is quite different from *dilatata*, de Haan, as evidenced by comparison with specimens of this species from Java in my collection.

Hyboella dilatata, Haan.

Acridium (Tetrix) dilatatum, de Haan, Bijdr., pp. 167, 169, pl. xxii, fig., 1843.

Habitat.—Carin Asciuii (*Brunner*).

Hyboella obesa, sp. nov.

Body strongly crassate, above granose, little rugose; head little retracted under the pronotum; face large; eyes strongly conoidal in profile. Colour fuscous, variegated with dull yellow, hind femora often bearing vertical bars of yellow. Vertex wider than one of the eyes, barely emarginate in front, median carina very small and low, slightly fossulate on each side; paired ocelli placed scarcely below the middle of the eyes; antennae rather short, and inserted barely between the lower margin of the eyes, apical articles of palpi compressed: frontal costa barely arcuate elevated between the antennae, and little depressed between the eyes: vertex in profile angulate. Pronotum anteriorly very obtuse angulate, when viewed in front the body quadrate, on either side of pronotum concave, and above tectiform. median carina of pronotum in profile elevated forward, and arcuate compressed from the front border to a point backward above the articulation of the hind femora, then strongly depressed irregularly sinuate backward; the dorsum backward planate, broadly fossulate opposite the elytral apices; hind process acute cuneate, the lateral carinae compressed, the apex reaching to the extremities of the hind femora or surpassing them; humeral angles widely arcuate, the humeral carinae subobliterated, not at all compressed; elytra sinus shallow: posterior angles of the lateral lobes of the pronotum widely dilated, obliquely truncate behind: elytra elongate, widened at the middle, acuminate toward the apices; wings shortened, not reaching to the pronotal apex, or little surpassing the apex. Anterior and middle femora elongate, granulose, margins subentire: middle femora little compressed bearing a row of rounded tubercles at the middle; hind femora strongly crassate, very wide, the superior carina entire, minutely subsulcate serrulate, below entire; hind tibiae black, pale annulate at the anterior fourth; margins narrow and armed with strong spines; the first joint of the hind tarsi distinctly longer than the third, the first and second pulvilli of the first joint acute, the third longer than the second.

Entire length of male and female 12·5–16 mm.; pronotum 12–15 mm.; posterior femora 8–9 mm.

Habitat.—Ghumti, Darjiling Dist., E. Himalayas, 4000 ft., July, 1911 (*F. H. Gravely*).

A series of adults and larvae in the Indian Museum. The nymphs have the dorsum of pronotum strongly compressed, the median carina rounded-cristate as in some of the adult Cladonotinae.

Hyboella angulifrons, sp. nov.

Body moderately crassate, above granose-rugose: head very little exserted, in profile angulate; face oblique; eyes conoidal; colour fuscous, variegated with yellow on the hind process and hind femora; vertex viewed from above little produced and rounded at the sides, granose, wider than one of the eyes, widened backward and subampliate, little ascendant forward: antennae yellow, inserted between the extreme lower part of the eyes; paired ocelli placed little below the middle of the eyes; frontal costa slightly elevated arcuate between the antennae. Pronotum very little convex at the front margin: dorsum deplanate, between the shoulders little convex, behind the shoulders depressed and broadly fossulate; hind process above planate, rugose and subulate extended backward beyond the hind femoral apices; median carina of the pronotum compressed slightly arcuate and crassate forward extending only as far backward as the sulci; posteriorly indistinct but irregularly compressed-tuberculate: lateral carinae of process compressed: humeral angles very obtuse; posterior angles of the lateral lobes of the pronotum oblique very little dilated, the apices obtuse, the margin behind obliquely truncate: elytra lanceolate, widest at the middle, acuminate toward the apices; wings extended backward to the pronotal apex; anterior and middle femora elongate; margins of the anterior entire; margins of the middle femora undulate; hind femora crassate, the superior carina minutely serrulate, at the anterior half strongly rounded, the inferior margins entire; the first article of the posterior tarsi longer than the third; the first and second pulvilli acute, the third distinctly longer than the second.

Entire length of female 17·5 mm.; pronotum 16·5 mm.; posterior femora 9 mm.

Habitat.—Dawna Hills, Lower Burma, "third camp to misty hollow" 400–2400 ft., Nov. 22-30, 1911 (*F. H. Gravely*). Ind. Mus. coll.

This species is not so robust in stature as *obesa*, the vertex is more angulate in profile; the median carina of the pronotum forward is lower; and the hind process is lengthily subulate, instead of shortened and cuneate.

Hyboella conioptica, sp. nov.

Body slightly robust; colour greyish, the hind femora obscurely marked with yellow; head not at all exserted, in profile subrounded; vertex nearly equal in width to one of the eyes, not at all advanced beyond the eyes, narrowed forward and fossulate on each side, little ampliate backward; antennae inserted between the lower part of the eyes; paired ocelli placed between the middle of the eyes; frontal costa viewed in front sulcate above the paired ocelli, in profile arcuate produced before the eyes. Pronotum anteriorly truncate or very slightly convex, posteriorly subulate extended beyond the hind femoral apices; dorsum rather smooth granulate-tuberculate, between the shoulders convex, but forward subtectiform, and behind the shoulders planate; median carina very low, elevated forward and little arcuate behind the anterior margin between the sulci; depressed between the humeral angles, and depressed sinuate backward, irregularly compressed; humeral angles obtuse and roundly depressed; posterior angles of the lateral lobes of the pronotum little dilated, obliquely truncate behind, the apices dull; elytra elongate, widened at the middle, narrowly rounded at the apices; wings extended barely beyond the pronotal apex; anterior and middle femora elongate, little compressed, margins entire; posterior femora stout, externally granose, margins entire; the four anterior tibiae and tarsi black and pale annulate; first joint of the hind tarsi longer than the third, the first and second pulvilli acute, the third longer than the second.

Entire length of male and female 11·5 mm.; pronotum 10·5–12·5 mm.; posterior femora 5·6–7 mm.

Habitat.—Singla, Darjiling Dist., 1500 ft., Mar. 1913 (*Lord Carmichael's coll.*); Assam (*H. H. Godwin-Austen*). Ind. Mus. coll.

Hyboella problematica, Bolivar.

Coptotettix problematica, Bol., Real Soc. Espan., ix, p. 401, 1909.

Habitat.—Upper Assam (*Bolivar*).

TETTIGINAE.

Genus Teredorus, Hancock.

Hancock, Gen. Ins. Orth. Acrid. Tetr., pp. 51, 52, 1906.

The genus *Teredorus*, Hancock, was based on a Peruvian (S. American) species, *stenofrons*. The new Indian species described below have the frontal costa somewhat depressed, whereas, in *stenofrons*, it is roundly protuberant between the antennae; the dorsum of pronotum is rounded between the shoulders instead of being convex as in *carmichaeli*.

Teredorus carmichaeli, sp. nov.

Near *stenofrons*, Hancock. Body wholly white, or grey and pale variegated, tibiae fuscous, annulate with white, wings black or infumate. Head not at all exserted, viewed from above very small; vertex strongly contracted forward drawing the eyes very near together, minutely tricarinate; frontal costa barely subproduced beyond the eyes, sinuate at the median ocellus; antennae inserted little below the eyes; eyes distinctly globose, pronotum smooth glabrous, minutely granose, subcylindrical forward, widened between the shoulders; median carina very indistinct; prozonal carinae behind the front border mere obsolete lines and parallel; antehumeral carinae indistinct; hind process long subulate surpassing the hind femoral apices; posterior angles of the lateral lobes of the pronotum turned down, the apices rounded or subtruncate; elytra moderately large, externally reticulate, rather wide forward and acuminate backward and narrowly rounded at the apices; wings extended to the pronotal apex; the four anterior femoral margins entire, minutely serrulate; middle femora little compressed and externally bicarinate; posterior femoral margins entire, minutely serrulate, the antegenicular denticle acute; the three pulvilli of the first joint of the hind tarsi equal in length, the first and third tarsal joints equal in length.

Entire length of female 17 mm.; pronotum 16 mm.; hind femora 7 mm.

Habitat.—Singla, Darjiling Dist., 1500 ft., Apr., June, 1913 (*Lord Carmichael's coll.*). Ind. Mus. coll.

Teredorus frontalis, sp. nov.

Near *carmichaeli*, differing as follows: stature smaller, body coloured fuscous, variegated with ochre, the underparts, fusco-variegated, lower half of the lateral lobes of the pronotum pale ochre; the four anterior femora little more compressed, subrugose: the hind femora stouter; elytra smaller and punctate, not so reticulate; the antennae inserted little lower; the hind process of the pronotum not so long subulate, but the wings extended to the pronotal apex.

Entire length of female 11 mm.; pronotum 10·5 mm.; posterior femora 5·8 mm.

Habitat.—Dharampur, Simla Hills, 5000 ft., May, 1907 (*N. Annandale*).

Teredorus ridleyi, Hancock.

Systolederus ridleyi, Hanc., Trans. Ent. Soc. Lond., p. 401, 1908.

Habitat.—Singapore, Botanical Gardens (*Hancock*).

EXPLANATION OF PLATE XIV.

Lateral lobes of the pronotum in *Criotettix*. Line indicates length of pronotum.

FIG. 1. *Criotettix montanus*, Hanc.
,, 2. *Criotettix rugosus*, Bol.
,, 3. *Criotettix aequalis*, Hanc.
,, 4. *Criotettix dohertyi*, sp. nov.
,, 5. *Criotettix pallidus*, sp. nov.
,, 6. *Criotettix annandalei*, sp. nov.
,, 7. *Criotettix maximus*, Hanc.
,, 8. *Criotettix grandis*, Hanc.
,, 9. *Criotettix gravelyi*, sp. nov.
,, 10. *Criotettix flavopictus*, Bol.
,, 11. *Criotettix extremus*, Hanc.
,, 12. *Criotettix orientalis*, Hanc.
,, 13. *Criotettix spinilobus*, Hanc.
,, 14. *Criotettix tricarinatus*, Bol.

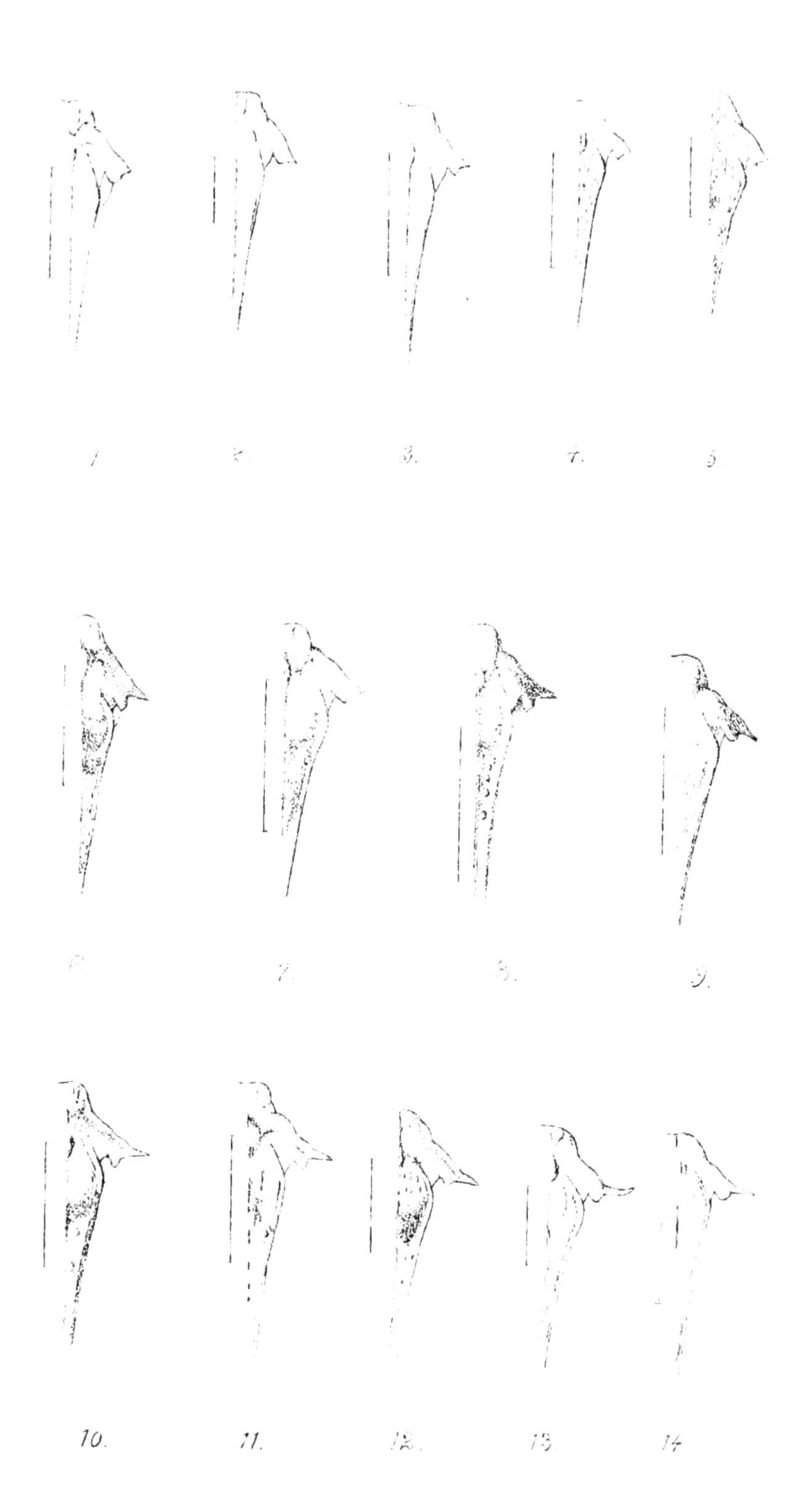

J. L. Hancock, del. A. Chowdhury, lith.

PRONOTUM OF CRIOTETTIX.

Genus **Paratettix,** Bolivar.

Table for separating the Indian species.

1. Wings strongly caudate, 3·5–5 mm. longer than the pronotal process.
2. Dorsum wide between the shoulders; body crassate; frontal costa little produced, arcuate.
3. Median carina of pronotum percurrent, not at all undulate; vertex equal in width to one of the eyes; dorsum of pronotum rugulose, slightly scabrous; posterior femora above strongly serrulate toward the apices, bearing an acute elevated denticle; pronotum of male and female 11·8–14 mm. .. *hirsutus*, Brunner.

3. 3. Median carina of pronotum compressed subgibbose forward behind the front border, and depressed just before the shoulders; vertex wider than one of the eyes; dorsum rugose-tuberculose, subtumid forward; the front border obtuse-angulate; hind femoral carinae above toward the knees minutely subserrulate, antegenicular denticle little prominent; pronotum of female 12 mm. .. .. *rotundatus*, sp. nov.

2. 2. Dorsum much narrower between the shoulders, body slender; vertex distinctly narrower than one of the eyes; median carina of pronotum percurrent, little compressed, and longitudinally low arcuate forward, highest between the shoulders, barely sinuate before the shoulders; head not at all exserted; pronotum of female 8·6 mm. *latipennis*, sp. nov.

1. 1. Wings caudate but less than 3 mm. longer than the pronotal apex, or abbreviated.

4. Median carina percurrent, undulate before the shoulders, but not behind; vertex equal in width to one of the eyes; posterior tibiae fuscous at the basal part, the apical part pale; pronotum of female 11 mm. .. .. *semihirsutus*, Brunner.

4. 4. Median carina of pronotum distinctly undulate, compressed gibbose forward between the sulci; vertex subnarrower or subequal to one of the eyes.

5. Hind process of pronotum acute, not ex-

tended beyond the hind femoral apices; pronotum of male and female 7–9·5 mm. .. .. *curtipennis*, Hancock.

5. 5. Hind process of pronotum long subulate, extended beyond the hind femoral apices; pronotum of female 12·8 mm.; wings 1·5 mm. longer than the pronotum *alatus*, sp. nov.

Paratettix hirsutus, Brunner.

Brunn., Ann. Mus. Genova, xxxiii, p. 110, fig. 43, 1893.

Habitat.—Sibsagar, N. E. Assam (*S. E. Peal*). Ind. Mus. coll.

Paratettix rotundatus, sp. nov.

Colour ferrugineous; body moderately stout, hirsute below, head not at all exserted, eyes and vertex not at all elevated; vertex wider than one of the eyes, not produced, on either side fossulate, little ampliate behind the fossae on the occiput, frontal carinulae laterally reflexed and little rounded compressed; paired ocelli placed little in advance of the eyes between the middle; antennae inserted barely on a line with the lower border of the eyes; frontal costa moderately arcuate produced beyond the eyes. Pronotum above rugose-tuberculose, anteriorly very obtuse angulate, posteriorly acute subulate, little surpassing the hind femoral apices; dorsum forward between the shoulders little tumid, backward planate; median carina behind the anterior border compressed, little arcuate subgibbose, between the shoulders depressed, backward very slightly subundulate compressed; humeral angles obtuse depressed; lateral carinae posteriorly little compressed on base of process; the two prozonal carinae behind the front margin parallel; posterior angles of the lateral lobes of the pronotum with the apices rounded, and somewhat narrowed; elytra widest near the middle, narrowed toward the apices and rounded; wings strongly caudate, extended 4·5 mm. beyond the pronotal apex in the type; anterior and middle femora elongate and compressed, hirsute, margins subparallel; posterior femora elongate, moderately broad, with a prominent antegenicular denticle, and a minute lobe anterior to it, the superior margin also being minutely serrulate-granose; hind tibiae with two light and two black annulations; the three pulvilli of the first joint of the hind tarsi spiculate, the third pulvillus little longer than the second.

Entire length of female 16·5 mm.; pronotum 12 mm.; posterior femora 7 mm.

Habitat.—Tezpur, Mangaldai Dist., Assam, Oct. 8, 1910 (*Kemp*).

This species resembles *hirsutus*, Brunn., differing in the wider vertex; the compressed subgibbose median carina of the pronotum forward; the wider sulcation of the frontal costa; and in the superior carina of the posterior femora being subentire, but minutely serrulate, whereas, in *hirsutus* the carina above is strongly serrulate toward the knees and the antegenicular denticle more strongly elevated and acute. Even with these differences however, this may be a variety of *hirsutus*, which fact can only be determined by a study of more extensive series.

Paratettix latipennis, sp. nov.

Stature slender; colour pale ferrugineous; body slightly rugose, little hirsute below; head not at all exserted; vertex distinctly narrower than one of the eyes; antennae inserted little below the eyes; frontal costa moderately arcuate produced, very lightly sinuate at the median ocellus. Pronotum anteriorly truncate, posteriorly acute subulate, produced about one millimetre beyond the hind femoral apices; dorsum between the shoulders convex; median carina of the pronotum percurrent, little compressed, and forward forming a low gentle arc, highest between the shoulders, and little compresso-elevated forward between the sulci; humeral angles obtuse angulate; the two short prozonal carinae behind the anterior margin indistinctly divergent backward; posterior angles of the lateral lobes of the pronotum rounded-truncate; elytra moderately small, elongate, apices narrowly rounded, externally reticulate; wings strongly caudate produced nearly 3·5 mm. beyond the pronotal apex in the type; anterior and middle femora little compressed, elongate; middle femoral margins in the female parallel, pilose; posterior femora stout, the apical half of the superior carina minutely serrulate; antegenicular denticle prominent, acute; first article of the hind tarsi longer than the third; the three pulvilli spiculate, the third pulvillus distinctly longer than the second.

Entire length of female 13·8 mm.; pronotum 8·6 mm.; posterior femora 4·8 mm.

Habitat.—Moradabad, U.P., July 21, 1911 (*T. B. Fletcher*). Author's coll.

Paratettix semihirsutus, Brunner.

Brunn., Ann. Mus. Genova, xxxiii, pp. 110, 111, 1893.

Habitat.—Carin Cheba (*Brunner*).

Paratettix curtipennis, Hancock.

Coptotettix curtipennis, Hanc., Mem. Dept. Agric. India, iv, pp. 146, 147, 1912.

Habitat.—Ghumti, Darjiling Dist., E. Himalayas, 4000 ft., July, 1911 (*F. H. Gravely*). Ind. Mus. coll.

Paratettix alatus, sp. nov.

Colour obscure fuscous; body moderately stout; head not at all exserted; eyes and vertex very little elevated; vertex subequal in width to one of the eyes, anteriorly truncate, frontal carinulae laterally reflexed and rounded-compressed, front not advanced beyond the eyes; antennae inserted below the eyes; paired ocelli placed little below the middle of the eyes: frontal costa widely sulcate, in profile not produced at the vertex between the eyes, but little arcuate produced between the antennae, sinuate at the median ocellus; face slightly oblique. Pronotum anteriorly truncate, posteriorly lengthily acute subulate: dorsum between the shoulders convex, rugose, backward depressed planate and somewhat rugulose, bearing small tubercles; median carina very thin, gibbose forward between the sulci, depressed and sub-straight between the shoulders, backward depressed and irregularly undulate; the two short prozonal carinae behind the front margin distinctly convergent backward; humeral angles carinate, little compressed; lateral carinae backwards compressed and towards the apex serrulate: posterior angles of the lateral lobes of the pronotum little obliquely dilated, the apices dull, rather widely obliquely truncate behind; elytra oblong, widest behind the middle, but the apices narrowly rounded; wings surpassing the pronotal apex 1·3 mm.; anterior femora elongate, little compressed, the superior carina strongly curved, outer face and below sparingly hirsute; middle femora rather broad, compressed, elongate, sparingly hirsute, margins subundulate; posterior femora moderately stout, externally subscabrous, the apical third above serrulate-lobate, antegenicular lobe acute spinate first article of the hind tarsi longer than the third; the first and second pulvilli acute, subspinate, the third distinctly longer than the second.

Entire length of female 15 mm.; pronotum 12·8 mm.; hind femora 6·3 mm.

Habitat.—Ghumti, Darjiling Dist., E. Himalayas, 4000 ft., July, 1912 (*F. H. Gravely*).

This species may be the long-wing form of *curtipennis*. It is near *semihirsutus*, but differs in the frontal costa not being produced above between the eyes, and in the distinctly undulate median carina of the pronotum.

Genus **Acrydium**, Goeff.

(*Tettix*; *Tetrix* of authors.)

Acrydium bipunctatum, Linn.

Kirby, Syn. Cat. Orth. Brit. Mus., iii, pp. 39–42, 1910.

Habitat.—Sikkim, Darjiling Dist.; Europe. Indian Mus. coll.

This species was previously reported from India by Bolivar as occurring at Kodaikanal, S. India. Two specimens in the Indian Museum bear perfect resemblance to specimens in my collection from different parts of Europe. Kirby gives the range of habitat as: Europe; North Africa, and North and West Africa.

Acrydium variegatum, Bolivar.

Paratettix variegatus, Bol., Ann. Soc. Ent. Belg., xxxi, p. 280, 1887; *Tettix atypicalis*, Hanc., Spol. Zeylan., ii,

p. 142, 1904; Hanc., Mem. Dept. Agric. India, iv, p. 149, 1912.

Habitat.—Moulmein, Lower Burma, Nov. 16, 1911 (*F. H. Gravely*); Peradeniya, Ceylon, Oct. 1905; Ratnaputa, Ceylon, Apr. 1905 (*E. E. Green*); Singla, Darjiling Dist., 1500 ft.

Acrydium ceylonicum, Hancock.

This is the short-wing form of *variegatum*, *l. c.*, 1904.

Habitat.—Peradeniya, Ceylon. Indian Mus. coll.

Acrydium indicum, Bolivar.

Paratettix indicum, Bol., Ann. Soc. Ent. Belg., xxxi, p. 281, 1887; Hanc., Mem. Dept. Agric. India, iv, p. 150, 1912.

Habitat.—Bihar, Chapra; Gorakhpur; Pupri, Muzaffarpur. Author's coll. "Indes Orientalis" (*Bolivar*).

Acrydium mundum, Walker.

Tettix mundum, Walk., Cat. Derm. Salt. Brit. Mus., p. 827, 1871.

Tettix umbriferum, Walk., *l. c.*, p. 824; *T. lineiferum*, Walk., *l. c*; *T. vittiferum*, Walk.; *T. dorsiferum*, Walk.; *T. oliquiferum*, Walk.; *T. nigricolle*, Walk.; *T. balteatum*, Walk.; *T. lineosum*, Walk., *l. c.* 1871.

Habitat.—N. India (*Walker*).

The above species appear to be mere colour varieties of one or two species which I find it impossible to place.[1]

Acrydium tectitergum, sp. nov.

Near *subulatum*, Linn. Fuscous or obscure fusco-ferrugineous, granulose; vertex wider than one of the eyes, distinctly produced, carinulae laterally subangularly reflexed, and little compressed, the front border convex, middle carinate, on each side elongate fossulate; eyes moderately small; head in profile obtuse angulate, the face strongly produced, convex, advanced equal to the width of one of the eyes; indistinctly sinuate at the median ocellus; antennae very slender filiform, inserted between the lower border of the eyes and partly below. Pronotum anteriorly very obtuse angulate, posteriorly acute subulate produced beyond the hind femoral apices; dorsum tectiform, median carina of the pronotum forward distinctly compressed elevated cristate, above substraight from the sulci to the shoulders, then gently concavely sloping backward, substraight on the hind process; humeral angles obtuse; lateral carinae little compressed; prozonal carinae behind the anterior border shortened; posterior angles of the lateral lobes

[1] See Kirby's Cat. Orth., III, 1910.

narrowed, apices hebitate; elytra elongate, widest at the middle, the apices narrowly rounded; wings caudate; anterior and middle femora elongate, compressed, margins below undulate, the middle femora below lobate; pulvilli of the first joint of the hind tarsi spinose, the third longer than the second.

Entire length of male and female 10·8–11·5 mm.; pronotum 8·5 mm.; hind femora 3 8 4 mm.

Habitat.—Hoshangabad, Sept. 14, 1911 (*T. B. Fletcher*); Surat, Bombay, July 8, 1904 (*T. B. Fletcher*). Author's collection.

One example from the former locality is distinctly larger in stature; entire length 13 mm.; pronotum 10·5 mm.; post. fem. 5·3 mm.; the colour obscure ferrugineous, with an elongate black spot on each side of the disc. In structure this example is near typical *tectitergum*. Until a larger series are obtained for study I hesitate to designate this form as new.

This species differs from *subulatum*, Linn., in the cristate median carina of the pronotum, in the slender antennae, in the lobate inferior carina of the middle femora, and somewhat shorter third pulvillus of the hind tarsi.

Genus **Coptotettix**, Bolivar.

Table for separating the Indian species.

1. Elytra very minute; stature very small.
2. Hind process of pronotum acute; median carina of pronotum anteriorly largely elevated, posteriorly pluri-interrupted; dorsum rugose tuberculate, pronotum of male 7 mm. .. *fossulatus*, Bolivar.

2. 2. Hind process in 3 and 3. 3. rounded-truncate at the apex.

3. Median carina of pronotum percurrent, anteriorly lightly elevated, rounded; posterior femora crassate, bearing a strong triangular antegenicular spine, the superior carina also serrulate; pronotum 3·8 mm. .. *parvulus*, Hancock.

3. 3. Median carina of pronotum percurrent, little elevated, subsinuate before the shoulders; posterior femora with the apical half slender; pronotum of female 8 mm. .. .. *capitatus*, Bolivar.

1. 1. Elytra larger; wings explicate, caudate, or abbreviated.

4. Pronotum with the dorsum deplanate, subsmooth, minutely granose.

5. Frontal costa distinctly sinuate at the median ocellus, arcuate or roundly produced before the eyes.

6. Posterior angles of the lateral lobes with the apices rounded; median carina of pronotum straight, not at all interrupted; wings long surpassing the pronotal apex; pronotum of male 11 mm. *interruptus*, Bolivar.

6. 6. Posterior angles of the lateral lobes little expanded, behind obliquely truncate, apices obtuse; hind process and wings either abbreviated or extended beyond the hind femoral knees, the wings then as long as the pronotum; pronotum between the shoulders bearing two short granulose lines; pronotum of male and female 8-9 mm.; male long-winged form 11 mm. .. *annandalei*, sp. nov.

4. 4. Pronotum with the dorsum deplanate, but with more or less distinct rugulae, or tubercles.

5. 5. Frontal costa not at all or indistinctly sinuate at the median ocellus; roundly or arcuate produced.

7. Dorsum barely subtectiform forward when viewed from behind, elevated forward; median carina percurrent, subdepressed in front of the shoulders, very low subundulate backward; elytra rather large, ovate; pronotum of male and female 11-13 mm. *indicus*, Hancock.

7. 7. Dorsum distinctly convex between the shoulders, lightly granulose-tuberculose, and provided with two short indistinct lineate carinae; backward on the process bearing minute irregular lineate tubercles; pronotum lengthily subulate; posterior angles of the lateral lobes with the apices barely prominent, subtruncate; male and female pronotum 12-14 mm. .. *conspersus*, sp. nov.

8. Posterior angles of the lateral lobes of pronotum obtuse, rounded; pronotum above little rugose; median carina low, little elevated arcuate behind the anterior border, backward irregularly undulate; body coloured grey palpi white; middle femora of male crassate; pronotum of male 10·5 mm. .. .. *retractus*, sp. nov.

8. 8. Posterior angles of the lateral lobes triangular, the apices narrowly subtruncate.

9. Dorsum of pronotum little rugose, bearing lineate tubercles; two short carinulae between the shoulders distinctly expressed; colour yellowish-ferrugineous; pronotum of male and female 13–14·8 mm. .. .. .. *artolobus*, sp. nov.

9. 9. Dorsum of pronotum distinctly rugose-tuberculose; median carina of pronotum low, pluri-interrupted; hind femora testaceous, four anterior femora fusco-fasciate; the lateral carinae of pronotum maculate with yellow and testaceous; pronotum of female 12·5 mm. *testaceous*, Bolivar.

Coptotettix fossulatus, Bolivar.

Bol., Ann. Soc. Ent. Belg., xxxi, p. 288, 1887; Hanc., Spol. Zeyl., ii, p. 153, 1904.

Habitat.—Balugaon, Puri Dist., Orissa, July 21, 1913 "on lichen-covered rocks" (*N. Annandale*). Ind. Mus. coll.

The type came from Ceylon, and this was the first species described in the genus *Coptotettix* by Bolivar, but the type for this genus was later fixed by Kirby in the selection of *C. asperatus*, Bol., which was figured in Bolivar's "Essai." In some respects it resembles in miniature, representatives of the genus *Hyboella*, but for the present I have included it here.

Coptotettix parvulus, Hancock.

Hanc., Mem. Dept. Agric. India, iv, pp. 145, 146, 1912.

Habitat.—Singla, Darjiling Dist., 1500 ft., June 1913 (*Lord Carmichael's coll.*); Kushtea, Bengal, Oct. 7, 1909 (*J. T. Jenkins*); Sikkim, Darjiling Dist. (*L. Mandelli*); Calcutta. Indian Mus. coll. Also: Kobo, "On rotten wood" (*Kemp*); Janakmukh, "Under bark"; Chapra, Bengal (*Mackenzie*). Author's coll.

Coptotettix capitatus, Bolivar.

Bol., Ann. Soc. Ent. Belg., xxxi, p. 289, 1887; Brunn., Ann. Mus. Genova, xxxiii, p. 111, 1893.

Habitat.—Burma (*Brunner*).

Coptotettix interruptus, Bolivar.

Bol., Ann. Soc. Ent. Belg., xxxi, pp. 291, 292, 1887; Brunn., Ann. Mus. Genova, xxxiii, p. 113, 1893.

Habitat.—Burma.

Coptotettix annandalei, sp. nov.

Colour dark grey, or paler grey sometimes marked with light yellow on the pronotum and hind femora. Head not at all exserted; vertex subequal in width to one of the eyes, slightly narrowed forward, fossulate on each side, not produced; antennae inserted between the lower part of the eyes: paired ocelli placed between the middle of the eyes; frontal costa arcuate produced, distinctly sinuate at the median ocellus. Pronotum anteriorly truncate, posteriorly shortened and acute produced, not at all reaching the knees of the hind femora, or longer subulate, surpassing the hind femoral apices: dorsum deplanate subsmooth, coarsely granulose, with scattered small tubercles, between the shoulders convex; median carina low, forward between the sulci little elevate and rounded crassate, posteriorly little irregularly compressed-elevated; humeral angles strongly obtuse and rounded-depressed; prozonal carinae behind the anterior margin very little expressed and convergent backward; two short carinulae between the shoulders merely granulate lines; posterior angles of the lateral lobes of the pronotum little dilated, obliquely truncate, apices little prominent; elytra elongate, rounded-truncate at the apices; wings extended little beyond the pronotal apex; middle and anterior femora elongate, margins of the middle femora subundulate; posterior femora moderately stout, superior carina minutely serrulate, the antegenicular denticle acute; first joints of the hind tarsi longer than the third, the first and second pulvilli acute spinose, the third longer than the second and flat below.

Entire length of male and female 9-10 (to end of pronotum); pronotum 8-9 mm.; posterior femora 5-6 mm.; male long-wing form 12 5 mm.

Habitat.—Singla, Darjiling Dist., 1500 ft., Apr. 1913 (*Lord Carmichael's coll.*); Northern Shan Hills, Upper Burma (*J. C. Brown*); Calcutta, India. Ind Mus coll.

Some examples from Calcutta are somewhat smaller in stature than the type from Singla. This species resembles *interruptus*, Bol., in the frontal costa being sinuate at the median ocellus, but differs in the posterior angles of lateral lobes; in *annandalei* they are obliquely truncate, the apices of the angles little prominent, whereas in *interruptus* they are rounded, and the median carina of the pronotum is straighter.

Coptotettix conspersus, sp. nov.

Colour yellowish-ferrugineous, subinfuscate; vertex narrower than one of the eyes, narrowed forward, not produced, fossulate on each side; frontal costa strongly roundly produced, barely sinuate at the median ocellus. Pronotum truncate anteriorly, lengthily subulate posteriorly, long surpassing the hind femoral apices; dorsum above granose-rugose, and tuberculose backward, between the shoulders convex: median carina of the pronotum very low forward, but between the sulci barely elevated, posteriorly irregularly undulate; posterior angles of the lateral lobes narrowed, barely prominent, subtruncate; elytra somewhat widely rounded at the apices, subovate; wings surpassing the pronotal apex; anterior femora elongate, margins entire; middle femora in the male little crassate toward the base, the lower carina with the basal two-thirds little dilated, in the female both margins subparallel: hind femoral margins minutely serrulate; first joint of the hind tarsi longer than the third, the third pulvillus longer than the second and straighter below.

Entire length of male and female 13-16 mm.; pronotum 12-14·5 mm.: posterior femora 6-7 mm.

Habitat.—Sibsagar, N. E. Assam (*S. E. Peal*): Bhim Tal, 4500 ft., Kumaon: Siliguri, base of E. Himalayas, Bengal, June, 1906.

This species is somewhat stouter in stature than *retractus*, and the posterior angles of the pronotum more narrowed, subtruncate; on the dorsum of pronotum between the shoulders there are vestiges of two short carinulae, or they are entirely wanting; on the base of process rugose tuberculose, but forward granose.

Coptotettix indicus, Hancock.

Hanc., Mem. Dept. Agric. India, iv, pp. 144, 145, 1912.

Habitat.—Vela, Koyna Valley, Satara Dist., 2100 ft., Apr. 26, 1912 (*F. H. Gravely*). Ind. Mus. coll.

This species has the general appearance of *Hedotettix gracilis*, but with the vertex characters and rugose pronotum of *Coptotettix*.

Coptotettix retractus, sp. nov.

Colour grey; body above granose, sparingly tuberculose; head not at all exserted; vertex subnarrower than one of the eyes, narrowed forward, not advanced quite as far as the eyes, carinula on either side subcurvate, little compressed, open in front, with minute fossa on each side: frontal costa strongly arcuate produced, not sinuate at the median ocellus: antennae inserted between the lower part of the eyes; palpi white. Pronotum anteriorly truncate, posteriorly lengthily subulate surpassing the apices of the hind femora; dorsum granose and scattered with small tubercles; median carina low, behind the anterior margin little compressed-elevated, between the humeral angles depressed, posteriorly pluri-undulate; prozonal carinae behind the front border very lightly expressed, convergent backward: posterior angles of the lateral lobes of the pronotum with the apices rounded, very slightly prominent; elytra widened at the middle, narrowed and rounded at the apices; wings caudate and pale; anterior femora elongate, margins entire; middle femora toward the base crassate in the male, lower margin undulate; hind femora moderately stout, margins entire, granose; hind tibiae pale yellow; first joint of the hind tarsi longer than the third, the first and second pulvilli acute, the third longer than the second and straighter below.

Entire length of male 12·5 mm.: pronotum 10·5 mm.; posterior femora 5·2 mm.

Habitat.—Pusa, Bihar, Aug. 28 (*T. B. Fletcher*). Author's coll.

Coptotettix artolobus, sp. nov.

Near *conspersus*, but differing in the narrower pronotum, the hind process being very slender subulate: the hind femora not so stout; the dorsum between the shoulders bearing two distinctly expressed carinulae, and backward many minute rugulae; the posterior angles of the lateral lobes of the pronotum distinctly narrowed, minutely obliquely truncate, and as viewed from above the apices little prominent; middle femora in the male very slightly or not at all crassate, the inferior carina very little elevated toward the base; wings caudate; colour yellowish-rufescent or ferrugineous.

Entire length of male and female 13-14·8 mm.: pronotum 11·5-13·5 mm.; hind femora 5·2-6·5 mm.

Habitat.—Ceylon. Indian Mus. coll.

Coptotettix testaceus, Bolivar.

Bol., Ann. Soc. Ent. Belg., xxxi, p. 291, 1887; Hanc., Spol. Zeylan., ii, p. 153, 1904.

Habitat.—Chota Nagpur, Pass between Chaibassa and Chakardharpur, Mar. 2, 1913 (*F. H. Gravely*), Ind. Mus. coll. Ceylon (*Bolivar*).

Genus **Hedotettix**, Bolivar.

Table for separating the Indian species.

1. Median carina of pronotum anteriorly between the shoulders strongly arcuate-cristate, very thin and translucent.

2. Pronotum at the front border distinctly angulate produced over the head ... *cristitergus*, sp. nov.

2, 2. Pronotum anteriorly not produced over the head, subangulate or truncate.

1, 1. Median carina of pronotum more or less compressed-arcuate forward before the shoulders, higher at the sulci; vertex and front margin of pronotum viewed from above subobtuse angulate, in profile rounded .. .. .. *gracilis*, de Haan.

3. Median carina of pronotum very low, not at all or little compressed; frontal costa strongly produced before the eyes; vertex in profile and viewed from above angulate; front of head oblique ... *costatus*, Hancock.

3. 3. Median carina of pronotum little compressed, longitudinally low arcuate forward.

4. Dorsum of pronotum between the shoulders subtectiform; frontal costa joined with the vertex in profile strongly arcuate produced; vertex viewed from above subwidened between the frontal carinulae; stature moderate; colour often obscure ferrugineous or infuscate, sometimes with a median light fascia on the dorsum .. *attennuatus*, Hancock.

4. 4. Dorsum between the shoulders distinctly tectiform; frontal costa strongly arcuate produced; blades of the female ovipositor long; the first and third pulvilli of the first joint of hind tarsi subequal in length; stature moderately large; colour grey .. .. *grossus*, sp. nov.

Hedotettix cristitergus, sp. nov.

Near *punctatus*, Hancock. Obscure yellowish-ferrugineous or somewhat infuscate; body above granulose. Vertex subequal or barely wider than one of the eyes, frontal carinulae formed in obtuse angle, roundly reflexed at the sides, middle carinate; frontal costa strongly advanced before the eyes and arcuate, barely subsinuate at the median ocellus; antennae inserted distinctly between the lower fourth of the eyes. Pronotum anteriorly distinctly angulate produced over the head; median carina strongly compressed, very thin punctate, arcuate forward, the crest highest above or barely behind the shoulders; hind process acute subulate extended beyond the hind femoral apices; prozonal carinae behind the anterior border abbreviated and parallel; dorsum between the shoulders on either side of the crest somewhat smooth; humeral angles obtuse, rounded, carinae little compressed; posterior angles of the lateral lobes narrowed, apices hebitate; elytra oblong, apices widely rounded; wings strongly caudate; anterior femora elongate, entire; middle femora elongate, margins little compressed in the female but subparallel, in the male arcuate dilated towards the base; hind femora elongate, margins entire, minutely serrulate; first joints of the posterior tarsi longer than the third, the first two pulvilli spinose, the third longer, substraight below.

Entire length of male and female 14·5–16·5 mm.; pronotum 10·8–13 mm.; hind femora 6–6·7 mm.

Habitat.—Hoshangabad, Nov. 14, 1911 (*T. B. Fletcher*). Several examples in author's collection.

This species resembles *punctatus*, Hanc. The habitat of the latter is unknown, the type being in the University Mus., Oxford.

Hedotettix gracilis, de Haan.

Acridium (Tettix) gracile, de Haan, Temminck, Verhandel. Orth., p. 167, 1842; *Hedotettix gracilis*, Bol., Ann. Soc. Ent. Belg., xxxi, p. 283, 1887; Hanc., Spol. Zeylan., ii, p. 156, pl. 3, fig. 19, 1904; *Hedotettix abortus*, Hanc., *l. c.*, p. 151, 1904, short-wing form; *Hedotettix festivus*, Bol., *l. c.*, p. 286, pl. 5, fig. 24, 1887; *Hedotettix diminutus*, Hanc., Mem. Dept. Agric. India, iv, pp. 149, 150, 1912.

Habitat.—Calcutta (*N. Annandale*; *C. A. Paiva*; *F. H. Gravely*; *Brunetti*); Madhupur, Bengal, Oct. 14, 1909 (*C. Paiva*) "at light": Kharagpur, Bengal, June 17, 1911 (*R. Hodgart*): Wellawaya, Ceylon, Nov. 1905; Peradeniya, Ceylon, Nov. 1910; Goalbathan, E. Bengal, July 9, 1909 (*R. Hodgart*); Berhampur, Murshidabad Dist., Bengal, Jan. 1, 1908 (*R. Lloyd*); Basanti F. S. Sunderbuns (*J. T. Jenkins*): Rangoon, Burma, Feb. 26, 1908 (*N. Annandale*); Balugaon, Lake Chilka, Orissa, Sept. 20, 1913 (*N. Annandale*); Marikuppam, S. India, 2500 ft., Oct. 1, 1911 (F. H. Gravely). Indian Museum coll. Also: Pusa, Bihar, Oct. 12, 1911 (*T. B. Fletcher*): Hoshangabad, Sept. 14, 1911 (*T. B. Fletcher*); Chapra, Bengal (*Mackenzie*). Author's coll.

This species has been reported from Java and Sumatra.

The stature of *Hedotettix diminutus*, Hanc., *l. c.*, p. 149, 1912, is smaller, yet from a study of the large series of specimens of *Hedotettix gracilis* from India, this species shows a wide range of variation that intergradate into *diminutus*. The latter I reported as found at Surat, Bombay.

Hedotettix costatus, Hancock.

Hanc., Mem. Dept. Agric. India, iv, pp. 147, 148, 1912.

Habitat.—Madhupur, Bengal, Oct. 16, 1910 (*C. Paiva*) "at light"; Nepal, Nov. 21, 1908; Allahabad, U. Prov., Oct. 25, 1910 (*Kemp*); Tirvani, Nepal Terai, Dec. 27, 1909 (*B. Warren*); Monghyr, Bengal, Sept. 23, 1909 (*J. T. Jenkins*). Indian Mus. coll. Also: Bengal, July 26, 1910 (*T. B. Fletcher*); Naraingani, Assam, Oct. 29, 1906 (*C. S. M.—Fletcher*); Chapra, Bengal (*Mackenzie*); Munshiganj, Assam, Oct 22, 1906 (*C. S. M.—Fletcher*); Durbhanga, Bengal, Jan. 5, 1903 (*T. V. R. A.—Fletcher*) "on grass" and at light. Author's collection.

Hedotettix attenuatus, Hancock.

Hanc., Spol. Zeylan., ii, p. 151, pl. 3, fig. 18, 1904; Hanc., Gen. Ins. Orth. Acrid. Tetr., p. 60, fig. 23, 1906.

Habitat.—Sur Lake, near Puri, Orissa coast, Aug. 19, 1911 (*N. Annandale*; *F. H. Gravely*): Balighai, near Puri, Orissa, Oct. 24, 1908 (*N. Annandale*); Victoria Gardens, Colombo (*C. Paiva*); Kesbewa, Ceylon, Apr. 1903; Trincomalee, Ceylon, Nov. 1906; Sibsagar, N. E. Assam; Assam-Bhutan Frontier, Mangaldai Dist., N.-E., Dec. 30, 1910 (*Kemp*).

Hedotettix grossus, sp. nov.

Greyish-cinereous: granulose; vertex equal or subnarrower in width to one of the eyes, narrowed forward, frontal carinula convex and subrounded reflexed at the sides; middle carinate; frontal costa strongly arcuate produced. Pronotum anteriorly truncate, dorsum tectiform: median carina low, little compressed, not at all cristate, low arcuate forward, substraight backward; hind process long acute subulate: between the shoulders on the dorsum presenting two abbreviated carinulae; prozonal carinae behind the anterior margin parallel and rather widely separated; posterior angles of the lateral lobes of the pronotum strongly narrowed; elytra rather large, elongate, apices widely rounded; wings caudate; femoral margins entire; first joint of the posterior tarsi longer than the third: the second and third pulvilli equal in length; valves of the female ovipositor long, acuminate, terminating in curvate spines, the margins above and below serrulate.

Entire length of female 16·5 mm.; pronotum 13·5 mm.; hind femora 6·8 mm.

Habitat.—Singla, Darjiling Dist., 1500 ft., May, 1913 (*Lord Carmichael's coll.*). Indian Mus. coll.

Somewhat larger in stature than *gracilis*; the frontal costa more narrowly sulcate; the pronotum anteriorly truncate; the median carina not at all compresso-elevated forward, though little compressed and forming a low longitudinal arc; the ovipositor much longer than in *gracilis*, and the colour grey, with very obscure dark oblique bands on the outer face of the hind femora.

Genus **Euparatettix**, Hancock.

In this genus the representatives have the antennae inserted partly between the lower margin of the eyes; the paired ocelli placed between the middle of the eyes; the vertex narrower than one of the eyes and truncate; the frontal costa rounded or arcuate produced; the head exserted; the body graceful, more or less slender, the pronotum posteriorly lengthily subulate and the wings caudate; the intermediate femora elongate, not at all crassate.

Table for separating the Indian species.

1. Hind tibiae dense black, white annulate just behind the knees; head distinctly exserted; pronotum of male and female 7·5–9·5 mm. .. .. *personatus*, Bolivar.
   Slightly smaller in stature, with additional white marking at the apical third of hind tibiae. .. *var. A. E. p. birmanicus*, nov.

1\. 1. Hind tibiae subunicoloured, or obscurely marked, but not dense black.
2\. Head exserted; pronotum narrow between the shoulders; stature very slender; median carina of pronotum substraight, percurrent; pronotum of male and female 11·5–13·5 mm. .. *tenuis*, Hancock.
2\. 2. Head very little exserted; pronotum moderately dilated between the shoulders: stature more robust; median carina of pronotum arcuate forward, often little undulate before the shoulders, subobliterated just behind the front border; pronotum of male and female 9–11 mm. .. .. *variabilis*, Bolivar.
Stature larger; pronotum of male and female 10–13 mm. .. *corpulentus*, Hancock.

Euparatettix personatus, Bolivar.

Paratettix personatus, Bol., Ann. Soc. Ent. Belg., xxxi, p. 278, 1887: *Euparatettix personatus*, Hanc., Spol. Zeylan., ii pp. 155, 156, pl. 2, fig. 10 and pl. 3, fig. 26, 1904; Hanc., Gen. Ins. Tetr., p. 55, pl. 3, fig. 32, 1906; Hanc., Mem. Dept. Agric. India, iv, p. 152, 1912; Brunn., Ann. Mus. Genova, xxxiii, p. 109, 1893.

Habitat.—Peradeniya, Ceylon, May 30, 1910; Kandy, Ceylon, July 1909; Calcutta, Oct. 27, 1911 (*F. H. Gravely*); Balugaon, Puri Dist., Orissa, July 21, 1913 (*N. Annandale*); Sigiriya, Ceylon, Aug. 8, 1901; Anuradhapura, Ceylon, "low country," Oct. 18, 1911 (*N. Annandale*); Sibsagar, N. E. Assam (*S. E. Peal*); Berhampur, Murshidabad Dist., Bengal, Jan. 1, 1908 (*R. Lloyd*); Waikam, Travancore, Coastal Region, Nov. 5, 1908 (*N. Annandale*). Indian Museum collection.

The colour pattern in this species varies greatly, and among these examples from Ceylon and India, there are some specimens that agree with the description of the colour as described by Bolivar, namely: "body fuscous, head in front and sides of pronotum cinereous or black." All the specimens have black hind tibiae with a white circular fascia behind the knees, and the antennae are rather long, and the four apical articles are little compressed and black.

Euparatettix personatus var. birmanicus, nov.

Besides the above mentioned specimens under *personatus*, there are a number of representatives which are distinguished by the slightly smaller stature, and the hind tibiae has, besides the usual white annulation behind the knees, a little white on the apical third of the black shaft.

Habitat.—Rangoon, Burma, Feb. 26, 1908 (*N. Annandale*): Assam-Bhutan Frontier, Mangaldai Dist., N.-E., Jan. 1. 1911 (*Kemp*).

Euparatettix tenuis, Hancock.

Hanc., Mem. Dept. Agric. India, iv, pp. 151, 152, 1912.

Habitat.—Calcutta, Oct. 1912 (*F. H. Gravely*); Thingannyin-aung to Sukli, Dawna Hills, 900-2100 ft., Nov. 23, 1911 (*F. H. Gravely*): Pusa, Bihar, June 15, 1911 (*T. B. Fletcher*; *Hancock coll.*); Monghyr, Bengal, Sept. 22, 1909 (*J. T. Jenkins*). Indian Mus. coll.

This species can be recognized by the slender and graceful stature, the narrow pronotum, and strongly caudate wings.

Euparatettix variabilis, Bolivar.

Paratettix variabilis, Bol., Ann. Soc Ent. Belg, xxxi, pp 276, 277, 1887: *Euparatettix variabilis*, Hanc., Mem. Dept. Agric. India, iv, p. 150, 1912.

Habitat.—Rajshai, E. Bengal, Feb. 6, 1907 (*N. Annandale*): Asansol, Bengal, Nov. 13, 1910 (*Paiva* and *Caunter*); Mandalay, U. Burma, Mar. 12, 1908 (*N. Annandale*); Chotajalla, Rajmahal, Bengal, Feb. 14, 1910 (*B. L. Chaudhuri*); Berhampur, Murshidabad Dist., Bengal, Jan. 2, 1908 (*R. Lloyd*); Puri, Orissa, Jan. 20, 1908; Anuradhapura, Ceylon, "low country", Oct. 1911 (*N. Annandale*). Ind. Mus. coll.

This species is smaller in stature than *corpulentus*, Hanc., and nearly allied, and it appears from some of the specimens that they cross, producing hybrids.

Euparatettix corpulentus, Hancock.

Hanc., Mem. Dept. Agric. India, iv, p 158, 1912.

Habitat.—Balugaon, Puri Dist., Orissa, July 21, 1913 (*N. Annandale*): Sur Lake, Puri, Orissa coast, July 19, 1911 (*N. Annandale* and *F. H. Gravely*); Bosondhur, Khulna Dist., Ganges Delta, Aug. 21, 1909, "On launch, at light" (*J. T. Jenkins*); Kalandhungi, Naini Tal Dist., U. P. Agra and Oudh, May 4, 1913 (*R. Hodgart*): Kasara, W. base of W. Ghats, Bombay, Nov. 23, 1901; Balighai, near Puri, Orissa, Oct. 24, 1909 (*N. Annandale*): S. India "On board ship four miles off Tuticorin, May 25, 1908 (*C. Paiva*): Berhampur, Murshidabad Dist., Bengal, Jan. 1, 1908 (*R. Lloyd*); Calcutta: Madhupur, Bengal, Oct. 18, 1909 (*C. Paiva*); Dhappa, nr. Calcutta (*N. Annandale*); Bangalore, S. India (*Cameron*); Purulia, Manbhum Dist., Chota Nagpur, Feb. 10, 1912; Chotajalla, Rajmahal, Bengal, Nov. 14, 1910 (*B. L. Chaudhuri*); Northern Shan Hills, Upper Burma (*J. C. Brown*); Neapalganj, Nepal Frontier, Nov. 22, 1911. Indian Mus. coll.

Genus **Indatettix**, nov. (*Euparatettix* in part)

Resembling *Euparatettix*, Hancock. Head exserted: vertex strongly narrower than one of the eyes; paired ocelli placed nearly between the lower third of the eyes; antennae inserted just below the eyes; frontal costa little arcuate produced between the antennae, but not above between the eyes as in *Euparatettix*; median carina of the pronotum often undulate, or sinuate, or interrupted; the hind process of pronotum with the lateral carinae toward the apex not at all, or more often, minutely crenulate, or with small dilated lobes the apex then often minutely subdilated-truncate; middle femora crassate, or the margins undulate; body bearing elytra and fully developed wings. The type is *Euparatettix nodulosus*, Hancock.

Table for separating the Indian species.

1. Hind tibiae white and black biannulate, or bifasciate, more or less intensely pigmented, or white annulate behind the knees, the shaft black or fuscous interrupted with white marking; head distinctly exserted

2. Stature small; pronotum of female not over 9·5 mm.; body above more or less rugose.

3. Middle femora compressed, margins above and below distinctly undulate lobate; pronotum above rugose; hind femora with the outer face bearing compressed prominent tubercles as viewed from above; median carina of pronotum little cristulate forward between the sulci, depressed between the shoulders and backward strongly sinuate, with small elevated nodes; hind process toward the apex bearing minute dilated-serrulate lobes, the apex often minutely expanded-truncate. *nodulosus*, Hancock.

3. 3. Middle femora compressed, margins above and below undulate; median carina of pronotum undulate, the median nodes backwards suppressed or not evident; hind process with the lateral carinae very indistinctly lobate toward the apex .. .. *parvus*, Hancock.

2. 2. Stature somewhat larger: pronotum of female 9·5–11 mm.; above plain or little rugose; hind process of prono

tum with the lateral carinae toward the apex bearing small more or less dilated lobes, or entire; pronotum light testaceous toward the apex, and often maculate.

4 Wings towards the apex dark, often pale maculate.

5. Median carina of pronotum behind the shoulders undulate, bearing obtuse crenules; hind process of pronotum with the lateral carinae toward the apex subentire; anterior and middle femora narrow; pronotum of female 10·6–11 mm. .. .. *interruptus*, Brunner.

5. 5. Median carina of pronotum backward behind the humeral angles more or less distinctly nodulose and sinuate; the minute lobes of the lateral carinae towards the apex more or less evident *var. A. aff.*

6. Pronotum above behind the shoulders backward subsmooth; small lobes of the lateral carinae toward the apex more or less distinct; median carina of pronotum backward very low, thin and substraight, barely undulate *var. B. lobulosus*, nov.

6. 6. Pronotum above behind the shoulders little rugose; the small lobes of lateral carinae evident .. .. *var. C.*

4. 4. Wings plain, not at all maculate; stature little larger; hind process of pronotum above little rugose, light and fusco-maculate: lateral carinae toward the apex with small serrulate lobes; median carina of pronotum backward subnodulose-sinuate; body pale, variegated with fuscous, legs pale, fusco-fasciate; frontal costa depressed, barely arcuate between the antennae; pronotum of female 11·6 mm. .. *callosus*, sp. nov.

1. 1. Hind tibiae subinornate, or sometimes bearing obscure fumate markings, but not distinctly annulate or fasciate with black.

7. Dorsum of pronotum above subsmooth.

8. Head distinctly exserted.

9. Median carina of pronotum percurrent, little compressed-elevated before the shoulders, little sinuate near the anterior border, between the shoulders and backward gently undulate or sub-

straight; highest between the shoulders; middle femora lightly hirsute below; in the male crassate-compressed, the margins dilated, in the female margins subparallel; stature slender .. .. .. *crassipes*, Hancock

9. 9. Median carina of pronotum behind the shoulders backward not at all undulate, little compressed; dorsum often bearing a pale longitudinal fascia; stature slender. .. .. .. *var. A. aff.* nov.

7. 7. Dorsum of pronotum above minutely rugose; body pale grey; median carina of pronotum marked with minute black spots; wings hyaline at the base, but toward the apex black or fumate, interrupted with white vertical venations; stature small .. .. *var. B. hybridus*, nov.

8. 8. Head little exserted: median carina of pronotum undulate: dorsum little rugose; colour brownish-rufescent; hind tibiae little infuscate toward the apices. *var. C. bengalensis*, Hancock

Indatettix nodulosus, Hancock.

Euparatettix nodulosus, Hanc.. Mem Dept Agric. India, pp. 155, 156, 1912.

Habitat.—Calcutta, India, Feb. 22, 1907; Puri, Orissa. Jan. 20, 1908: Purulia, Manbhum Dist., Chota Nagpur, Nov 10, 1912 (*N. Annandale*); Kaladhugi, Naini Tal Dist., May 4 1913 (*R. Hodgart*); Kiari, Naini Tal Dist., W. Himalayas, Dec. 24, 1910; Vela, Koyna Valley, Satara Dist., 2100 ft. Apr. 26, 1912 (*F. H. Gravely*): Jalaban, Naini Tal Dist., base of W. Himalayas. Mar. 22, 1910; Motisal, Gharwal Dist., W. Himalayas, Mar. 5, 1910; Amangarh, Bijnor Dist., U.P., Feb. 24, 1910; Raxaul, Nepal Frontier, Nov. 10, 1911. Indian Mus. coll.

Indatettix parvus, Hancock.

Euparatettix parvus, Hanc., Spol. Zeylan., ii, p. 145, 1904; *Euparatettix pilosus*, Hanc.. Trans. Ent. Soc. London, p. 410, 1909.

Habitat.—Bhim Tal, 4450 ft., Kumaon, W. Himalayas, May 9, 1911 (*Kemp*); Calcutta; Puri, Orissa, Jan. 20, 1908; Amangar, Bijnor Dist.; Dhampur, Bijnor Dist., U.P., Nov. 11, 1907; Igatpuri, W. Ghats, Bombay Pres., Nov. 20, 1901 (*N. Annandale*). Indian Mus. coll.

Indatettix interruptus, Brunner.

Paratettix interruptus, Brunn., Ann. Mus. Genova, xxxiii, p. 109, 1893.

Habitat.—Marikuppam, S. India, 2500 ft., Oct. 19, 1910, Batticaloa, Ceylon, July 1907; Calcutta, Nov. 2, 1907; Siripur, Saran, N. Bengal, Sept. 25, 1910; Madhupur, Bengal Oct. 16, 1909 (*C. Paiva*); Sur Lake, near Puri, Orissa coast, Aug. 19, 1911 (*N. Annandale*); Ayaramtengu, S. end of L. Kayangulam, Travancore, Nov. 6, 1908 (*N. Annandale*); Damukdia, E. Bengal, June 7, 1908. Indian Mus. coll.

Indatettix interruptus, *var. A. aff.*

Habitat.—Satara Dist., 2050 ft., May 3, 1913. Ind. Mus. coll.

Indatettix interruptus, *var. B.* **lobulatus**, nov. and *var. C.*

Habitat.—Hoshangabad, Sept. 14, 1911 (*T. B. Fletcher*). Ind. Mus. coll. Also in author's coll.

Indatettix callosus, sp. nov.

Body above very little rugose, pale, disc ornate with two black spots, and variegated with fuscous, legs pale, fasciate with fuscous. Head distinctly exserted; vertex and upper part of the eyes elevated higher than the dorsum; vertex not produced, strongly narrower than one of the eyes, narrowed forward; antennae inserted below the eyes; paired ocelli placed between the lower third of the eyes; frontal costa slightly prominent, little arcuate between the antennae, and depressed above between the eyes. Pronotum somewhat wide between the shoulders; posteriorly subulate much surpassing the hind femoral apices; dorsum little rugose and callose; median carina little cristulate forward between the sulci, between the shoulders and backward strongly undulate, and nodulose on base of process; lateral carinae toward the apex bearing small serrulate lobes; hind process white and black maculate; middle femoral margins undulate-sublobate; tibial margins subdilated toward the base; posterior angles of the lateral lobes of pronotum little expanded, apices obtusely-rounded; the three pulvilli of the first joint of the hind tarsi spinose, the third barely longer than the second.

Entire length of female 13·4 mm.; pronotum 11·6 mm.; hind femora 5 mm.

Habitat.—Darjiling Dist., Singla, 1500 ft., May 1913 (*Lord Carmichael's coll.*).

Indatettix crassipes, Hancock.

Euparatettix crassipes, Hanc., Mem. Dept. Agric. India, pp. 153, 157, 1912.

Habitat.—Siripur, N. Bengal, Sept 27, 1910; Madhupur, Bengal, Oct. 14, 1913 (*C. Paiva*): Kaladhungi, Naini Tal Dist., Unit Prov. Agra and Oudh, May 4, 1913 (*R. Hodgart*); Bhim Tal, 4500 ft., Kumaon, Sept. 27, 1907; Damukdia, Bengal, June 7, 1908; Rajmahal, Bengal, July 6, 1909 (*N. Annandale*); Sicktan, Nepal, Nov. 13, 1908; Ghumti, Darjiling Dist. (*F. H. Gravely*); Pusa, Bihar (*T. B. Fletcher*). Indian Mus. coll. Also: Bankipur and Muzafferpur. Author's coll

Indatettix crassipes, *var A. aff.*

Habitat.—Chapra, Bihar (*Mackenzie*); Pusa, Bihar, July 6, 1911 (*T. B. Fletcher*). From Hanc. coll. in the Ind. Mus. coll.

This variety often bears a longitudinal fascia on the pronotum.

Indatettix crassipes *var. B.* **hybridus**, nov.

Habitat.—Dibrugarh, N.-E. Assam, Abor Exp., Nov. 11, 1911; Madhupur, Bengal, Oct. 16, 1909 (*C. Paiva*); Sukhwani, Nepal, Nov. 15, 1908. Ind. Mus. coll.

Indatettix crassipes *var. C.* **bengalensis**, Hanc.

Hanc., Mem. Dept. Agric. India, pp. 155, 156, 1912.

Habitat.—Bengal.

BATRACHIDINAE.

Genus **Saussurella**, Bolivar.

Saussurella indica, Hancock.

Mem. Dept. Agric. India, iv, pp. 156, 157, 1912.

Habitat.—Lebong, India. Author's coll.

Saussurella curticornu, Hancock.

Mem Dept. Agric. India, iv, pp. 158, 159, 1912.

Habitat.—?. Ind. Mus. coll. Bihar, Pusa. Author's coll.

Saussurella decurva, Brunner.

Habitat.—Dejoo, North Lakhimpur, base of hills, Upper Assam, June 29, 1910 (*H. Stevens*); Kawkareik to third camp, Amherst Dist., Lower Burma, Nov.-Jan. 21, 1911 (*F. H. Gravely*). Ind. Mus. coll.

Length of pronotum of male 19 mm.; female 19·6 mm.

Saussurella brunneri, Hancock.

Hanc., Mem. Dept. Agric. India, iv, p. 156, 1912; *Saussurella cornuta*, Brunn., Ann. Mus. Genova, xxxiii, p. 113, pl. 5, fig. 45, 1893.

This species was referred by Brunner to *S. cornuta*, de Haan, but as I have previously pointed out, *l. c.*, p. 156, foot note, that it is distinct.

The pronotum of male 13 mm.; female 16 mm.
Habitat.—Sibsagar, N. E. Assam. Indian Mus. coll.

Identifications of Indian Specimens in the Museum.

Kirby's identification.	Hancock's identification.
Coptotettix latifrons, Brunner.	Mazarredia ghumtiana, Hanc. (Type).
,, ,,	Hyboella obesa, Hanc.
Ergatettix tarsalis, Kirby.	Indatettix interruptus, Brunn.
,, ,,	Indatettix parvus, Hanc.
,, ,,	Indatettix nodulosus, Hanc.
Systolederus cinereus, Brunn.	Indatettix interruptus, Brunn.
,, ,,	Bolotettix lobatus, Hanc.
Euparatettix interruptus, Brunn.	Indatettix interruptus, Brunn.
,, ,,	Indatettix parvus, Hanc.
Hedotettix gracilis, de Haan.	Hedotettix costatus, Hanc.
,, ,,	Euparatettix corpulentus, Hanc.
,, ,,	Euparatettix variabilis, Bol.
Hybotettix, sp.	Xistrella dromadaria, Bol.
Apterotettix obtusus, Hanc.	Scelimena harpago, Serv.
Acantholabus cunecatus, Hanc.	Thoradonta spiculoba, Hanc.
Euparatettix personatus, Bol. var. longicornis, Krby.	Criotettix rugosus, Bol.
,, ,,	Acrydium polypictum, Hanc.
Coptotettix acuteterminatus, Brunn.	Mazarredia ghumtiana, Hanc. (? larva).
Mazarredia lugubris, sp. n.	Bolotettix lobatus, Hanc.
Criotettix obscurus, Krby. (Type).	Criotettix spinolobus, Hanc.
(on label) Eugavialidium hastatum, Kb. (Type) (in F.B.I. *E. hastulatum*, Kb.).	Tettitelum hastatum, Hanc. (Type).
Scelimena producta (Serv.)	Eugavialidium multidentatum, Hanc. (Type).
Scelimena gavialis, Ssr.	Scelimena spinata, Hanc. (Type).
Criotettix exsertus, Bol.	Criotettix annandalei, Hanc. (Type).
Ergatettix tarsalis, Kby. (n. g. & sp.)	Indatettix parvus, Hanc.
,, ,, ,,	Indatettix nodulosus, Hanc.
,, ,, ,,	Indatettix crassipes hybridus, Hanc.
,, ,, ,,	Eupartettix personatus, Bol.
,, ,, ,,	Indatettix interruptus, Brunn.
Euparatettix personatus, Bol.	Euparatettix corpulentus, Hanc.
,, ,,	Indatettix interruptus, Brunn.
,, ,,	Euparatettix tenuis, Hanc.
,, ,,	Euparatettix variabilis, Bol.
,, ,,	Euparatettix personatus, Bol.
,, ,,	,, ,, burmanicus, Hanc.
,, ,,	Indatettix crassipes, Hanc.
,, ,,	Hedotettix attenuatus, Hanc.
,, ,,	Indatettix crassipes hybridus, Hanc.
,, ,,	Indatettix nodulosus, Hanc.
,, ,,	Indatettix parvus, Hanc.
,, ,,	Acrydium variegatum, Bol.
Hedotettix gracilis, de Haan.	Euparatettix variabilis, Bol.
,, ,,	Euparatettix corpulentus, Hanc.
,, ,,	Indatettix crassipes, Hanc.
,, ,,	Hedotettix costatus, Hanc.
,, ,,	Hedotettix attenuatus, Hanc.
,, ,,	Coptotettix artolobus, Hanc. (Type).

SPECIES FROM OUTSIDE THE INDIAN EMPIRE IN THE INDIAN MUSEUM.

Discotettix belzebuth, Serville.

Habitat.—Kuching, Sarawak, Borneo, June 27, 1910 (*Beebe*).

Tripetalocera ferruginea, Westwood.

Habitat.—Kuching, Sarawak, Borneo, June 27, 1910 (*Beebe*).

Dasyleurotettix curriei, Rehn

S. Africa.

Genus **Hexocera**, nov.

Allied to *Eugavialidium*, Hancock. Pronotum ornate with spines and gibbose. Face slightly oblique; frontal costa obsolete or nearly so above the paired ocelli, protuberant between the antennae and sulcate; antennae long slender filiform, inserted far below the eyes; eyes globose, slightly sessile; vertex wider than one of the eyes, on either side little elevated or subacute terminated, not higher than the eyes Pronotum anteriorly truncate, dorsum concave between the shoulders, the humeral angles produced in a spine on each side; median carina of pronotum often bigibbose or obtuse spined behind the shoulders; posterior angles of the lateral lobes of the pronotum laminate outwards. and produced in a sharp spine, at the lateral margin in front produced on each side in a spiniform tubercle; hind process lengthily produced backward beyond the hind femoral apices. Elytra acuminate backward toward the apices, wings perfectly explicate, and extended backward nearly to the apex of the pronotal process. Anterior and middle femora narrow, and strongly elongate, margins not compressed, but often subtuberculate; posterior femora elongate, inferior margin obtuse dentate; hind tibiae and first article of the posterior tarsi marginate, entire. Type *Acridium (Tetrix) hexodon*, de Haan.

Hexocera sexspicata, sp. nov.

Near *H. hexodon*, de Haan. Body minutely punctate, coloured ferrugineous-fuscous, hind femora obscurely marked with alternating light and fuscous bars. Vertex wider than one of the eyes, concave forward, frontal carinulae rounded, abruptly terminated at the sides, and slightly elevated nearly as high as the eyes; eyes prominent, strongly globose, subsessile; paired ocelli placed on a line with the lower margin of the eyes; antennae long and very thin, articles strongly elongate, inserted far below the eyes; frontal costa strongly protuberant between the antennae, sulcate, depressed and thin near the vertex above the paired ocelli. Pronotum truncate anteriorly, dorsum inequal, concave between the shoulders; median carina obliterated near the front border, unigibbulate forward before the shoulders, and behind the shoulders elevated in the form of a dull conical spine or gibbosity, black at the apex: dorsum depressed just behind the spine; humeral angles broadly elevated, crenulate, and strongly produced on each side in an acute spine, directed obliquely upwards: before the shoulders appears a small tubercle on each side on the lateral carinae; dorsum above the elytral apices on each side bearing an obtuse rounded node, and farther backward the base of the process provided with two elongate nodes nearly joining in front, and fossulate between them; hind process lengthily produced and smooth above: lateral lobes of pronotum on each side in front below the eyes outwardly produced in a subspine, longer than a tubercle; posterior angles of the lateral lobes outwardly laminate and produced in a single transverse strong spine on each side, little curvate forward toward the apex; elytra externally impresso-punctate, elongate, widest near the middle, rounded below or acuminate toward the apices: wings extended backward nearly to the apex of the hind process; anterior femora strongly elongate, margins entire or subundulate above; middle femoral margins above subtrilobate, below nearly bituberculate; posterior femoral carinae above thinly compressed bearing two minute tubercles, below dentate: hind tibiae mutilated in the type.

Entire length of male 20·7 mm.; pronotum 19·8 mm.; post. femora 6 mm.; antennae 7 5 mm.

Habitat.—Saudakan, N. Borneo (*Pryer*). Ind. Mus. coll.

The type of de Haan's species *hexodon* is from Sakoenbang, Sumatra, and this species is closely allied to *hexodon*. The spined shoulders are more pronounced than in *dentifer*, Bolivar.

The type specimen bears a label on which is written "*Scelimena hexodon*, De Haan." and appears to be in Saussure's handwriting.

Genus **Eugavialidium**, Hancock.

Eugavialidium chinensis, sp. nov.

Body greyish, light and fusco-maculate. Vertex wider than one of the eyes, on either side dentate, but not elevated above the eyes; dorsum of the pronotum strongly rugose, reticulate, deplanate, depressed and fossulate behind the shoulders; median carina multi interrupted, gibbulate forward between the sulci and just back of the shoulders, at the front margin produced over the occiput in a tubercle; humeral angles armed with an obtuse, slightly produced tubercle on each side: front margin of pronotum on each side of the lateral lobes bearing a small subacute denticle: posterior angles of the lateral lobes produced in an uncinate spine on each side, which is rather slender; hind process more or less maculate above, very strongly produced, much longer than the apices of the outstretched hind tibiae, subulate, and toward the apex cylindrical, and rugose; wings fully explicate, pellucid, reaching nearly to the pronotal apex; anterior femora compressed, elongate, margins above distinctly lobate, toward the base acute lobate, below bidentate; middle femoral margins above trilobate, below strongly bidentate; hind femora elongate bearing a large very obtuse lobe at the middle, and two acute denticles backward before the antegenicular lobe; below armed with very small tubercles; hind tibiae marked with black and white, black at the base and with two black annulations on the shaft, the margins very moderately expanded and armed with small spines: first joint of the posterior tarsi very slightly expanded.

Entire length of male and female 26–29 mm.; pronotum 25–28 mm.; posterior femora 7·5–8·5 mm.

Habitat.—Phuc Son, Annam (*R. Rolle*). Author's coll.

Acrydium hancocki, Morse.

Habitat.—United States. Ind. Mus. coll.

Acrydium subulatum, Linn.

Habitat.—Europe; Siberia. Ind. Mus. coll.

Acrydium depressum, Bris.

Habitat.—Europe (*de Saussure*). Ind. Mus. coll.

Acrydium variegatum, Bolivar, *aff.*

Habitat.—Kuching, Sarawak, Borneo (*Beebe*): Tiberias, Palestine (*N. Annandale*).

Acrydium polypictum, Hanc.

Habitat.—10 miles south of Kuching, Borneo. Ind. Mus. coll.

Nomotettix tartarus, Saussure.

Habitat.—Turkestan (*Saussure*). Ind. Mus. coll.

This is the species described as *Tettix tartarus* by Bolivar, in Ann. Soc. Ent. Belg., xxxi, pp. 262, 263, 1887.

Nomotettix compressus, Morse.

Habitat.—Georgia, U.S. America. Ind. Mus. coll.

Paratettix singularis, Shiraki.

Habitat.—Japan. Ind. Mus. coll.

Paratettix meridionalis, Ramb.

Habitat.—Greece; Teneriffe: Europe. Ind. Mus. coll.

Paratettix texanus, Hancock.

Habitat.—Texas, U.S.A. Ind. Mus. coll.

Paratettix similis, Bol.

Habitat.—10 miles south of Kuching, Sarawak, Borneo. Ind. Mus. coll.

Paratettix toltecus, Saussure.

Habitat.—Mexico (*Saussure*). Ind. Mus. coll.

Euparatettix personatus, Bolivar, *aff.*

Habitat.—Borneo. Ind. Mus. coll.

Telmatettix aztecus, Saussure.

Habitat.—Mexico (*Saussure*). Ind. Mus. coll.

Tettigidea prorsa, Scudder.

Habitat.—Georgia, U.S.A. Ind. Mus. coll.

Tettigidea lateralis, Say.

Habitat.—Texas, United States (*Saussure*). Ind. Mus. coll.

Tettigidea parvipennis, form **pennata,** Morse.

Habitat.—Chicago, Illinois, U.S.A., Sept. 29, 1892 (*Hancock*). Ind. Mus. coll.

Tettigidea medialis, Hancock.

Habitat.—Georgia, U.S.A. (*Saussure*). Ind. Mus. coll.

Tettigidea mexicana, sp. nov.

Near *nigra*, Morse. Vertex wider than one of the eyes, not advanced beyond the eyes, median carina small, the crown subhorizontal, or barely convex in profile, and with the vertex obtuse angulate; frontal costa advanced beyond the eyes nearly one-half their breadth, and convex, little sinuate at median ocellus, viewed in front widely sulcate, and forked above the paired ocelli, and little divergent forward. Pronotum tectiform, plain granulose above, posteriorly cuneate reaching to the knees of the hind femora; median carina of the pronotum compressed, elevated, little arcuate before the shoulders, and little depressed between the shoulders, the hind process little depressed toward the apex; posterior angles of the lateral lobes of the pronotum angulate, the inferior margins little outwardly deflexed; the superior elytral sinus right angulate and very shallow; elytra very small, elliptical, bearing a subapical white mark; wings wanting; colour reddish brown, hind femora ferrugineous.

Entire length of female 12 mm.: pronotum 10 mm.; posterior femora 7 mm.

Habitat.—Orizaba, Sumuhran, Mexico (*Saussure*). Ind. Mus. coll.

The printed name "*Tettigidea polymorpha*, Burm." is on the label attached to this specimen, probably placed by Saussure.

MISCELLANEA.

INSECTS.

Additional Mallophaga from the Indian Museum (Calcutta).

In addition to a large collection of Mallophaga from birds of India and S. Asia generally, received from the Indian Museum, and recently reported on by Kellogg and Paine (*Rec. Ind. Mus.* Vol. X, pp. 217-243, 1914), we have received a small sending composed of the species noted in this paper. Although no new species are included in this collection, the new host and locality records are worth recording.

Docophorus rostratus, Nitzsch. Juvenile specimens from *Scops* sp. (taken at sea, off Aden).

Docophorus gonorhynchus, Nitzsch. Specimens taken from *Milvus melanotis* (Kurseong, E. Himalayas).

Nirmus rufus, Nitzsch. Specimens from the Brahmini Kite *Haliaster indicus* (Calcutta).

Lipeurus longus, Piaget. Specimens from the pheasant *Tragopan satyra* (Zool. Gardens, Calcutta).

Lipeurus antilogus, Nitzsch. Males and females of this well-marked and interesting parasite of the bustards from *Houbara (Otis) macqueeni* (in captivity, Lahore, Punjab, and also wild, Bhawalnagar, Punjab).

Goniodes bicuspidatus, Piaget. Males and females from the pheasant *Tragopan satyra* (Zool. Gardens, Calcutta).

Colpocephalum flavescens, Nitzsch. Specimens from the Brahmini Kite *Haliaster indicus* (Calcutta).

Colpocephalum subpachygaster, Piaget. Specimens from *Scops* sp. (at sea, off Aden).

Colpocephalum miandrium, Kellogg. Specimens from the African Brown Crane *Balearica pavonica* (in captivity, Calcutta). This species was originally described from specimens taken from a crane of the same genus collected by Sjöstedt's Kilimanjaro Meru Expedition in E. Africa in 1907.

Menopon gonophaeum, Nitzsch. Specimens from the Raven *Corvus corax* (Zool. Gardens, Calcutta, recently received from Nepal).

Menopon nigrum, Kellogg and Paine. Specimens from *Corvus splendens* (Calcutta). This species was described in 1911 from specimens taken from the White-Necked Raven *Corvultur albicollis*, shot at Oshogbo, Southern Nigeria, by J. J. Simpson. The species, though closely related to others of the genus found on crows and ravens, is a well-marked one.

Nitzschia minor, Kellogg and Paine. Specimens taken from the Swift *Cypselus affinis* (Calcutta). The species was described in 1914 from specimens from the same host taken in the same locality and included in the earlier sending from the Indian Museum.

Laemobothrium titan, Piaget. Male, female and young specimens from a Baza, *Baza jerdoni* (Kurseong, E. Himalayas).

V. L. KELLOGG and S. NAKAYAMA,
Stanford University, California.

REPTILES.

An abnormal specimen of *Naia bungarus*, Schleg.

Dr. Boulenger in the "Fauna" volume on "Reptilia and Batrachia" shows a rhomboidal shield, in between the *occipitals* anteriorly in fig. 114 on page 390, but in the description he says that the *parietals* are followed by a pair of large shields (*occipitals*), no mention being made of this shield.

Major Wall has also in his book on the "Poisonons Snakes of India and how to recognize them" (1913) shown the *parietals* followed by a pair of large *occipitals*; and he says that these (*occipitals*) are in contact with one another throughout.

Sir J. Fayrer, K C.S.I., in the "Thanatophidia of India" does not show any shield in between the *occipitals* which are shown in contact throughout. In some specimens examined the condition is exactly as shown by Wall or Fayrer, but in the singular specimen about which this note has been written the condition is exactly as shown in fig. 114, on page 390 of the "Fauna" volume.

BAINI PARSHAD, B.Sc.,
Alfred Patiala Research Student,
Zoological Laboratory.

Government College,
Lahore.

BATRACHIA.

A South Indian Flying Frog: RHACOPHORUS MALABARICUS (Jerdon).

(*Extract from a letter*). I have the honour to state that I have collected a specimen of a flying tree-frog near Sagar, a place in the Malnad forest regions, or the Western Ghats portion, of Mysore Province, some twenty miles from the famous Gersoppa Falls. I happened to catch it in this way. I was collecting and photographing natural science specimens in the locality for my College. As I approached a big tree with my camera, my attention was suddenly drawn by a rustling noise in the leaves above and, as I looked up, I found a beautifully coloured little animal having all the appearance of a small bird, falling from the top of the tree in a slanting direction. Its flight was curious, inasmuch as it did not flap its "wings". All the same, a sort of a whir was audible as it flew slantingly. It alighted on the ground a good distance from the tree it darted from. It is a pity I failed to measure the distance travelled by the animal. It may, I think, be somewhere between thirty to forty yards. My attendant happening to be close to where the creature alighted ran and caught it by throwing his cloth over it. When I went to see what it was, I found to my intense surprise and delight that it was not a bird, but a gaily coloured flying tree-frog.

Its upper side was coloured a beautiful grass-green, the webs bright red, and its underside a bright yellow. It possessed well-developed adhesive discs at the ends of its fingers and toes with which it could attach itself to any surface easily. It could with ease attach itself to the wet slippery sides of the glass bottle in which I carried it home. It was crouching in the bottle in such a way that it looked a lump of dull green. When among the green leaves, it could, I think, escape detection most efficiently. The brilliancy of its colour was to be seen only during its flight and might serve for purposes of recognition by others of its kind. I already mentioned that a kind of whir was audible during its flight. This whir and the sudden flash of colour as it darted from the tree brings to my mind certain grasshoppers with criptic colouring which make a sort of sound as they leap and take short-flights and at the same time display their brilliantly coloured nether garments of inner wings—a sort of warning to the effect that "danger is near; follow my lead."

I filled the bottle three-fourths with water, but the frog did not much tolerate the water. It climbed as high as it could up the sides of the bottle and avoided the water. Evidently it did not live in water. Its home and even nesting place probably were always the tree-tops, like those of some of its relatives.

Colouration.

Upper portion of the body, bright grass-green (dull steel-blue in spirit), obscurely dark dotted all over. Finely tuberculated, almost smooth, except at the sides of the upper jaw which are a bit coarser in granulation.

Underside golden yellow; granular a little from the arm pits downwards. Underside of the thighs interspersed with bigger granules. Two streaks of dull yellow at the sides speckled all over with dark brown spots and tinged with red. Upper portion of the arms, the legs, and the last digits of the limbs coloured green like the upper portion of the body. The underside of the legs yellow. The upper arms are coloured yellow, but bear a red streak on the side towards the body. The undersides of the thighs are coloured yellow and bear a reddish blotch which increases in redness towards the knee joint, the redness continuing lightly on the inner unexposed side of the leg towards the calf which is mainly yellow in colour. Web between 1st and 2nd finger, yellow; between 2nd and 3rd finger, yellow towards the distal end only, but the rest bright red; between 3rd and 4th finger, red throughout, except a little at the two corners near the discs. The web extends only a little beyond half way between the 1st and 2nd finger in each hand, but right up to the discs between the other fingers.

Measurement of the specimen (in spirit).

Length of the body	..	..	$3\frac{1}{4}$ inches.
,, ,, hand	..	..	2 ,,
,, ,, leg	..	..	5 ,,

Area of expanded fore web about $\frac{1}{2}$ sq. inch.
,, ,, hind ,, $1\frac{1}{8}$ sq. inch.
Total area of the four webs about $3\frac{1}{4}$ sq. inches.

I wish to express my thanks to my student, Mr. H. Channapayya for the help he gave me, and to Mr. N. P. Muniswami Naidu, Drawing-master, Teachers' College, Saidapet, for the excellent coloured sketch of the frog he has made.

M. O. PARTHASARATHY AYYANGAR,
Teachers' College, Saidapet, Madras.

VI. CONTRIBUTIONS TO A KNOWLEDGE OF THE TERRESTRIAL ISOPODA OF INDIA.

PART I.—ON A COLLECTION FROM THE MADRAS PROVINCE AND SOUTHERN INDIA.

By WALTER E. COLLINGE, *M.Sc., F.L.S., F.E.S.*

(Plates IV—XII.)

The majority of the species here described were collected by Dr. Annandale and Mr. S. W. Kemp in the Ganjam district in the north-eastern corner of the Madras Presidency. Unfortunately in a number of instances there are only single or imperfect specimens, these are not described. I have reluctantly been compelled to establish two new genera, viz. *Ennurensis* for an interesting species collected at Ennur, near Madras, and also at Mandapam, Southern India; and *Hemiporcellio* for two new species allied to both *Porcellio*, Latreille, and *Porcellionides*, Miers. Of the remainder there is a new species of *Arhina*, Budde-Lund, one each of *Philoscia*, Latreille, *Periscyphis*, Gerst., and three new species of *Cubaris*, Brandt. The complete list is as follows:—

Ennurensis hispidus, gen. et sp. nov.
Philoscia tenuissima, n. sp.
Porcellio sp.
Hemiporcellio carinatus, gen. et sp. nov.
,, *hispidus*, n. sp.
Arhina barkulensis, n. sp.
Periscyphis gigas, n. sp.
Cubaris solidulus, n. sp.
,, *nacrum*, n. sp.
,, *granulatus*, n. sp.

Ennurensis hispidus, gen. et sp. nov.

(Plate iv, figs. 1–10.)

Body oblong oval, convex, covered with small setae. Cephalon (figs. 1 and 2) convex, fairly long, almost smooth excepting for numerous fine setae; lateral lobes small, no median lobe or definite anterior margin; epistoma convex, smooth and setaceous.

Eyes large, dorso-lateral. Antennulae small, 3-jointed. Antennae (fig. 3), first four joints short, fifth joint long, 2nd–4th characterised by deep groove on their inner border: flagellum 2-jointed, proximal joint slightly longer than the distal one. First maxillae (fig. 4), the outer lobe terminates in four stout, slightly incurved spines and seven smaller finer ones, inner lobe thin and spoon-shaped terminally, proximally thickened. Second maxillae (fig. 5) thin, plate-like, bilobed distally and setaceous. Segments of the mesosome convex and almost sub-equal, the lateral plates well developed on the 1st segment, but small on the remaining ones, the posterior angles of segments 1–4 produced backwardly, overlapping the succeeding segments. Maxillipedes (fig. 6) with elongated palps, outer one terminating in outer multispinous process and two inner spines, inner palp with single spine. Thoracic appendages (fig. 7) rather short, fringed on the inner side with few stout spines, claws long. Uropoda (figs. 8 and 9) extending beyond the telson, basal plate convex both sides with lateral expansions, dorsally there is a lateral process with which the endopodite articulates; exopodite and endopodite small and almost sub-equal in length, setose and each terminating in a fine spine. Telson (fig. 10) small, sub-equal with basal plates of uropoda, triangular with antero-lateral portions extended, depressed in the median line, apex sub-acute. Length 7 mm. Colour (in alcohol) very variable, some a creamy white with posterior margins and lateral plates of all segments a fuscous-brown, others a light brown with darker markings laterally and in the median line.

Habitat.—Ennur, nr. Madras, under stones on sand, 19-x-13, No. 8671/10: Mandapam, Pamben Passage, S. India, No. 8605/10. "In both cases the specimens were on bare sand close to the sea-shore". (*N. Annandale.*)

Type.—In the collection of the Indian Museum.

This interesting species will, I believe, be found to have a wide distribution in India. The form of the head, antennae, telson, and uropoda at once separate it from any other genus I know of, while there are a number of minor, but pronounced characters in the mouth parts. In colour it is exceedingly variable, particularly so in specimens under 7 mm. in length. To those who attach any great importance to the mouth parts, the form of the inner lobe of the 1st maxilla should prove of interest.

Porcellio sp.

Habitat.—Marikuppam, S. India, 2500 feet, 21-x-10, No. 8588. (*R. Hodgart.*)

Two examples, both without their antennae or uropoda, I am referring to the genus *Porcellio.* In colour they are a deep blackish-brown, with the posterior angles of the lateral plates of the mesosomatic segments a yellowish-brown. The head and all the segments of the body are richly tuberculated. The lateral and median lobes of the head are well-developed, epistoma convex.

Philoscia tenuissima, n. sp.

(Plate v, figs. 1–10.)

Body oblong oval, slightly convex, lateral plates of mesosome but slightly expanded; metasome abruptly narrower than mesosome. Cephalon (figs. 1 and 2) convex above, slightly rounded in front, medium and lateral lobes absent: frontal margin ill-defined and bent downward laterally; epistoma flattened with transverse ridge. Eyes prominent, dorso-lateral, ocelli large. Antennulae small, 3-jointed, basal joint large. Antennae (fig. 3) long and slender, the distal joint being the longest; flagellum 3-jointed, with deep groove on the anterior border, terminating in long spinous style, setaceous. First maxillae (fig. 4), outer lobe terminates in three stout, curved spines and five smaller inner ones: inner lobe terminates in two setaceous spines. Second maxillae (fig. 5) thin, plate-like, bilobed distally, inner division terminating in fairly long setae. Segments of the mesosome almost subequal, posterior angle of lateral plates not produced backwards. Maxillipedes (fig. 6) with elongated palps, outer one terminating in multispinous process and a single long spine: inner palp somewhat cone-shaped, sunken at the apex, with tooth-like spine on the inner border and a long pointed one arising from the base of the concavity, and four small tooth-like spines on the outer border. Thoracic appendages (fig. 7) comparatively short, setaceous, 5th joint and claw elongated, 4th joint with two spines on the inner border with obtuse plumose apices. Uropoda (figs. 8 and 9) extending beyond the telson, basal plate small with deep groove on the under side which also extends along the inner border of the exopodite, the endopodite is also grooved on its ventral side. Telson (fig. 10) short and broad, produced to a blunt point in the median line posteriorly. Length 6·5 mm. Colour (in alcohol) horny-brown with light greyish markings.

Habitat.—Museum compound, Madras (town). No. 8668/10. (*N. Annandale.*)

Type.—In the collection of the Indian Museum.

Hemiporcellio carinatus, gen. et sp. nov.

(Plate vi, figs. 1–10.)

Body (fig. 1) oblong oval, flattened, with irregular tuberculations, and tooth-like tubercles on the posterior margin of the metosomatic segments; metasome narrower than mesosome. Cephalon (figs. 2 and 3) narrow, tuberculated, setose, lateral lobes cup-shaped, median lobe formed by a dipping forward and downward of the anterior margin which is continuous; epistoma convex with numerous small setae. Eyes large, sub-lateral. Antennulae very small. Antennae (fig. 4) elongated, with well-marked carination on the dorsal side of 3rd, 4th and 5th joints; flagellum 2-jointed, the proximal joint being the longer. First maxillae (fig. 5), outer palp terminating in four stout curved spines and four smaller ones with bifid terminations, inner lobe with short blunt spine on the outer side and two setaceous spines on the inner side. Second maxillae (fig. 6) thin, plate-like, bilobed, inner lobe setaceous. The segments of the mesosome somewhat depressed, lateral plates small, slightly deflected outwards, the posterior angle of 5-7 produced backwardly, as also those of the three last metasomatic segments. Maxillipedes (fig. 7), outer palp with multispinous process on the outer side and three long spines internal to this, the inner palp has two spines and four small tooth-like processes on the margins. Thoracic appendages (fig. 8) fringed with stout spines on the inner side of the three distal joints. Uropoda (fig. 9) extending beyond the telson, basal plate with lateral extensions dorsally and ventrally, with the former the slightly curved endopodite articulates, the exopodite which is cuniform articulates at the base of the basal plate. Exopodite nearly twice the length of the basal plate, and of the endopodite, both covered with fine setae. Telson (fig. 10) short, not extending beyond the basal plates of the uropoda, triangular, apex subacute. Length 7·5 mm. Colour (in alcohol) greyish-brown, cephalon and metasome usually darker, lateral plates of mesosome with dark patch uniformly a blackish-brown, very variable.

Habitat.—Under stones and dead water weeds at edge of Chilka Lake, Rambha, Ganjam, Madras Pres., 27-xii-13, No. 8692/10. "Apparently an amphibious species". (*N. Annandale.*)

Type.—In the collection of the Indian Museum.

This species is closely allied to *Porcellio immsi*, Cllge.,[1] which latter must now be referred to the genus *Hemiporcellio*; it differs, however, from *immsi* in the form of the antennae, in the anterior margin of the cephalon, which is continuous, and in the form of the uropoda.

The colour is exceedingly variable. Examples measuring 5·5 and 6 mm. invariably have the first three and last mesosomatic segments a reddish-brown colour and the whole of the metasome a deep blackish-brown. Similar variations obtain in *H. immsi* judging from an immature specimen.

Hemiporcellio hispidus, n. sp.

(Plate vii, figs. 1–9.)

Body oblong-oval, flattened, tuberculated and covered with fine setae; metasome narrower than the mesosome. Cephalon (figs. 1 and 2) small, tuberculated, lateral lobes slightly cup-shaped, median lobe absent, anterior margin distinct and continuous, epistoma convex and covered with stellate setae. Eyes small, sub-dorsal. Antennulae small, 3-jointed, basal joint prominent. Antennae (fig. 3) elongated, with well-marked carination on the

[1] Ann. Mag. Nat. Hist. 1914, s. 8, vol. xiv, p. 207, pl. ix.

3rd, 4th and 5th joints; flagellum 2-jointed, the distal joint being the longer. First maxillae (fig. 4), outer lobe terminates in three long and one short stout spine on the outer side, on the inner side are four smaller ones and two fine spines; the inner lobe terminates in two setaceous spines. The lateral plates of the mesosomatic segments are small and slightly overlap one another, posterior angle slightly produced. Maxillipedes (fig. 5), outer palp with multispinous process on the outer side and three spines internal to this, inner palp has a single spine and three tooth-like marginal processes. Thoracic appendages (fig. 6) short, with the 1st and 2nd joints grooved on their outer side, the three terminal joints are fringed with stout spines with trifid terminations. The whole of the appendages are covered with fine hair-like setae. Uropoda (figs. 7 and 8) extending beyond the telson and covered with fine setae, basal plate with lateral extensions, exopodite cuniform and grooved on the outer side, endopodite triangular in section. Telson (fig. 9) short, extending beyond the basal plates of the uropoda, triangular, apex sub-acute. Length 5 mm. Colour (in alcohol) greenish-grey with few irregular blackish blotches.

Habitat.—Satpara, Lake Chilka, Orissa, 17-ix-13, No. 8635/10. "A terrestrial species". (*N. Annandale*).

Type.—In the collection of the Indian Museum.

Arhina barkulensis, n. sp.

(Plate viii, figs. 1-10.)

Body oblong-oval, strongly convex, surface shiny, minutely punctate. Cephalon (figs. 1 and 2) covered with minute raised tubercles, lateral lobes well developed, median lobe small, epistoma slightly raised in median line, concave laterally with transverse ridge above the antennules. Eyes large, sub-lateral. Antennulae (fig. 3) 3-jointed, the terminal joint having a number of bristle-like setae at the apex and side. Antennae (fig. 4) characterised by the shortness of the three first joints, 3rd and 4th joints together equal in length to the 5th which is as long as the flagellum, the three joints of which are almost sub-equal; terminal stylet slender; whole of the appendage covered with short setae, those on the flagellum rather longer. First maxillae (fig. 5), the outer lobe is very solid and terminates in eight stout spines; the inner lobe is scroll-like, the inner border partly overlapping the flat outer portion, terminally there are two long setaceous spines. Second maxillae (fig. 6) a thin bilobed plate terminating in an inner dense tuft of setae and an outer tuft. The segments of the mesosome almost sub-equal, posterior margins of 1–4 almost straight, lateral angles rounded, of 5–7 slightly produced backward, sub-acute. Lateral plates of metasomatic segments greatly prolonged backwards (fig. 10). Maxillipedes (fig. 7), the outer palp terminates in a strong spine with two tufts of setae, at the base of this are two further tufts arising from a slight eminence, and a third pair still more inwardly: there are no spines on the inner palp, which is fringed with short setae. Thoracic appendages (fig. 8) stout and comparatively short, first joints almost equal in length to the next three, stout claw with lateral spines; the 3rd, 4th and 5th joints have on their inner side a dense mass of long setae, with paired stouter spines on the outer side of the 2nd, 3rd, and 4th and on the inner side of the 5th joints. Uropoda (fig. 9) extending beyond the telson; basal plate sparsely covered with setae, there is a short, blunt spine on the outer side and a raised portion extending across the proximal end to the inner side, beneath which the endopodite articulates; exopodite somewhat conical in shape, more globose on the inner side, endopodite slender, terminating in two long setae. Telson (fig. 10) triangular, flat, sides straight, apex sub-acute. Length 11·5 mm. Colour (in alcohol) greenish-brown with yellow flecks on the head and mesosomatic segments.

Habitat.—Under stones at edge of lake, Barkul, Lake Chilka, Orissa, 22-vii-13, No. 8670/10. "Apparently an amphibious species". (*N. Annandale.*)

Type.—In the collection of the Indian Museum.

The genus *Arhina* was constituted by Budde-Lund (Rev. Crust. Isop. Terr., 1904, p. 44) for a species, *A. porcellioides*, found "in a warehouse at Copenhagen, perhaps imported from East India." The genus is placed by Budde-Lund under the sub-family Spherillioninae, which includes *Pseudophiloscia*, *Suarezia*, *Scleropactes*, *Sunniva*, *Saidjahus*, *Ambounia* and *Spherillo*. Neither *Pseudophiloscia* or *Arhina* are closely related to any of the above mentioned genera, and whilst I differ strongly from Budde-Lund in his views on classification, they would, in my opinion, have found a more natural position in his Tribe Alloniscoidea, (*op. cit.*, p. 37).

The form of the antennae, 1st maxillae, maxillipedes and uropoda clearly indicate the relationship of this species to *Arhina porcellioides*.

Periscyphis gigas, n. sp.

(Plate ix, figs. 1-10.)

Body (fig 1) oblong oval, dorsal face strongly convex, sloping downwards posteriorly, almost smooth. Cephalon (figs. 2 and 3) small with median depression, flanked laterally by the lateral plates of the 1st segment of the mesosome, the anterior border of which extends slightly beyond the cephalon; lateral lobes well developed, median lobe absent. Ventrally there is a strong median carination. Eyes prominent, sub-dorsal. Antennulae (fig. 4) short and stout, 3-jointed. Mandibles (fig. 5), the outer cutting edge has three blunt teeth and a blunt process on the inner edge, beneath which is a tuft of setae. First maxillae (fig. 6), the outer lobe terminates in four curved spines and five finer and straighter ones on the inner side. Second maxillae (fig. 7) thin and plate-like; the inner lobe terminates in a mass of setae, whilst the outer lobe is more robust and tooth-like. The segments of the mesosome are strongly convex, excepting those of the 1st the lateral plates are only slightly produced backwards. Maxillipedes (fig.

8), inner lobe palp-like with setae on the inner side and two rows across the dorsal face, the outer palp terminates in three multispinous processes. Basally there is a raised portion studded with numerous small setae. Thoracic appendages (fig. 9) robust and fringed with numerous spines with trifid terminal portions (fig. 9*a*) and smaller spines, 2nd appendages having on the apical border of the fifth joint two with obtuse plumose apices. Uropoda (fig. 10), basal plate large, extending beyond the telson: outer margin sub-crenate; exopodite articulating on the middle inner border, endopodite slightly longer than the exopodite and articulating at the top of the inner border of the basal plate. Telson (fig. 1), dorsal face strongly convex, obtusely triangular, almost smooth. Length 20·5 × 13 mm. Colour (in alcohol) horny-brown with the lateral plates of the 1st, 5th and 6th mesosomatic and the 3rd and 4th metasomatic segments yellow.

Habitat.—Ponmudi, Travancore, September, 1893, No. 8626/10. (*H. S. Ferguson.*)

Type.—In the collection of the Indian Museum.

This interesting species is, I believe, the largest yet described of this genus. Unfortunately there were no antennae on either of the two specimens. Owing to the strong convexity of the 1st segment of the mesosome the dorsal surface of the head is almost vertically disposed.

Cubaris solidulus, n. sp.

(Plate x, figs. 1–12)

Body oblong oval, moderately convex, minutely punctate with lateral rugosities. Cephalon (figs. 1 and 2) small with posterior margin slightly raised, lateral lobes small, median lobe absent, epistome flat. Eyes small, situated dorso-laterally. Antennulae (fig. 3) small, 3-jointed, with numerous short setae on the terminal joint. Antennae (fig. 4) short, covered with fine setae, 2nd to 5th joints grooved on their inner side; flagellum 2-jointed, the distal joint being a little over one and a half times as long as the proximal one. First maxillae (fig. 5), outer lobe terminates in four stout incurved spines and six smaller, almost straight ones; inner lobe terminally rounded, with two setose spines. Second maxillae (fig. 6) thin and plate-like, terminating distally in two setaceous lobes, the inner of which is jointed. Segments of the mesosome with posterior angles of 1–4 produced backwards, overlapping the succeeding segments, fitting into a slight groove in segments 2–5, segments 5–7 almost straight. Segments 1 and 2 notched on their lower inner margins for reception of succeeding segments (figs. 7 and 8). Lateral plates of metasomatic segments 3–5 elongated. Maxillipedes (fig. 9), the outer palp terminates in a multispinous process on the outer side, with two small spines below it, internal to the large process are two long fine spines, the inner palp possesses five tooth-like spines and one larger one. Thoracic appendages (fig. 10) comparatively short, setaceous, with few stout spines on the inner side of the three distal joints. Uropoda (fig. 11) not extending beyond the telson, basal plate somewhat triangular, posterior margin almost straight; exopodite small, situated on the inner lower border of the basal plate, endopodite nearly twice as long, setaceous and terminating in two long setae, situated at the top of the inner margin of the basal plate. Telson (fig. 12) slightly longer than broad, concave laterally, posterior margin straight, anteriorly with slight median depression. Lenghth 11 mm. Colour (in alcohol) horny-brown with greyish rugosities.

Habitat.—Oorgaum, Kolar District, S. India, 20-x-10, No. 8598/10.

Type—In the collection of the Indian Museum.

The rugosities on the mesosomatic segments are more distinct in some specimens than in others. The form of the uropoda, antennae and telson serve to separate it from its allies, and minor differences are also present in the mouth parts and thoracic appendages.

Cubaris nacrum, n. sp.

(Plate xi, figs. 1-10.)

Body oblong oval, strongly convex, smooth and shiny. Cephalon (figs. 1 and 2) small, with sloping anterior half, lateral lobes very small, median lobe absent, epistome almost flat. Eyes small, situated antero-dorsally. Antennulae (fig. 3) small, 3-jointed, with numerous short, thick setae on the terminal joint. Antennae (fig. 4) short, covered with fine setae, flagellum 2-jointed, distal joint almost twice as long as the proximal one. First maxillae (fig. 5), outer lobe terminates distally in four incurved spines and five smaller almost straight ones, short, simple, hair-like setae on the outer margin; inner lobe terminally rounded, thin, and with two setose spines, broader at the base. Second maxillae (fig 6) thin and plate-like, terminating distally in a bi-lobed manner, the inner lobe having a row of strong setae on the inner face. The segments of the mesosome are strongly convex, with the lateral plates not expanded excepting in the 1st mesosomatic segment. Segments 1 and 2 notched on their lower inner margins for reception of succeeding segments Lateral plates of metasomatic segments 3-5 greatly elongated. Maxillipedes (fig. 7), the outer palp terminates in a multispinous process on the outer side and two small spines inwardly, the inner palp possesses a single terminal spine and two smaller ones on the outer border. Thoracic appendages (fig. 8) comparatively short, setaceous, the three distal joints being fringed on the inner side with numerous stout spines. Uropoda (fig. 10) not extending beyond the telson, basal plate triangular in shape, posterior margin pointed, outer half raised above the flat, inner half; exopodite very small, situated towards the base of the raised outer half, endopodite large and situated at top of the inner margin of the basal plate, but not extending beyond it. Telson (fig. 9) slightly longer than broad, contracted laterally, posterior margin almost straight. Length 16·5 mm. Colour (in alcohol) slaty-grey with lighter coloured lateral markings on the mesosome.

Habitat.—Under stones on hill near Rambha, Ganjam District. No. 8690/10. (*N. Annandale.*)

Type.—In the collection of the Indian Museum.

The form of the uropoda at once serve to separate this species from any other known form. Considerable variation was noticed in the mouth parts. In alcohol it is a slaty-grey colour, but when dry the specimens look like large pearls.

Cubaris granulatus, n. sp.

(Plate xii, figs. 1–11.)

Body oblong-oval, moderately convex, finely granulated with few irregular rugosities on the cephalon. Cephalon (figs. 1 and 2) small, anterior margin slightly raised, lateral lobes small, median lobe absent, posterior margin distinct, irregularly rugose, epistome with triangular convexity, deeply sunken around base of antennae. Eyes moderately large, situated dorso-laterally. Antennulae small, 3-jointed. Antennae (fig. 3) short, covered with fine setae, 2nd to 4th joints grooved on their inner side: flagellum 2-jointed, the distal joint being nearly three times as long as the proximal one. First maxillae (fig. 4), outer lobe terminates in four stout incurved spines and six smaller, almost straight ones: inner lobe terminally rounder, with two long setose spines. Segments of the mesosome with posterior angles of 1–4 produced backwards, overlapping the succeeding segments, fitting into a slight groove in segments 2–5, lateral plates of segments 6 and 7 slightly expanded. Segments 1 and 2 notched on their inner margins for reception of succeeding segments (figs. 5 and 6). Lateral plates of metasomatic segments 3–5 elongated. Maxillipedes (fig. 7), outer palp terminates in a multispinous process on the outer side and two long inner spines, at the base of the innermost are two very small spines: the inner palp has a single spine and a small tooth-like process. Thoracic appendages (fig. 8) comparatively short, setaceous with dense mass on the inner side of the 3rd and 4th joints. Uropoda (figs. 9 and 10) not extending beyond the telson, basal plate somewhat triangular, posterior margin almost straight, plicated on the ventral side; exopodite small, situated on the inner border of the basal plate, endopodite two-and-a-half times as long as the exopodite, setaceous, situated at the top of the inner border of the basal plate. Telson (fig. 11) slightly longer than broad posteriorly, expanded anteriorly, posterior margin almost straight with concavity anteriorly in the median line. Length 5·5 mm. Colour (in alcohol) dark olive brown.

Habitat.—Rambha, L. Chilka, Ganjam Dist., Madras, 22-ix-13, No. 8639-10. "Probably a terrestrial species". (*N. Annandale* and *S. W. Kemp.*)

Type.—In the collection of the Indian Museum.

EXPLANATION OF PLATE IV.

Ennurensis hispidus, n. sp.

FIG. 1.—Dorsal view of the cephalon.
,, 2.—Anterior view of the cephalon.
,, 3.—Antenna.
,, 4.—First maxilla, terminal portions of inner and outer lobes
,, 5.—Second maxilla, terminal portion.
,, 6.—Maxillipede, terminal portion.
,, 7.—Second thoracic appendage
,, 8.—Left uropod, dorsal view.
,, 9.—Left uropod, ventral view.
,, 10.—Last abdominal segment and telson.

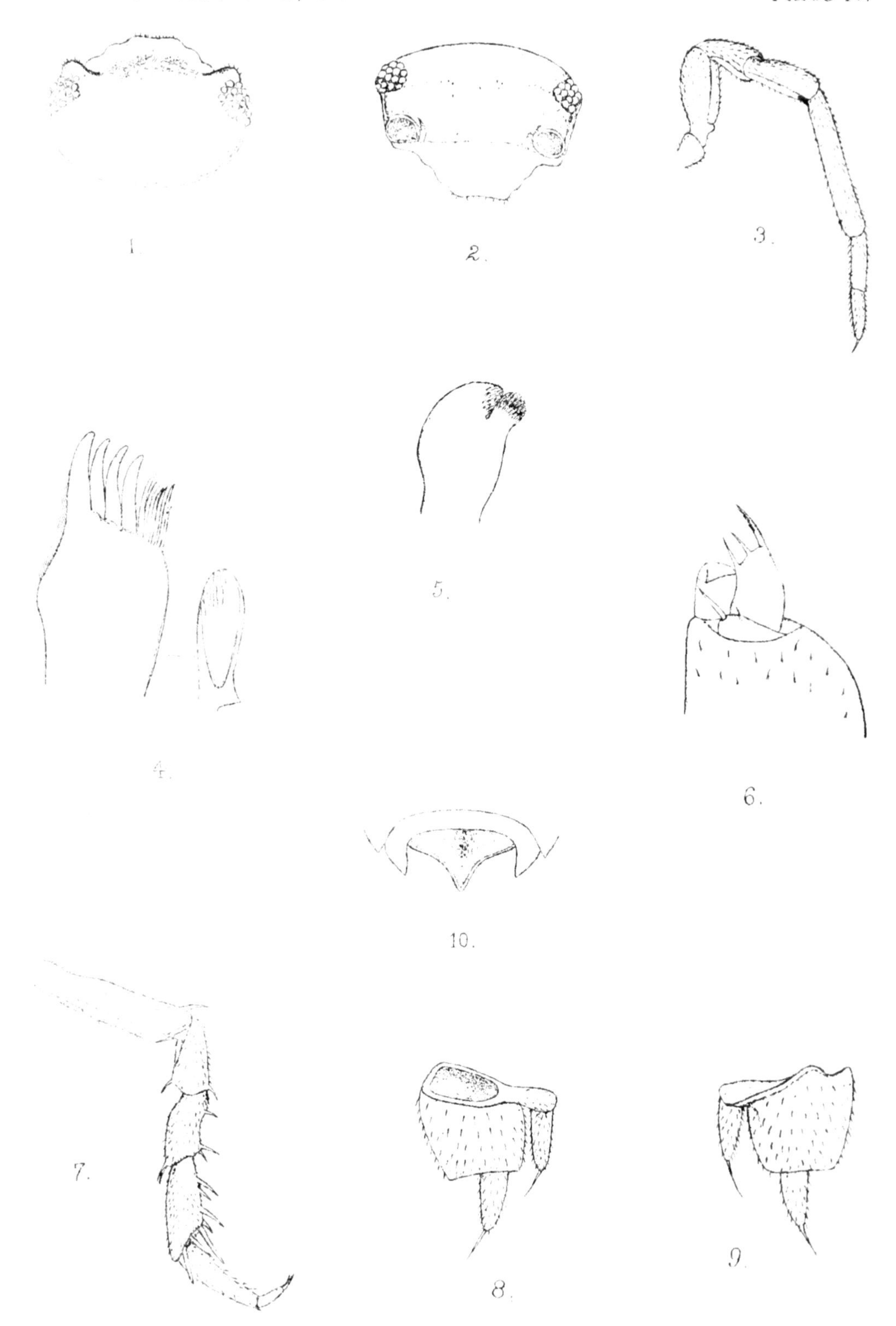

E. W. E. del.

A. Chowdhary, lith.

ENNURENSIS HISPIDUS, n.sp.

EXPLANATION OF PLATE V.

Philoscia tenuissima, n. sp.

FIG. 1.—Dorsal view of the cephalon.
,, 2.—Anterior view of the cephalon.
,, 3.—Antenna.
,, 4.—First maxilla, terminal portions of inner and outer lobes.
,, 5.—Second maxilla, terminal portion.
,, 6.—Maxillipede, terminal portion.
,, 7.—Second thoracic appendage.
,, 8.—Left uropod, dorsal view.
,, 9.—Left uropod, ventral view.
,, 10.—Telson and last abdominal segment.

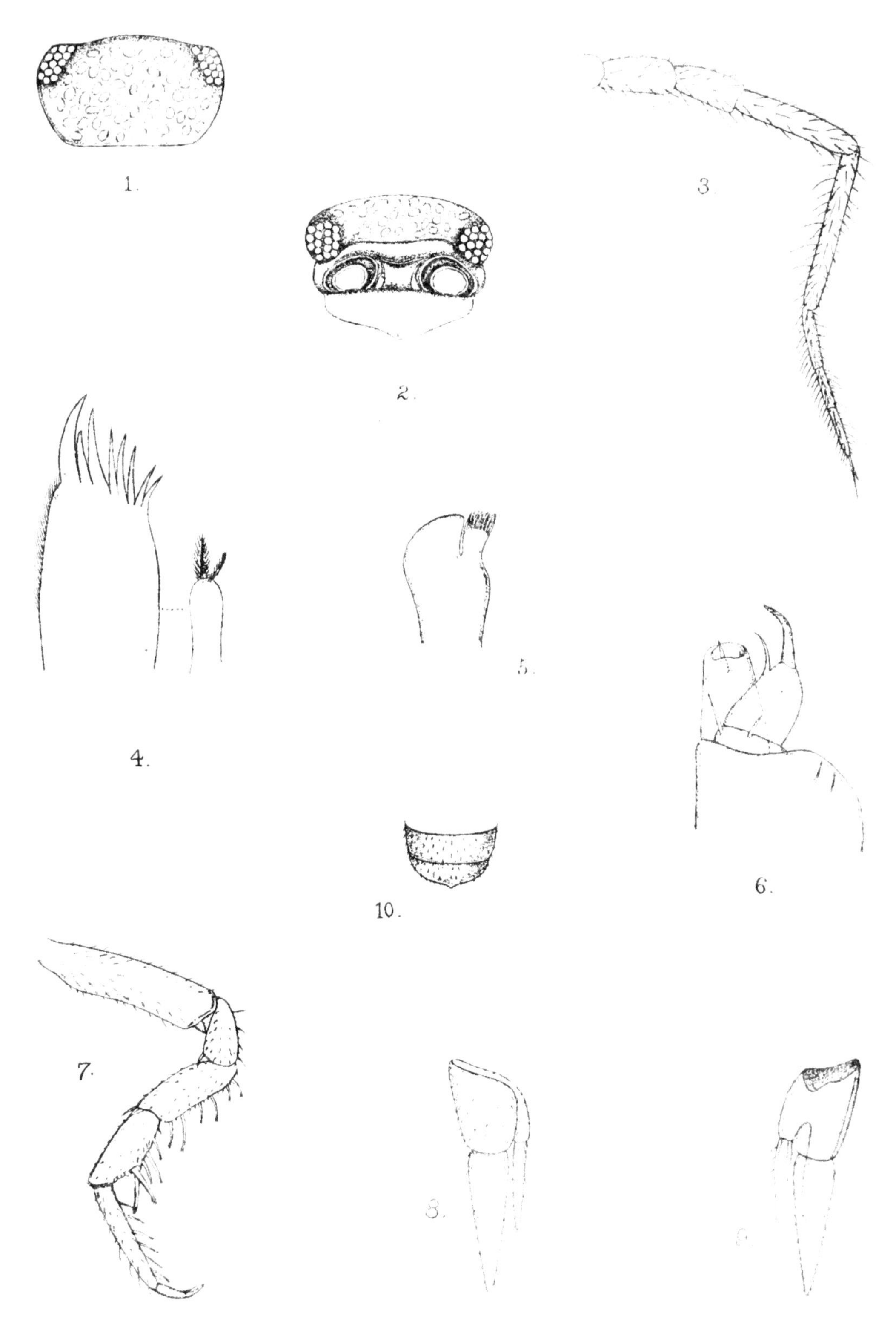

E.W.E. del. A. Chowdhary, lith.

PHILOSCIA TENUISSIMA, n. sp.

EXPLANATION OF PLATE VI.

Hemiporcellio carinatus, n. sp.

FIG. 1.—Dorsal view, × $8\frac{1}{2}$.
,, 2.—Dorsal view of the cephalon.
,, 3.—Anterior view of the cephalon.
,, 4.—Antenna.
,, 5.—First maxilla, terminal portions of inner and outer lobes.
,, 6.—Second maxilla.
,, 7.—Maxillipede, terminal portion.
,, 8.—Second thoracic appendage.
,, 9.—Left uropod.
,, 10.—Last abdominal segment and telson.

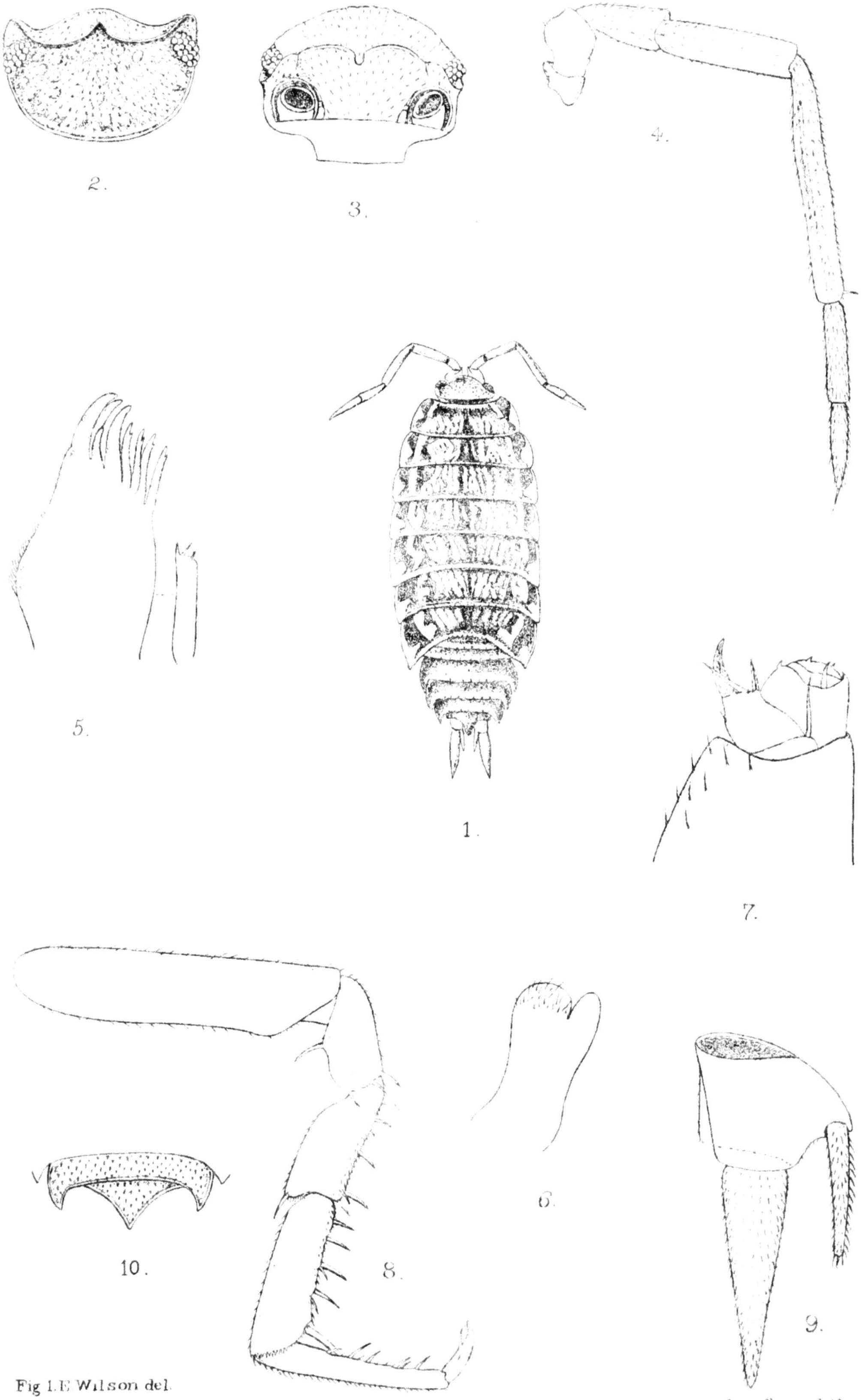

2. 3. 4.

5. 1. 7.

6. 9. 10. 8.

Fig 1. E Wilson del.
E. W. E. del.

A Chowdhary, lith.

HEMIPORCELLIO CARINATUS, n. sp.

EXPLANATION OF PLATE VII.

Hemiporcellio hispidus, n. sp.

FIG. 1.—Dorsal view of the cephalon.
,, 2.—Anterior view of the cephalon.
,, 3.—Antenna.
,, 4.—First maxilla, terminal portions of inner and outer lobes.
,, 5.—Maxillipede, terminal portion.
,, 6.—Second thoracic appendage.
,, 7.—Left uropod, dorsal view.
,, 8.—Left uropod, ventral view.
,, 9—Telson and part of last abdominal segment.

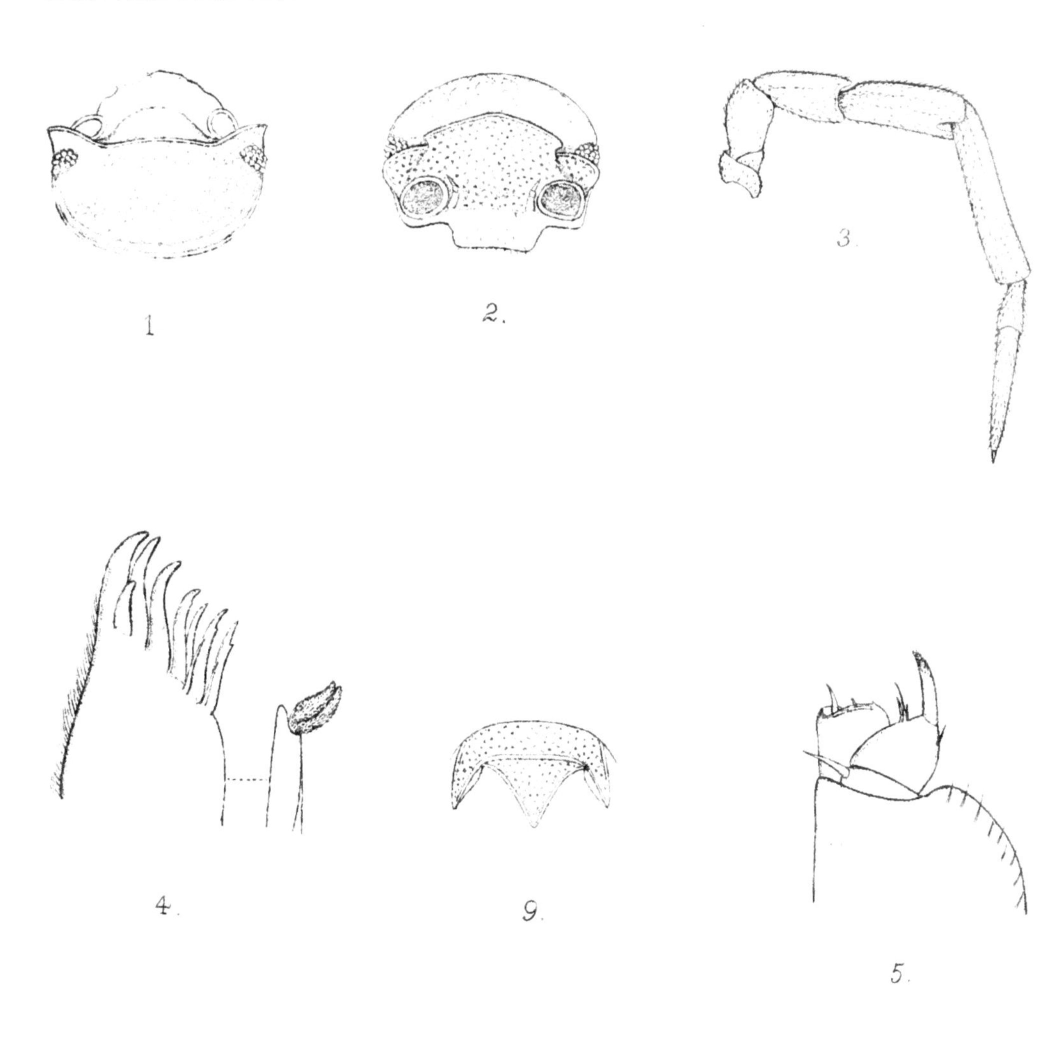

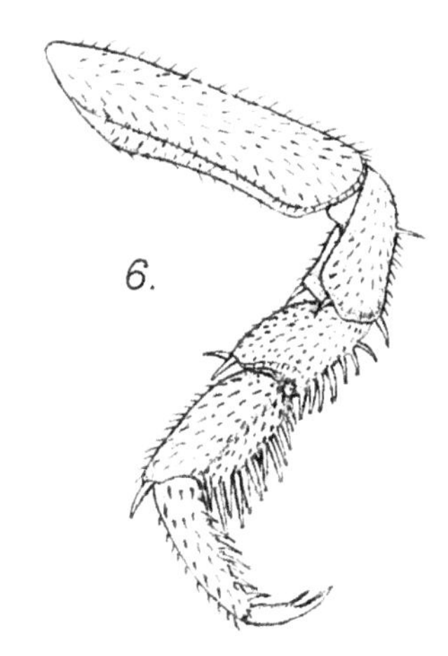

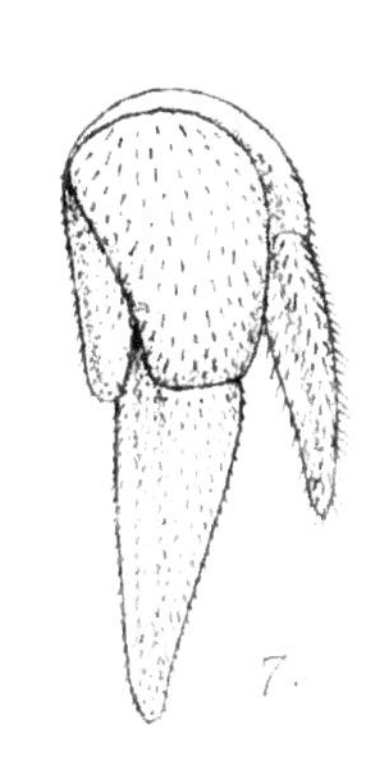

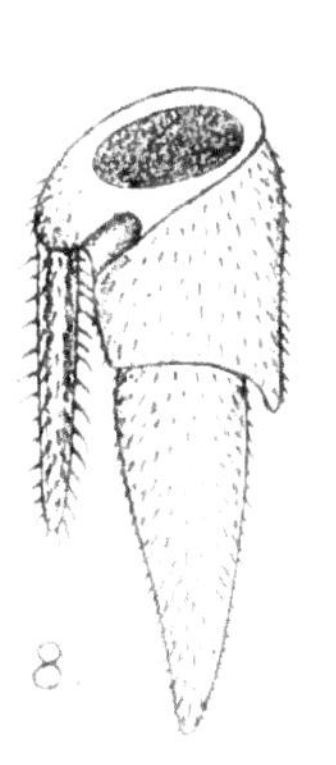

E. W. E. del. A. Chowdhary, lith.

HEMIPORCELLIO HISPIDUS, n. sp.

EXPLANATION OF PLATE VIII.

Arhina barkulensis, n. sp.

FIG. 1.—Dorsal view of cephalon.
,, 2.—Anterior view of cephalon.
,, 3.—Antennule.
,, 4.—Antenna.
,, 5.—First maxilla, terminal portion of inner and outer lobes.
,, 6.—Second maxilla, terminal portion.
,, 7.—Maxillipede, terminal portion.
,, 8.—Second thoracic appendage.
,, 9.—Left uropod, dorsal view.
,, 10.—Telson and part of last abdominal segment.

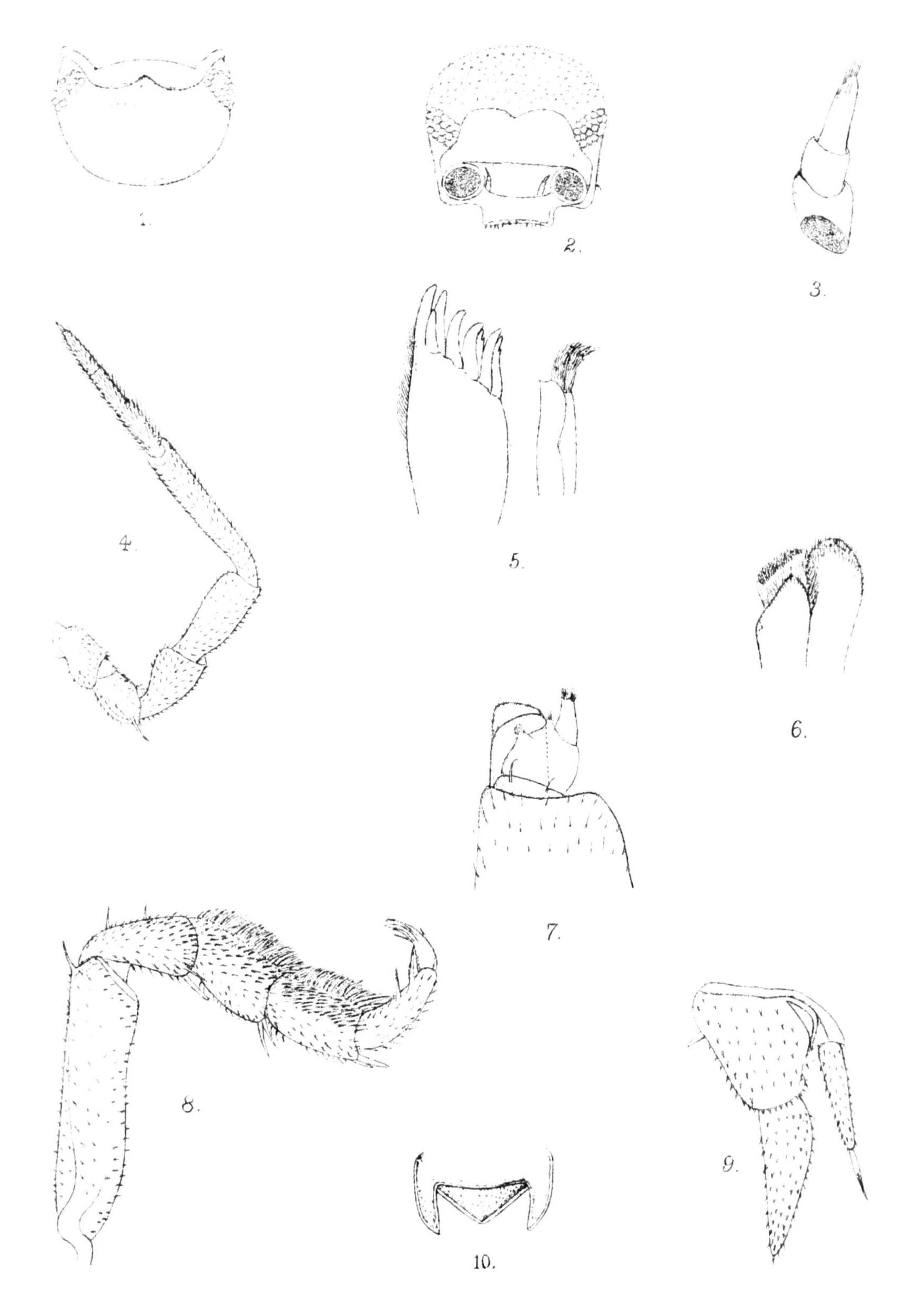

E. W. F. del. A. Chowdhary, lith.

ARHINA BARKULENSIS, n.sp.

EXPLANATION OF PLATE IX.

Periscyphis gigas, n. sp.

FIG. 1.—Dorsal view, × $2\frac{1}{2}$.
,, 2.—Dorsal view of the cephalon.
,, 3.—Anterior view of the cephalon.
,, 4.—Antennule.
,, 5.—Right mandible.
,, 6.—Outer lobe of 1st maxilla.
,, 7.—Second maxilla.
,, 8.—Maxillipede.
,, 9.—Second thoracic appendage.
,, 9*a*.—Terminal portion of spine, much enlarged.
,, 10.—Left uropod.

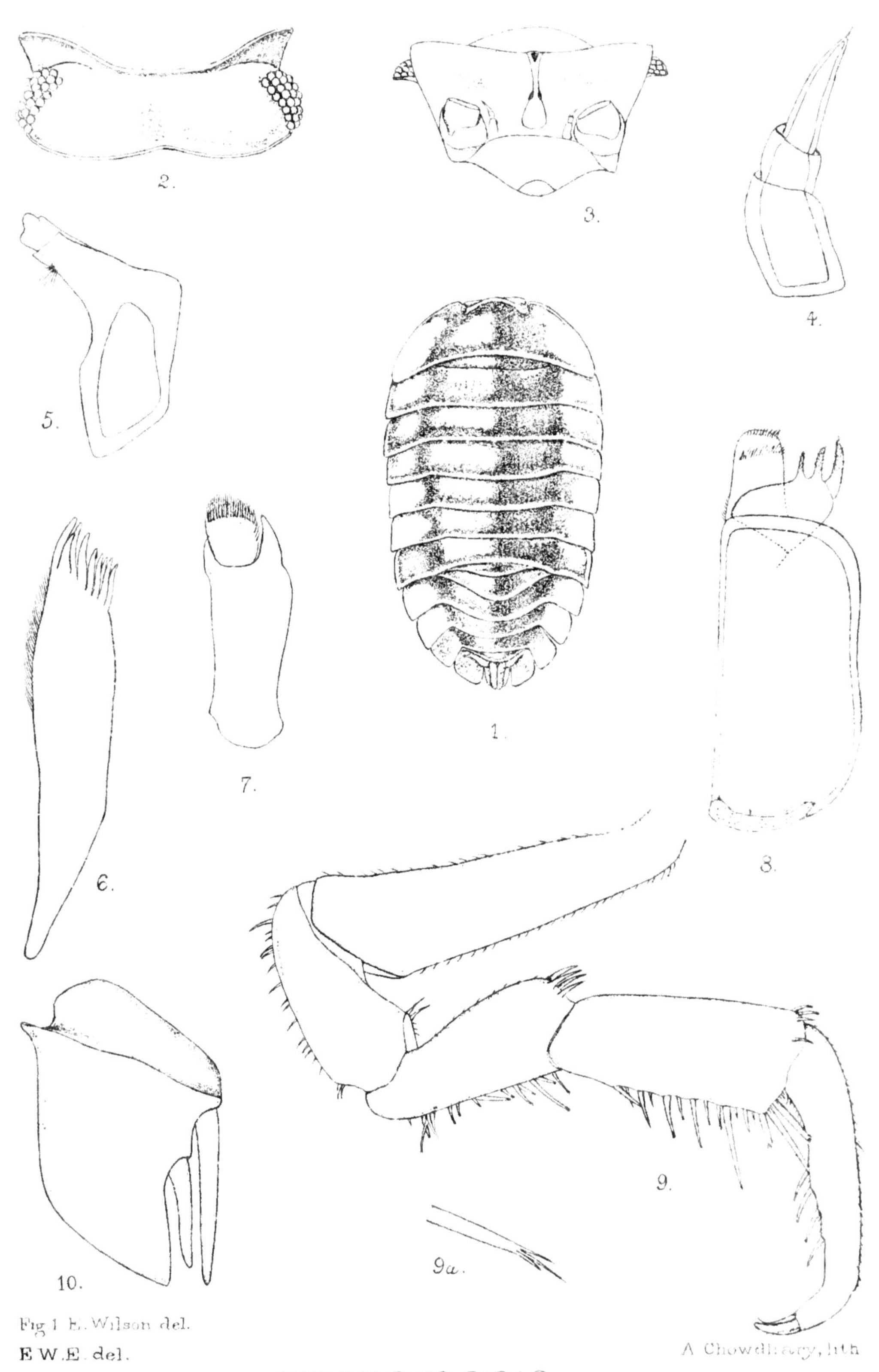

Fig 1 E. Wilson del.
E.W.E. del.

A Chowdhary, lith.

PERISCYPHIS GIGAS, n. sp.

EXPLANATION OF PLATE X.

Cubaris solidulus, n. sp.

FIG. 1.—Dorsal view of the cephalon.
,, 2.—Anterior view of the cephalon.
,, 3.—Antennule.
,, 4.—Antenna.
,, 5.—First maxilla, terminal portion of inner and outer lobes.
,, 6.—Second maxilla, terminal portion.
,, 7.—Lateral portion of 1st mesosomatic segment, showing notch on the under side.
,, 8.—The same on the 2nd mesosomatic segment.
,, 9.—Maxillipede, terminal portion.
,, 10.—Second thoracic appendage.
,, 11.—Right uropod.
,, 12.—Last abdominal segment and telson.

1.

2.

3.

5.

4.

6.

7.

8.

9.

10.

12.

11.

E.W.B., del. A Chowdhary, lith.

CUBARIS SOLIDULUS, n. sp.

EXPLANATION OF PLATE XI.

Cubaris nacrum, n. sp.

FIG. 1.—Antero-dorsal view of the cephalon.
,, 2.—Anterior view of the cephalon.
,, 3.—Antennule.
,, 4.—Antenna.
,, 5.—First maxilla, terminal portion of inner and outer lobes.
,, 6.—Second maxilla, terminal portion.
,, 7.—Maxillipede, terminal portion.
,, 8.—Second thoracic appendage.
,, 9.—Last abdominal segment and telson.
,, 10.—Right uropod

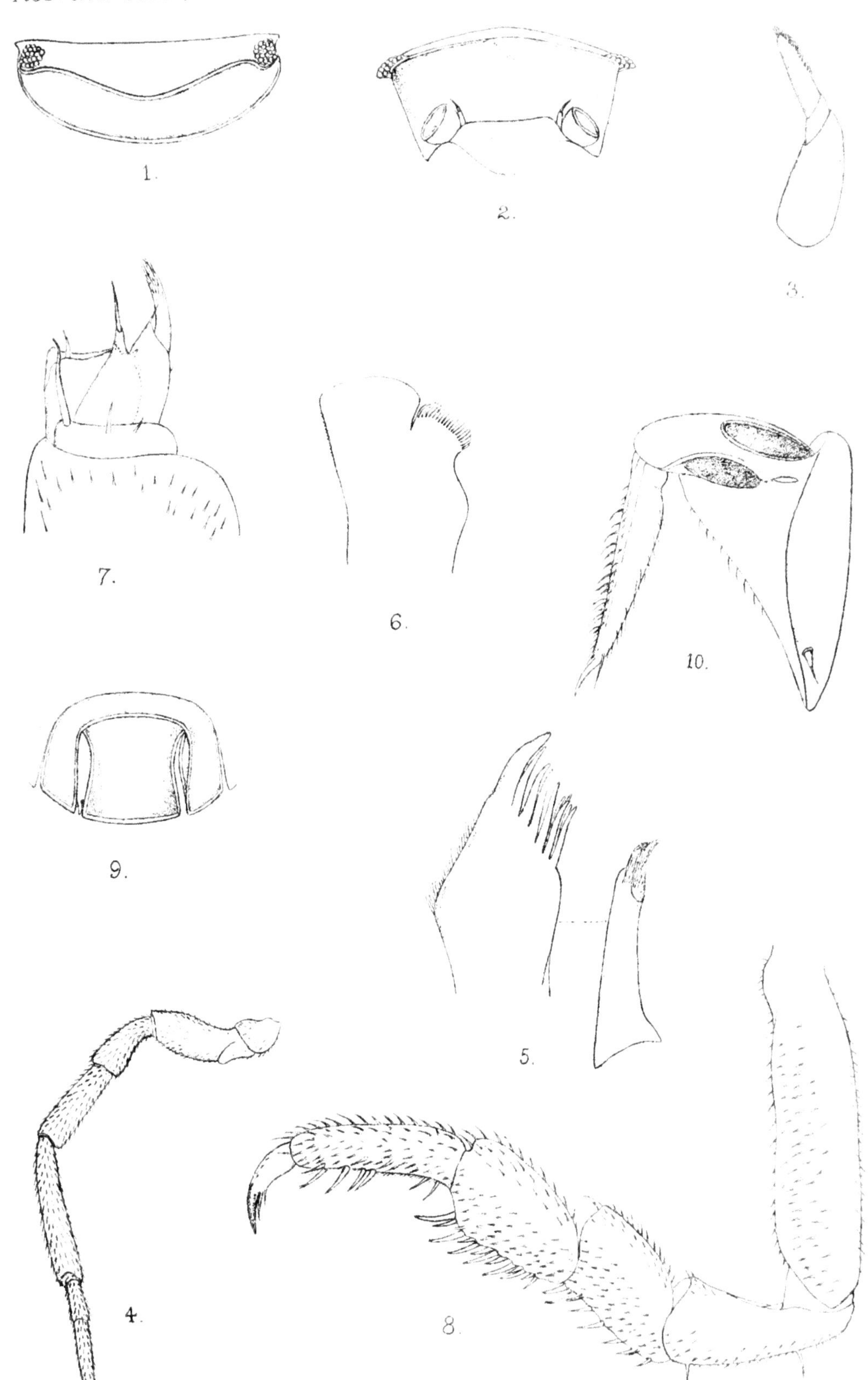

E. W. E. del.

A. Chowdhary, lith.

CUBARIS NACRUM, n. sp.

EXPLANATION OF PLATE XII.

Cubaris granulatus, n. sp.

FIG. 1.—Dorsal view of the cephalon.
,, 2.—Anterior view of the cephalon.
,, 3.—Antenna.
,, 4.—First maxilla, terminal portion of inner and outer lobes.
,, 5.—Lateral portion of the 1st mesosomatic segment, showing notch on the under side.
,, 6.—The same on the 2nd mesosomatic segment.
,, 7.—Maxillipede, terminal portion.
,, 8.—Second thoracic appendage.
,, 9.—Right uropod, dorsal view.
,, 10.—Right uropod, ventral view.
,, 11.—Last abdominal segment and telson.

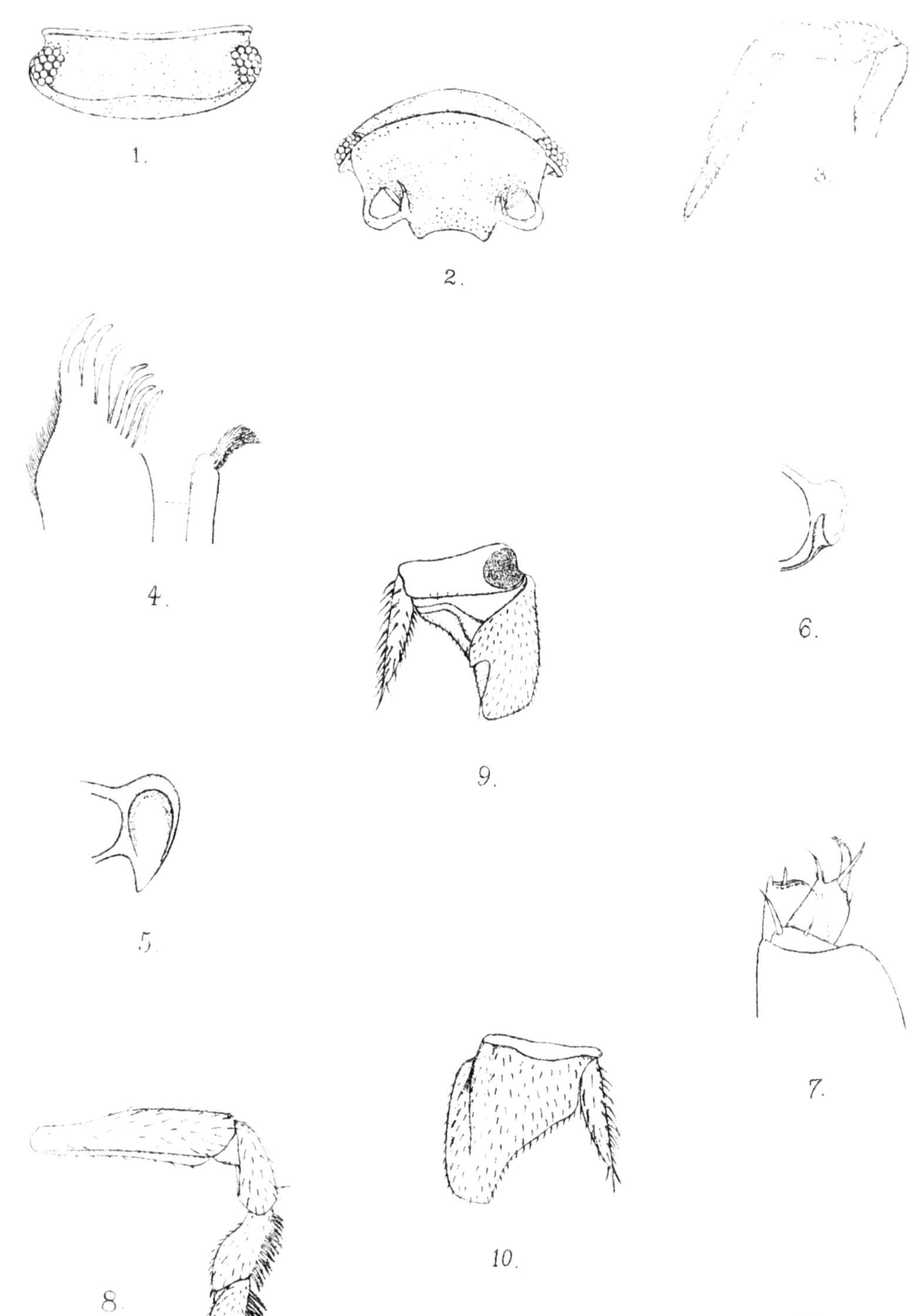

E. W. E. del. A. Chowdhary, lith.

CUBARIS GRANULATUS, n. sp.

VII. ON THE ANATOMY OF A BURMESE SLUG OF THE GENUS *ATOPOS*.

By EKENDRANATH GHOSH, *M.Sc., Assistant Professor of Biology, Medical College, Calcutta.*

(Plates xvi—xix.)

Two specimens of this slug were collected by Mr. F. H. Gravely, Assistant Superintendent, Indian Museum, from the base of Dawna Hills near Thingannyinaung, about 900 ft. above sea-level, on the 27th November, 1911.

They belong to the genus *Atopos*, Simroth, and are made the type of a new subgenus (*Parapodangia*) under the name of *A.* (*P.*) *gravelyi*.

Subgenus **Parapodangia**, nov.

An anterior portion of the mantle (notum) (about $\frac{1}{6}$th the body length) along the middle line separate from the thin dorsal wall of the body beneath. An **H**-shaped fold of the integument beneath the mouth between the precephalic flap and the foot. The lower tentacles fused with the precephalic flap as in *Podangia*.

Atopos (Parapodangia) gravelyi, n. sp.

Colour of notum sepia (with a slight brownish tint) with blotches and pinpoint dots of dark brown above, and of slaty gray (bluish) with blotches and spots of slaty black below; in the middle, a longitudinal row of pale buff blotches above and a continuous dark clove-brown band below. The anterior portion of the head (with ommatophores, lower tentacles and precephalic flaps) is slaty blue, while the posterior portion is ochraceous yellow behind. Foot sole pale yellow Keel prominent, rounded and of dark clove-brown colour.

Length of notum (along mid-dorsal line) 14·7 cm. Height of notum 1·4 cm. Breadth 1·3 cm. Female genital aperture 1·35 cm. from male genital aperture.

ANATOMY.

I. Body wall.

The inner surface of the thick mantle in its anterior portion where it is separate from the dorsal integument beneath, is very vascular and appears to share a prominent part in respiration.

The vessels are connected to the dorsal sinus in the mid-dorsal line of the body wall. In other respects, the body wall is quite similar to that in other species.

II. Pallial Complex.

The *pallial complex* forms a circular area extending from the right side at the junction of the inner surface of the mantle and the dorsal surface of the foot to within 6 cm. of the margin of the mantle on the left side. It lies at about 2·4 cm. distant from the anterior margin of the mantle. The *pericardium* lies in the anterior two-thirds of the pallial complex and to the right, the *kidney* occupying the remainder. There is no distinct pulmonary chamber at all. The whole pallial complex is adherent to the thick mantle and is roofed by a thin membrane which is fused with the inner surface of the latter. The ventral wall lies over the anterior end of the liver.

The *heart* is placed in the long axis of the pericardium. The auricle is placed behind the ventricle.

Minute structure of the ventricle.—The outer surface is lined by a single layer of cubical-cells with oval nuclei. There is no distinct epithelial lining of the cavity. The superficial layers form a thin continuous coat of transversely-arranged muscle fibres. Beneath this, the muscle fibres form thick bundles which are disposed irregularly in different directions. Just beneath the superficial layer, the thick bundles are arranged circularly in a transverse direction, being separated from the former widely in many places by thick bundles which pass inwards, some radially and others obliquely, from the superficial bundles to these circular ones with which they seem to unite. More internally the disposition of the fibres are mainly longitudinal with a few oblique ones. The cavity of the ventricle is traversed by these muscle bands which extend through the cavity in various directions.

Minute structure of the auricle.—The wall is lined externally by a layer of rectangular cells with their long axis parallel to the surface. The superficial muscles form a transversely circular layer. The inner bundles form a longitudinally circular layer. Between these two layers, there are a few bundles which are arranged obliquely and seem to pass from one layer to another.

III. Digestive System.

(i) The *buccal bulb* forms a protrudable proboscis which, when retracted, is placed inside a proboscis sheath having a narrow funnel-like shape at its outer aspect. The proboscis forms the acrecbolic (pleurembolic) introvert of Sir E. R. Lankester.

As seen in a longitudinal section, the proboscis, when retracted, lies in the tubular space formed inside the proboscis sheath, which is folded a little behind its middle in such a way that its posterior portion is invaginated into the anterior portion;

the morphologically inner surface of the posterior portion of the sheath now becomes external and lies in contact with the inner surface of the anterior portion. On the dorsal surface and the lateral aspects of the anterior portion of the proboscis sheath are numerous flattened muscular strands which pass upwards and backwards to be inserted into the thin dorsal integument. The presence of these strands prevents this tubular anterior portion of the proboscis sheath from being protruded with the proboscis. Again the posterior division of the sheath, which, owing to the doubling of its wall, is placed inside the anterior division, is prevented from being straightened out behind by the presence of fine strands of connective tissue, which extend from its morphologically outer side (inner side in the present condition) to the outer surface of the proboscis in a direction backwards and inwards from the wall of the sheath.

It is remarkable to note that in *A.* (*P.*) *kempii*, Ghosh, owing to the absence of special muscle strands from the outer surface of the proboscis sheath to the dorsal integument, the proboscis can be extended to its full extent so that the wall of the sheath is seen to become continuous with the anterior end of the head in the position of the mouth. A partial protrusion of the proboscis is only possible in the present case. Again a simpler condition exists in *A.* (*P.*) *sanguinolenta* (Stol. MS.). Here the proboscis is attached behind to the posterior end of the sheath surrounding it, just in front of the beginning of the oesophagus, so that after the proboscis has been fully protruded it drags from behind the sheath which then becomes gradually everted and forms a covering of the radular portion of the proboscis, so that the proboscis sheath becomes continuous with the proboscis in front.

Hence the present species shows an intermediate condition as regards the structure of its protrudable buccal bulb.

Minute structure of the proboscis.—The inner surface is raised into transverse folds: the ridges on the upper half fit into depressions on the lower half like the teeth on the blades of a pair of forceps. The inner surface of the organ is lined by a single layer of cubical epithelium which secretes a layer of hard homogenous cuticle, as thick as the cells themselves. On the outer side of the epithelium are placed the muscular layer, the bundles being arranged in various directions.

(ii) The *radula sac* is a club-shaped body curved somewhat like the letter S, the narrow end of which is attached to the posterior end of the proboscis. The sac is surrounded by a thin outer coat of muscular tissue within which is a thick muscular coat. Both these two coats are continued behind to form the retractor muscle. Beneath the thick inner coat and lying in the ventral and lateral aspect of the cavity of the sac, is a thick flap of muscular tissue which is attached to the inner surface of the muscular sheath behind and laterally, but projects anteriorly into the cavity about half the length of the sac from its posterior end; this free anterior end of flap gives attachment to the radula. Inside the muscular sheath and lying over the flap is a bilobed hollow fibrous cushion with a thick and hard wall. This cushion is free at its round anterior end and ventral aspect, but is attached to the muscular sheath behind and laterally towards the dorsal aspect. Ventrally between the two lobes is a deep transverse fissure into which the anterior end of the ventral flap fits. The dorsal surface of the cushion is convex in the middle line, but concave at the sides, where it is continuous with the inner surface of the muscular sheath. A little anterior to the middle of its length is a transversely placed crescentic aperture leading into the radula sac proper. The *radula* is attached to the anterior end of the ventral flap and passes over the anterior rounded end of the cushion to its dorsal surface over which it is traced backwards into the radula sac proper. The portion of the radula lying over the dorsal surface of the cushion extends laterally on the concave lateral portions of the surface and the inner surface of the muscular sheath, so that in a longitudinal section a little to one side of the middle line we get two sections of the radula—one lying on the dorsal surface of the sac, and another beneath the inner surface of the inner muscular sheath at a higher level than the first.

Minute structure of the radula and the radula sac proper.—The *radula* consists of the following layers :—

(1) A thin fibrous membrane lined beneath by a single layer of pavement epithelium. The membrane consists of white fibres alternating with single rows of connective tissue corpuscles.

(2) A single layer of cubical epithelium over the fibrous layer.

(3) A thick homogeneous corneous layer with fine longitudinal striation. To this are attached the bases of the teeth which are all unicuspid and are arranged in V-shaped rows.

The *radula sac* lies in the middle of the bilobed cushion in its dorsal aspect. The crescentic aperture (mentioned above) leads into the narrow cavity which is directed downwards and backwards, and which ends blindly after curving a little backwards and outwards. At the sides the cavity extends downwards and outwards and then upwards and inwards again for a short distance, where it ends blindly abutting on the wall of the hollow mass on the dorsal aspect. In a longitudinal section of the sac, a little to the side of the middle line, we get a sort of horse-shoe-shaped appearance as the cavity extends for some distance on the anterior aspect where the two portions become continuous. In a transverse section through the middle of the sac we get a reniform outline with the middle third of the convex side absent.

The sac is surrounded by a sheath of connective tissue. The ventral and the outer walls of the sac are thin; the lower surface of the radula is applied on these surfaces. A little in front of the posterior blind end of the cavity, the lower and outer walls of the sac is lined by a single layer of large granular cells with round or oval nuclei. This layer is continuous with the cubical epithelium of the radula. At the extreme posterior end of the cavity of the sac lies a mass of cells arranged obliquely and probably in several rows. These are placed on a thin layer of connective tissue, and seem to be continuous in front with the layer of granular cells just mentioned, while the connective tissue layer is continued in front to that of the radula.

The corneous layer on which the teeth are placed becomes suddenly narrowed down, just behind the point where the cubical epithelium ends, and is continued backwards as a thin layer to the tip of the cavity where it ends above the upper tiers of cell of the cellular mass just mentioned. The dorsal and inner lining of the cavity is convex and have the teeth of the radula embedded in them. The cavity of the sac is thus so narrow as to keep the radula between its two surfaces, there being no space left between the radula and the lining of the cavity of the sac. The postero-dorsal wall of the sac is thick and projects into the cavity of the sac. The base of the projecting mass consists of a curved stratum of connective tissue in front of which lies an oval mass of large muscular fibres, arranged transversely and separated widely from each other by connective tissue fibres and cells. Still in front lies an elongated mass of connective tissue cells embedded in a gelatinous matrix traversed by a few fine fibres, on the lower aspect of this wall lies a row of much elongated obliquely-placed cells, continuous round the blind end of the cavity to the cellular mass on the upper and outer aspect of the cavity. On the lower surface of this cellular layer are seen two or three transverse rows of teeth one before another and placed flatly on the homogeneous layer. In front of these rows, the teeth are arranged obliquely on the thick homogeneous layer between rows of cells continued to the posterior and inner wall of the sac, and filling up the space between the successive rows and probably between the individual teeth of the same row. The cells are probably concerned in the secretion of the teeth of the radula.

(iii) The two *salivary glands* are fused to form a single oval mass, but there are two salivary ducts which pass to their destination as usual.

(iv) The *oesophagus* ends in the substance of the liver. Its course is exactly similar to that in other species.

(v) The *digestive gland* is elongately conical in shape and rounded in front. It ends about 1·5 cm. in front of the posterior end of the mantle. The upper surface is convex from side to side; it presents the S-shaped curve of the rectum in its anterior portion about $\frac{1}{4}$th the length of the gland from this end. The cavity of the liver is a C-shaped slit in transverse section.

(vi) The *intestine* forms a S-shaped curve lying partially embedded in the substance of the liver on the dorsal aspect. It begins on the left lateral aspect of the liver and, taking the curve just mentioned, emerges out of its wall from near the anterior end to the left. It then passes along the anterior border of the liver forwards and outwards to the right, being surrounded by a coil of the oviduct in its course, to end in the anus in the groove between the foot and the mantle. The portion which lies beyond the liver may be conveniently named *rectum*.

IV. Reproductive System.

(i) The *hermaphrodite gland* is an oval mass—more or less flattened from side to side and placed on the right side of the anterior end of the liver. The organ is connected to the liver by a flat strand of fibrous tissue. The *retractor penis* muscle passes over the outer side of the gland.

Minute structure.—Under the low power the true glandular portion of the body lies in its distal end. It consists of a flattened mass of more or less rounded acini held together by loose connective tissue. Each acinus gives rise to a duct which unites with others to form the oviduct. The oviduct is coiled and looped in various ways as it passes on, and then emerges from the glandular mass after having received the vas deferens in the same. The vas deferens seems to arise from the centre of the glandular mass, and passes outwards nearly to its proximal end where it opens into the oviduct.

(ii) The *albumen gland* forms an elongated mass along the upper border of the hermaphrodite gland with which it is inseparably fused from the distal end.

Minute structure.—The gland consists of numerous irregular lobules separated somewhat widely from each other by loose connective tissue. Each lobule consists of an irregular mass of acini held together by a thin layer of connective tissue. The acini open together into a short duct which ends in the main duct of the gland. The main duct passes along the upper border of the hermaphrodite gland and then turns downwards between the glandular mass and the coiled mass of oviduct to the lower border, where it seems to open into the oviduct. It also receives several ducts of lobules scattered along its course.

(iii) The *hermaphrodite duct* is a short tube which forms a V-shaped loop as it emerges from the glandular mass and coils itself round the intestine to pass outwards, downwards and a little backwards, where it ends in the external genital aperture placed just behind and internal to the anus.

(iv) The *penis* is enclosed in a penial sheath, which opens on the outer side of the base of the lower tentacles. There is one *simrothian gland*, on the right side only. The *penial sheath* as usual gives attachment to a retractor penis muscle. A fine thread-like tubule, the *flagellum*, also opens into the penis, at its distal end.

The *penial sheath* consists of two portions: (1) A stout reniform mass containing the penis when the latter is fully retracted, the wall of the penis being continuous with that of the penial sheath at their distal ends. (2) A narrow tubular portion, which ends in the external aperture with the *right simrothian gland.*

(v) The *right simrothian gland* consists of two portions: (1) A narrow tubular portion coiled in various ways in its distal two-thirds. (2) A stout portion (also looped once) ending in the external aperture. At the junction of the two, is a fine tubular blind-sac (coecum) directed towards the outer end. The base of the sac gives attachment to a few muscle fibres. These correspond to the first and fourth portions of the simrothian glands of *A. (P.) sanguinolenta.*

V. *Nervous System.*

The general arrangement of the ganglia is similar to that in other species. As the system has not yet been studied in detail in other species, it is convenient to deal with them rather fully in the present species.

(i) *Cerebral ganglia.* Each ganglion contains numerous groups of ganglionic cells arranged correspondingly to the origin of nerves from it. Three such rows can be recognized as follows:—

(1) An elongated row of cell-group in the inner third of the ganglion along the whole length.
(2) A similar row in the outer third.
(3) A narrow elongated group in the middle third in its anterior fourth.

The nerves (C 1-5) from the cerebral ganglion:—

(1) A stout nerve on the inner side lying close to the nerve of the opposite side. It arises from the ventral aspect and supplies the ommatophore.
(2) A stout nerve dividing into three branches immediately after its origin. These supply the cephalic flap and its retractor muscle. One of these communicate with the buccal ganglion of the same side.
(3) A stout nerve from the antero-external corner of the ganglion; it supplies the outer side of the head.
(4) A fine nerve on the dorsal aspect of the nerve (3) supplying the dorsal integument of the head.
(5) Several small nerves on the outerside supplying the sides of the head and the muscular strands in connection with the proboscis and proboscis-sheath.
(6) A few fine nerves from the ventral aspect to the body wall at the base of the proboscis, one of which supplies the retractor muscle of the tentacle (C 6).

(ii) *Buccal ganglia.*—Each ganglion gives off (1) a number of nerves which spread over the proboscis and its radular portion; (2) one long nerve which passes along the oesophagus and serves to supply this portion of the alimentary canal. It gives off a fine nerve to the salivary gland.

(iii) *Viscero-pleural ganglia.*—The nerves (VP 1-5) are:—

(1) Two stout nerves arising side by side from the outer side. The outer one can be traced to the V-shaped process at the anterior end of the foot of the same side. The inner one seems to supply the anterior end of the foot. These two nerves arise from the oval group of ganglionic cells on the outer and anterior portion of the ganglia.

(2) A fine nerve arising from the outer aspect of the left ganglion at its posterior end. It passes along the gullet and ends in the liver a little behind the antero-inferior surface of the liver.

(3) A nerve from the right ganglion which supplies the female genital organs.

(4) A fine nerve from the right ganglion to the penis, penial sheath and the simrothian gland.

(5) Nerves to the side of the mouth above the pedal groove (VP 5).

(iv) *Pedal ganglia* (P 1-3).—The nerves are:—

(1) A nerve to the pedal gland.

(2) Nerves to the lateral wall of the mantle just above the groove round the foot.

(3) The long pedal cord which passes backwards along the dorsal surface of the foot to its posterior end. It gives off numerous nerves from its outer side to supply the foot.

VI. Eyes and Head Appendages.

The *eyes* are of rhipidoglossate type. Each forms a vesicle which is closed anteriorly forming a cornea composed of an external layer of epithelial cells, continuous with the tegumentary epithelium, an internal layer of epithelial cells (continuous with retina) and an intervening layer of transparent connective tissue. There is an oval crystalline lens with a surrounding layer of vitreous humour.

Ommatophore.—The cylindrical body of the ommatophores is hollow with a thick wall. Just behind the optic vesicle is a thin septum of connective tissue stretching transversely across the cavity. The wall consists of the following layers:—

(1) A single layer of cubical epithelium on the outer side.

(2) A thick layer of connective tissue with numerous cells. This layer contains some mucous glands placed at distant intervals. There are numerous pigment granules along the course of the connective tissue fibres.

(3) A layer of circular muscle fibres with a few radial fibres from the next internal coat.

(4) A layer of longitudinal muscle fibres.

VII. Pedal Gland.

(i) The *pedal gland* is a tubular body. The anterior portion is somewhat flattened from above downwards, while the posterior portion is triangular in transverse section.

Minute structure.—The pedal gland agrees in minute structure with that in *Atopos (Podangia) kempii*, Ghosh, except in the following points:—

(1) In the present species there is a blood-sinus on the dorsal aspect of the lumen of the gland in the middle line. Its wall consists of longitudinal muscle fibres bounded on the outer side by a layer of connective tissue.

(2) Owing to the interposition of a blood-sinus, the lumen of the gland comes to lie more or less in the centre and has become flattened out a little, instead of lying more towards the dorsal aspect and of being circular in transverse section as in *A. (P.) kempii*.

(ii) The *supra-pedal gland* is a small tongue-shaped body lying between the proboscis sheath and the pedal gland, and opening into the exterior just above the aperture of the pedal gland.

Minute structure.—The anterior portion forms a wide U-shaped cavity with the curve of the U continued in front to open into the exterior. The posterior two-thirds form a glandular mass, which consists of numerous lobules held together by connective tissue. Each lobule consists of a number of many-sided cells with spherical nuclei placed on one side. The ducts of these glands seem to open into the cavity of the body. On the outer aspects of the cavity, there are also numerous glands of similar structure with ordinary mucus-secreting cells in addition. The cavity of the gland is lined by a single layer of cubical cells. Immediately on the outer side of the epithelium is a thin layer of connective tissue which is prolonged outwards between the lobules.

LITERATURE

For references see my paper "Mollusca, I", in the Zoological Results of the Abor Expedition (*Records of the Indian Museum*, VIII, part III, No. 15).

EXPLANATION OF PLATE XVI.

Atopos (Parapodangia) gravelyi, n. sp.

FIG. 1.—Side view of the mature specimen (nat. size); ♂, male aperture.

,, 1*a*.—Side view of the smaller specimen (nat. size); the proboscis is protruded.

,, 1*b*.—Ventral view of the head.

,, 1*c*.—Side view of the head.

,, 2.—Transverse sections of the body; *a*, through the middle; *b*, about 1·4 cm. in front of the posterior end; *c*, about 6 cm. from the posterior end.

,, 3.—Inner surface of the anterior portion of the mantle; 1, attachment of the foot; 2, line of attachment of the dorsal integument to the mantle; 3, the dorsal blood sinus.

,, 4.—Pallial complex seen from the inner side, × 2. 1, rectum; 2, renal aperture; 3, ventricle; 4, auricle; 5, kidney; 6, pericardium.

,, 5.—Longitudinal section of the ventricle.

,, 6.—Longitudinal section of proboscis, diagrammatic.

,, 7.—Longitudinal section of the radular portion of the proboscis (a litcle to the side of the middle line); 1, wall of proboscis sheath; 2, 3, radula; 4, radula sac; 5, salivaiy duct.

1. 1a. 1b. 1a.

a. b. c.

2.

3. 6.

5.

4. 7.

E. N. Ghosh, del. A. Chowdhary, lith.

EXPLANATION OF PLATE XVII.

Atopos (Parapodangia) gravelyi, n. sp.

FIG. 8.—Longitudinal section of the radular portion of the proboscis (diagrammatic) in the middle line.

,, 9.—Transverse section of the same, *a-b*.

,, 10.—Dorsal view of the aperture of the radula sac.

,, 11.—A tooth of radula [Zeiss ocular vi, Leitz obj. 6].

,, 12.—Longitudinal section of radula sac. [Zeiss ocular vi, Leitz obj. 3].

,, 13.—A portion of the same, marked B roughly in fig. 12.

,, 14.—A portion of the same as in fig. 12, marked A.

,, 15.—Digestive gland, × 1. 1, oesophagus; 2, rectum.

,, 16.—Anterior portion of digestive gland, ventral view, × 1.

,, 17.—Transverse sections of the digestive gland at various levels shown in fig 15.

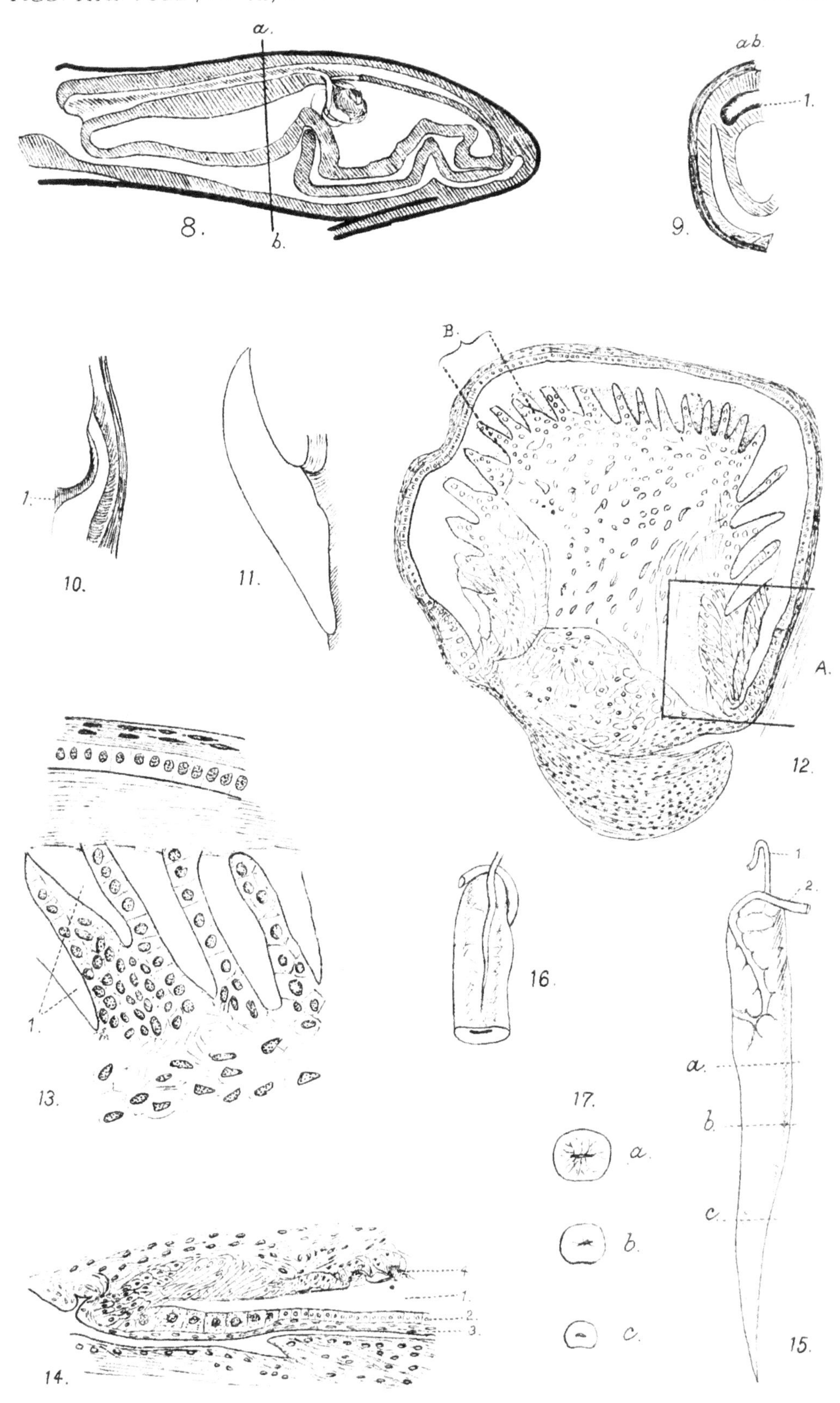

E. N. Ghosh, del. A. Chowdhary, lith.

EXPLANATION OF PLATE XVIII.

Atopos (Parapodangia) gravelyi, n. sp.

FIG. 18.—Proboscis, its radular portion (*a*), and salivary gland, (*b*), × 1.

,, 18*a*.—Ventral view of the proboscis with the ends of the salivary ducts, × 1.

,, 18*b*.—Longitudinal section of proboscis, × 2.

,, 19.—Salivary gland with the two ducts (*a*), × 1.

,, 20.—Genital organs; 1, hermaphrodite duct; 2, rectum; 3, hermaphrodite gland.

,, 21.—Inner surface of the gland.

,, 22.—Pedal gland, × 1.

,, 23.—Suprapedal gland, × 1.

,, 24.—Transverse section of the pedal gland, × 103.

,, 25.—Longitudinal section of the suprapedal gland.

,, 26.—A portion of the same marked *a-b* in fig. 25.

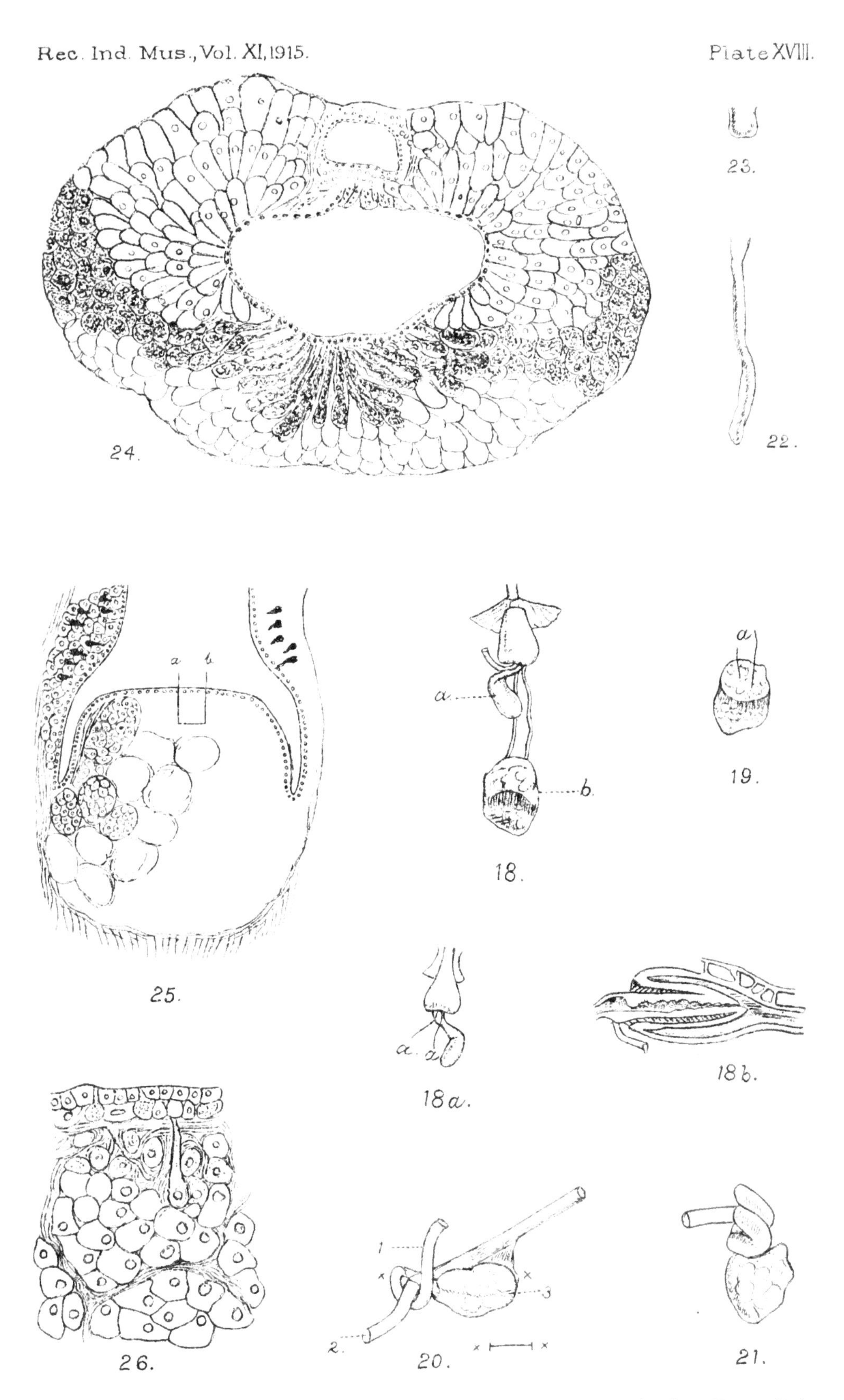

E N. Ghosh, del. A. Chowdhary, lith.

EXPLANATION OF PLATE XIX.

Atopos (Parapodangia) gravelyi, n. sp.

FIG. 27.—Male genital organs; 1, flagellum; 2, penis: 3, simrothian gland (right).

,, 28.—Diagrammatic enlarged drawing of the hermophrodite gland of the smaller specimen. *a*. Albumen gland: *b*, acini; *c*, coiled mass of oviduct; *d*, spermduct.

,, 29.—Nervous system; *a*, nat. size; *b*, enlarged drawing.

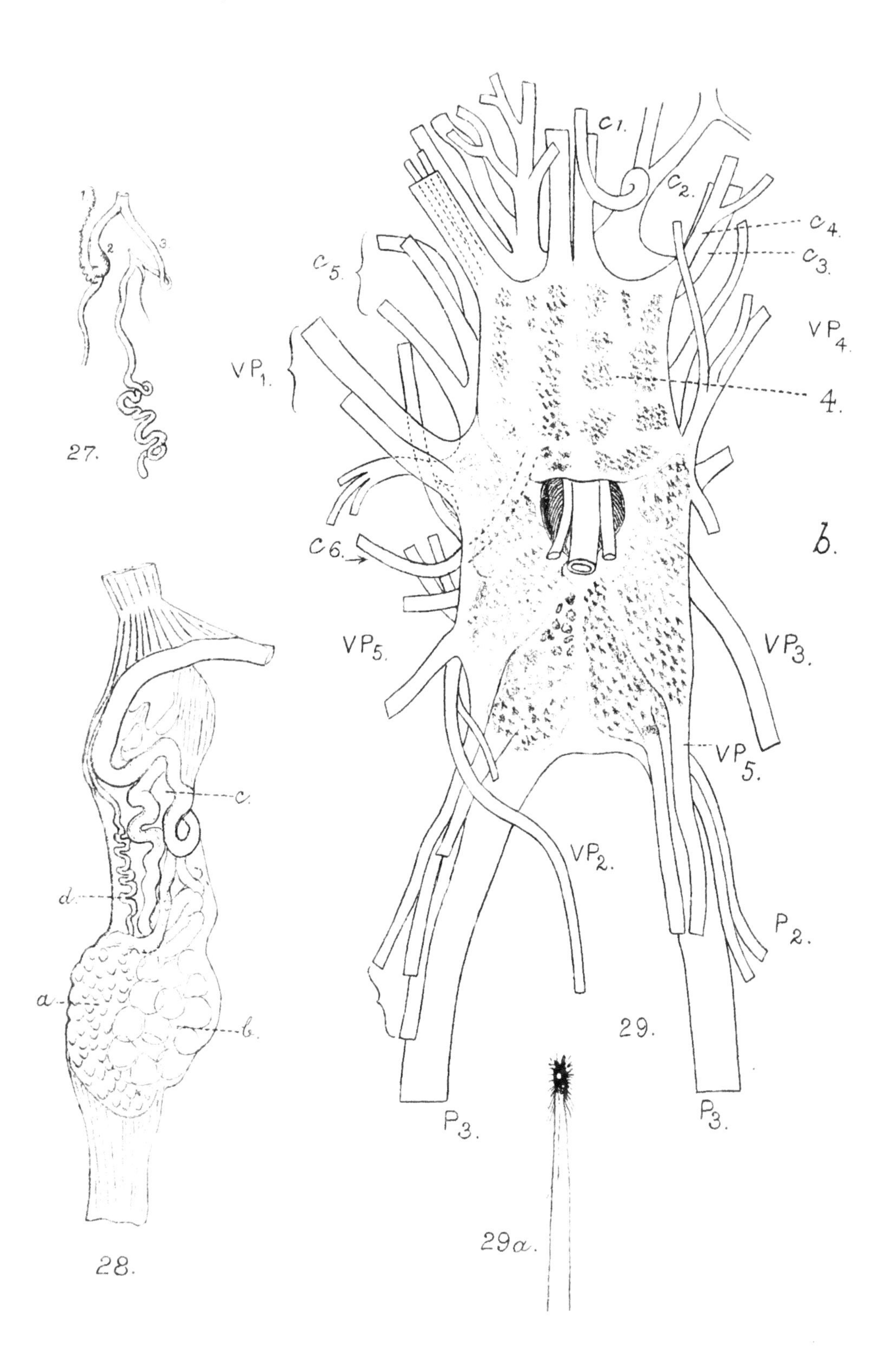

E. N. Ghosh, del.

A. Chowdhary, lith.

VIII. THE GENUS *AUSTRALELLA* AND SOME ALLIED SPECIES OF PHYLACTOLAEMATOUS POLYZOA.

By N. ANNANDALE, *D.Sc., F.A.S.B., Superintendent, Indian Museum.*

(Plates II, III).

I described the genus *Australella* in 1910[1] in a footnote to a paper on Indian Phylactolaemata, but the type-species, which was the only one then distinguished, was an Australian form (*Lophopus lendenfeldi*, Ridley[2]) known from the original description only. The type-specimen is in the British Museum, whence I have been able to obtain a fragment through the kind offices of Mr. R. Kirkpatrick. This schizotype was before me when I described the genus, but the shrivelled condition of the colony rendered it necessary to rely on Ridley's diagnosis and figures rather than on direct observation. Relying, therefore, on this description, I placed *Australella* in the subfamily Lophopinae. An examination of admirably preserved material of a new species leaves no doubt now that it belongs to the Plumatellinae, as is indicated by Kraepelin's[3] recent note on *Lophopus jheringi*, Meissner, which he regards as a congener.

We owe the discovery of the new species to Mr. Baini Prasad, Patiala Research Scholar in the Government College, Lahore, whose keenness as a collector and observer is already beginning to cast light on obscure places in our knowledge of the aquatic fauna of the Punjab.

The genus *Australella* may now be defined as follows:—

Genus **Australella**, Annandale.

Plumatellinae in which the colonies are recumbent and dendritic but enclosed in a uniform apparently structureless jelly that fills up the interstices between individual zooecia and branches. There is no stolon; the zooecia arise directly one from another. Individually they are semirecumbent, the proximal part of each resting, when the branch to which it belongs is fully formed, on the object to which the colony is fixed, while the distal part is almost vertical. The polypide

---

[1] *Rec. Ind. Mus.* V, p. 40 (1910).

[2] *Journ. Linn. Soc. London* (Zool.) XX, p. 62 (1890).

[3] Michaelsen's *Land und Susswasserfauna Deutsch-Sudwestafrikas* I, Bryozoa, p. 61 (1914).

is normal; it has some 40 to 60 tentacles, which are moderately or very long. The lophophore generally resembles that of *Plumatella.* The statoblasts are large (0·4 mm. to 1 mm. long), but as a rule smaller than those of the Lophopinae. They resemble the free statoblasts of *Plumatella* in structure and have neither marginal processes nor terminal prolongations.

Apart from the synoecial jelly, the structure of the colony in this genus is very like that of *Plumatella*, but the order of branching is not quite the same. In the younger parts each zooecium normally produces a single bud, but the precise stage at which this bud is produced differs in different species and even in different parts of the same colony; in *A. lendenfeldi* it probably does not appear as a rule until the mother-zooecium is well developed, whereas in *A. indica* it develops while the latter is still small. As a rule, in both species, it arises on the left and the right side respectively of alternating zooecia, so that a zig-zag stem is produced, consisting of a linear series of zooecia pointing alternately in different directions. As the colony grows older a secondary bud is often produced on the opposite side of the mother-zooecium to that on which the primary bud was formed. These secondary buds are the mother-zooecia of lateral branches that pursue a similar course to that of the stem from which they orginated, but at an acute angle to it. The figure may be further complicated by the production of secondary buds, and ultimately of secondary branches, from zooecia of the primary branches, and as a matter of fact this often takes place at an early stage in the development of the colony.

The result is the formation of a solid encrusting body closely compacted and agglutinated together by the synoecial jelly, but increasing in bulk mainly in one plane and without vertical branches.

It sometimes happens that branches or parts of branches die off or are killed by injury. In such cases the synoecial jelly remains intact. New branches may arise in vacant masses of jelly by budding from isolated fragments of the polyparium and are thus found separated from the remainder of the colony expect in so far as they are united by the jelly. This fact sometimes gives the whole structure the false appearance of being a compound colony like that of *Pectinatella.*

The genus *Australella* has now been found in Australia, India and South America.

Key to the species of AUSTRALELLA.

I. Synoecial jelly cartilaginous, scanty.
   Statoblasts oval, rounded at the ends .. *A. indica.*

II. Synoecial jelly soft, very copious.
   A. Statoblasts oval, subtruncate .. *A. lendenfeldi.*
   B. Statoblasts subcircular or polygonal *A. jheringi.*

Australella indica, sp. nov.

(Plate II.)

Zoarium.—The zoarium forms a massive, somewhat nodular structure growing round the stems of water-plants. It has an opaline gelatinous appearance and (preserved in formalin or spirit) a hard but elastic consistency. Even when the polypides are completely retracted the individual zooecia are distinctly visible, each having the appearance of being enclosed in a separate cell-like compartment. The surface, apart from the larger nodulosity, is otherwise smooth.

As is usually the case in the Plumatellinae, the precise organization of the colony is best seen in its terminal parts. There it is quite evident that the zoarium consists essentially of a main stem giving off lateral branches symmetrically in pairs, one branch at each side. The branches join the main stem at an acute angle and those that form each pair are given off almost simultaneously at the same level. The main stem is, as a whole, recumbent and adherent, but the lateral branches, although they are horizontal, at first run in the synoecial jelly, parallel to rather than in contact with the object to which the colony is attached. As they develop further, they become adherent and themselves give off lateral branches. Both the main stem and the main branches have actually a zig-zag course, because they are composed of zooecia which point alternately in two directions, this can only be seen clearly in the younger parts; for in the older parts interdigitation of the secondary branches takes place to such an extent that it is difficult to follow the course of any one branch, and the whole mass of zooecia seems to have a practically homogeneous honeycomb-like structure. Although the phase "main stem" is a convenient one, it must be understood that there are actually several or many stems of the kind in a single large colony such as the one figured on plate II, and that each is actually a unit or ray in a radiate dendritic whole.

The jelly which fills the interstices between the zooecia and between the stems and branches occupies a relatively small space. It is colourless and hyaline and, preserved in spirit or formalin, has the consistency of cartilage. I can detect no cells either in it or on its surface except, on the surface, those of unicellular algae. It is easily removed from the zooecia.

Zooecia.—The individual zooecia are distinctly J-shaped. The horizontal arm is more slender than the vertical one, which is sometimes constricted at its base in such a way that it assumes an outline like that of an egg-cup. The soft tissues are very delicate and easily torn and there seems to be no horny or other non-cellular layer between them and the jelly, which, indeed, is itself the homologue of such a layer.

Polypides.—The polypides closely resemble those of *Plumatella.* The lophophore is slender and bears between 40 and 50 slender and moderately elongate tentacles which have a narrow but distinctly festooned web at their base. The epistome is large and rather broad. The whole polypide is slender, the stomach particularly so. The colour of the latter in formalin is pale yellowish green.

Statoblasts.—The statoblast is of moderate size, elongate and rounded at the extremities; the capsule is broadly oval and often sometimes eccentric (fig. 1). The swim-ring is broad. It is remarkable for its vertical curvature (figs. 1*a* and 1*b*), surrounding the capsule like the rim of a dish in such a way that one surface of the statoblast is distinctly concave and the other convex, although the capsule itself is perfectly symmetrical. The convex surface is the one by which it is fastened to the funiculus.

The average length of the statoblast is about 0·46 mm. and the average breadth about 0·29 mm., the corresponding dimensions of the capsule being about 0·25 and 0·187 mm.; but owing

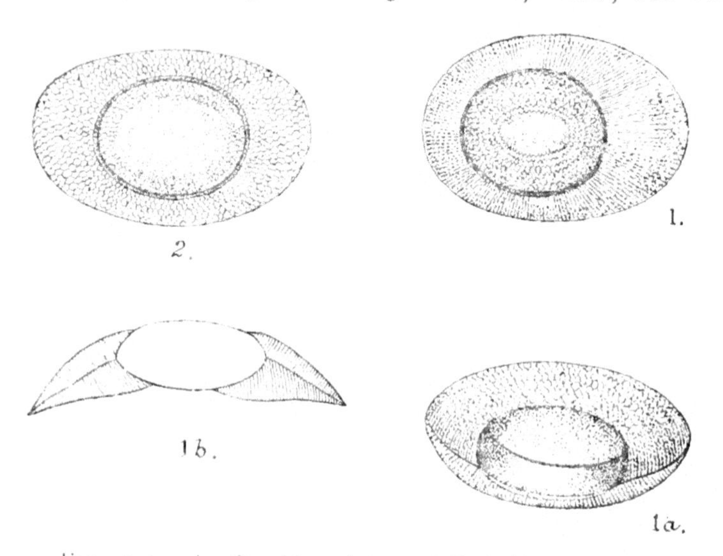

FIGS. 1, 1*a*, 1*b*.—Statoblast of *Australella indica*, sp. nov.
FIG. 2.—Statoblast of *Plumatella punctata* var. *longigemmis*, nov.

to the curvature of the rim it is difficult to obtain exact measurements.

Type.—No. 6629/7 Z.E.V., *Ind. Mus.* A cotype has been sent to the British Museum.

Locality.—Lahore, Punjab (12-x-14).

The points in which this species differs from *A. lendenfeldi* are discussed under the heading of that species.

In one of the specimens part of the colony is undergoing regeneration. The polypides have apparently been killed or injured but the jelly remains intact. New branches are arising from single polypides or pieces of polypides that have not perished. They consist of single or double rows of zooecia which have not yet produced lateral branches. The precise structure of such rows has already been discussed (p. 164).

One colony examined seems to have overwhelmed a colony of *Plumatella emarginata*. The latter, which has assumed the Alcyonelloid form, has managed to keep a small space clear for its more vigorous branches in the midst of the *Australella*.

Mr. Baini Prasad has given me the following notes on the occurrence of *A. indica* and on some species found in the same environment.

"On the occasion of a recent visit to Ferozpore I collected some material from the stagnant rainwater pools that abound on the banks of the river Sutlej mostly near the Kaiser-i-Hind railway bridge. The following representatives of the three classes, Sponges, Hydrozoa and Polyzoa were found.

SPONGES.

Spongilla carteri, Carter (Bowerbank *in litt.*).

Large masses of this sponge were found in two ponds, some measured more than a foot in length. Large dried-up masses consisting of spicules and gemmules only were found in another place where the water had quite dried up.

Spongilla lacustris subsp. *reticulata*, Annandale.

Large flat masses of this sponge of a bright green colour were found attached to the stems and leaves of *Potamogeton*. The pond was in an open place with no shade at all.

HYDROZOA.

Hydra oligactis, Pallas.

Only a single specimen of this form was found. It was attached to a *Potamogeton* leaf from the same pond in which *Spongilla lacustris* subsp. *reticulata* had been found. The specimen has only four tentacles and one bud, which shows the rudiments of the tentacles.

POLYZOA.

Australella indica, Annandale.

Large masses of this polyzoon were found in the same pond: these were covering the *Potamogeton* stems and leaves. Most of the individuals, however, were dead and large numbers of statoblasts had developed in them."

Australella lendenfeldi (Ridley).

1890. *Lophopus lendenfeldi*, Ridley, *Journ. Linn. Soc. London* (Zool.) XX, p. 62.

So far as I can judge from the specimen before me, this species differs from *A. indica* mainly in the following characters:—

1. The synoecial jelly is much softer and more copious.
2. The colony is broken up into a number of short branches lying separated in the jelly.
3. The tentacles of the polypide are longer and more numerous.
4. The statoblast is larger and has the sides more nearly parallel and the extremities subtruncate; the swim-ring is not curved in cross-section.

Ridley's statement that the different parts of the polyparium are joined together by a stolon is due to a misunderstanding: a stolon is indeed present at the base of the jelly, but it is that of a hydroid (*Cordylophora whiteleggi*, v. Lendenfeld), of which I have found a single hydranth projecting from the surface of the colony in the schizotype of the polyzoon.

Australella jheringhi (Meissner).

1893. *Lophopus jheringhi*. Meissner, *Zool., Anz.*, p. 290.
1914. *Australella jheringhi*, Kraepelin, Michaelsen's *Land und Süsswasserfauna Deutsch-Südwestafrikas* 1, Bryozoa, p. 61, pl. i, fig. 9.

I have not seen this species, which is only known from Brazil. It may be readily distinguished from the other two by its nearly circular statoblasts.

Genus **Plumatella**, Lamarck.

Plumatella punctata, Hancock.

1887. *Plumatella punctata*, Kraepelin, *Deutsch. Susswasserbryozoen* 1, p. 126 (numerous figures).
1911. *Plumatella punctata*, Annandale, *Faun. Brit. Ind., Freshw. Sponges*, etc., p. 227, p. 213, figs. 42 G and G' and pl. iv, fig. 5.
1914. *Plumatella punctata*, Kraepelin, *op. cit. supra*, p. 60, pl. i, fig. 10.

Since 1911 I have found this species fairly abundant, with *Fredericella sultana indica* and *Plumatella tanganyikae bombayensis*, in the canal at Cuttack in Orissa. Kraepelin has recently recorded its occurrence in South-West Africa.

var. longigemmis, nov.

(Plate III, fig. 2).

A closely allied form grows luxuriantly in a small pool of practically fresh water on Barkuda Island in the Chilka Lake (Ganjam district, Madras Presidency). It agrees with *P punctata* in every respect except that the gemmules are uniformly more elongate and have relatively smaller capsules than is usually the case (fig. 2, p. 166). They have the swim-ring slightly curved in cross-section, in this respect somewhat resembling those of *Australella indica,* though the feature is less marked. The average length of the gemmule is 0·42 mm., the average breadth 0·25 mm., but the difficulty in exact measurement noted in the case of *A. indica* (p. 166) also occurs with reference to them, though not to the same extent. The average measurements of the capsule are about 0·24×0·17 mm.

The colonies of this new variety were found on stones and rushes in July, 1914. They exhibited among themselves, so far as the zoarium was concerned, a complete transition between the two seasonal phases of the species found in Europe. In several instances the freshwater sponge *Spongilla alba*, Carter had already begun to grow over them; by November of the same year it seemed to have exterminated them altogether.

Genus **Stolella**, Annandale.

Stolella himalayana, Annandale.

(Plate III, fig. 1).

1911. *Stolella himalayana*, Annandale, *Faun. Brit. Ind., Freshw. Sponges,* etc., p. 246, fig. 49.

For some reason all reference to this species is omitted from the *Zoological Record*. I take the opportunity to reproduce an enlarged photograph of one of the types, a young colony from Malwa Tal in the Western Himalayas.

Stolella indica, Annandale.

1911. *Stolella indica*, Annandale, *op. cit.*, p. 299, fig. 45, pl. v, figs. 3, 4.

Professor K. Ramunni Menon has sent me specimens of this species from the town of Madras. It thus occurs in the Main Area of the Indian Peninsula as well as in the Indo-Gangetic Plain.

EXPLANATION OF PLATE II.

Australella indica, sp. nov.

FIG. 1.—The type-specimen, × 2.

,, 2.—Terminal part of the same specimen further enlarged.

,, 3.—Surface of a piece of the central part of the colony, showing the honeycomb-like arrangement of the zooecia.

,, 4.—Part of another colony in which regeneration of the branches is taking place in a mass of dead synoecial jelly.

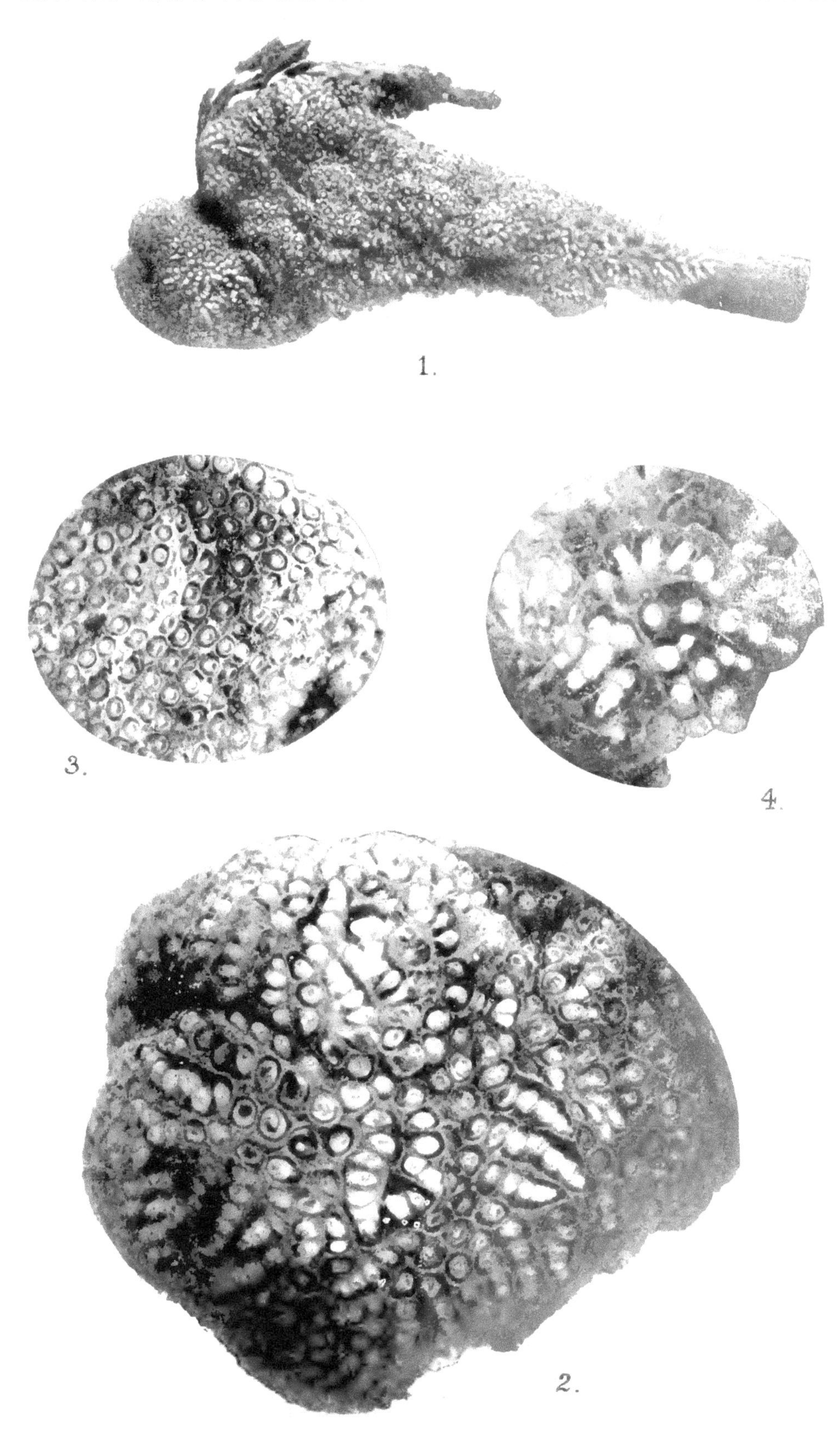

S. C. Mondul, *Phot.*

Australella indica.

EXPLANATION OF PLATE III.

FIG. 1.—*Stolella himalayana*, Annandale.
Part of one of the type-specimens, a young colony, enlarged. × = the terminal branch of a colony of *Fredericella sultana* (Blumenbach).

" 2.—*Plumatella punctata*, Hancock var. *longigemmis*, nov.
Part of the type-specimen, enlarged.

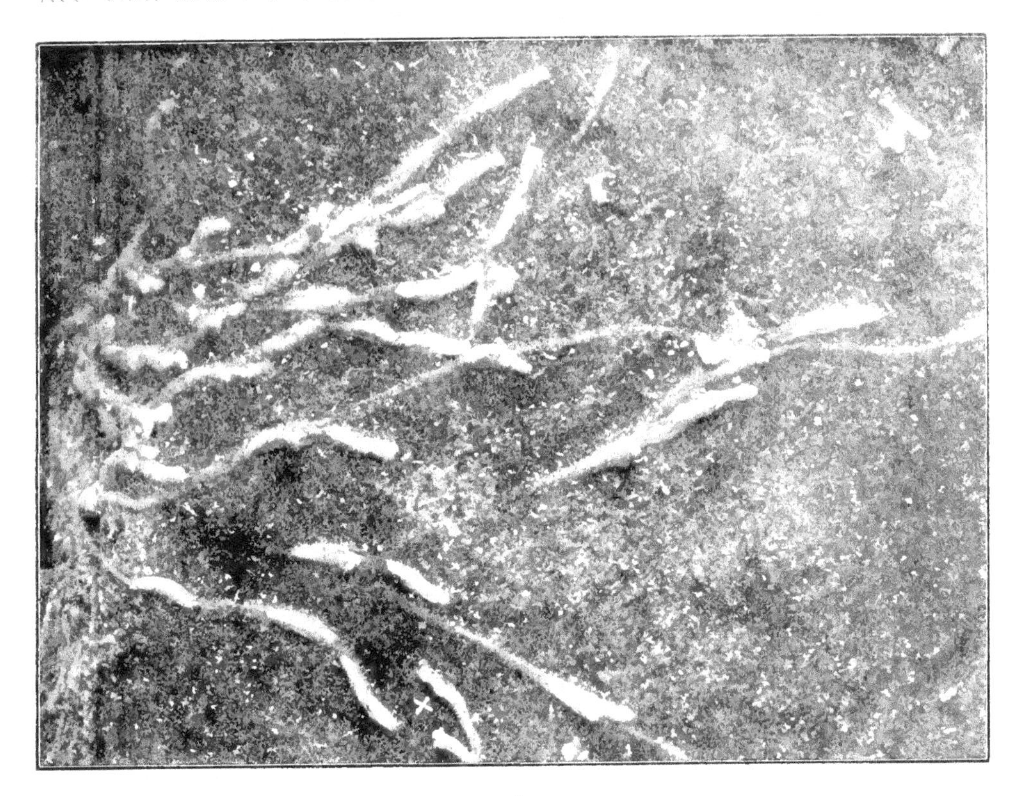

1. Stolella himalayana.

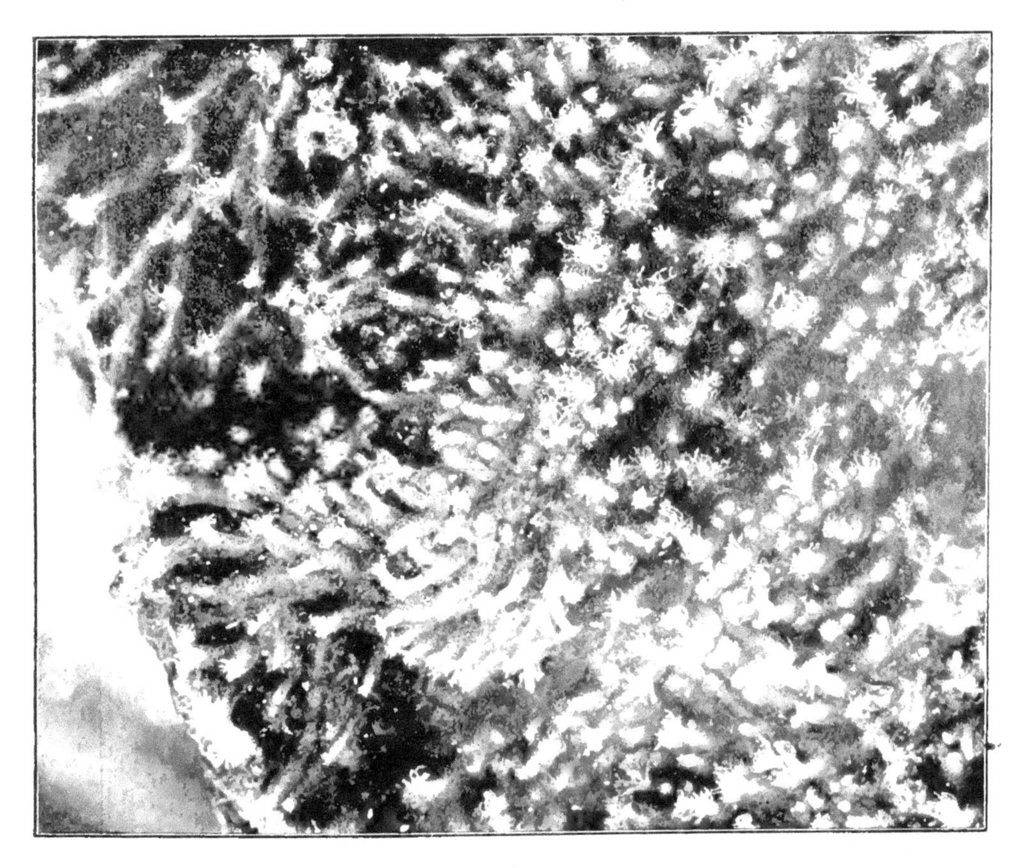

S. C. Mondul, *Phot.*

2. Plumatella punctata *var.* longigemmis.

IX. NOTES ON FRESHWATER SPONGES.

By N. ANNANDALE, *D.Sc., F.A.S.B., Superintendent, Indian Museum.*

No. XVI.—THE GENUS *Pectispongilla* AND ITS ALLIES.

The genus *Pectispongilla* was described in 1909 (*Rec. Ind. Mus.* III, p. 103) for the reception of a single species (*P. aurea*) from Travancore in the south-west of the Indian Peninsular Area; the subsequent account in the "Fauna" (*Freshwater Sponges*, etc., p. 106: 1911) added nothing to the generic diagnosis, but included the description of another form (*subspinosa*) from Cochin in the same part of India. This form was then regarded as a variety of *P. aurea*. The receipt of fresh material from Cochin has resulted in a re-examination of the original specimens and in the detection of an error in the generic diagnosis, *viz.* the statement that free microscleres were absent. It has also been found necessary to recognize at least three distinct species.

The genus may now be redefined as follows :—

> Small Spongillinae of massive or encrusting habit, of soft and friable consistency, with delicate skeletons in which the vertical fibres, though well-defined and not devoid of horny substance, are always very slender. Dermal membrane aspiculous. Skeleton-spicules rough or smooth amphioxi; free microscleres present in the flesh of the sponge, often of more than one type ; gemmule-spicules with the extremities flattened and expanded in the main axis, the terminal expansions bearing, on one face only, large spines arranged longitudinally in parallel comb-like rows.

Type-species.—Pectispongilla aurea, Annandale.

Geographical Distribution.—The plains of Travancore and Cochin in the southern part of the Malabar Zone of Peninsular India.

Affinities.—In the original description of the genus I suggested that the peculiar gemmule-spicule had been derived from that of *Ephydatia* by a rotation of the terminal rotules. Dr. W. Weltner wrote to me shortly afterwards expressing the opinion that this type of spicule had more probably been produced from one like that of *Spongilla* by a specialization of the extremities. A consideration of the form of the gemmule-spicules in the species of *Spongilla* most nearly related in general structure to *Pectispongilla* has induced me to accept Dr. Weltner's views. These species of *Spongilla* constitute a little group in the sub-

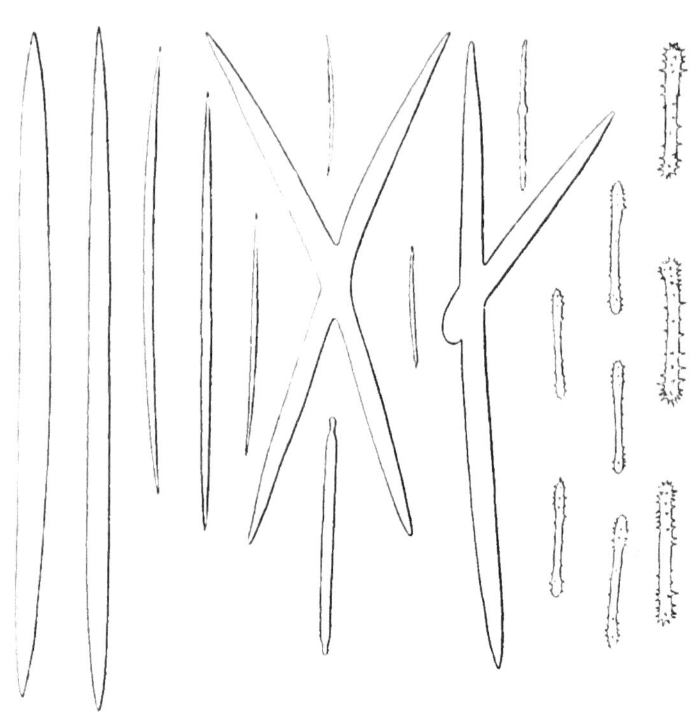

FIG. 1.—Spicules of *Spongilla hemephydatia*, Annandale.

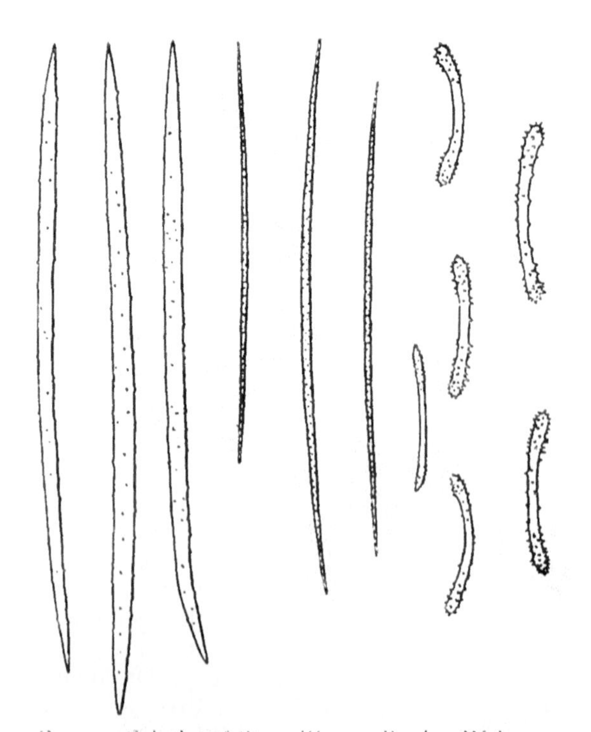

FIG. 2.—Spicules of *Spongilla zanzibarica*, Weltner.

genus *Euspongilla* typified by *S. crateriformis* (Potts), and distinguished from other members of the subgenus by the erect or semi-erect posture of all or most of their gemmule-spicules and by the fact that the terminations of these spicules are clearly specialized. The specialization may, however, take one or other of two directions, for the ends of the spicule may (as in *S. crateriformis* and the closely allied *S. biseriata*, Weltner[1]) bear an imperfect horizontal rotule of large recurved spines, or they may (as in my own *S. hemephydatia*, in *S. sansibarica*, Weltner[1] and apparently in Haswell's imperfectly known *S. botryoides*) be inflated, so that the spicule is technically tornote. The group is, therefore, of particular interest as representing the ancestral form, at any rate so far as the gemmule-spicule is concerned, of both *Ephydatia* and *Pectispongilla*.

I give figures here, in both cases from the type-specimens, of *S. hemephydatia* (fig. 1) and *S. sansibarica* (fig. 2). The tetraxon spicules that occur not uncommonly among the macroscleres of the type-specimen of the former are of course abnormal, but they have some interest as possible examples of reversion of the type by no means uncommon in the Spongillidae.

The general structure of the skeleton in *Pectispongilla* is identical in the different species and does not differ in any important feature from that found as a rule in *Euspongilla*. In particular it agrees closely with that which can be readily demonstrated by the use of pyrogallic acid in *S. hemephydatia*, *S. crateriformis* and *S. sansibarica*. Weltner (*op. cit.*, p. 127), indeed, states that the skeleton of the last species corresponds precisely with that of the Chalininae, in that the fibres are enclosed in a sheath of horny substance. That this substance is present in amount much greater than can be seen in unstained preparations or might be argued from the thinness of the fibres, is certainly a fact; but its arrangement seems to me to be quite different from that I recently demonstrated in *Lubomirskia*[2], for although it permeates the fibre, cementing together the component spicules and occupying the spaces between them, I can detect no external fibre-sheath. Where, as is often the case, it forms veil-like films at the nodes of the skeleton, it has the appearance of a perfectly homogeneous film.

The geographical distribution of *Pectispongilla* is peculiar. It is apparently the only genus of the Spongillinae that has so limited a range, for even *Asteromeyenia*,[3] which is confined, so far as we know, to the southern part of the United States of America, considerably surpasses it in this respect.

The two species of *Spongilla* most closely allied to *Pectispongilla* (*S. hemephydatia* and *S. sansibarica*) occur in the main area

[1] *Mitt. Naturh. Museum Hamburg* XV, pp. 121, 127, pl.—, figs. 1-5, 13-17 (1897).

[2] *Rec. Ind. Mus.*, X, p. 144, pl. ix, fig. 1*a* (1914).

[3] Annandale, *Proc. U.S. Nat. Mus.*, XL, p. 593 (1911).

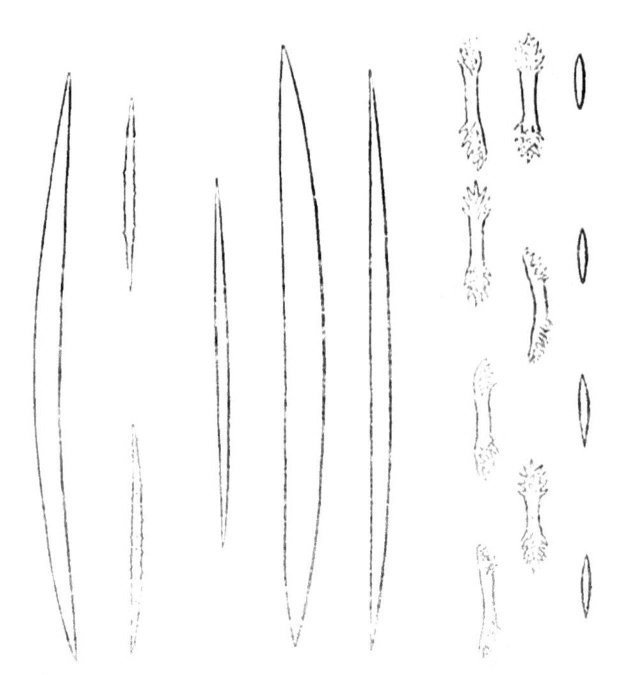

FIG. 3.—Spicules of *Pectispongilla aurea*, Annandale.

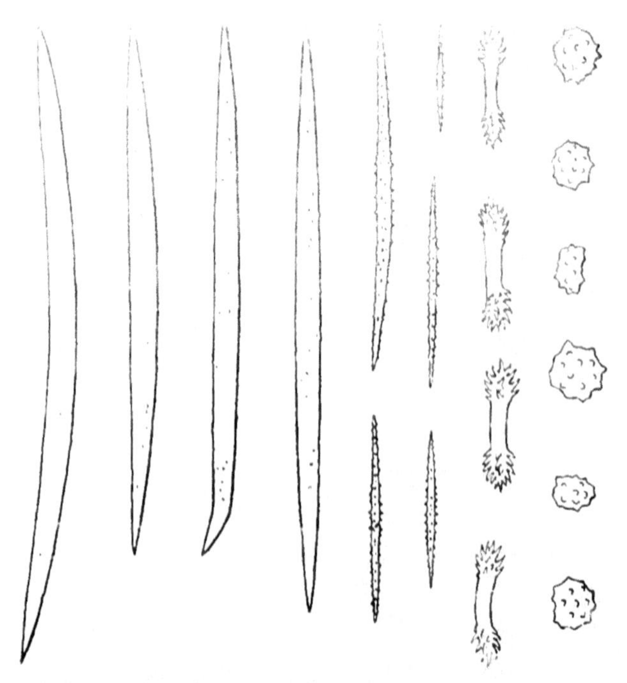

FIG. 4.—Spicules of *Pectispongilla stellifera*, sp. nov.

of the Indian Peninsula and at Zanzibar off the East Coast of Africa respectively, while *S. botryoides*, which may also be related, has been found only in New South Wales.

Key to the known species of PECTISPONGILLA.

1. Skeleton-spicules quite smooth.
   Free microscleres of two types: (*a*) minute, smooth, rhomboidal, and (*b*) moderately large, slender, spindle-shaped, bearing scattered spines .. *P. aurea.*
2. Skeleton-spicules rough or spiny.
   A. Free microscleres of two types: (*a*) amphioxous, spindle-shaped, somewhat closely spined, and (*b*) subspherical with scattered tubercles. *P. stellifera.*
   B. Free microscleres less distinctly of two kinds, spindle-shaped or cylindrical, amphioxous or truncated, all definitely spiny .. *P. subspinosa.*

Pectispongilla aurea, Annandale.

(Fig. 3).

1909. *Rec. Ind. Mus.*, III, p. 103, pl. xii, fig. 2.
1911. *Faun. Brit Ind., Freshw. Sponges*, etc., p. 106, fig. 20.

In describing this species I neglected to observe the free microscleres, or rather confused them with immature macroscleres and with the skeletons of diatoms. The free spicules are actually of two kinds: (1) small, slender, straight or nearly straight, spindle-shaped, sparsely spiny amphioxi on an average about 0·084 mm. long, and (2) minute, smooth, relatively thick amphioxi rhomboidal in outline and on an average about 0·024 mm. long. The former (1) are extremely scarce, the latter (2) abundant. Both types of spicules are confined to the flesh of the sponge. *P. aurea* is only known from Tenmalai on the western side of the Western Ghats in Travancore.

Pectispongilla stellifera, sp. nov.

(Fig. 4).

The sponge apparently forms thin films encrusting bodies such as the fibres of cocoanut-husks that have fallen or been thrown into the water, but my specimens are dry and not in very good condition. They have a brownish colour. The skeleton resembles that of *P. aurea*, but is rather stouter.

The macroscleres are slender and sharply pointed; they have minute rounded spines or tubercles scattered almost uni-

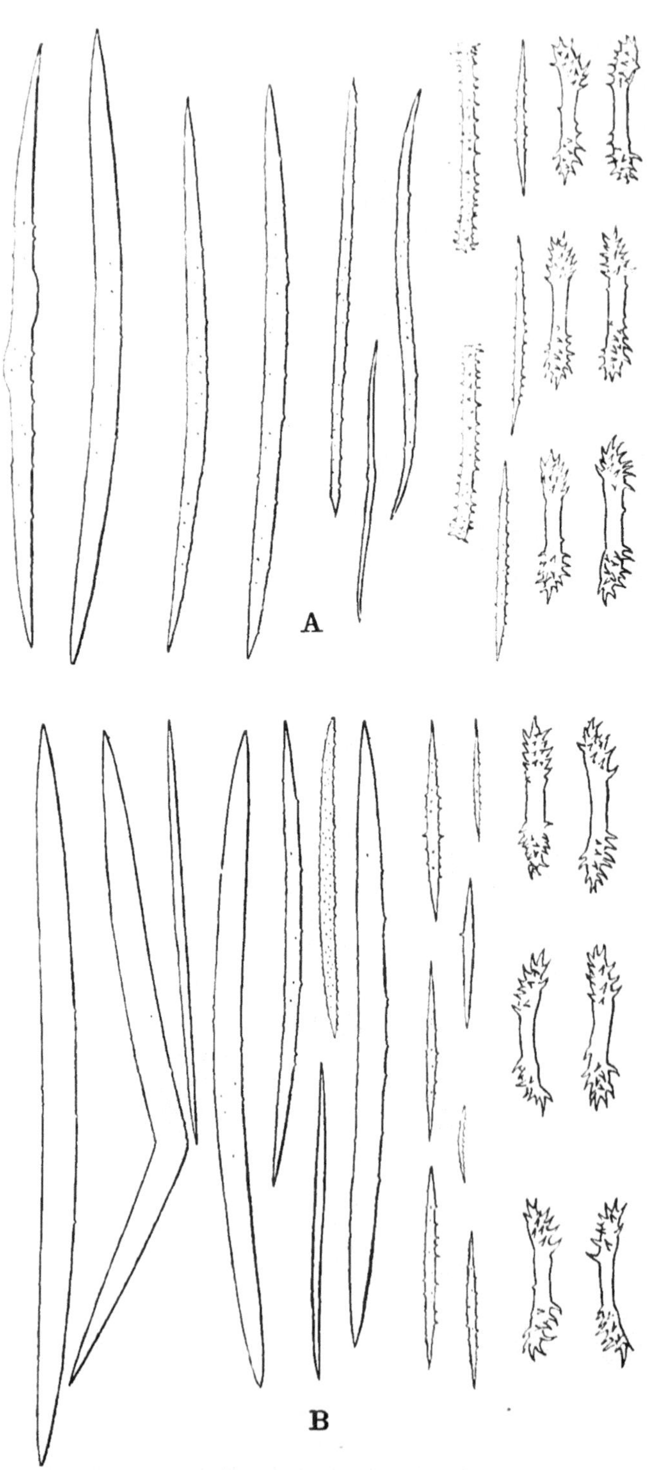

FIG. 5.—Spicules of *Pectispongilla subspinosa*, Annandale.

formly, though sparsely, over their surface, but their extremities are as a rule smooth.

The gemmule spicules resemble those of *P. aurea* but are a little stouter.

The free microscleres are of two quite distinct kinds: (1) slender, spiny, spindle-shaped, straight or nearly straight amphioxi, and (2) short, stout, cylindrical or subspherical tuberculate bodies of very characteristic form. The former vary greatly in size and proportions; their spines, which are scattered less sparsely than those of the macroscleres, are short and not very sharp. The free microscleres of the second type are, so far as I am aware, unique in the Spongillidae; their form is shown in the figure.

Diameter of gemmule	0·201	mm.
Length of macrosclere (average) ..	0·273	,,
Diameter of macrosclere (greatest average)	0·0084	,,
Length of gemmule spicule (average) ..	0·0336	,,
Length of free microsclere of type 1	0·0546—0·1554	,,
Diameter of free microsclere of type 1	0·0021—0·0063	,,
Length of free microsclere of type 2 (average)	0·0126	,,
Diameter of free microsclere of type 2 ..	0·0084	,,

Locality.—Trichur, Cochin State, Malabar Zone.
Type-specimen.—Z E.V. 3790/7, *Ind. Mus.*

Pectispongilla subspinosa, Annandale.

(Fig. 5, *A-B*).

1911. *Pectispongilla aurea* var. *subspinosa*, Annandale, *Faun. Brit. Ind., Freshw. Sponges*, etc., p 107.

This species is closely related to *P. stellifera*, with which it was at first confused, but lacks the aster-like microscleres characteristic of the latter.

The free microscleres are not so definitely separated in o two kinds as in the other two species of the genus but, in the type-specimen from Ernakulam at any rate (fig 5*A*), there are a few spicules that closely resemble the gemmule-spicules of *Spongilla crateriformis* in shape, being truncate at the extremities, and having rudimentary rotules thereat These spicules are, however, lacking in sponges recently obtained by Mr F. H Gravely at Trichur (fig 5*B*) in which the amphioxous free microscleres are also more variable. The truncate spicules may possibly be adventitious and until further specimens are obtained it seems inadvisable to separate the form discovered by Mr Gravely as a species or variety. His specimens, which were growing on rocks in a small pool connected with a sluggish stream, are (in spirit) of a dull brown colour and form an irregular crust some 2 to 5 mm. thick. The external apertures are small and inconspicuous, the subdermal cavity is relatively small and the external surface smooth. The skeleton resembles that of *P. stellifera.*

P. subspinosa is known only from Trichur and Ernakulam in the plains of Cochin.

X. LAGRIIDAE UND ALLECULIDAE DES "INDIAN MUSEUM."

Von F. BORCHMANN, *Hamburg.*

LAGRIIDAE.

1. Lagria ventralis, Reitt.

Deutsche Ent. Zeitschr. 1880, p. 255.

Viele Exemplare. Sikkim, E. Himalayas; Sukna, E. Himalayas, 500 ft.; Mazbat, Mangaldai distr., Assam; Dejoo, N. Lakhimpur, Upper Assam; Wan-hsaung, near Myitkyina, N.O. Burma, 600 ft; betw. Mongwan and Nan Tien, Yunnan.

2. Lagria hirticollis, Borchm.

Bull. Soc. Ent. Italiana 1909 (1910), p. 201.

Kawkareik, Amherst distr., L. Burma; Khayon, nr. Moulmein, L. Burma.

Die Art wurde von Pegu, Borneo, beschrieben.

3. Lagria concolor, Blanch.

Voyage au Pole Sud IV, 1853, p. 104, t. xii, f. 10.

Mazbat, Mangaldai, Assam; Assam-Bhutan Frontier, Mangaldai distr., N.E.; Burdwar, Nepal Terai.

Die Art ist sehr weit verbreitet.

4. Lagria ruficollis, Hope.

Gray's Zool. Misc. 1831, p. 32.

Viele Exemplare. Kurseong, 5000 ft., E. Himalayas; Bim Tal, 4500 ft., Kumaon; Pussumbing, Darjeeling, 4700 ft.; Siliguri, base of E. Himalayas; Ghumti, Darjiling distr., 4000 ft.

5. Lagria foveifrons, n. sp.

Länge: 6½—7 mm., Schulterbreite 2½—3 mm. Nach hinten schwach erweitert, mässig gewölbt, mässig glänzend, oben ziemlich dicht, unten spärlicher lang weisslich behaart; schwarz mit bronzenem und violettem Metallschimmer oder dunkelblau, Vorderkörper oben dunkelblau, Flügeldecken grünlich bronzefarbig, Beine schwarzblau, Fühler schwarz; Kopf rundlich, flach und undicht punktiert, Mundteile typisch, Endglied der Maxillartaster kurz und sehr breit; Oberlippe kurz, vorn etwas ausgerandet,

Vorderrand glatt, sonst dicht, fein und tief punktiert, lang beborstet; Clypeus stark quer, nach vorn verengt, Vorderrand so breit wie die Oberlippe, stark ausgerandet, von der Stirn durch eine starke, fast gerade Querfurche geschieden; Stirn im hintern Teile mit einem starken, hufeisenförmigen Eindrucke; Schläfen grob punktiert, etwa so lang wie ein Auge; Fühler schlank, die Schultern überragend, mit Ausnahme des 2 alle Glieder länger als breit, 3. Glied etwas länger als das 4., Endglied so lang wie das 10., spitz, 10. Glied wenig länger als breit; Augen stark ausgerandet, gewölbt, Abstand auf der Stirn gleich 1 Augendurchmesser von oben gesehen; Halsschild sehr wenig quer oder quadratisch, vorn etwas breiter als der Kopf mit den Augen, undicht mit flachen, tuberkelartigen Punkten besetzt, Vorderrand gerade, schmal gerandet, Hinterrand in der Mitte etwas eingezogen, breiter gerandet, Seiten vor der Basis eingezogen, Vorderecken abgerundet, Hinterecken etwas vortretend, Scheibe uneben vorn in der Mitte und vor der Basis beiderseits mit einem breiten, flachen Eindrucke; Schildchen kurz, rundlich, dicht punktiert; Flügeldecken doppelt so breit wie der Halsschild, Schultern nach vorn etwas vorgezogen, Scheibe dicht, grob und ziemlich stark querrunzlig punktiert, im 1. Viertel flach quer eingedrückt, Decken einzeln zugespitzt, etwas vorgezogen, Epipleuren breit, skulptiert wie die Flügeldecken; Unterseite sehr fein punktiert, Hinterleibsringe an den Seiten mit ringförmigen Eindrücken, Spitze des Analsegments rund, Intercoxalfortsatz des 1. Segments so lang wie breit, spitz, breit gerandet; Beine mittel, Schenkel wenig verdickt, Schienen schwach gebogen, Schienenspitze innen dicht gelb behaart, Hinterschenkel erreichen kaum den Hinterrand des 3. Segments, Metatarsus der Hinterfüsse so lang wie Glied 2 und 3 zusammen.

2 ♂♂ von Dibrugarh, N.E. Assam, 17—19-xi-1911. Die Art ist der *Lagria concolor*, Blanch. ähnlich, unterscheidet sich aber leicht durch die abweichende Färbung und durch den Mangel des quergerunzelten Eindruckes auf dem Halsschilde.

1 ♂ von N.O. Sumatra, Tebing-tinggi (gesammelt von Dr. Schultheiss) unterscheidet sich durch die viel schlankere Form und die stark abweichende Färbung: dunkelbraun mit starkem blauen Scheine, Basis der Oberschenkel braun, Vorderkörper grün bronzefarbig, Schildchen blau, Flügeldecken rötlich metallisch, Naht schmal grün. Ich benenne die Varietät **sumatrana**, n.v.

6. Lagria nigrita, n. sp.

Länge: 8 mm. Form wie *L. concolor*, Blanch., Flügeldecken hinter den Schultern etwas flachgedrückt: mässig glänzend, lang, abstehend, weisslich behaart; tiefschwarz, Spitze des letzten Hinterleibssegmentes rot; Kopf gewöhnlich, grob, undicht unpunktiert; Oberlippe stark quer, gewölbt, ausgerandet; Clypeus ebenfalls stark quer, nach vorn verengt, stark ausgerandet, von der Stirn durch eine tiefe, gerade Furche getrennt; Stirn mit hufeisenförmigem Eindrucke, Schläfen länger als ein Auge, Hals deutlich; Augen schmal, stark ausgerandet, weit getrennt; Fühler mittel, die Schultern überragend, nach aussen etwas verdickt, Glieder kürzer werdend, Grundglied dick, 3. Glied etwas länger als das 4. Endglied so lang wie Glied 9 und 10 zusammen, stumpf zugespitzt; Mundteile gewöhnlich. Halsschild so breit wie der Kopf mit den Augen, fast quadratisch, mässig grob, nicht dicht, etwas querrunzlig punktiert, in der Mitte eine rundliche Fläche mit feinerer und dichterer Punktierung, Vorderrand nicht, Hinterrand deutlich gerandet, Seiten in der Mitte eingeschnürt, Vorderecken abgerundet, Hinterecken stumpf, Scheibe in der Mitte an der Seite beiderseits mit einem Quereindrucke, Schildchen gross, rundlich, fein und dicht punktiert; Flügeldecken zweimal so breit wie der Halsschild, ziemlich dicht und grob runzlig punktiert, Schultern etwas eckig gefaltet, Decken einzeln zugespitzt, Epipleuren breit, skulptiert wie die Flügeldecken, vor der Spitze nach aussen gewendet; Beine schlank, Schienen wenig gebogen; Metatarsus der Hinterfüsse so lang wie die folgenden Glieder zusammen; Unterseite feiner punktiert, Seiten gröber; Fortsatz des Abdomens breit, zugespitzt, gerandet.

4 ♀♀ von Burdwar, Nepal Terai, 1-i-1910; Thamaspur, Nepal, 18—20-ii-1908; Noalpur, Nepal, 21-ii-1908; Paresnath, W. Bengal, 4000 ft., 9-iv-1909.

Die Art ist nahe verwandt mit *L. concolor*, Blanch., unterscheidet sich aber leicht durch die Farbe und die abweichende Halsschildbildung.

7. Cerogria nepalensis, Hope.

Gray's Zool. Misc. 1831, p. 32.—Dohrn, Stett. Ent. Zeit. XLVII, 1886, p. 354.

Viele Exemplare. Kurseong, E. Himalayas, 5000 ft.; Cheerapunji, Khasi Hills, Assam.

8. Cerogria basalis, Hope.

Gray's Zool. Misc. 1831, p. 32.—Dohrn, Stett. Ent. Zeit. XLVII, 1886, p. 353.

Kurseong, E. Himalayas, 5000 ft.; Karak, 3000 ft., 10-iii-1912.

9. Cerogria quadrimaculata, Hope.

Gray's Zool. Misc. 1831, p. 32.

Siliguri, base of E. Himalayas; Kurseong, E. Himalayas, 4700-5000 ft.; Pussumbing, Darjiling, 4700 ft.; Mazbat, Mangaldai, Assam.

10. Cerogria flavicornis, Borchm.

Bull. Soc. Ent. Ital. 1909 (1910), p. 210.

Shan Hills, Upp. Burma.

11. Lagriocera cavicornis, Fairm.

Ann. Soc. Ent. Belg. XL, 1896, p. 41.

1 Exemplar. Shan Hills, Upper Burma.

12. Nemostira hirta, n. sp.

Länge: 6½ mm. Gestreckt, nach hinten sehr wenig erweitert, mässig glänzend, undicht, sehr lang, abstehend braun behaart; glänzend schwarz, Flügeldecken mit grünlichem Metallschimmer, Fühler braunschwarz. Kopf gestreckt; Oberlippe gewölbt, quer herzförmig, fein punktiert, lang beborstet, vorn nicht ausgerandet; Clypeus nach vorn verengt, gewölbt, fast glatt, an der Basis mit wenigen Borstenpunkten, nicht ausgerandet, von der Stirn durch eine gebogene Furche getrennt, Stirn grob, zerstreut punktiert, zwischen den Augen mit einem tiefen Quereindrucke, Hinterkopf glatt, Schläfen länger als ein Auge, grob punktiert, beborstet, Hals dünn; Fühler gleich der halben Körperlänge, kräftig, alle Glieder mit Ausnahme des 2. länger als breit, Glieder gegen die Spitze etwas breiter, 3. Glied kürzer als das 4., Endglied dünn, etwas gebogen, länger als die 3 vorhergehenden Glieder zusammen; Augen schmal, stark gewölbt, vorn wenig ausgerandet, Stirnabstand bedeutend grösser als ein Augendurchmesser: Mundteile typisch. Halsschild kaum breiter als der Kopf mit den Augen, so lang wie breit, grob und ziemlich dicht punktiert, beborstet, gewölbt, Vorderrand gerade, fein gerandet, Hinterrand in der Mitte etwas eingezogen, breiter gerandet, Seiten gerundet, grösste Breite in der Mitte, vor der Basis eingezogen, Vorderecken stumpf, Hinterecken vortretend, Seitenrand geschwunden. Schildchen länglich, abgerundet, fast glatt. Flügeldecken doppelt so breit wie die Halsschildbasis, mit groben Punktreihen, Streifen wenig vertieft, Punkte gegen die Spitze flacher, Zwischenräume fast eben, Schultern vorgezogen, Spitzen abgestutzt, Nahtwinkel zähnchenartig, Epipleuren schmal, verkürzt, ausgehöhlt. Unterseite lackartig glänzend, Seiten der Brust grob punktiert; Intercoxalfortsatz der Vorderbrust so hoch wie die Hüften, gerandet, in der Mitte vertieft, nicht hinter die Hüften verlängert; Beine kräftig, Schenkel keulig, Hinterschenkelspitze fast die Spitze des Hinterleibes erreichend, Seiten des glatten Abdomens mit flachen Eindrücken, Intercoxalfortsatz des 1. Segments abgerundet, schmal gerandet, die Mitte mit einer starken dreieckigen Erhebung, die einen Längskiel bildet, 2. Segment mit einer ähnlichen Bildung; Schienen schwach gebogen; Metatarsus der Hinterfüsse fast so lang wie die folgenden Glieder zusammen.

1 ♂. Siionbari, North Lakhimpur (base of hills), Upper Assam (*H. Stevens*), 31-v-1911.

Die kleine Art unterscheidet sich leicht von ihren Verwandten durch die eigentümliche Bildung der beiden ersten Abdominalsegmente.

13. **Nemostira ceylanica**, n. sp.

Länge: 8½-9 mm. Form und Grösse der *N. terminata*, Fairm. und der *Casnonidea brevicollis*, Fairm.; mässig glänzend, gewölbt, nach hinten wenig erweitert, ziemlich dicht, lang, abstehend, rotgelb behaart; schwärzlich braun, Basis der Beine wenig heller, Kopf, Halsschild und Vorderbrust und die Flügeldecken bräunlich rotgelb, Flügeldecken mit schwarzen Epipleuren, schwarzer Naht und je einer grossen schwarzen Makel, die das letzte Viertel einnimmt; Kopf rundlich, zerstreut punktiert; Oberlippe quer, flach ausgerandet, fein punktiert; Clypeus quer, nach vorn verengt, von der Stirn durch eine gebogene Furche getrennt, breit und flach ausgerandet; Stirn mit 2 grösseren Punkten zwischen den Augen, Scheitel mit Grübchen; Augen gross, ausgerandet, Abstand auf der Stirn gleich ½ Augendurchmesser (♂); Schläfen kurz, Hals sehr deutlich; Fühler gleich der halben Körperlänge, kräftig, nach aussen etwas verdickt, nicht gesägt, 3. Glied gleich dem 4., Endglied etwas kürzer als die 3 vorhergehenden Glieder zusammen (♂), beim ♀ kürzer; Endglied der Kiefertaster schmal, aber nicht messerförmig; Halsschild etwas länger als breit, von typischer Form, gerstreut punktiert, etwas breiter als der Kopf mit den Augen; Schildchen klein, rundlich; Flügeldecken doppelt so breit wie die Halsschildbasis, mit starken Punktstreifen, Punkte in den Streifen gross, dicht, rund, nach hinten feiner, Zwischenräume gewölbt, mit zahlreichen feinen Borstenpunkten, Spitzen der Flügeldecken zusammen abgerundet; Unterseite feiner, die Seiten grob punktiert; Abdominalfortsatz kurz, rund, breit gerandet; Prosternalfortsatz breit, gerandet, hinter den Hüften nicht erweitert; Beine kräftig, lang behaart, ohne Geschlechtsmerkmale; Metatarsus der Hinterfüsse wenig kürzer als die folgenden Glieder zusammen.

2 Exemplare Paradeniya, Ceylon, 8-viii und 18-v-1910.

Die Art ist mit den oben genannten nahe verwandt, unterscheidet sich aber leicht schon durch die Färbung.

ALLECULIDAE.

1. **Allecula indica**, Borchm.?

Deutsche Ent. Zeitschr. 1909, p. 714.—Fairm., Ann. Soc. Ent. Belg., XL, 1896, p. 38 (brachydera Fairm.).

1 Exemplar. Kurseong, E. Himalayas, 4700-5000 ft.

2. **Allecula arthritica**, Fairm.

Ann. Soc. Ent. France LXII, 1893, p. 36.

Sikkim, E. Himalayas, v—vi-1912.

3. **Allecula femorata**, n. sp.

Länge: 12½-14½ mm.; Breite: 3½-4 mm. Gestreckt, ♂ schlanker als ♀, ♂ nach hinten verengt, ♀ nicht; dunkel pechbraun,

Mundteile rötlich, Fühler hell rötlich, gegen die Spitze dunkler, Beine gelblich, Kniee und Füsse dunkler; mässig glänzend; ziemlich dicht, anliegend, nicht sehr dicht gelblich behaart. Kopf verlängert, Mundteile vortretend; Oberlippe quer, dicht und ziemlich fein punktiert, mit langen gelben Borsten, vorn stark ausgeschnitten, Seiten gerundet, gegen die Basis verengt; Clypeus ebenso skulptiert, Vorderrand gerade, so breit wie die Basis der Oberlippe, von der Stirn durch eine gerade Furche getrennt; Stirn ebenso skulptiert wie die Oberlippe; Schläfen

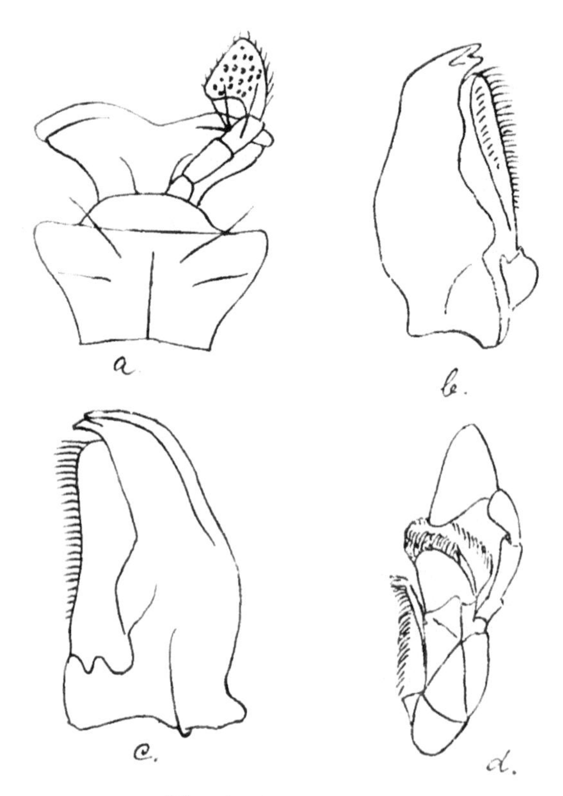

Allecula femorata, n. sp.
a. Unterlippe und Mentum; *b*. Mandibel von unten; *c*. Mandibel von oben; *d*. Maxille.

sehr kurz, Hals durch eine scharfe Furche abgesetzt; Augen gross, gewölbt, ausgerandet, Abstand auf der Stirn ½ Augendurchmesser ♀, ♂ noch geringer. Mandibeln zweispitzig, schlank, mit kräftigem Sacke; innere Lade der Maxillen sehr klein, äussere gross, auf der Oberseite filzartig behaart, Endglied der Taster quer, dreieckig; Unterlippe stark quer, vorn ausgeschnitten, Vorderecken spitz, Seiten gegen die Basis stark verengt, Endglied der Taster dreieckig; Mentum stark quer, vorn fast gerade, Vorderecken scharf, Seiten zweimal gebogen verengt; Fühler wenig kürzer als der Körper, dünn, fadenförmig, Glieder nehmen nach der Spitze an Länge ab, 3. Glied kaum länger als das 4; Halsschild breiter als der Kopf mit den Augen, wenig kürzer als breit, wenig gewölbt dicht und ziemlich fein punktiert, nach vorn verengt, nahe der Basis oder an der Basis jederseits nahe der Ecke mit einem flachen Grübchen, zuweilen auch die Scheibe nahe der Mitte mit 2 Grübchen, Vorderrand schmal, Hinterrand etwas breiter gerandet, flach zweibuchtig, in der Mitte etwas eingedrückt, Vorderecken gerundet, Hinterecken fast rechtwinklig, Seiten etwas gerundet, Seitenrand vollständig. Schildchen klein, rundlich. Flügeldecken an den Schultern kaum breiter als der Halsschild, mit Punktstreifen, Punkte grob, gegen die Spitze schwindend. Streifen kaum flacher, Schulter schräge, Spitzen zusammen abgerundet, Zwischenräume der Streifen etwas gewölbt, sehr dicht, sehr fein punktiert; Epipleuren schmal, an der Spitze nach aussen gewendet. Abdomen ziemlich dicht und fein, die Seiten der Brust gröber punktiert, Analsegment ♂ gerundet; Fortsatz der Vorderbrust hinter den Hüften etwas vorragend, stumpf. Beine kräftig, lang, Schienen gebogen, 2-4. Glied der Vorder- und Mittelfüsse, 3. Glied der Hinterfüsse gelappt, 1. Glied der Vorderfüsse beim ♂ stark erweitert und unten ausgehöhlt.

1 ♂, 2 ♀♀. Kurseong, E. Himalayas, 4700-5000 ft., 24-vi-1910 (*Annandale*).

Die Art ist durch die Behaarung leicht kenntlich.

4. Allecula sukliensis, n. sp.

Länge: 14 mm. Körper etwas kräftiger als bei der vorigen Art, nach hinten fast gleichmässig erweitert, mässig gewölbt, wenig glänzend, überall fein, kurz, anliegend, gelb behaart; pechbraun, Vorderkörper, Schieldchen und Flügeldecken rötlich, Fühler und Beine gelb, Schenkelspitze und Tarsen gebräunt. Kopf gewöhnlich, fein und dicht punktiert; Oberlippe stark quer vorn weniger ausgerandet als bei der vorigen Art, beborstet, Clypeus quer, vorn gerade, von der Stirn durch eine breite, schlecht begrenzte Furche getrennt, Schläfen sehr kurz, Hals dick; Augen schmal, gewölbt, stark ausgerandet, Abstand auf der Stirn geringer als ein Augendurchmesser; Fühler fadenförmig, (beide beschädigt), 3. Glied kürzer als das 4.; Endglied der Lippentaster nach innen stark eckig erweitert. Halsschild bedeutend breiter als der Kopf mit den Augen, leicht quer, fein, mässig dicht punktiert, vorn und hinten fein gerandet, etwas 2 buchtig, Mitte schwach vorgezogen, Scheibe mit schwacher Mittelfurche, Vorderecken gerundet, Hinterecken fast rechteckig, Seiten nach vorn schwach gerundet verengt, Seitenrand deutlich. Schildchen kurz, rundlich. Flügeldecken nicht doppelt so breit wie die Halsschildbasis, mit Punktstreifen, Punkte ziemlich fein, nach hinten schwindend, Zwischenräume wenig gewölbt, mit sehr feinen Borstenpunkten, Schultern schräge, Spitzen einzeln abgerundet, Epipleuren erst sehr breit, dann stark verschmälert; Unterseite stark glänzend, fein und dicht, Seiten der Brust grob punktiert, Analsegment gerundet, Intercoxalfortsatz des 1. Segments schmal, spitz, gerandet. Hinterbrust zweilappig vorgezogen, Fortsatz der Vorderbrust wie bei der vorigen Art. Beine kräftig, Schenkel keulig, Vorder- und Mittelschienen gebogen, Füsse breit, an den Vorder- und Mittelfüssen Glied 1–4 gelappt (an den Mittelfüssen Glied 1 schwach), an den Hinterfüssen deutlich nur Glied 2 und 3. Metatarsus der Hinterfüsse etwas kürzer als die folgenden Glieder zusammen.

1 ♂ von Sukli, Ostseite der Dawna Hills, 2100 ft., 22—29-xi-1911, gesammelt von Herrn F. H Gravely.

Die Art unterscheidet sich von ihren Verwandten durch die Bildung der Füsse, der Hinterbrust und der Lippentaster.

5. Allecula sobrina, n. sp.

Länge: 11 mm. Form wie *Allecula femorata*, wenig gewölbt, wenig glänzend, ziemlich lang, abstehend, undicht gelblich behaart; dunkelbraun, Flügeldecken rotbraun, Beine gelb, Spitzen der Schenkel und Basis der Schienen oder die ganzen Schienen dunkelbraun, Fühler rotbraun, die einzelnen Glieder an der Spitze dunkler. Kopf gewöhnlich, ziemlich dicht, nicht grob punktiert; Oberlippe kürzer, weniger ausgerandet als bei *A. femorata*, Augen schmäler, weniger genähert, Endglied der Lippentaster weniger breit; Fühler fadenförmig, (beide beschädigt) 3. Glied so lang wie das 4; Halsschild so lang wie breit, glanzlos; Schildchen kurz, rundlich; Flügeldecken mit stärker eingestochenen Punkten in den Streifen; Beine wie bei der genannten Art, aber an den Vorderfüssen Glied 1 erweitert, unten gekielt, Glied 1–4, an den Mittelfüssen Glied 3–4 und an Hinterfüssen Glied 2 und 3 gelappt; Metatarsus der Hinterfüsse länger als Glied 2 und 3 zusammen; das Übrige wie bei *A. geniculata* m.

1 ♂ von Kurseong, Ost-Himalaya, 4700–5000 ft., 21-vi-1910, gesammelt von Annandale.

6. Borboresthes suturalis, n. sp.

Länge: 6–7 mm. Form etwas schlanker als *B. fuliginosus*, Fairm.; Brust braun, Bauch hellbraun, Beine, Fühler und Flügeldecken hell gelbbraun, letztere mit schmaler dunkler Naht und schmalem, dunklem Seitenrande, Vorderkörper rotbraun: mässig glänzend, mässig gewölbt; fein, anliegend, wenig dicht, ziemlich kurz gelb behaart. Kopf fein punktiert; Oberlippe stark quer, vorn gerade, Clypeus ebenso, von der Stirn undeutlich geschieden: Augen gewölbt, weit getrennt, Schläfen sehr kurz; Fühler halb so lang wie der Körper, fadenförmig, 3. Glied kürzer als das 4., Endglied kürzer als das 10.; Halsschild stark quer, Vorderrand gerade, Hinterrand zweimal gebuchtet, Seitenrand scharf, Seiten gerundet verengt, Hinterecken fast rechtwinklig: Schildchen rundlich; Flügeldecken mit starken Punktstreifen, die hinten nicht flacher werden, Punkte schwinden gegen die Spitze, Spitzen zusammen abgerundet; Epipleuren schmal, ganz, glatt. Unterseite und Beine gewöhnlich: Metatarsus der Hinterfüsse viel länger als die folgenden Glieder zusammen.

2 Ex. Kurseong, E. Himalayas, 4700–5000 feet 24-vi-1910 (*Annandale*). Unterscheidet sich durch die Färbung und die schlankere Gestalt leicht von ihren Verwandten.

7. Cistelopsis (?) aborensis, n. sp.

Länge: 7 mm. Form typisch. Rotbraun; mässig glänzend; anliegend, ziemlich dicht gelblich behaart. Kopf verlängert, ziemlich fein und dicht punktiert; Oberlippe und Clypeus stark quer, vorn gerade; Clypeus von der Stirn durch eine wenig gebogene Furche getrennt; Augen stark gewölbt, ausgerandet, Abstand auf der Stirn gleich einem Augendurchmesser: Fühler typisch, so lang wie der halbe Körper, 3. Glied halb so lang wie das 4., Endglied kürzer als das 10. Glied; Endglied der Kiefertaster fast messerförmig, der Lippentaster dreieckig: Form des Halsschildes typisch, rings um scharf gerandet, dicht mit Nabelpunkten besetzt, Basis in der Mitte breit vorgezogen, Hinterecken etwas abgerundet; Schildchen breit, spitz; Flügeldecken mit feinen Punktstreifen, die an der Spitze vertieft sind, Zwischenräume flach, dicht, etwas querrissig punktiert, Epipleuren mässig breit, vollständig; Unterseite feiner, an den Seiten gröber punktiert. Prosternalfortsatz hinter den Hüften nicht verlängert; Abdominalfortsatz spitz; Hinterhüften hinten scharf gerandet. Beine gewöhnlich; das vorletzte Tarsenglied aller Füsse gelappt; Metatarsus der Hinterfüsse so lang wie die folgenden Glieder zusammen.

1 Exemplar. Kobo, Abor country, 400 ft., 8-xii-1911 (gesammelt von Mr. Kemp), unter Baumrinde.

Cistelopsis rufa, n. sp.

Länge: 7 mm. Form sehr ähnlich wie *C. validicornis*, Fairm., aber etwas schlanker, die Schultern weniger gerundet; rotbraun. Hinterleib dunkel, letztes Fühlerglied gebräunt; kurz, anliegend, gelb behaart; Halsschild fein, wenig dicht, nicht runzlig punktiert, Hinterrand in der Mitte breit vorgezogen, sonst gerade, Hinterecken rechtwinklig; Schildchen dreieckig; Flügeldecken wenig erweitert, sehr fein punktiertgestreift, Streifen nicht vertieft, nur gegen die Spitze 3 vertiefte Streifen, nicht querrunzlig, Spitzen zusammen abgerundet; Epipleuren schmal; Metatarsus der Hinterfüsse so lang wie die folgenden Glieder zusammen. Alles Übrige wie bei *C. validicornis*, Fairm.

1 Ex. von Pattipola, Ceylon, 6000 ft., 13-x-1911.

8. Cistelomorpha andrewesi, Fairm.

Ann. Soc. Ent. Belg. XL, 1896, p. 58.

Phagu, Simla Hills, 9000 ft.

9. Cistelomorpha alternans, Fairm.

Ann. Soc. Ent. Belg. XXXVIII. 1894, p. 40.

Kurseong, E. Himalayas, 4700–5000 ft.; Sikkim, E. Himalayas; Mungphu, Darjiling distr.

10. Cistelomorpha trabeata, Fairm.

Ann. Soc. Ent. Belg. XXXVIII, 1894, p. 40.

Bangalore.

XI. NOTES ON SOME INDIAN CHELONIA.

By N. Annandale, *D.Sc., F.A.S.B., Superintendent, Indian Museum.*

The majority of the specimens mentioned in these notes have been sent me by Dr J. R. Henderson of the Madras Museum, who has been at great pains, in this and other respects, to assist us in the Indian Museum with specimens from the Madras Presidency.

Family TRIONYCHIDAE.

Trionyx leithii, Gray.

1873. *Trionyx leithii*, Gray, *Proc. Zool. Soc. London*, p. 49, fig. 3 (skull).
1873. *Trionyx gangeticus*, *id.*, *ibid.*, pl. viii.
1889. *Trionyx leithii*, Boulenger, *Cat. Chel. Brit. Mus.*, p. 249.
1912. *Trionyx leithii*, Annandale, *Rec. Ind. Mus.*, VII, p. 150, fig. 2 (plastron of young).

Boulenger regards this species as intermediate between *T. gangeticus*, which it somewhat resembles in colouration, and *T. hurum*, with which the shape of its skull is to some extent in agreement; but the structure of the skull is nearer that of *T. formosus*. The only differences that I can detect are that the snout is a little less declivous and slightly longer, the horizontal groove on the palate broader, the post-cranial spine less dilated and the proximal articular part of the lower jaw more slender in *T. leithii* than in the Burmese species. The former has much the same relationship to the latter as *T. nigricans* has to *T. phayrei*, and the existence in *T. leithii* of two neural bones between the first pair of costals is a modification probably of slight importance though it serves to separate all the Indian forms from their Burmese allies. The branchial skeleton of *T. leithii* also resembles that of *T. formosus*, in particular in that two additional ossifications are present at the tip of the hypobranchial bone. In the adult animal the hypoplastra approach one another in the middle line of the plastron, though they do not actually meet, and the internal projections practically disappear.

The natural colouration and the external appearance of the disk do not appear to have been observed in the adult living animal. The following notes are based on two individuals which Dr. Henderson has been kind enough to send to Calcutta for examination.

The dorsal surface of the disk is greenish black obscurely vermiculated and marbled with olive-green; that of the limbs and tail is also blackish, while that of the head is variegated with dark olive-green and a much paler olivaceous brown. In one specimen[1] dark green predominated on the head and the paler markings were not of a very definite nature, but on that of the other (fig. 1) there was a much larger proportion of the paler shade, a dark line extended backwards and downwards from each eye and there were two distinct forwardly-directed **V**-shaped black bars on the temporal and occipital regions, interrupted somewhat at their apices in the middle line; the ends of the bars extended backwards more faintly on to the neck.

The cartilaginous disk is long and relatively narrow, expanding slightly behind. In front of the bony carapace there is a conspicuous projecting pad of coarsely tuberculate cartilage, and

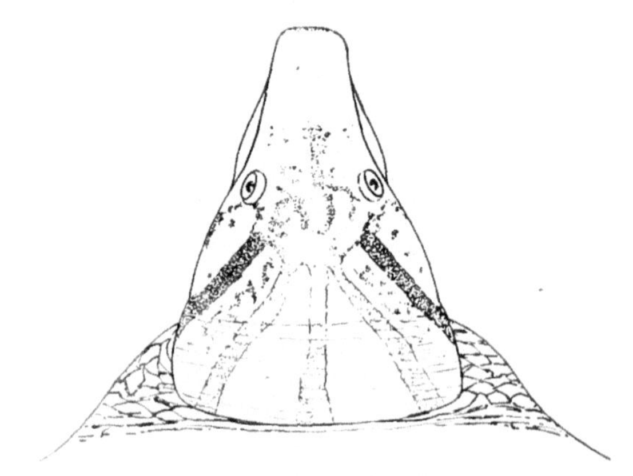

FIG. 1.—*Trionyx leithii*, Gray.
Head of a living specimen from the Kurnool district (× ½).

behind there is a group of large tubercles in the central region. The anterior part of the carapace itself bears a prominent rounded boss; there is no middorsal ridge or groove

The disk of the larger of the two specimens is 49 cm. long and 41 cm. broad, that of the smaller one 47 cm. × 42 cm. In the former the breadth of the bony carapace, which is broadly emarginate behind, is 29 cm. and the breadth 32 cm. The length of the skull, which agrees well with Gray's figure of 1873 except that the lower jaw is not cleft at the tip, is 91 mm. and the zygomatic breadth 58 mm.

Dr. Henderson's specimens, of which the larger has been retained in the Indian Museum and the smaller returned to Madras, were taken by his assistant Mr. Sundara Raj in a small stream in

---

[1] The snout had been injured in this specimen and possibly the dark colouration was to some extent due to inflammation or congestion.

the Nallamalai range of the Eastern Ghats, where they were dug from the mud in the bed of a pool in January or February.

The evidence for the occurrence of this species in the Ganges or the Indus is not satisfactory. Murray's specimens from Sind assigned provisionally to it by Boulenger, apparently on the grounds of probability only, were almost certainly representatives of *T. gangeticus*, while those from which Hardwicke's Ms. figures (reproduced by Gray in 1873) were drawn, though said to be from "Futtegurh", may have been either introduced or ascribed to an incorrect locality. All definite records of specimens now in existence refer to places in the Indian Peninsular Area south of the Indo-Gangetic Plain.

Trionyx hurum, Gray.

1912. *Trionyx hurum*, Annandale, *Rec Ind. Mus.*, VII, pp 160, 180, pl. v, fig. 3.

Mr. F M. Howlett Imperial Pathological Entomologist, has recently sent me a young specimen taken in the Little Gundak River near Pusa in Bihar Its colouration is normal agreeing closely with that of a young turtle from Dacca in Eastern Bengal. It is thus clear that the normal *T hurum* has made its way in the Gangetic system far above the delta.

Family TESTUDINIDAE.

Testudo travancorica, Boulenger.

1906. *Testudo travancorica*, Boulenger, *Journ. Bombay Nat. Hist. Soc.*, XVII, p. 560, 2 pls

Mr. F H. Gravely recently obtained further specimens in the forests on the western slopes of the Western Ghats in Cochin while Mr. F. Hannyngton, I.C.S., has presented to the Indian Museum one from Coorg on the eastern side of the Ghats. The known range of the species may, therefore, now be stated thus:—Travancore and Cochin on the western slopes of the Western Ghats and Coorg on the eastern slopes. It is probable that the tortoise also occurs on the western side of the hills in the western districts of the Madras Presidency, and also in parts of Travancore and the adjacent districts situated on the eastern side, but no records as yet exist. The common land-tortoise over the greater part of the Presidency is certainly *T. elegans*, which as Mr. Sundara Raj informs me, occurs in the Eastern Ghats in the Kurnool district

Geoemyda trijuga (Schweigg.)

1913. *Geoemyda trijuga*, Annandale, *Rec. Ind. Mus.*, IX, p 67.

It is probable that the range of the typical form of this species is confined to the east-central part of the Madras Presidency, but specimens from Mysore, the northern part of the Presidency and other parts of the Peninsular Area must be examined before this question is settled. I have here to describe a new race from Coorg on the east side of the Western Ghats considerably east and a little south of the Madras district.

The following key to the known races may be useful:—

Key to the races of G. trijuga.

I. Head with conspicuous red or yellow markings.
   A. Temporal regions pale yellow, snout black; shell uniform black or very dark brown .. .. .. *coronata.*
   B. Whole of the head black with orange-red streaks and spots; shell dark brown plastron as a rule bordered with yellow .. .. .. *thermalis.*

II. Head without conspicuous red or yellow markings.
   A. Head plumbeous grey, obscurely vermiculated on the temporal regions; shell brown, the plastron with yellow borders . .. *plumbea.*
   B. Head olivaceous with inconspicuous yellowish or greenish spots and streaks; shell brown or blackish, the plastron as a rule with yellow borders. *forma typica (madraspatana).*
   C. Head uniform brown or with an obscure reticulation of olive-brown and orange-yellow; shell of adult black, the plastral borders and the dorsal keels yellow .. .. .. *edeniana.*

Subsp. **plumbea**, nov.

The carapace is dark brown and when wet appears to be obscurely marbled with a still darker shade; there are as a rule yellow markings on each or some of the costal shields and the dorsal ridges are for the most part tinged with yellow. The central part of the plastron is dark brown with broad yellow margins; the bridge is dark brown with yellow spots along the outer edge. In life the dorsal surface of the limbs, head and tail is leaden grey; that of the snout and of a triangle extending backwards from the eyes has a brownish tinge and is devoid of markings. The temporal region is obscurely vermiculated with a paler shade of grey and two pale lines extend backwards from the eye above the tympanum, which is somewhat darkened; the beak and the ventral surface of the head, neck and limbs are pale grey; on the chin and neck there are obscure dark horizontal lines.

After some weeks in spirit the markings on the head have become obscure and the whole has a livid greyish tinge very different from the colour of that of specimens of the typical form of the species that have been even longer in alcohol.

The iris is pale chestnut. There are a number of small tubercles on the side of the head between the eye and the tympanum.

I can detect no constant peculiarity in the skull or in the shell, except that the dorsal keels appear to be blunter than in specimens of the same size from Madras. Possibly this is correlated with the fact that the race is a very small one and that shells of small size are therefore more worn and belong to older individuals than their dimensions would suggest. The concentric rings on the dorsal shields are, however, very distinct.

	Reg. No. 17712 (sk.)	Reg. No. 17715(sp.)
Carapace.		
Total length with the callipers .	155 mm.	162 mm.
Total length with the tape ..	170 ,,	175 ,,
Total breadth with the callipers	108 ,,	113 ,,
Total breadth with the tape ..	125 ,,	144 ,,
Depth of the shell .. .	60 ,,	57 ,,
Plastron.		
Total length with the callipers.	139 ,,	145 ,,
Length of the bridge ..	58 ,,	60 ,,
Skull.		
Total length	38 ,,	..
Zygomatic breadth	23 ,,	..

Types. No 17712 (skeleton) and No. 17715 (spirit), *Ind. Mus.*

I have examined three living specimens which Dr. Henderson has been kind enough to send me. They were collected by his assistant Mr. Sundara Raj in a pond. One has been returned to the Madras Museum, one skeletonized and one preserved in alcohol. All are apparently adult females of approximately the same size: they are very uniform as regards their racial characters.

Subsp. **coronata** (Anderson).

1913. *Geoemyda trijuga coronata*, Annandale, *op. cit.*, p. 68, pl. vi, figs. 3, *3a*.

It is strange that there is no reference to this very distinct race in the "Fauna", but, to judge from the labels on specimens in the British Museum, it seems possible that it was regarded by Dr. Boulenger as the fully adult or possibly aged phase of the typical Madras form, to which Anderson gave the name *madraspatana*.

We have received from Dr. Henderson specimens of this race from Chalakudi in Cochin and from a locality about 25 miles N.E. of Calicut in the Malabar district of the Madras Presidency.

The Malabar specimen is a large female, of which the measurements are given below. The shell, dry, is practically black all over. The central dorsal keel remains distinct throughout its length, but the lateral keels are obsolescent. The colour of the head was typical though slightly less brilliant than in smaller individuals.

Carapace.

		Reg. No. 17437 (sk.)
Total length with the callipers	..	.. 233 mm.
,, ,, ,, ,, tape	..	.. 260 ,,
Total breadth with the callipers	..	.. 158 ,,
,, ,, ,, ,, tape	..	.. 215 ,,
Depth of the shell ..	..	.. 81 ,,

Plastron.

Total length with the callipers	..	.. 200 ,,
Total length of the bridge	..	.. 85 ,,

Subsp. **thermalis** (Lesson).

1913. *Geoemyda trijuga thermalis*, Annandale, *op. cit*, p. 68, pl. vi, figs. 4, 4*a*.

Further specimens of this race were recently obtained in the Ramnad district by Dr. Henderson and Mr Kemp

Geoemyda tricarinata, Blyth

1913. *Geoemyda tricarinata*, Annandale, *op. cit.*, p. 73, pl. vi, figs. 6, 6*a*, 6*b*.

In a footnote to the paper cited (p. 74) I have recorded the occurrence of this species in the Jalpaiguri district of northern Bengal. Possibly it is one of those Assamese reptiles whose western range along the base of the Himalayas has been limited or practically limited by the R. Tista. If so, its occurrence in Chota Nagpur is all the more remarkable.

Geoemyda silvatica, Henderson.

1912. *Geoemyda silvatica*, Henderson, *Rec. Ind. Mus.*, VII, p. 217.

The type-specimen of this species has been presented by the Madras Museum to the Indian Museum. It is now preserved in spirit and is numbered 17115 in our register of reptiles. A good watercolour sketch of the living animal was made by Babu Abhoya Charan Chowdhary and is available for reference.

Bellia crassicollis, (Gray)

1906 *Bellia crassicollis*, Annandale, *Journ. As. Soc. Bengal* (n.s.) II, p. 205.

The specimen said to be from Travancore and referred to in the paper cited had, it is now evident, suffered from an accidental transposition of labels. There is, therefore, no reason to think that this Malayan species occurs in South India. The authentic in the collection of the Indian Museum are from Burma and Penang.

XII. NOTES ON ORIENTAL DRAGONFLIES IN THE INDIAN MUSEUM.

By F. F. LAIDLAW.

No. I. THE GENUS *OROGOMPHUS*.

Order ANISOPTERA.

AESCHNIDAE.

CHLOROGOMPHINAE.

So far as at present known only the genus *Orogomphus* occurs in the Indian area. The other genus of the family, *Chlorogomphus* is found in Sumatra, Java and Tonkin, whilst *Orogomphus* has representatives in Bengal, Burma and the Himalayas, as well as in Borneo, the Philippine Islands and Formosa. The subfamily is the only one confined to the Oriental region.

Four species of *Orogomphus* are known. They are—

Orogomphus atkinsoni, de Selys. Bengal, Assam.
,, *speciosus*, de Selys. Burma, Darjiling.
,, *splendidus*, de Selys. Philippine Islands, Borneo, Formosa.
,, *dyak*, Laidlaw. Borneo.

For figures of *O. splendidus* see Ris in *Supplementa Entomologica*, No. I, 1912: text-fig. 15 *a-b*; taf. iii, fig. 1-6; taf. v. fig. 5 In this paper Dr. Ris also discusses the venation and characters of the male previously unknown (*loc. cit.* pp. 77-79).

The wings of a female presumed to belong to this species, collected in Borneo, are figured by myself (*Proc. Zool. Soc. London*, 1914, pl. i, fig. 8). Selys's original account of the type female from Luzon is given in his 4me additions. Synops. Gomph *Bull. Acad. Roy. Belg.* xlvi (2), 1878, pp. 681-682.

Orogomphus atkinsoni, de Selys.

O. atkinsoni. Selys, 4me add. *Synops. Gomph.*, p. 682 (1878).
Kirby, *Cat. Odonata*, p. 79 (1890).
Selys, *Ann. Mus. Civ. Genova* (2) x, pp. 481-482 (1891).
Williamson, *Proc. U.S. Nat. Mus.* xxxii, pp. 278-279, fig. 5-6 (1907).
Laidlaw, *Proc. Zool. Soc. Lond.*, pp. 61-62 (1914).

I ♂ $\frac{5114}{20}$ I ♀ $\frac{5413}{20}$. Both specimens are named; the male is without a locality, the female from Sibsagar, Assam. The specimen recorded by me (*loc. cit.*) is from Kumaon, de Selys's type is from 'Bengal.' He was not acquainted with the male at the date of publication of the species (1878). Both the present specimens are in poor condition, but fortunately the abdomen of the male is complete. I take the opportunity of figuring the anal appendages of the male. These bear a close resemblance to the corresponding structures of *O. dyak*.

Orogomphus speciosus, de Selys.

O. speciosus, Selys. *Ann. Mus. Civ. Genova* (2) x, pp. 481-482 (1891). Kirby, *Cat. Odonata*, p. 79 (1890).

1 ♂ [illegible]. Lord Carmichael's collection, Darjiling District, 1000-3000 ft., May, 1912.

The male of this species has not been described.

The dimensions of the specimen are as follows:—

Length of abdomen 54 mm., of hind-wing 40 mm., breadth of h. w. 12·5 mm.; of the type female, length of abdomen 57 mm., of hind-wing 46 mm., breadth of h. w. 15 mm.

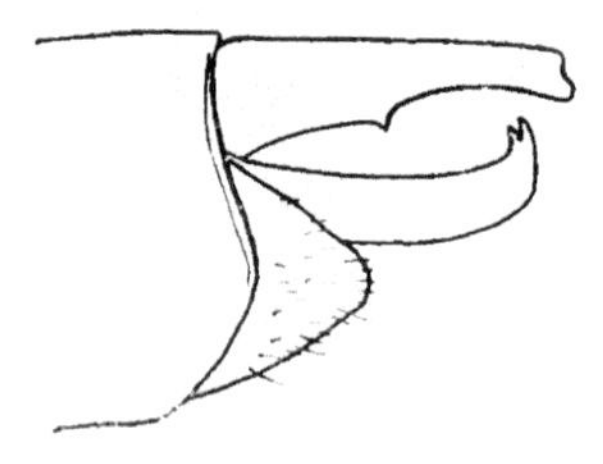

FIG. 1.—Profile of anal appendages of *O. atkinsoni* ♂.

There is thus a considerable difference in size, scarcely greater than occurs in other species of the genus.

In colouring the present example shows the following points of disagreement with the account of the type.

i. The occiput is black not yellow.

ii. Abdominal segment 2 is largely yellow above, with a transverse black band not touching either extremity.

iii. Segment 8 shows no lateral yellow spot.

Venation-formula:

	Anal loop.	An. n.	Pn. n.	M.	Cu.	t. (cells)	supra. t.
basal post-costal		21—22	11—12	1—1	7—7	2—2	4—3
+	0—0	20—22	16—15	1—1	5—6	2—2	3—3
♀ type		23—23	13—13	2	7—8	2—2	4—4
		17—19	17—17	2		3—3	4—4

In other respects there is close agreement between the characters of the male here recorded and those of the female as described by de Selys, so that I have little hesitation in referring them both to the same species.

The wing of the male is colourless and is very like that of the male *O. atkinsoni*, broadly speaking.

The anal appendages differ markedly in detail from those of the allied species. The upper pair are stout, a trifle shorter than the lower appendage, and are curved inwards towards each other; bifid at the extremity, and with a very small ventro-

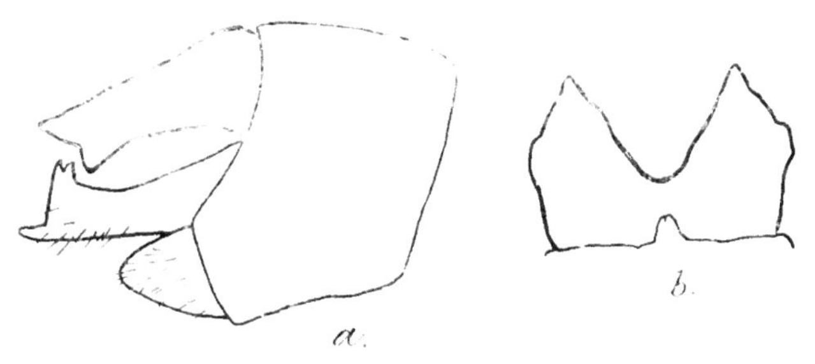

FIG. 2a.—Profile of anal appendages of *O. speciosus* ♂.
„ 2b. Lower anal appendage of *O. speciosus* ♂, viewed from below.

internal tubercle on each just beyond its middle and scarcely visible in profile.

The lower appendage likewise is stout; its two limbs each carry a projection directed straight upwards. This terminates in a doubly toothed point. Beyond the projection each limb ends in a pointed, backwardly directed spur. When looked at from below the lower appendage appears as ending in a pair of triangular processes not divaricated from each other.

XIII. NOTES ON ORIENTAL SYRPHIDAE; WITH DESCRIPTIONS OF NEW SPECIES.

PART II.

By E. BRUNETTI.

(Plate xiii.)

My previous paper on this family appeared in April, 1908 and revised our knowledge of certain oriental genera up to that date, including descriptions of thirty-nine new species.

In the present paper thirty-five additional species and some new varieties or "forms" are described, and those set up by other authors recorded, with such synonymical and other notes as appear of interest.

Two or three genera are, perforce, treated herein tentatively, such as *Sphaerophoria* and *Eumerus*, whilst many species of *Syrphus* and *Eristalis* are still imperfectly understood. Dr. Meijere has made much progress in identifying and redescribing several of the older authors' species of *Eristalis* and offers a valuable tabulation of those known to him.

Subfamily *SYRPHINAE*.

PARAGUS.

One new species *rufiventris* recently described by me (Rec. Ind. Mus., viii, 157, (r 1913) from Assam, the Western Himalayas and Ceylon. *Type* in Indian Museum.

Paragus serratus, F.

This common and widely distributed species extends to Assam; Sadiya, 23-xi-11, and Dibrugarh, 17-19 xi-11. I have it in my own collection, taken by myself from Cawnpore 29 xi 04, Calcutta 1-ii 07 and Rangoon 9 ii-06. It is common at Pusa in Bihar.

Paragus indica, Brun.

Pipizella indica, Brun., Rec. Ind. Mus. II, 52.

This species was wrongly placed by me in *Pipizella*. Further specimens in the Indian Museum are from Darjiling, Matiana, and Tenmalai (Travancore), 21-xi-08. It is perhaps identical with *Paragus politus*, W. described from China. The sides of the thorax are not whitish as in Wiedemann's species, but bear some long white hair anteriorly.

Paragus atratus, Meij.

One ♂ specimen from Bijrani, Naini Tal District, 19 iii-10 in the Indian Museum, agrees exactly with a specimen in the collection from Java, sent by Dr. Meijere. He records further specimens of both sexes from Java.

Pipizella rufiventris, mihi, sp. nov.

♂ Western Himalayas. Long. 7 mm.

Head.—Vertex aeneous black, with violet reflections, and black hairs; ocelli concolorous Frons and upper part of antennal prominence shining blue black, with black hairs; a broad grey dust band from eye to eye across middle of former. Tip of antennal prominence, just between the antennae, pale. Antennae black, 3rd joint large and elongate, arista black. Face yellowish, with whitish pubescence, a little darker about mouth opening; a black narrow median stripe. Eyes brown, with short distinct grey pubescence; occiput black, with yellow hairs around margin.

Thorax aeneous black, shining, with soft yellowish grey pubescence, which is more whitish and ragged on the sides. Scutellum luteous, semi-translucent, a little darker in middle, pale yellow pubescent.

Abdomen reddish yellow, basal segment, central basal part of 2nd segment, a moderately narrow band on hind borders of 2nd and 3rd segments, apical half of 4th and all the 5th segment, black. Dorsum of abdomen with pale yellowish grey pubescence, which is longer and thicker about sides of 2nd segment. Belly yellowish, a broad black transverse band on 2nd segment, apical part of abdomen black.

Legs.—Coxae, basal half of anterior femora and basal three-fourths of hind femora, black; also median half of hind tibiae, though less well defined; and upper sides of all tarsi. Rest of legs brownish yellow; underside of hind tarsi with golden brown pubescence.

Wings clear, subcostal cell pale yellowish; squamae pale yellow; halteres yellow.

Described from a perfect ♂ in the Indian Museum, presented by Col. Tytler, taken by him at Kousanie, 6075 ft., Kumaon District, 22-vii-14.

Psilota cyanea, mihi, sp. nov.

(Plate xiii, fig. 1.)

♀ Eastern Himalayas. Long. 4½ mm.

Head.—Frons and face brilliantly shining violet black; the vertical triangle demarcated by an impressed line: antennal prominence very slight. Some erect brown hairs on vertex. Frons and face with sparse whitish soft hairs. Viewed from above, a just perceptible whitish dust spot about the middle of the frons, contiguous on the eye margin each side, and the edges of the face with vague whitish reflections in certain lights. Upper mouth border distinctly produced[1]; proboscis short, brownish black. Antennae brownish yellow, under side of 3rd joint much paler. Eyes of exactly the same vertical height as the head, dark brown, very shortly but rather thickly pubescent; occiput slightly produced behind the vertex, aeneous black, with a fringe of short white hairs.

Thorax shining cyaneous black with a faint violet tinge, scutellum concolorous, both with sparse very short whitish pubescence: some rather long whitish hairs at sides, anteriorly.

Abdomen of three obvious segments only, the first very narrow, the 2nd, 3rd and 4th subequal, the 5th barely visible; all wholly cyaneous black, with short soft pale pubescence, belly similar.

Legs black; knees, tibiae tips and anterior tarsi brownish orange, hind tarsi darkened above. Hind femora and all tibiae with a little pale pubescence.

Wings almost clear; stigma pale yellow; halteres reddish brown.

Described from two ♀♀; Gangtok, Sikkim, 6150 ft., 9-ix-09, *type*; and Kurseong 10–26-ix-09. In Indian Museum

Chrysogaster (Orthoneura) indica, mihi, sp. nov.

♂ Punjab. Long. 6 mm.

Head.—Vertex very small, with a little dark brown hair. Eyes contiguous for a moderate space, about half the height of the frons, dark brown, bare. Frons shining blue black, with light brown or greyish hairs; antennal prominence slight. Face shining blue black with a little whitish hair, mouth border well produced. Proboscis and palp blackish brown. Antennae rather dark brown, 3rd joint ovate, arista almost basal. Occiput blackish, not at all produced beyond upper half of eye and only slightly so on lower half, which bears a fringe of white hairs.

Thorax cupreous, with brownish or yellowish hairs; scutellum aeneous, similarly pubescent. Sides of thorax cupreous, with a little greyish hair.

Abdomen cupreous, with moderately thick soft whitish pubescence, which is thicker at the sides and on the belly.

Legs wholly black, with the usual amount of greyish pubescence, undersides of tarsi reddish brown.

Wings pale grey, stigma yellowish, halteres orange.

Described from a unique ♂ in the Indian Museum from the Kangra Valley, 4500 ft., xi-09 [*Dudgeon*].

[1] A generic character according to Verrall.

CHILOSIA, Mg.

One new species *apicalis* ♀ recently described by me (Rec. Ind. Mus. viii, 158, 1913) from Rotung, 1400 ft. (N. E. Front. Ind.) 4—13-iii-12. *Type* in Indian Museum.

Chilosia hirticincta, mihi, sp. nov.

♂ Darjiling. Long. 9-10 mm.

Head.—Eyes covered with dense brownish yellow pubescence, touching for a considerable space, leaving a very small vertical triangle, blackish, with three or four long black hairs intermixed with the long brownish yellow ones. Frons sharply demarcated from face, very convex, aeneous blackish, with only a very narrow grey-dusted eye border, and an indistinct median similar line: the whole frons covered with long black hair.

Face moderately prominent, central bump small, mouth border not very prominent, extremely narrowly orange, the face blackish, with very short, almost microscopic pubescence. Eye margins greyish, with long yellowish grey hair. Occiput and lower part of head grey, with yellowish grey hairs. Proboscis black. Antennal first two joints, black, 3rd black, with, visible in certain lights, a greyish dust; rounded, but with rather truncate tip: arista bare, black. The depression in which the antennae are set, brownish yellow.

Thorax and scutellum shining aeneous with rather long and thick brownish yellow hair with which some black hair is intermixed. No stiff bristly hairs on either thorax or scutellum. Sides of thorax cinereous grey (the colour extending almost over the shoulders), with rather long brownish yellow hair.

Abdomen aeneous black, shining, with long yellowish hair; the third segment with all black hairs on the dorsum.

Legs.—Femora blackish with long yellowish hair, extreme tips orange. Tibiae black, the base broadly, the tips less broadly orange or brownish yellow; with yellow or golden yellow short pubescence on front side of front pair. Tarsi blackish above, with a little yellowish brown hair, under side with rich golden brown or golden yellow close pubescence, the first two joints of the middle pair brownish yellow.

Wings pale grey, a little yellowish on basal half anteriorly, in one specimen slightly yellowish in the neighbourhood of the veins. Halteres and tegulae brownish yellow.

Described from 3 ♂ ♂ in the Indian Museum from the Darjiling District [*Lynch*].

This species is easily known by the conspicuous, wholly black haired 3rd abdominal segment.

Chilosia nigroaenea, mihi, sp. nov.

♂ ♀ Simla District. Long. ♂ 7 ♀ 6 mm.

Head.—Eyes in ♂ contiguous for about one-third of the distance from extreme vertex to root of antennae, vertical triangle blackish, with some long black hairs. Eyes with rather thick short yellowish grey pubescence, which when viewed from certain directions appears quite white. Frons as in *hirticincta*, the dust on the eye margins less distinct. Face with the central knob somewhat large, conspicuous and rounded; aeneous black, shining, with very sparse and short, almost microscopic greyish pubescence, the central knob and the space immediately below it very shining black. Mouth border narrowly orange, moderately produced, with a small bump on each side of it. Eye margins dull blackish, with sparse rather short greyish hairs. Occiput grey with short greyish hairs. Antennae dull dirty brownish grey, arista bare, black.

In the ♀ the frons not much narrowed on vertex, and about one-third the width of the head just above the antennae is shining black with a little grey pubescence.

Thorax and scutellum shining black, with rather thick brownish yellow pubescence, a few black bristly hairs intermixed in front of wings, and on hind margin of scutellum, where these black hairs are much longer than the general pubescence. Sides moderately dark shining greyish, with yellowish grey hair.

Abdomen all shining black, with close, pale yellowish grey pubescence; belly similar.

Legs blackish, with pale yellowish grey pubescence. Extreme tips of femora, base of tibiae rather narrowly in ♂ and to the extent of basal third in ♀, brownish yellow. Tarsi blackish, with pale hairs above and thick rich golden brown or golden yellow pubescence below; base of middle tarsi above more or less brownish yellow.

Wings very pale grey, ♂, practically clear, ♀; stigma pale yellow, halteres yellowish.

Described from a single ♂, Matiana and ♀, Simla 7-v-10 in the Indian Museum [both *Annandale*].

Chilosia plumbiventris, mihi, sp. nov.

♀ Simla. Long. 8 mm.

Head.—Frons and face shining aeneous black, almost with a deep indigo tinge, the frons widening gradually from vertex to about one-third the width of the head above the antennae. Frons slightly prominent above the antennae from eye to eye, giving the appearance of an elongate transverse callus. Above this callus-like prominence is an oval, yellowish grey dust spot each side touching the eye margins. Frons, except the dust spots, wholly covered with thick black hairs. Face shining black, the central knob large and prominent, the mouth border but slightly produced, very narrowly orange; a little almost microscopic pubescence at the sides of the mouth but not extending just below the eyes. Eye margins distinct, grey, widened immediately below antennal prominence, and on this wider part on each side of the face are three elongate notches as though impressed with a knife.

Vertical margin, occiput and lower part of head aeneous black, but the facial eye margins are continued narrowly round the eyes to the vertex, bearing a fringe of yellowish grey hairs, and similarly coloured hairs also cover the lower part of the head below and behind the eyes. Antennal first two joints brownish yellow, 3rd large, broadened, rounded, darker and duskier brown: arista concolorous, bare.

Thorax dark aeneous black, shining, covered rather closely with very short brownish yellow pubescence, which, viewed at a low angle from in front, appears uniform and continuous over the whole dorsum, but viewed from behind appears to form three longitudinal stripes, the median one narrowly divided in the middle and attaining the front margin; the exterior ones foreshortened. Sides of thorax concolorous, with very sparse and short, brownish yellow hair. Scutellum concolorous, with short, brownish yellow pubescence and a single pair of well separated apical long black bristles.

Abdomen shining lead colour with almost microscopic pale yellow pubescence; a dull black broad band, half the length of the segment, on the hind margins of 2nd and 3rd segments, narrowed to a point at the sides of the segment, and very slightly notched in the middle in front. These transverse black bands are best seen from behind.

Legs dark reddish brown: femora with a moderate amount of pale yellow pubescence; extreme tips of all femora, basal half of all tibiae, tips of middle tibiae and first three joints of middle tarsi, brownish yellow, the posterior margins of these three joints blackish. Under sides of fore and hind tarsi with golden brown pubescence.

Wings very pale grey, stigma pale yellow, halteres bright orange.

Described from a single ♀ in the Indian Museum from Simla 7-v-10 [*Annandale*].

The shining lead colour of the abdomen will easily distinguish this species.

Chilosia ? grossa, Fln.

A ♂ and ♀ taken at Binsar, Kumaon District, 28-v-12, by Dr. A. D. Imms, sent to me for examination appear to be this rather widely distributed European species.

Unfortunately no specimens are at hand for comparison, but the only discrepancies from Verrall's description are as follows. The antennae are dull dark reddish brown, not blackish; the vertex and frons have an admixture of black hairs in the pubescence, which is not the case in *grossa*; the tibiae are mainly black (not orange) in both sexes, with the base broadly, and the tip much less broadly orange, the black part beginning always distinctly before the middle, whilst of *grossa* Verrall says "blackish ring just below the middle." The halteres are wholly orange yellow, not with blackish knobs. In the ♀ the 4th and

5th abdominal segment shew no trace of black hairs (though the pubescence is considerably worn off).

In *grossa* the whole of the 5th segment, and the major part of the 4th segment are entirely black haired.

On the other hand, the special points of similarity, in addition to a very close general agreement with Verrall's description, are the shape of the face in profile, the three faint channels on the frons in the ♀, and the distinctly more reddish colour of the pubescence on the head and thorax in the ♀ specimen. The size also agrees, ♂ 10 ♀ 11 mm.; Verrall giving "about 11 mm."

MELANOSTOMA, Sch.

Melanostoma ambiguum, Fln.

Melanostoma dubium, Zett.

These two European species were introduced in my previous paper on Oriental Syrphidae, on a single example of each from the Simla District. The former, represented by a ♂ from Matiana, is truly identified, agreeing in every particular with Verrall's very faithful description, but the specimen referred by me to *dubium* proves on a closer examination to be only a melanoid *Platychirus albimanus*, F. There is the less excuse for this error, seeing that I knew this species to occur in the Himalayas.

Melanostoma orientale, W.

(Plate xiii, fig. 2.)

In my notes on diptera from Simla (Rec. Ind. Mus. i, 168) were included *M. mellinum*, L. and *M. scalare*, F., both common European species. The examples referred to *scalare* are only *orientale*, and as regards those supposed to be *mellinum* there is ample room for doubt as to their identity. In fact *mellinum* in typical form may possibly not occur in the East at all, although as it is so abundant throughout the whole of Europe it will be curious if it is not found in the Himalayas.

However, it seems to me highly probable that *orientale* is not specifically distinct from *mellinum*, a species it is more akin to than *scalare*.

The principal alleged difference is the grey-dusted frons and face in *orientale*, but numerous specimens occur in which this is much less conspicuous than usual, thereby closely approximating to *mellinum*. Among the males, specimens occur which are hard to definitely assign to either species, and three Darjiling specimens in the Museum taken by me may really be true *mellinum*.

The females in *orientale* are more easily recognised by the dust spots on the frons being more closely approximate, so that the vertex and the lower part of the frons are more clearly demarcated, but a near approximation to this is not infrequently met with in *mellinum* ♀. Meijere's redescription of the species is wholly applicable to the specimens referred by me to *orientale*. It was my impression at first that the facial bump was not so large or conspicuous as in *mellinum*, but an examination of a large number of specimens shews that there is no difference. Moreover, such examination has revealed the existence of an apparently undescribed form (pl. xiii, fig. 3) with a facial profile intermediate between *orientale* and *univittatum*, in which the central bump though distinct is much less conspicuous than in *orientale*. This form is represented by a dozen females from the Simla and Darjiling districts, the United Provinces, Bengal and Bangalore. It is further distinguished from the specimens representing my final view of *orientale* ♀ by the 1st pair of abdominal spots being larger than in *orientale*, oval, and carried over the side of the 2nd segment below the base. Also the hind femora are all yellow, the hind tibiae bearing only an indistinct median dark band which is frequently absent.

Meijere reports the 1st pair of spots in *orientale* as smaller, more rounded and "petty" as compared with *mellinum*, whilst Wiedemann describes them as obliquely placed

These twelve specimens approach my *univittatum* ♀, but the presence of the small though perfectly distinct facial bump at once separates them. When all the specimens are examined in conjunction with a series of *univittatum* ♀ they are seen to be almost certainly specifically distinct. I am at a loss to satisfactorily dispose of them, but as there are no males with the same characters, to set them up as a new species would be premature.

Melanostoma univittatum, W. ♂ ♀.

(Plate xiii, figs. 4-6.)

? *Syrphus planifacies*, Macq.

Wiedemann described only the ♂ of this species, nor have I seen any mention of the ♀ having been described. Nine specimens in the Indian Museum can hardly fail to be that sex of this species. They possess the smooth face without any trace of a central bump so characteristic of *univittatum*, and the peculiarity of the 1st pair of spots being fully as large as the others, with their bases on the anterior border of the segment or enclosing the anterior angle of it, or carried over the side just below the base. These front spots are sometimes whitish in colour, and occasionally occupy the whole of the segment, the colour extending well over the base of the 3rd segment also. The 1st pair of spots in *univittatum* ♂ also occupy nearly all the 2nd segment, and have their bases on the anterior border of that segment; although a more suitable description would be to regard the abdomen as reddish yellow, with a narrow black median line and the posterior borders of the segments narrowly black, the colour extending slightly forward towards the sides. The hind legs are wholly yellow except for an indistinct median dark band on the tibiae, and this is often absent.

As regards *planifacies*, Macq. I think it may also be regarded as the ♀ of *univittatum*. The sole disagreement in Macquart's description is the colour of the thorax and frons, which he says is greenish black. Although in the nine ♀ ♀ that I refer to *univittatum* the thorax and frons are aeneous black as it normally is in the ♂, some ♂ ♂ in the collection exhibit a distinctly greenish tinge. One of the ♀ examples (from Bangalore) agrees exactly with Macquart's plate, and his remark that the pale colour at the base of the abdomen extends to the side borders agrees with the nine specimens referred to. The legs in these specimens agree with those of my male *univittatum*.

Meijere records three *planifacies* from Singapore, Sumatra and Queensland respectively, but no ♂.

The ♂ *univittatum* specimens in the Indian Museum come from Darjiling, Katmandu, Dibrugarh, the Assam-Bhutan Frontier, Mergui, Travancore, Bangalore and Coromandel; whilst the 9 ♀ ♀ hail from Bhim Tal, the Assam-Bhutan Frontier, Sadiya, Travancore, Bangalore, Coromandel and Sarawak, the localities of both sexes thus supporting the view that they are the same species. Its range of distribution is evidently very wide.

Melanostoma cingulatum, Big.

This can hardly be a *Melanostoma*, the yellow scutellum and side stripes to the thorax throwing it out of this genus altogether. Bigot says it resembles *Syrphus consequens*, Walk., which latter has been referred to *Asarcina*, a totally different group of species. Bigot, in fact, did not understand the genus *Melanostoma* and introduced, with a query quite a number of species. In the Indian Museum are two specimens marked "*Melanostoma, hemiptera*, Big." in that author's handwriting which are merely the common *Syrphus* (*Asarcina*) *aegrotus* F.

Platychirus manicatus, Mg. var. himalayensis, mihi, nov. var.

Three ♂ ♂ from Garhwal differ from the European *manicatus* sufficiently to rank them as at least a very distinct variety, if not a distinct species. The dilatation of the first two joints of the front tarsi is more conspicuous, and more produced forwards on the inner side of the 1st joint. The hind metatarsus is distinctly less thickened in the middle though obviously broader throughout than the femur or the remaining tarsal joints.

The present form is 11·5 millimetres long, as against 9 to at most 10 millimetres in *manicatus*, and the abdominal yellow spots are smaller, more quadrate and of uniform size, the first pair being as large and as square as the others.

The close similarity in all other characters causes me to refrain from considering this form distinct, at least until further specimens are available.

Dideoides ovata, Brun.

One ♂ Sikkim v-1912; one ♀ Shillong 10—12-x-14 [*Kemp*].

ASARCINA, Macq.

This is not a good genus but I collect under this heading the species referred to it. Meijere regards it as a subgenus, Bezzi as a valid genus.

Syrphus (Asarcina) aegrotus, F.

One of the commonest species in the East, and easily recognized by the broad blackish band across the middle of the wings. This band sometimes extends to the base of the wing, and a specimen of this nature in the Indian Museum bears a label *Melanostoma hemiptera*, Big. Meijere records it from several places in Java and the Indian Museum has it from a wide range of localities.

Syrphus (Asarcina) ericetorum, F.

S. salviae, Wied.

S. salviae, W., is identical with *ericetorum*, F., described originally from Africa, and the latter name will have to be used for it. Meijere records it from Java, the Indian Museum has it from many localities and I took two at Colombo in June, 1904. Two were taken at Simla viii-14 by Capt. Evans, R.E., and two at Cherrapunji, Assam, 4400 ft., 2—8-x-14 [*Kemp*].

Syrphus (Asarcina) consequens, Walk.

Meijere records this species from Sumatra, Java and Papua, and confirms Osten Sacken's suggestion that *striatus*, Wulp, is synonymous.

The following two species have been recently described as belonging to *Asarcina*.

A. biroi, Bezzi, Ann. Mus Hung. vi, 902 (1908).
A. morokaensis, Meij., Tijd. v. Ent. li, 308 ♂ ♀, pl. viii, 33 (1908), Papua.

Meijere records *biroi* from several localities in Papua.

SYRPHUS, F.

Dr. Meijere gives a table of a number of species of *Syrphus* and records *serarius*, Wied., from Pattipola, Ceylon (200 metres), [*Biro*].

Syrphus balteatus, DeGeer.

Very common in the Himalayas and also in the plains of India and Assam, extending to Java, China and Japan.

The following new species were described by Meijere recently (Tijd. v. Ent. li, 1908):—

luteifrons, p. 304, ♂, pl. viii, 37, Moroka (1300 metres), Papua [*Loria*]. Type in Genoa Museum, a unique specimen.
triangulifrons, p. 305, ♂ ♀, pl. viii, 36, Moroka, Papua [*Loria*]. Types in Genoa Museum, a unique pair.
circumdatus, p. 306, ♂ ♀, pl. viii, 35, Moroka Papua [*Loria*]. Types in Genoa Museum.
longirostris, p. 307, ♂, pl. viii, 34, Moroka, Papua [*Loria*]. Type in Genoa Museum.
morokaensis, p. 308, ♂ ♀, pl. viii, 33, Moroka, Papua [*Loria*]. Types in Genoa Museum. Referred to the subgenus *Asarcina*.
elongatus, p. 309, ♂ ♀, pl. viii, 32, Moroka, Papua [*Loria*]. Types in Genoa Museum.

Four species taken on the Abor Expedition were described as new by me in Rec. Ind. Mus., viii (1913). Types in Indian Museum.

aeneifrons, ♂, p. 159, N.E Front. India, 1100 ft, 17-iii-12; 4000 ft., 18-iii-12.
transversus, ♀, p. 160, Sadiya, 28-xi-11, a unique specimen.
fulvifacies, ♀, p. 161, Rotung (N.E. Front. Ind.), 26-ix-11, a unique specimen.
maculipleura, ♀, p. 162, Rotung, 25-xii-11, a unique specimen.

Syrphus distinctus, mihi, sp. nov.

(Plate xiii, fig. 7).

♂ Western Himalayas. Long. 14—15 mm.

Head.—Frons, face and under side of head covered with pale orange yellow tomentum, being more dusky towards the frons. A broad median blackish stripe. Frons with black hairs. Vertex blackish with black hairs. Antennae and arista wholly black. Back of head dark grey with short yellow hairs, some black ones behind the vertex.

Thorax.—Blackish on dorsum, yellowish grey at sides, mainly covered with brownish yellow pubescence. Scutellum orange yellow with black hairs in the middle and yellowish white ones on anterior and posterior margin and below the latter.

Abdomen.—Blackish, 1st segment yellowish, hind margins of 2nd, 3rd and 4th segments pinkish grey, with a rather narrow cross band of the same colour across the middle of each; that of the 4th segment lying just before the middle. Dorsum of abdomen with rather thickly placed black hairs except on the 2nd segment, on the pale band on the 3rd and at the sides of the whole abdomen where the pubescence is whitish yellow. Belly blackish, yellowish at base and along the hind borders of the segments, covered with yellow or black pubescence according to the colour of the surface.

Legs.—Anterior pairs principally orange yellow; anterior femora black on about the basal half; hind legs principally black, knees broadly brownish yellow as are the last four tarsal joints. Anterior femora with some moderately long yellowish or brownish yellow hairs on under side, with black hairs intermixed towards tips of fore pair and generally on underside of middle pair. Conspicuous thick but short black hairs on hind femora, longest on underside, and on front and hind sides of hind tibiae.

Wings yellowish grey, stigma brown; squamae yellowish brown with fringe of the same colour.

Described from 3 ♂♂ from Tolpani, Garhwal District, 9500 ft., 23-iv-14 to 13-v-14

The unbroken pinkish grey bands on the abdomen easily separate this species from all other Oriental ones, and from all European or North American species known to me.

SPHAEROPHORIA, St. Farg.

Few genera offer more complexities than this, as regards the limits of the species.

The present notes must therefore be regarded as simply a contribution towards a better knowledge of the Oriental species; and apart from the two perfectly good species *scutellaris*, F., and *javana*, W., the four forms recognized and described herein are termed and understood as "forms" only, although it seems probable that *viridaenea* will eventually prove specifically distinct.

In working out the fairly good series of specimens in the Indian Museum I adopted the plan of dividing them into "forms" previous to consulting any of the descriptions, treating the ♂♂ first and the ♀♀ subsequently, moreover in each case without any reference to the localities of the specimens.

By this method one avoids being prejudiced in favour of pairing off ♂♂ and ♀♀ according to the localities, and a more trustworthy result is likely to ensue.

In the present instance the six male forms sorted themselves readily enough and were backed up in every case by females from the same localities; proving to be the two well marked and known species *scutellaris*, F., and *javana*, W., with four remaining forms of which I have ventured to give names to three.

One point noticeable about them all is that the yellow markings of the abdomen are almost always definite bands and not pairs of spots more or less resolving into bands as in the European species.

Apart from *scutellaris*, F. (with *aegyptius*, W., *longicornis*, Macq., *splendens*, Dol., and *Melithreptus novaeguineae*, Kert., as synonyms), and *javana*, W. (with *Melithreptus distinctus*, Kert., as a synonym) the only other two described species are *bengalensis*, Macq., and *indianus*, Big.

Macquart's *bengalensis* may be anything; he separates it from *taeniata*, Mg., on the shorter abdomen, with wider (yellow) bands, the 4th segment being tawny with a dorsal line. Though no individual specimen answers to this amongst those before me it may very well be my "Form 1."

Bigot's description of his *indiana* (♂), from Bengal, though more lengthy is very inconclusive and may easily be the same form again. His "derniers segments des tarses un peu brunatres" cannot be regarded as pointing to my *nigritarsis*, in the three ♂♂ of which the tarsi are very distinctly wholly deep blackish or blackish brown.

The characters studied in the present differentiation of forms are as follows:—

(1) Length of 3rd antennal joint. This is always simply elongate or almost rounded, except in *scutellaris*, F., a very distinct species which may be recognized at once by this character alone.

(2) Markings on frons and face, or absence of same.

(3) Thoracic dorsum with pale grey stripes or not.

(4) Scutellum with yellow or black hairs, or both. The best way to decide this is to view this part from behind and slightly above. If wholly yellow haired, hardly any pubescence can be seen at all at this angle, whereas any dark hairs are visible at once.

(5) Abdominal markings.

(6) Comparative length of wing and abdomen. One or two of the older writers spoke of the wing being longer than or equal in length to, the abdomen. It is the abdomen, which in some species (*scriptus*, L., of Europe, for instance) is abnormally long, that varies, the proportional length of the wings to that of the rest of the body being the same in all the forms now treated of.[1] Moreover it appears to vary within reasonable limits, and if "form 2" is the same species as my "*flavoabdominalis*" form, the proportionate length of wing to abdomen will prove to be of less value still.

(7) Coxae all yellow, or black marked. This hardly affects the Indian forms at all.

(8) Hind tibiae mainly yellow or mainly black. This character only serves to separate the second well known and distinct species *javana*, W., which has them wholly black except for a clear cut median yellow band of some little width; all the other forms possessing entirely yellow hind tibiae. It is true that *scutellaris* often has an indistinct obscure ring about the middle, but the very elongate 3rd antennal joint will always distinguish that species.

[1] If there is any exception to this it is in my "*flavoabdominalis*" form amongst the Indian ones, and in *scriptus*, L., with its varieties, amongst the European ones.

(9) Hind tarsi distinctly black or dark brown above, or mainly yellow. This only separates my form *nigritarsis* from the remainder, after eliminating *scutellaris* and *javana*. Occasional individuals of various forms may have them a *little* brownish, or a deeper orange yellow, but never sufficiently darkened to be mistaken for *nigritarsis*.

Table of Oriental species and "forms" of SPHAEROPHORIA.

A Antennal 3rd joint very elongate, about twice as long as broad.	sp. 1. *scutellaris*, F., ♂ ♀.
AA Antennal 3rd joint simply oval or rounded.	
B Hind tibiae wholly black except a well marked yellow median not very wide ring.	sp. 2. *javana*, W., ♂ ♀.
BB Hind tibae wholly yellow.	
C Face without black stripe. Thorax blackish with two obvious though faint grey stripes.	
D Hind tarsi all yellow.	
E Abdomen all yellow after 2nd segment, shorter than wings in ♂, generally also in ♀. "Form 1."	*flavoabdominalis*, mihi, ♂ ♀.
EE Abdomen with 3rd and 4th segments black at base and tip to a varying extent. Abdomen as long as wings ♂ ♀. "Form 2."	mihi, ♂ ♀
DD Hind tarsi all black above, anterior tarsi more or less so. "Form 3".	*nigritarsis*, mihi, ♂ ♀.
CC Face with distinct black median stripe. Thorax aeneous green, absolutely unstriped. "Form 4."	*viridaenea*, mihi, ♂ ♀

It will be seen that four "forms" are recognized in addition to the two well marked species *scutellaris*, F., and *javana*, W., which have been known for nearly a century. To three of the "forms" I have ventured to give names tentatively, to facilitate reference to them, and it seems probable that *viridaenea* will prove specifically distinct.

"Form 1, flavoabdominalis", mihi.

♂ ♀ Baluchistan, Persia, Simla, Nepal, Punjab, Bushire.

♂ *Frons* and face all yellow, rarely a very small black mark on or near central knob or mouth border: an individual aberration only. *Thorax* normally distinctly though faintly striped on at least anterior half, but occasionally the dorsum is quite dull and the stripes absent even in perfect specimens. *Scutellum* all yellow haired.

Abdomen with 1st segment shining aeneous, often appearing like a prominent triangle on each side of the base of the 2nd segment. The 2nd segment black, with a broad, clear cut, bright yellow transverse uninterrupted band forming about one-third of the segment; remainder of abdomen normally orange yellow, unmarked, and though there are generally a few irregular obscure markings there is nothing in the nature of transverse black bands or pairs of spots. Abdomen distinctly shorter than the wings. *Coxae* all yellow (in only one specimen the hind pair show a slight darkening); remainder of legs wholly yellow, the hind tarsi a little deeper orange.

Long. 6—7 mm.

Baluchistan, Bushire, Katmandu (Nepal), Dharampur (Simla Hills), 5000 ft., 6—8-v-07 [*Annandale*]; Agra, 4-iv-05; Ferozepore (Punjab), 28-iv-05 [*Brunetti*].

♀ Differing from the ♂ only as follows. Vertex shining black or dark aeneous, with a concolorous stripe, narrowing considerably and approximately reaching the antennae.

Abdomen about as long as the wings, 3rd and 4th segments with a wide black band on posterior margin.

I took this form in abundance at both Agra and Ferozepore, in company with the ♂ ♂ referred to, during April, 1905 in fields of dry grass, stubble and general vegetation.

One ♂ from Purneah (Bengal), 8—9-iii-09 [*Paiva*] agrees technically, but the wings and abdomen are equally long, and it is a little more robust. Long. 7 mm

It is difficult to differentiate this form from *scriptus*, L., yet it seems quite a distinct local race. Verrall notes the partiality of this species to form local races. Apart from size the ♂ in the present form is exactly like *scriptus* with all yellow abdomen after the 2nd segment, a form that species very often takes in European specimens, but on the other hand the ♀ does *not* so closely resemble the ♀ *scriptus*, the abdomen being mainly yellow, with black bands, instead of mainly black, with interrupted narrow yellow bands.

"Form 2", mihi.

♂ ♀ Shanghai, Simla, Nepal, Bengal.

♂ This differs from Form 1 only in the 3rd and 4th abdominal segments in the ♂ having a narrow black band at the base and a broad one at the tip of each. The wings are as long as the abdomen.

Long. 6 mm. Shanghai, 17-iv-06 [*Brunetti*]: Songara, Bengal, 3—5-iii-07.

♀ Agreeing with ♂ but the scutellum sometimes has some black hairs on the hinder part. The abdominal black bands are broader. Vertex shining aeneous black, frons with a broad black stripe to the antennae, this stripe sometimes of uniform width, sometimes narrowing anteriorly.

Shanghai, in company with ♂; Noalpur (Nepal), 21-ii-08; Dharampur, Simla, 5000 ft., 6—8-v-07 [*Annandale*]; Katihar, N. Bengal, 8—9-iii-09 [*Paiva*]; Bhanwar, Bengal, 26-ii-07.

This form seems to me practically identical with *S. menthrastri*, L. (*taeniata*, Mg.), the females agreeing exactly, but unfortunately there are no ♂ *menthrastri* specimens available for comparison, and there are several minor discrepancies between Verrall's descriptions of this sex and my ♂ ♂.

"Form 3, nigritarsis", mihi

♂ ♀ Simla, Kurseong.

♂ Differing from Form 1 as follows.

Scutellum with distinct blackish hairs on hinder part; these black hairs being longer than the yellow ones or than the yellow hairs on the hinder part of the scutellum in Form 1. *Abdomen* with 3rd and 4th segments each normally with a rather narrow black basal band and hind border of broader, but varying width. In one specimen these black bands are indistinct on the 3rd segment and altogether absent on the 4th. *Wings* and abdomen equal in length. *Coxae* wholly yellow in two specimens, in the other two, the front coxae are dusky on the anterior side. *Legs* yellow; hind tarsi wholly distinctly black or dark brown on upper side; anterior tarsi always distinctly brown or dark brown, always much deeper coloured than in Forms 1 and 2. The middle pair of tarsi the least deeply coloured of the three.

Long. 6 mm. Matiana, 28—30-iv-07, Theog, 27-iv-07, Simla Distr., 8000 ft. [*Annandale*].

Two ♀ ♀ from Simla and Kurseong respectively, agree with the ♂ except that the hind tarsi are a little less dark brown on the upper side, and the anterior tarsi are lighter brown but distinctly darker than the tibiae, yet not so dark as in the ♂. The frons has a very broad aeneous black stripe from the similarly coloured vertex to the antennae.

Kurseong, 7-ix-09, Kodiala, Simla Distr. [*Annandale*].

The black or nearly black upper side of the tarsi (always at least the hind pair) will distinguish this form from all the others. It is impossible to identify it with any recognized European variety of which an adequate description is open to me.

"Form 4, viridaenea", mihi.

♂ ♀ Simla, Kurseong.

♂ This form varies from Form 1 very materially and will probably prove a good species.

Frons with a very small frontal triangle; face with distinct *black median stripe*, not very regular in width. *Thorax* with wholly *aeneous green shining* dorsum, clothed with close yellow pubescence, *without any trace of stripes*; scutellum wholly, or at least mainly black-haired. *Abdomen* with 3rd and 4th segments orange yellow, the posterior border with a moderately broad black band, the anterior border generally black also. *Wings* and abdomen subequal. *Legs* all yellow, but hind tarsi rather darker orange.

Long. 6-7 mm. Simla, 16-v-09, Theog, 2-v-07 [*Annandale*]; Kurseong.

♀ A single specimen from Kurseong, 4-ix-09 [*Annandale*] agrees absolutely with the ♂♂. The vertex is broadly shining dark aeneous green with a broad stripe similarly coloured reaching the antennae.

I feel convinced this is a good species on the strength of the unstriped greenish aeneous thorax and very distinct black facial stripe, yet it seems preferable to rank it for the present as a "form" only.

Sphaerophoria scuttellaris, F.

In the Indian Museum from Maho, base of Nepalese Himalayas, 17-iii-09; Ferozepore, 28-iv-05; Agra, 3-iv-05 [both *Brunetti*]; Paresnath, W. Bengal, 4300 ft., 15-iv-09 [*Annandale*]; Bhanwar, 26-ii-07; Bettiah, Champaran, 8-iii-08: Dhampur, 24-ii-07: Rajmahal, 6-vii-09; Kulti Sitarampore, 10-viii-09 [*Lord*]; on launch off Coconada, Madras coast, 15-iv-08 [*Paiva*]; Calcutta, iii, x, xi, common. All the above localities in India. Base of Dawna Hills, Tennasserim, 4-iii-08 [*Annandale*]. I also took it myself at many places in India and the East but exact data are not available.

Sphaerophoria javana, W.

In the Indian Museum from Ukhrul, Manipur, 6400 ft. [*Pettigrew*]; Dawna Hills, 2000-3000 ft., 2-3-iii-08 [*Annandale*]; base of Dawna Hills, 1-iii-08 [*Annandale*]; Sukli, 2100 ft., 22-29-xi-11 [*Gravely*]; Sukna, 500 ft., 1-vii-08 [*Annandale*]; Burma-Siam Frontier, 900 ft., 24-26-xi-11 [*Gravely*].

This species was, by a clerical error of my own, recorded in my paper on the Diptera of the Abor Expedition (Rec. Ind. Mus., viii, 164), as *S. scutellaris*, F. Specimens were taken at Sadiya, Assam, 23-28-xi-11; Rotung, 1400 ft., 29-xii-11, and Kobo, 400 ft., 30-xi-11, the last two places being on the north-eastern Frontier of India.

Eriozona himalayensis, mihi, sp. nov.

♂ Western Himalayas. Long. 13-14 mm.

Head wholly moderately shining black. Frons with a pale yellowish grey tomentum when viewed in certain lights. Face with more obvious similarly coloured tomentum or minute pubescence; a median rather broad space being bare; some longer black hairs on the cheeks Proboscis black. Eyes with thick dark brown pubescence. Antennae black, 3rd joint dull, arista black. Occiput blackish grey with yellow hairs around the margins, with which are intermixed some black hairs behind the vertex.

Thorax moderately shining black, with, in certain lights, a slight aeneous tinge: prothorax dull aeneous, covered with light brownish yellow, rather thick pubescence, rest of dorsum covered with black pubescence; scutellum with long thick black pubescence, lower posterior margin with a fringe of short yellowish hairs. Mesopleura and sternopleura with thick yellowish pubescence, rest of sides of thorax with sparser black hairs.

Abdomen shining black, covered thickly with bright red pubescence, which becomes more yellowish on 1st segment and on sides of 2nd. Margins of 3rd and 4th segments, and whole of belly with black pubescence. Genitals dark grey with black hairs.

Legs black, with short black pubescence, which is rather long on under side of femora, the hind pair having in addition two diverging rows of long widely separated hairs.

Wings grey, anterior margin slightly darker; a broad brownish infuscation from around the stigmatic region across the middle of the wing, extending half way to the posterior margin. Halteres yellow, clubs black.

Described from several ♂ ♂ from the Kumaon District, 20-6-14 to 20-7-14.

This species evidently mimics the bee *Bombus haemorrhoidalis*, Smith.

BACCHA, F.

Meijere tabulates and notes a number of oriental species (Tijd. v. Ent. li, 316) and records the following previously known species: *pulchrifrons*, Aust., from Depok, W. Java, Singapore and Tsushima; *pedicellata*, Dol., from Semarang and Tandjong Priok, Java [*Jacobson*] also ♂ ♂ ♀ from Krakatua, and *purpuricola*, Walk., from two Papuan localities and the Key Islands.

I have myself received males of *nubilipennis*, Aust., and *pulchrifrons*, Aust., from Kandy and Peradeniya respectively.

Since Van der Wulp's catalogue, quite a number of new species of this genus have been set up. These are listed here.

rubella, Wulp, Termes. Fuzet. xxi. 423 (1898), Papua.
Meijere notes both sexes from Papua.

mundula, Wulp, *loc. cit.*, 423, ♀. Papua.
Meijere records the ♂ from Sukabumi, Java [*Kramer*], and a ♀ from Dilo, Papua [*Loria*].

circumcincta, Meij., Tijd. v. Ent., li, 320 (1908). ♀, Buitenzorg, Java [*Jacobson*]. Type in Amsterdam, a unique specimen.

pallida, *id.*, *loc. cit.*, 322, ♂ Stephansort, Papua [*Biro*]. Type, a unique specimen, in Hungarian Museum.

loriae, *id.*, *loc. cit.*, 324. ♀ Paumomu, Papua [*Loria*]. Type, a unique specimen, in Genoa Museum.

austeni, *id.*, *loc. cit.*, 325, ♂ ♀, environs of Buitenzorg, Java [*Jacobson*].

bicincta, *id*, *loc. cit.*, liii, 104 (1910), ♂ ♀, Batavia, Tandjong Priok (Java), Bekassi and from Krakatua [*Jacobson*]. Types in Amsterdam Museum.

chalybea, *id.*, *loc. cit.*, 105, ♂ ♀, Pasuruan, Java and Krakatua [*Jacobson*]. Types in Amsterdam Museum.

Baccha dispar, Walk.

A ♀ specimen in the Indian Museum, without data, identified by Bigot, is certainly this species.

Baccha robusta, mihi.

Three (♂ ♀) have been seen by me from Dehra Dun, sent by Dr. Imms; one ♂ in the Indian Museum from the base of the Dawna Hills, 4-iii-08 [*Annandale*]; and four (♂ ♀) from Sikkim, v-1912.

Baccha flavopunctata, mihi.

The specimens referred to in my description of this species, with the exception of the type, appear to be a different species, which is here described as *elegans*.

Of true *flavopunctata* further specimens have been acquired from Sibpur, Bengal, 4-iv-13 [*Gravely*]: Cherrapunji, Assam, 4400 ft., 2—8-x-14 [*Kemp*]; with an additional apparently immature one from the same locality. All are females.

It is possible that this species is synonymous with *pedicillata*, Dol., though that author's species is described as blackish brown with two semilunar pale bands; and mine as yellowish with black bands on the 3rd and 4th segments. Two specimens in the Indian Museum from Sibpur, Bengal, have the ground colour brown and the black bands a little larger and more extended at the sides of the abdomen, thus making a very close approximation to Doleschall's figure.

His description agrees exactly, except that he does not mention the conspicuous perpendicular yellow stripe on the mesopleura, with the adjoining spot on the sternopleura. His "metathorax luteo cincto" may refer to the conspicuous elongate yellow spot on the metapleura.

The specimens of this species with conspicuous yellow abdomens must bear some resemblance to *vespaeformis*, Dol. *Flavopunctata* differs from Doleschall's species by the presence of the metapleural and sternopleural yellow spots; by the black band at the base of the 3rd abdominal segment: the black rings on the hind legs and the blackish subcostal cell. Doleschall says the wings are clear except for a brownish red fore-border. There is little doubt the two forms are distinct. In the four examples present of *flavopunctata*, two (including the type) have the ground colour of the abdomen yellow, in the other two it is brownish, and in these the shape of the abdomen is also slightly different, the breadth of the 3rd, 4th and 5th segments being greater, and the widening of the 3rd segment more sudden than in the type and the Cherrapunji specimen, in both of which the greatest width of the abdomen is proportionately less, and the widening more gradual However, I include all under one species as in every other particular they agree with one another and it is no uncommon thing for the yellow parts of a species of Syrphidae to be replaced in individuals by brownish.

Baccha elegans, mihi, sp. nov.

♂ North Bengal; Lower Burma. Long. 11—12 mm.

Head.—Eyes absolutely contiguous for about half the distance from frontal triangle to vertex; (in one example they are quite distinctly though very narrowly separated). Frons shining violet black, frontal triangle and face wholly deep chrome yellow, with a very distinct median black stripe, broader on upper part, from below antennae to mouth border. No obvious bump on face. Antennae wholly bright orange yellow, antennal prominence hardly noticeable. Occiput whitish grey, cut away in profile behind upper part of eyes; a fringe of short white hairs round entire ocular orbit. Proboscis brownish yellow.

Thorax.—Dorsum shining deep blue, with very short whitish depressed pubescence. Sides dark blue black. Pale callus-like yellow spots are placed as follows: a large one on the shoulder contiguous to a lateral oblong one along the side, just below the dorsum and just touching a large perpendicular oblong one on the mesopleura, which in its turn is sub-contiguous to a round one on the sternopleura. A more or less oval one behind the wings. Scutellum mainly blackish brown, with a well marked pale lemon yellow base, this colour extending over the sides. Metanotum dark bluish black.

Abdomen.—First segment very short, sub-triangular; 2nd exceedingly narrow and elongate; 3rd equally narrow on basal third, thence suddenly widening to three times that width, the whole segment less long than 2nd; 4th distinctly shorter than 3rd, 5th less than half as long as 4th. The enlargement of the abdomen continues to the tip of the 4th segment, the 5th narrowing. The 1st segment is wholly brownish yellow, the rest of the abdomen is shining dark brown, with a vague violet tint, and there is a pale narrow space at the junction of the 2nd and 3rd segments, also broadly at tip of 3rd segment Genitalia shining brownish yellow, with some obscure markings and a small process below. Belly mainly a replica of upper side The whole abdomen with a little very short blackish pubescence, some longer, though still short, whitish pubescence at sides of first two segments.

Legs.—Anterior pairs bright brownish yellow, bare except for a little pale hair below the femora; hind legs with coxae obscure above, femora darker brownish yellow, tibiae pale yellow on basal half, black on remainder, as is also the metanotum; hind tarsi tips brownish.

Wings clear, except subcostal cell, from base to tip blackish or blackish brown, the colour carried narrowly along the front margin to tip of 3rd longitudinal vein. Halteres brownish yellow.

Described from several ♂♂ in the Indian Museum from Sukna, 500 ft. 1 and 2-vii 08 ; and from jungle at base of Dawna Hills, 1-iii-08 [both *Annandale*] ; Rungpo, Sikkim, 1400 ft. 6-ix-09. In the latter specimen the face is wholly pale, but it is undoubtedly this species.

This is evidently distinct from *flavopunctata*, though bearing a close resemblance, and at one time I thought it the ♂ of that species. It differs in the distinct blue tinge to the whole thorax instead of the almost cupreous dorsum in *flavopunctata*. Also in the metapleural stripe which is shorter and more truncate at its lower end, instead of longer, elongate oval and sometimes divided transversely. The femora are only slightly browner apically than basally, the tips seldom paler ; instead of a deeper brown middle part, the base distinctly pale and the tips always more or less so.

The whole hind metatarsus is black, instead of only at the base; the costal cell quite clear, not yellowish. The abdominal marks appear constant in *elegans* in the five specimens seen, except that the 3rd segment in one of them is all black.

Baccha apicenotata, mihi, sp. nov.

(Plate xiii, fig. 8).

♀ Western Himalayas. Long. 10 mm.

Head.—Frons shining aeneous black, with a dark blue tinge, narrowest immediately below vertex, thence gradually widening to double that width just above antennae. An elongate grey dust spot at about the middle of the frons each side, contiguous to eye margin. Face grey at sides, leaving a broad median blue black shining stripe ; the central bump rather large. Antennal prominence rather large, antennae bright orange. Proboscis brownish yellow. Occiput greyish, ocular orbit with a fringe of whitish hairs.

Thorax.—Dorsum shining dark blue, with sparse short brownish grey pubescence. Sides obscurely brownish, a small greyish shoulder spot, another similar spot half way between the latter and the wing root, contiguous to a perpendicular, similarly coloured oblong spot on the mesopleura. Scutellum shining dark blue with very sparse short pale hairs.

Abdomen only very slightly widened,[1] dark brownish, posterior margins of 2nd, 3rd and 4th segments broadly black, 5th segment mainly so.

Legs.—Anterior pairs wholly yellow ; hind pair a little more obscure ; coxae darkened, a subapical light brownish ring on

[1] A little may be allowed for the sides curling underneath, but the species is evidently nearly linear in form.

femora and a broad apical band on tibiae, neither of the rings very definite, tarsi wholly yellow.

Wings clear; subcostal cell wholly blackish brown, and beyond tip of cell the colour spreads into an apical wing spot, contiguous to front margin and limited posteriorly by the 3rd longitudinal vein. Halteres brownish yellow.

Described from a single ♀ from Bhowali, 5700 ft., vii-09 [*Imms*], the specimen presented by him to the Indian Museum.

This might easily be taken for the ♀ of *elegans*, but a good structural difference exists. In *apicenotata* the antennae are seen to be set on a rather conspicuous prominence, and the facial bump is also distinct, but in *elegans* there is no obvious antennal prominence and the facial bump is barely noticeable. Other differences consist of the absence of the yellow stripe on the metapleura, the wholly blue scutellum, the wholly yellow hind metatarsus and the more conspicuous wing-tip spot.

There are two further examples in the Museum collection which are apparently additional ♀ ♀ of the present species. The differences in the first are: (1) the frons is a little broader, (2) the abdomen enlarges very suddenly at the base of the 3rd segment and reaches its greatest width at the tip of that segment. The abdomen is black, except for the 1st segment, the extreme base of the 2nd and (indistinctly) the basal half of the 3rd. The black streak on the costa reaches the tip of the 3rd vein, but only weakly, and shows no sign of enlargement into an apical spot as in *apicenotata*. The specimen is from "Jungle at base of Dawna Hills", 1-iii-08 [*Annandale*].

The second specimen is an obviously immature one from Cherrapunji, Assam, 4400 ft., 2—8-x-14 [*Kemp*], and differs only in the wholly clear wings.

Baccha plumbicincta, mihi, sp. nov.

♀ Assam. Long. 8½ mm.

Head.—Frons broad, distinctly but not greatly broader above antennal prominence, where it is nearly one-fourth the width of the head; bluish black, the colour sharply demarcated behind vertex; a little whitish tomentum about the middle of inner orbit of eyes. Face, down to a little above mouth opening, bluish black, slightly grey-dusted, with a central conspicuous black bump. Remainder of lower part of head, including buccal region, uniformly bright yellow. Antennae black, 3rd joint broad, arista black. Occiput grey.

Thorax.—Dorsum and scutellum almost lead colour, shining, with slight coloured reflections when viewed from different angles; minute yellow pubescence; remainder of thorax bright yellow.

Abdomen.—Only slightly contracted on 2nd segment, remainder of segments barely wider, the abdomen at no point quite so wide as thorax, shining bluish black with very short inconspicuous pubescence, base of 3rd, 4th and 5th segments with a moderately broad lead coloured band.

Legs yellow; an indistinct broad brownish ring on apical half of hind femora; the apical half of hind tibiae blackish except broadly at tip; upper side of hind metatarsi brown, rest of hind tarsi black.

Wings clear; subcostal cell dark brown except on the narrow basal part; halteres yellow.

Described from one perfect ♀ in the Indian Museum from Cherrapunji, 2—8-x-14 [*Kemp*].

SPHEGINA, Mg.

One species described by me recently, *tristriata*, ♀, from a unique specimen from Rotung (N.-E. Front India), 6—13-iii-12 (Rec. Ind. Mus., viii. 165, ♀, pl. vi. 1913). *Type* in Indian Museum.

Sphegina bispinosa, mihi, sp. nov.

♂ ♀ Assam, E. and W. Himalayas. Long 5½ mm.

This species is remarkably close to the tolerably common and very widely distributed *S. clunipes*, Fln of Europe, but differs in two essential characters.

In the first place there is a short tooth-like black spine on the side of the basal abdominal joint lying immediately behind the halter. Three or four stiff black bristles lie behind the spine.

The second specific character is that the costa is a little brownish about the middle, the colour spreading slightly over the base of the 2nd longitudinal vein. The turned-up portions of the 4th and 5th longitudinal veins, with the posterior cross vein, are all distinctly brown suffused.

Described from a ♂ (*type*) from Margherita, Assam, a ♀ (*type*) from Darjiling (7000 ft.) taken by me, 29-v-10: also two ♀ ♀ taken by Mr. Imms near Bhowali, Kumaon, Western Himalayas (5700 ft.) in July, 1909.

Type ♂ and ♀ in Indian Museum.

Sphegina asciiformis, mihi, sp. nov.

♀ Darjiling. Long. 4 mm.

Head.—Frons aeneous black, with a little yellowish grey tomentose dust along the eye margins. Antennae with 1st and 2nd joints dark brown, 3rd joint black with long dorsal arista placed at the base of the joint. Mouth parts reddish brown. Occiput dark grey.

Thorax.—Yellowish grey-dusted, a little lighter on the shoulders: three moderately wide dorsal infuscated stripes, separated from each other by less than their own width. Scutellum shining black, with a little hoary dust. Sides of thorax blackish, with a little greyish dust on upper parts.

Abdomen.—The 1st segment narrow, 2nd very much contracted at base, thence suddenly widened; rich shining deep mahogany brown, nearly black, with very sparse and almost microscopic whitish hairs. Belly yellow ochre; two small black spots in a dorsal line near the base, and a median well marked black line on the apical half.

Legs.—Anterior four bright yellow. Hind femora much incrassated, yellow, a blackish band in the middle (incomplete below), and a complete broad black ring at the tip Under side with two rows of minute black spines; hind tibiae pale yellow, a long black streak below at base, and a blackish ring (incomplete on upper side) at tip. Hind tarsi brown, their metatarsi distinctly thicker than the tibiae, nearly half as long and about as wide as rest of tarsi.

Wings absolutely clear, brilliantly iridescent: halteres blackish.

Described from a unique ♀ taken by me, 29-v-10, at Darjiling. In the Indian Museum.

From the small size and very contracted base of the abdomen, this species closely resembles an *Ascia*.

Sphegina tenuis, mihi, sp. nov.

♀ Darjiling. Long. 4½ mm.

Head.—Frons dull black, with grey dust, ocelli distinct, red; the concavity in profile below the antennae well marked. Antennae black, a little dull grey-dusted, arista very curved; mouth parts reddish brown. Occiput grey.

Thorax black, with yellowish grey dust, and three dorsal infuscated stripes, the median one the widest, the outer ones slightly interrupted at the suture, and not reaching the shoulders. A pale grey spot on the latter can be seen if viewed from behind. Sides of thorax blackish, with yellowish grey hair.

Abdomen black, 2nd segment much attenuated and very long, 3rd with a broad yellowish sub-basal band. Genital organs large, globular apparently. Belly black, greater part of 3rd segment brownish yellow.

Legs.—The two first pairs pale yellow with the two last tarsal joints black. Hind coxae black, hind femora considerably incrassate; basal half pale yellow, apical half black. Hind tibiae mainly dark brown, pale at tips, and a narrow band just beyond the middle (which band appears as if in some examples it might be interrupted). Hind tarsi blackish brown, the hind metatarsi thickened, but only one-third as long as the tibiae.

Wings yellowish grey, brilliantly iridescent; stigma long, brown, halteres brownish yellow.

Described from one ♀ from Darjiling, taken by me, 29-v-10. In the Indian Museum.

Sphegina tricoloripes, mihi, sp. nov.

(Plate xiii, fig. 9).

♀ Western Himalayas. Long. 7 mm.

Head.—Frons blackish grey, nearly one-third the width of the head, uniform in width, vertical triangle not very distinct; face blackish grey. Upper mouth border well produced, proboscis moderately long, brownish yellow. Antennal prominence distinct but small, antennae blackish; 3rd joint slightly produced above at base; occiput blackish grey.

Thorax dull blackish, with two rather narrow, well separated, greyish dorsal stripes from anterior margin to scutellum; shoulders a little greyish. Scutellum rather shining black, with a pair of apical pale bristles, convergent and weak.

Abdomen.—Tawny brown, much contracted at base, widening rapidly from middle of 2nd segment to tip of 3rd, thence gradually narrowing. Upper side of last segment a little obscure. A few long whitish hairs at sides at base of abdomen, the remainder of the dorsal and ventral surfaces practically bare. Belly tawny brown.

Legs.—Front pair with coxae, base and tip of femora, basal half of tibiae and the metatarsi yellow, the remainder black. Middle pair similar, but the very short coxae obscure. Hind pair much enlarged, with obscure coxae. Of the hind femora the basal fourth is bright lemon yellow, the remaining portion having the proximal half black and the distal half reddish brown; the extreme tip is black. Under side beset with several rows of very short spines, and an additional row of about 8 or 9 longer ones. Tibiae distinctly but not greatly curved, pale yellow, rather less than the apical half black; tarsi all black, metatarsi distinctly enlarged and lengthened.

Wings pale grey; subcostal cell yellowish from tip of auxiliary vein; 4th longitudinal vein curved upwards to 3rd in a very rounded loop; 5th vein bent upwards at a slightly obtuse angle; halteres yellow.

Described from a single ♀ in the Indian Museum presented by Dr. A. D. Imms, taken by him at Bhowali, Kumaon District, 5700 ft., 2-vii-10.

Rhinobaccha gracilis, Meij.

One specimen in the Indian Museum taken at Pattipola, Ceylon, 3-vii-10, the exact locality from which the type came, agrees with every generic and specific character as given in Meijere's description.

I am uncertain as to its sex having seen only the one, but it is apparently a ♂.

The genus was described by Meijere in the Tijd. v. Ent. li, 315 (1908).

SPHEGINOBACCHA, Meij.

Tijd. v. Ent. li, 327 (1908).

Near *Ascia* and *Sphegina*. One species is referred here, *Sphegina macropoda*, Big. Meijere figures this, *l.c.* pl. viii, 43, and records a ♂ and ♀ from Semarang [*Jacobson*].

RHINGIA.

Rhingia binotata, Brun.

Only the ♂ was described originally. The ♀ has appeared from the banks of the Siyom River, near Yekshi (N.-E. Front. India), 3-ii-12, taken by Mr. Kemp on the Abor Expedition. In the Indian Museum.

R. sexmaculata, sp. nov. ♀, described by me (Rec. Ind. Mus., viii, 163, 1913) from a single ♀ from Dibrugarh, Assam, 17-xi-11 [*Kemp*]. *Type* in Indian Museum.

Subfamily *VOLUCELLINAE.*

Volucella pellucens, L.

One ♂ of this very common European species from Takula, Kumaon District, Western Himalayas. Not previously recorded from India. In the Forest Zoology Coll My *basalis* is very near it, but the distinctions stated in my description of the species hold good.

Meijere records *V. trifasciata*, Wied , from Semarang and *discolor*, Brun. from Japan.

GRAPTOMYZA, Wied.

Meijere records *G. longirostris*, W., and *brevirostris*, W., from Java, and adds a note on *G. atripes*, Big.; whilst *brevirostris* was taken by Mr. Kemp at Rotung, 1400 ft. (N.-E. Front. India), 25-xii-11. It also occurs in the Nilgiri Hills.

Graptomyza ventralis, W. var. nigripes, Brun.

Gangtok, Sikkim, 6150 ft., 9-ix-09, one ♀, and Kurseong, 3-vii-08 and 9-ix-09 [*Annandale*]. In Indian Museum. Meijere records *ventralis*, Wied., from near Buitenzorg, Java

One ♀ from Sadiya, Assam, 27 xi-11 [*Kemp*]. Of the typical form Mr. Kemp took a ♀ at Rotung, 26-xii-11.

Five new species have been recently described by Meijere:

punctata (Tijd v. Ent., li, p. 280, pl. viii, 28, 1908). Erima, Astrolabe Bay, Papua [*Biro*].

Type in Hungarian Museum, a unique specimen

longicornis, *l.c.,* p. 281. Sattelberg, Huon Gulf, Papua [*Biro*].
Type in Hungarian Museum, a unique specimen.
trilineata, *l.c.,* p. 282, ♂, Paumomu-Fluss, Papua [*Loria*].
Type in Genoa Museum.
jacobsoni, *loc. cit.,* liv, 343 (1911) Telaga Mendjer and Gunung Ungaren, Java [*Jacobson*].
flavipes, p. 344. Gunung Ungaren [*Jacobson*].

Graptomyza tinctovittata, mihi, sp. nov.

(Plate xiii, fig. 10).

♀ N. Bengal. Long. 3 mm.

Head.—Pale lemon yellow, face with a shining brown median stripe from antennae to mouth border. Occiput black, the colour encroaching narrowly on the vertex. Frons with a very large subquadrate blackish brown, moderately shining spot, which occupies nearly all the surface, not contiguous to the eyes, but extending downwards to the root of the antennae; this square spot joined to the vertex by a short, broad stripe embracing the ocelli. Eyes sparsely and microscopically hairy. Antennae brownish yellow, upper side a little brownish, arista bare.

Thorax.—Shining black, with short yellowish grey pubescence; side margins and posterior margin of dorsum narrowly pale yellow. Shoulders with a yellow callus, and there is an elongate perpendicular yellow spot on the mesopleura, just before the wing and united to the yellow margin of the thorax. Scutellum shining black, with two long bristles on each side of margin near the base and a pair of similar, widely separated apical ones.

Abdomen.—Bright yellow, 2nd segment with a broad black band on posterior border, widest in the middle, where it extends nearly to the base of the short and very narrow 1st segment. A similar band on 3rd segment, rest of abdomen black. Belly yellow, with a few blackish marks.

Legs.—Wholly yellow, except the hind coxae rather obscure, a broad dark brown band on hind femora leaving the knees narrowly pale, and hind tibiae blackish brown, with base and tips narrowly yellow.

Wings.—Very pale grey. A brownish very short stripe from tip of auxiliary vein to 2nd longitudinal vein, a second stripe from tip of 1st vein to (and indistinctly including) the upturned end of lower branch of 4th vein, and a 3rd stripe from tip of 2nd vein to, and including, the upturned end of upper branch of 4th vein; all these stripes being narrow and indistinct yet perfectly obvious. The closed anal cell very slightly infuscated at tip. Halteres brownish yellow.

Described from one specimen in the Indian Museum, sex uncertain but probably ♀, from Sukna, 500 ft., 1-vii-08 [*Annandale*].

Subfamily *ERISTALINAE.*

ERISTALIS, Latr.

This genus was not dealt with in my first paper owing to reluctance to identify closely allied species from descriptions alone. A certain number of interesting notes on some of the species are now added.

Eristalis tenax, L.

This very cosmopolitan species occurs freely in the Himalayas during the summer, the specimens in no way differing from European and North American ones.

The var. *campestris*, Mg., is also common, ♀ ♀ only. *E. tenax* occurs sparingly in the plains (Meerut, 8—14-iii-07; Bareilly, 15—22-iii-07); and it is in the Indian Museum, from Yunnan, China. I have taken it freely at Mussoorie and Darjiling.

Eristalis sepulchralis, F.

This common European species was taken by me at Shanghai, i-v-06 and at Hankow, 22—26-iv-06, at both places being common. The dark spot on the 2nd abdominal segment in the ♂ instead of being of the usual shape takes the form of a broad stripe with a transverse line at base and apex, whilst in the ♀ the spot on the 1st segment is almost reduced to a broad stripe, and that on the 2nd segment to a narrow streak. The antennae in the ♀ are apparently a little darker.

On a ♀ specimen in the Indian Museum from Yange-Hissar taken on the Yarkand Expedition, the abdominal spots are quite normal.

Eristalis arvorum, F.

Meijere makes *E. quadrilineatus*, F., a synonym.

The species is the commonest of the Indian ones and occurs apparently all over the country from the Himalayas to the south; extending also to the East Indian Islands and China. It has been found by Dr. Annandale breeding in rotting seaweed in brackish water at Lake Chilka, Orissa, in February and November.

Eristalis quinquestriatus, F.

Meijere records it from various localities in Java and redescribes both sexes.

Eristalis obliquus, Wied.

Meijere records, figures (pl. vii, 17-18), and notes both sexes, the ♂ from Papua, the ♀ (hitherto unknown) from Batavia. It is closely allied to *arvorum*, F.

Eristalis orientalis, Wied.

Wiedeman described the ♂ only. Meijere records and describes the ♀ from Tosari, Java [*Kobus*]. Some of both sexes are in the Indian Museum from Sikkim and the Darjiling District.

Eristalis niger, Wied.

The ♂ redescribed by Meijere from Sukabumi, Java [*Kramer*]. A ♀, without data, is under this name in the Indian Museum, identified by Bigot, but I cannot be sure that it is this species.

Eristalis sinensis, Wied.

Two specimens from Assam are in my collection purchased some years ago in a miscellaneous lot of diptera at a sale.

Eristalis taphicus, Wied.

A few in the Indian Museum from Karachi, both sexes. Verrall claims this to be a variety of the European *aeneus*, Scop., from a series taken at Aden, and this may probably be the case.

Eristalis splendens, Le Guillou.

Apparently generally distributed in the East, Meijere recording it from Erima, Papua [*Biro*]. I possess one specimen from Key Island.

Eristalis tortuosa, Walk.

This species, described in Proc. Linn. Soc Lond., v, 266 (1861), was omitted from Van der Wulp's Catalogue. The ♂ only is mentioned, coming from Tondano. There is no indications as to where the type is located.

Eristalis suavissimus, Walk.

Meijere records from Meranke, South Papua [*Koch*].

Eristalis postcriptus, Walk.

One in my collection from Papua, but I do not know if the identification is correct.

Eristalis resolutus, Walk.

Recorded from several localities in Papua by Meijere. He redescribes both sexes.

Eristalis muscoides, Walk.

Meranke, Etna Bay, South Papua [*Koch*]. Recorded and noted by Meijere, both sexes.

Eristalis externus, Walk.

A ♂ and ♀ under this name exist in the Indian Museum collection. They were identified by Bigot but, I think, incorrectly, owing to discrepancies in the size, the length of the abdomen and the marks of the latter.

Eristalis nitidulus, Wulp.

Meijere records a ♂ from Semarang, July [*Jacobson*].

Eristalis solitus, Walk.

This species is common in Himalayan localities occurring freely at Darjiling during my two last visits (13—18-ix and 1—11-x-13) and I have taken it as far north-east as Yokohama, 24-v-06. In the Indian Museum from Shillong, Darjiling, Mussoorie, Naini Tal, Simla and Gangtok.

Eristalis inscriptus, Dol.

Meijere records this from Paumomu-Fluss, Papua [*Loria*], noting that it is very near *muscoides*, Walk.

Eristalis saphirina, Big.

This species, placed in the sub-genus *Eristalomyia* and described in Ann. Soc. Ent. Fr. (5) x, 230 (1880) from Papua, was omitted from Van der Wulp's Catalogue. Type in the Bigot collection.

The following new species are set up by Dr. Meijere in the paper from which the above notes by him are taken (Tijd. v. Ent. li, 1908). They are preceded by a very valuable analytical table of over twenty species known to him.

obscuritarsis, p. 250, ♂ ♀, pl. vii, 19, 20. Semarang [*Jacobson*]; Singapore and Bombay [*Biro*].
kobusi, p. 252, ♂ ♀. Tosari, Java [*Kobus*].
kochi, p. 255, ♂ ♀, Meranke, South Papua [*Koch*].
collaris, p. 258, ♂ ♀, Papua, several localities.
Types in Hungarian Museum.
maculipennis, p. 261, ♂, Lawang, Java [*Fruhstorfer*].
Type in Hungarian Museum, a unique specimen.
lunatus, p. 264, ♀, Astrolabe Bay, Papua [*Biro*].
Type in Hungarian Museum.
fenestratus, p. 269, ♀, Friedrich Wilhelmshagen, Papua.
Type, a unique specimen, in Hungarian Museum.
cupreus, p. 271, ♂ ♀, Simbang, Huon Gulf, Papua [*Biro*]; Meranke, Papua [*Koch*].
Types in Hungarian and Amsterdam Museums.
heterothrix, p. 273, ♂ ♀, Tami, Cretin Is., Mahakkam, Borneo [*Nieuwenhuis*].

From the context it is to be gathered that the type ♂ and ♀ are in the Hungarian National Museum and a further specimen in the Leyden Museum.

In *loc. cit.*, liv (1911), the same author describes the following:—

nigroscutatus, p. 337, ♂, Tandjong Priok and Batavia environs [*Jacobson*].

ferrugineus, p. 339, ♂ ♀, Batavia environs [*Jacobson*], a unique pair.

neptunus, p. 340, ♂, a unique specimen, Batavia environs [*Jacobson*].

lucilia, p. 341, ♂, a unique specimen, Semarang [*Jacobson*].

tristriatus, p. 342, ♂, Semarang, Batavia [*Jacobson*].

The types of these species are in the Amsterdam Museum.

MEGASPIS, Macq.

Meijere records *M. chrysopygus*, W., *errans*, F., *zonalis*, F., and *crassus*, F., all from Java and relegates my *transversus* to a synonym of *argyrocephalus*, Macq (*Eristalis*). He adds a table to five species, including *sculptatus*, Wulp. I have seen *M. crassus* and *zonalis* recently from Darjiling—and an *errans* ♀ from Cochin State, 1700-3200 ft., 16—24-ix-14 [*Gravely*].

Mr. Austen writes me that *Megaspis* is antedated by *Phytomyia*, Guer. (1833), in Belanger's Voyage aux Indes orientales, 509, with *chrysopygus*, Wied., as type, but I do not like to change the name after it has stood so long.

HELOPHILUS, Mg.

Meijere gives a table embracing eight species, including the following new ones. in Tijd. v. Ent. li (1908):—

niveiceps, p. 236, ♂, pl. vii, 16, Java [*Piepers*].
Type in Amsterdam Museum.

fulvus, p. 237, ♂, Moroka, Papua, 1300 metres [*Loria*].
Type in Genoa Museum.

scutatus, p. 238, ♂, Paumomu Fluss, Papua [*Loria*].
Type in Genoa Museum.

Dr. Meijere redescribes *H. quadrivittatus*, Wied., ♂ ♀, and records it from Semarang; also adding notes on *curvigaster*, Macq., and *vestitus*, W. (recording it from Sumatra).

AXONA, Walk.

To this interesting genus I have been able to add a second species, *cyanea* (Rec Ind. Mus., ix, 272 (♂) and 277 ♀, pl. xiv, fig. 3, full insect, 1913), from Darjiling, iv-1913, sent to the Indian Museum by Lord Carmichael. Only one species, *chalcopygus*, W., was previously known.

MALLOTA.

Mallota rufipes, Brun.

Described from a unique ♂ (Rec. Ind. Mus., ix, 271, 1913) from Singla, Darjiling District, April 1913.

Merodon ornatus, mihi, sp. nov.

(Plate xiii, fig. 11).

♂ Western Himalayas. Long. 10 mm.

Head.—Vertex wholly occupied by a moderately elevated aeneous black tubercle, bearing the three reddish ocelli. Frontal triangle small, black, with a little yellowish grey tomentum. The eyes contiguous for barely one-third of their total height, as viewed from in front. Whole under side of head yellowish, with whitish reflections, except the projecting face, which is shining black: the oral margin very narrowly reddish brown. Antennae pale brownish yellow, the 1st joint the darkest, the 3rd with whitish dust and a pale yellowish, basal, bare arista. Proboscis blackish. Back of head aeneous black, the upper ocular orbit with short yellow hair, the outer and lower ocular orbits with whitish hair.

Thorax.—Dorsum dull aeneous black, mainly covered with short yellowish hair, but which takes a golden brown hue where it forms two moderately broad dorsal stripes. The yellow hair is a little more prominent below the broadly whitish shoulders, behind the wings and on the entire hind margin of the concolorous aeneous scutellum.

In an indistinct manner, the dorsum of the thorax bears three broad blackish stripes: a median one, and one on each side of it, well separated, commencing just behind the whitish shoulders and continued to the posterior margin, the median dark stripe attaining the anterior margin of the dorsum.

Between these three indistinct dark stripes, the aeneous ground colour is more pronounced, and these spaces bear deeper golden brown hairs. Under side of thorax blackish, slightly aeneous, a patch of white hair between the anterior pairs of legs, immediately below the end of the transverse suture; and a little white hair generally distributed over the ventral surface.

Abdomen.—Black, moderately shining, with an aeneous tinge, which latter is most conspicuous on the unicolorous 1st segment; the 2nd segment has a pair of large yellowish spots, separated by a moderately wide space, and enlarged laterally to the full length of the segment. A similar pair of spots on the 3rd segment, but narrower at the sides, the colour not there reaching the hind margin. The whole surface of the abdomen is covered with short bright yellow hairs. At the sides, the hair is more whitish, especially towards the base, where it is also longer.

Belly mainly black, except on 2nd and 3rd segments, which are yellowish.

Legs.—Coxae black, with a little pale greyish hair, anterior pair grey-dusted; remainder of anterior legs wholly bright pale yellow. Hind femora greatly incrassated as usual, reddish brown; a few short black spines of unequal length on under side towards the tip, the whole limb with short soft yellow hair. Hind tibiae well curved, yellowish, with a tolerably distinct subapical black band with ill defined edges, and a tendency to a sub-basal narrower and still less definitely marked band. The whole limb with very short yellowish hair, but on the inner side is a thick row of very short and stiff black hairs; hind tarsi yellow. Claws, basal half bright yellow, apical half black.

Wings.—Pale yellowish grey, stigma brownish yellow; halteres bright yellow.

Described from one ♂ in the Indian Museum from Bhowali (5700 ft.), Kumaon District, taken by Mr. A. D. Imms, June 1909.

Subfamily *MILESINAE.*

Myiolepta himalayana, mihi, sp nov.

(Plate xiii, figs. 12, 13).

♂ ♀ West Himalayas. Long 7—8 mm.

Head, ♂.—Eyes bare; contiguous for only a short space, leaving a rather small vertical triangle, which is shining black, with some yellowish grey hairs. Sides of frons narrowly grey-dusted, the whole of the upper part of the face also, that is to say, the part immediately below the rather conspicuously produced antennal prominence, which latter is shining black, the extreme frontal edge narrowly orange. Facial bump very large and conspicuous, the central knob distinct, not cut away below (in profile), but the mouth opening less projecting. The whole protuberance shining black. The lower sides of the face with a little grey dust, and a few stiff long hairs near lower corner of eyes. Antennal third joint rounded, the whole organ pale vinaceous, with a hoary bloom, arista bare, orange at base. Back of head shining black, ash grey behind lower part of eyes, where it is considerably developed, and bearing there a fringe of yellowish hairs. An arc of short bristly brownish black hairs behind the vertex.

In the ♀ the frons, at the level of the antennal prominence, is one-third the width of the head, the frons and face being mainly shining black but narrowly grey-dusted at the sides, and with a little stripe of very short greyish pubescence along the sides from the cheeks to the mouth opening. There is a little grey hair in front of the lower corner of the eyes as in the ♂. Eye margins are present in both sexes as in *Chilosia.*

Thorax and scutellum aeneous black, with short yellowish grey pubescence; anterior margin of dorsum, including humeri, a little ash greyish. A fringe of long yellowish grey wavy hairs placed transversely in front of the wings.

Abdomen blackish ; 2nd segment nearly wholly orange reddish, the colour encroaching on base of 3rd segment, whilst in the ♀ the posterior border is also reddish. Extreme tip of abdomen orange red. Whole abdomen with short greyish pubescence, which is a little longer at the sides. Belly blackish, with grey pubescence, dull orange reddish for a considerable space about the 2nd segment.

Legs simple but somewhat strong, the femora having small spines below, towards the tips ; black, with fairly dense greyish pubescence. Trochanters, base and tips of tibiae, orange yellow. The underside of the hind tarsi (of which the metatarsus is distinctly though not greatly enlarged), brownish yellow, and the upper side of the 2nd and 3rd joints is brown in the ♂. In the ♂ the first three joints of the middle tarsi are orange yellow, as is the whole middle tarsus in the ♀. The exact limits of the pale colour in the tarsi is probably variable.

Wings pale yellowish grey, stigma yellowish, subcostal cell up to the stigma, brownish; a barely obvious suffusion immediately before and below the stigma. Halteres pale orange.

Described from a single ♂ and ♀ in the Indian Museum from Matiana taken by Dr. Annandale.

It has been rather difficult satisfactorily to place the present species generically. It has every appearance of a *Chilosia*, even to the eye margins, which are quite as distinct as in many species of that genus. But *Chilosia* should have no trace of pale markings, so that the nearly all orange red 2nd abdominal segment would throw it out. Considering the species as of the Syrphinae, it works down by Verrall's table of genera to *Chrysochlamys*, a genus which it is totally unlike in facies, colour, the shape of the closed 1st posterior cell and in the absence of the thoracic and scutellar bristly hairs.

If the exact position of the anterior cross-vein is not regarded as an absolute character, and Verrall doubted its inviolability,[1] it becomes a *Myiolepta*, which that author puts in the Milesinae, considering its affinities with *Tropidia* greater than those with Syrphinae, and he speaks of the genus as of "rather doubtful location." He says the femora are all swollen, and serrate near the tips below, but as Schiner gives the femora as simply "rather thickened" and there seems to be no further discrepancy, the new species is placed here.

XYLOTA, Mg.

One new species described, *X. aeneimaculata*, Meij., in Tijd. v. Ent. li, 227, 1913 from Moroka, Papua, 1300 metres [*Loria*]; one ♂ in the Genoa Museum. Dr. Meijere adds notes on some of the

[1] British Flies, *Syrphidae*, 572, footnote.

other known species, and I have described an additional species from Darjiling, *annulata* ♂ ♀, v-12 and iv-13 (Rec. Ind. Mus., ix, 270, ♂ ♀, pl. xiv, 11-15, 1913).

Xylota bistriata, mihi, sp. nov.

♂ ♀ Cochin. Long. 11—13 mm.

Head.—Eyes in ♂ practically contiguous for about lower third of distance from vertex to base of frons. Width of vertex about one-eighth that of head, vertex blackish aeneous with a little pale hair, the small ocelli distinct, reddish. Eyes in ♀ separated by a frons about one-eighth the breadth of the head, widening a little at base of antennae.

Face and frons blackish, covered with yellowish white tomentum; antennae covered with yellowish grey dust, arista black, base brownish yellow. Occiput blackish grey, with whitish dust; some bright yellow short hairs along top of head, intermixed behind vertex with black ones. Ocular orbit with a fringe of short white hairs which are longest on under side of head.

Thorax.—Dorsum greenish aeneous, with short and rather thick bright yellow pubescence. A pair of well separated pale median longitudinal stripes bearing short bright yellow hairs, becoming indistinct posteriorly but just attaining the scutellum, which latter is also greenish aeneous with short yellow pubescence and a fringe of short yellow hairs below hind margin. Sides of thorax blackish aeneous, nearly bare; sternopleura and mesopleura with a grey tinge and bearing some short yellow pubescence. Humeri apparently bare; but if viewed from behind they are seen to bear some short yellow pubescence.

Abdomen.—Blackish aeneous with a dull steel tinge, which on the 2nd segment in the ♀ may occasionally shew, seen from behind, a pale violet reflection; basal segment a little darker; on hind margin of both 2nd and 3rd segments a large dull black (seen from behind) sub-triangular spot, the apex reaching nearly to the base on the 2nd segment, but only to the middle on the 3rd segment. Dorsum of abdomen with microscopic dark hairs, sides with short pubescence, which is longer towards the base and is yellowish in the ♂ and white in the ♀. Genitalia in ♂ globular, of a dull steel colour, with some yellow hairs; ovipositor brownish yellow.

Legs.—Coxae aeneous, grey-dusted; hind pair with soft pale hair below. Anterior legs yellowish with short concolorous pubescence, which is longest on inner side of middle tibiae; tips of middle femora narrowly brown. Anterior tibiae longitudinally streaked irregularly with brown on inner and outer sides, last tarsal joint brown. Hind femora considerably incrassate, brownish yellow with a broad blackish brown median band, and the tips dark brown; a moderately long distinct black spine below at base and on the under side towards tip, an outer row of 6 to 8 black spines of moderate size, gradually diminishing in length posteriorly, and also an inner row of about four shorter ones of uniform length. A little long soft pale yellow hair on middle of underside; remainder of hind femora with very short yellow pubescence, which is longest about the middle on the outer side. Hind tibiae considerably curved yellow, with yellow pubescence, inner side mainly black; hind tarsi blackish brown with pale yellow pubescence; golden brown minute pubescence below.

Wings pale grey; subcostal cell pale yellow; halteres pale lemon yellow; anterior cross-vein barely beyond middle of discal cell.

Described from 3 ♂♂ and ♀♀ in perfect condition in the Indian Museum from Parambikulam, Cochin, 1700—3200 ft., 16—24-ix-14 [*Gravely*].

Criorhina imitator mihi, sp. nov.

(Plate xiii, fig. 14).

♀ Western Himalayas. Long. 17 mm.

Head produced downwards to a greater length than height of eyes. Frons and vertex blackish, with yellowish grey dust and dark brown hairs, the vertex with long brownish yellow hairs. Antennal prominence shining black, with yellow dust about the sides, and covered with some sparse brownish yellow hair. Face and lower part of head shining black, face with yellowish brown tomentum on each side up to end of snout, leaving an irregular median bare stripe; a few yellow hairs along inner orbit of eyes. Proboscis considerably longer than head, blackish, labella rather large; palpi more than half as long as proboscis, blackish. Antennal 1st and 2nd joints black, 3rd reddish brown, blackish at tip, arista black. Back of head dull shining black with brownish yellow hair, which extends to the vicinity of the cheeks, where it is longer.

Thorax moderately shining black, with a grey tinge anteriorly, covered with thick pubescence, which is mainly black, but is yellow on about the anterior half, and again for a narrow space along the hind border. The shoulders, posterior corners and scutellum are covered thickly with yellow pubescence which extends to the pleura below the shoulders.

Abdomen moderately shining black, with black pubescence. On 2nd segment the pubescence is yellowish, on posterior margins of 3rd and whole surface of 4th and 5th, bright red, long and conspicuous. Belly black, with pale yellow hairs on basal half.

Legs black, some yellow pubescence about the base and sides of all the femora.

Wings pale grey, pale brown tinged on anterior half; a slight infuscation about the stigmatic region, origin of 3rd vein, the posterior cross vein, and most of the veins being just perceptibly infuscated.

Described from one ♀ in the Indian Museum from Onari, Garhwal Distr., 11,000 ft., W. Himalayas, 27-vi-14 (*Tytler*).

In connection with this species may be noted an interesting case of mimicry. *C. imitator* itself, in the pale pubescence on the anterior part of the thorax and on the scutellum, in the colo-ration of the abdomen, the black legs and grey wings, distinctly resembles the bee *Bombus trifasciatus*, Smith; but the protective resemblance accorded to a large *Echinomyia*-like Tachinid fly (though not belonging to that genus), 20 mm. long, by the similarity of its appearance to that of the bee, is even more striking. The pubescence of the fly is tolerably dense, black, except for a broad yellowish grey band on the anterior margin of the thorax, and on the scutellum. The apical third of the abdomen bears rather bright red pubescence. No strong bristles are present any-where, the eyes are bare, the antennae short, the 3rd joint much broadened vertically, notched at the truncate apex. Five speci-mens are present, taken in company with the Syrphid and one specimen of the bee.

Lycastris cornutus, Enderl.

Described in Stett Ent. Zeit. lxii, 136 (1910), from Formosa. Type in Stettin Zoological Museum.

SYRITTA, St. Farg.

In my previous paper on Syrphidae my impression that there were only three Indian species of this genus was noted, and the further examination of a good number of specimens increases that impression. One of these is the common *S. pipiens*, L., of Europe and North America which occurs commonly in the Himalayas and also more rarely in the Indian plains. One specimen is in the Indian Museum from Mergui.

Of *orientalis*, Macq.[1] and *rufifacies*, Big., I prefer to speak at present, as *forms* only, for two reasons. Firstly because there is primarily *S. indica*, Wied., to be disposed of as the earliest des-cribed oriental species; but as his description is so meagre, it is unidentifiable. Still he says "very like *pipiens*, L." from which it may be inferred that the hind femora are practically wholly black. Now in *pipiens* there is normally a pale transverse streak in the middle, on the underside, which is often of considerable width and length, but which also is sometimes barely traceable, so that specimens may quite possibly occur which are practically wholly black. Wiedemann's type, moreover, may have not been in the best condition so that the presence or absence of such a pale streak may not have been easily ascertained, nor, inciden-tally, considered of much consequence in those days. Therefore,

---

[1] See Tijd. v. Ent. li, 224 for redescription ♂ ♀.

if *indica* should have wholly black hind femora there can be little doubt of its identity with *orientalis*, Macq., the former name taking priority.[1]

This form *orientalis* (I call it so until the synonymy is established) is quite a good one and is mainly distinguished by the wholly black hind femora.

Dr. Meijere sinks *Senogaster lutescens*, Dol., as synonymous, whilst *laticincta*, Big., *nom. nud.*, in the Indian Museum from Karachi and Calcutta, is certainly so; moreover *illucida*, Walk., from Celebes is likely to be also identical, the expression "vertex black with an elongated white point on each side" reading as though reference was made to the small portion of the whitish grey occiput visible on each side from above.

S. amboinensis, Dol., from Amboina may or may not be distinct; the anterior legs are obscurely ringed, which may mean anything, and as occasional specimens of both *pipiens* and *orientalis* have a dark streak on the anterior femora, it may be only a variety of the latter.

The form *rufifacies*, Big., is as well marked as *orientalis* and is distinguished by its bright reddish orange hind femora, the apical third being black. Though Dr. Meijere records it as synonymous with *orientalis*, the form is as distinct as that one, several of each sex in the Indian Museum answering exactly to Bigot's description. I have taken it myself at Agra, 4-iv-05.

There are, however, 2 ♂♂ in the Museum collection which appear intermediate between *orientalis* and *rufifacies*, and which may break down the barrier between them. These have dark brown or reddish brown femora and one has the tips more or less darker still. I have one in my own collection taken by me at Agra.

The abdominal markings are but a slight guide, as in both *orientalis* and *rufifacies* the pairs of spots on the 2nd and 3rd segments[2] are sometimes quite separate and sometimes merged into a transverse band. This happens with each pair of spots independently of one another and is equally variable in both forms.

There appears to be no other character offering any solution of the number of forms existing.

At present my own opinion is towards the following synonymy, regarding them taxonomically as forms only, except *pipiens* and my supposed *indica* of Wiedemann.

[1] There is certainly the possibility that *indica* may be simply *pipiens* after all, but it is hardly to be supposed that Wiedemann would not have recognized it as such, although probably in those days species were not thought to have so wide a distribution.

[2] Macquart speaks of the spots on the second segment being united into a band, but as it is more usually those on the third segment which are contiguous, I think he must have overlooked the very short 1st segment and was really referring to the 3rd segment.

1. **pipiens**, L.

2. **indica**, Wied.

orientalis, Macq.
lutescens, Dol (*Senogaster*).
illucida, Walk.
laticincta, Big. *nom. nud.*

3. **amboinensis**, Dol.

4. **rufifacies**, Big.

(Possibly synonymous with *orientalis*).

5. **luteinervis**, Meij.

The latter species, recently described (Tijd. v. Ent. li, 226, ♂, 1908), from Papua, is distinguished from *orientalis* by the pale yellow veins, which seems at best a very slender character.

EUMERUS, Mg.

Meijere describes four new species in the Tijd. v. Ent. li (1908).

flavicinctus, p. . 15, ♂, Semarang, Java; Medan, Sumatra
parallelus, p. 217, pl. vii, 12, ♀, environs of Batavia.
niveipes, p. 220, ♂, Batavia: (♀ described by him in *loc. cit.* liv, 335, from Semarang).
peltatus, p. 223, ♂, Friedrich Wilhelmshafen, Papua

Types of the first three species in Amsterdam Museum, type of the last one in the Hungarian Museum.

I have myself described a new species from Darjiling, *E. rufoscutellatus*, ♂ (Rec. Ind. Mus. ix, 269, ♂, pl. xiv, 13).

I had anticipated drawing up a table of oriental species in this genus, but from the descriptions only this is quite impracticable, the species being very closely allied, whilst the few characters that appear most useful taxonomically, *viz.* the width and shape of the frons, the structure of the hind tarsi and the degree of pubescence or bareness of the eyes, are ignored by all the older writers. The presence or absence of an infuscation at the wing tip, the intensity or entire absence of the pale stripes on the thorax, and the proportion of tawny colour in the legs are all characters subject to considerable variation.

It is probable that my *nepalensis* will sink to synonymy, but it is not certain which species it is identical with, as three or four appear very closely allied if allowances for variation are made. These are *macrocerus*, W., *aurifrons*, W., (*splendens*, W.), *nicobarensis*, Sch., and *niveipes*, Meij. Specimens agreeing with the description of my *nepalensis* are in the Indian Museum from Mergui, Margherita, Pallode and Travancore, 15-xi-08 [*Annandale*], these being four males, and from Mergui, Nepal (the type specimen of *nepalensis*), and Sibu, Sarawak, 2-vii-10 [*Beebe*], three females, that is seven specimens altogether.

All these appear to come within the range of a single species possessing the following variations of character. The frons in the female from shining black to rich blue black: the antennal 3rd joint may be black on upperside or unicolorous; the dorsal thoracic stripes vary in intensity and the 3rd pair of abdominal spots are wanting in one specimen; the wing tip varies from quite clear to distinctly and broadly brown infuscated; the hind tarsi vary from white to brownish yellow, the upper side of the metatarsus (and sometimes also the basal half of the succeeding joint) may be wholly or partly brown.

Taking all things into consideration the chances are in favour of *aurifrons*, W., being the species at present referred to.

A description of the specimen from Borneo is added, simply as such, as an augmentation of that of my *nepalensis*. In the ♂ of the species under discussion, whatever it may be, the frons is two to three times as broad on the vertex as at the point of nearest contiguity of the eyes. There was an error in my description, the frons not being black but brilliantly shining blue black.

Eumerus aurifrons, Wied.

Dr. Meijere makes *splendens*, W., a synonym of this and redescribes the ♂, recording the species from Batavia, Semarang, Ceylon and the Dammer Is. (Tijd. v. Ent. li, 218). This may be the species described by me as *nepalensis* (*infra*).

Eumerus nepalensis, Brun.

(*Description of a specimen from Borneo*).

♀ Borneo. Long. 5—6 mm.

Head.—Frons distinctly narrowed at vertex, measuring at the greatest width, just above the antennae, one-fourth of the head; shining black, with a grey-dusted spot each side about the middle of the eyes and contiguous to these latter, the spots nearly meeting one another in the middle of the frons. Vertex with brown hairs. Back of head behind vertex and upper part of eyes, shining black, narrow, occipital margin imperceptible below middle of eyes, occiput dark grey or blackish. The margins of the face from opposite the base of the antennae, a little grey-dusted, and the face itself with a little yellowish hair. Antennae bright brownish yellow, upper margin of 3rd joint blackish, arista black, base a little pale.

Thorax.—Shining black, with two well separated narrow whitish median stripes from anterior margin to behind transverse suture. Anterior part of dorsum a little aeneous in certain lights. Dorsum with yellow hair which becomes greyish about the shoulders and pleurae. Sides of thorax dull black; scutellum shining black, with yellowish grey hairs.

Abdomen.—Shining black, with almost microscopic greyish pubescence except towards the sides where it is quite distinct. The 2nd segment with two oval yellowish, diagonally placed good sized spots. The 3rd and 4th segments each with two narrow greyish, barely curved lunules, diagonally placed, beginning in the middle of the segment, well separated from one another, and lying towards the posterior corners of the segments. Belly yellowish with a median black stripe.

Legs.—Coxae and anterior femora black, the latter narrowly but very distinctly brownish yellow at tip; anterior tibiae mainly brownish yellow with a more or less distinct wide blackish band beyond the middle; anterior tarsi brownish yellow with whitish reflections. Hind femora much larger than anterior pairs but not incrassated, with a row of about twelve small spines on apical half of underside; and a second row towards the outerside of a less number; hind tibiae mainly black, narrowly brownish yellow at base and tip; hind tarsi brownish yellow with whitish reflections, basal half of upper and underside of hind metatarsi black. Anterior femora with grey hair below; anterior tibiae with similar hair but more extensive; hind femora and tibiae covered with moderately short greyish hairs; all tarsi moderately grey pubescent.

Wings.—Very pale grey; stigma yellowish brown, a very slight suffusion over upper part of the upturned section of the 4th longitudinal vein; there is also the suspicion of an appendix in the middle of the outer side of the anterior cross vein. Halteres very pale lemon yellow.

Described from a single perfect ♀ from Sibu, Sarawak, 2-vii-10 [*Beebe*], in the Indian Museum.

Eumerus flavipes, mihi, sp. nov.

♀ Borneo. Long. 5 mm.

A single example, taken by Mr. Beebe 10 miles south of Kuching, Sarawak, 24-vi-10, appears to be a closely allied species to the above. The principal difference is in the anterior legs which are all wholly bright orange yellow. The other differences are as follows: the 2nd pair of abdominal spots are yellow, not white; the 3rd antennal joint is wholly bright yellow, without trace of darkening on the upper edge; the greyish stripes on the thorax are absent; the wing tips are distinctly, though not deeply darkened as far inwards as to encroach on the 1st posterior cell, and there is no sign of an appendix to the anterior cross-vein.

Eumerus halictiformis, mihi, sp. nov.

♂ ♀ Bengal. Long. 5 mm.

Head.—In ♂ eyes quite bare, touching for a short distance only, the front facets a little larger than the others. Frons shining black with greyish dust except for a space bearing the two upper ocelli a little below the vertex, and a space lower on the frons bearing the 3rd ocellus. Blackish hairs on the frons rather thickly placed. Face and the narrow occipital margin wholly ash grey-dusted, the former with whitish hair. Antennae blackish, 2nd joint wholly and the 3rd joint more or less, dull reddish brown on the basal part.

In the ♀ the frons is barely narrowed at the vertex, and at the level of the antennae is equal to one-fourth the width of the head; the lowest ocellus less far removed from the others than in the ♂. The frons is considerably covered with yellowish grey hair.

Thorax. The general impression of the dorsum is that of a bluish grey background with four dark spots, one pair of which are more or less rounded ones on the anterior half occupying the greater part of the space, with a second pair, produced posteriorly, behind the suture, less in size than the others; whilst there are two median narrow black stripes from the anterior margin in about the middle. Sides grey with whitish grey hair on pleurae; scutellum aeneous with rather thick brownish yellow hair.

Abdomen.—Shining black, with, on each of the 2nd, 3rd and 4th segments, a pair of diagonally placed grey lunule-like, barely curved spots, beginning almost contiguous to one another in the middle of the segment near the anterior margin, and extending to the posterior corners, which they attain. The whole abdomen covered with very short yellow socketed hairs. Belly dark.

Legs.—Anterior femora and tibiae black, both brownish yellow at both base and tip, the former with greyish white hair behind, and the latter more extensively covered with similar hair. Hind femora considerably incrassated, aeneous, covered with grey hair; hind tibiae aeneous, covered with grey hair; knees and base of hind femora brownish yellow. Anterior tarsi brownish yellow with whitish reflections viewed in certain lights; hind tarsi brown, the hind metatarsi much enlarged, black. The hind tarsi with yellowish grey hair above and rich golden brown pubescence below.

Wings.—Nearly clear; stigma pale brownish yellow; halteres pale brownish yellow.

Described from one ♂ and one ♀ from Puri, Orissa Coast, 1—5-viii-10 [*Annandale*]. In the Indian Museum.

Eumerus halictoides, mihi, sp. nov.

♂ ♀ E. and W. Himalayas. Long. 5—6 mm.

Very near *halictiformis* but certainly distinct. The differences are as follows:—

The 3rd antennal joint is rounded above at the tip, instead of being broadly truncate; the thorax is a little, but obviously, cupreous, with two widely separated whitish dorsal lines; the frontal triangle in the ♂ is distinctly yellow, with yellow hairs, in complete contrast to the whitish face; the hind metatarsus is not greatly thicker than the rest of the hind tarsus and is longer proportionately than in *halictiformis*, in which the hind metatarsus is twice as broad as the other joints, and apparently flatter; lastly the tibiae and tarsi are nearly wholly black except the reddish brown underside of the hind tarsi.

The species is also slightly larger and more robust.

Described from a type ♂ from Darjiling, 2-x-08 [*Brunetti*], and a type ♀ from Simla, 9-v-09 [*Annandale*]; both in the Indian Museum.

Eumerus pulcherrima, mihi, sp. nov.

♀ Darjiling. Long. 7 mm.

Head.—Frons one-sixth the width of the head, aeneous, darker on vertex, a slightly greenish tinge in front, minutely punctured. Ocelli small, red, well separated from one another and from the eye margins. At each side of the frons, along the eye margins, from the lowest ocellus to just above the antennae, a little yellowish pollinose dusting, which becomes white at the level of the antennae, where it merges in the white-dusted face covered with yellowish white hair.

The frons is covered with a moderate amount of light yellowish hair, which on the vertex is replaced by dark brown hair. Posterior orbits of eyes rather narrow, yellow-dusted, with bright yellow hair behind the vertex. Eyes with dense short brownish grey hairs. Antennae black, with a little hoary bloom, if viewed from in front; the dorsal arista black, curved upwards, a little pale at the base. Proboscis dark brown.

Thorax.—Aeneous, with brilliant cupreous and violet reflections; a little but conspicuously hoary below the anterior margin in front. Three very narrow whitish dorsal lines from the anterior margin, but not reaching the posterior margin; a transverse narrow whitish line follows the transverse suture. Sides below shoulders yellowish white with rather shaggy yellowish white hair. Humeral calli small, aeneous; remainder of thorax below dorsum, grey. Scutellum very conspicuous, bright shining cupreous with dense long reddish orange hair.

Abdomen.—Aeneous violet; a large triangular cupreous spot with yellow hairs in front and with whitish hairs behind, on each side of the 2nd segment. In certain lights the sides of the abdomen towards the tip, and the whole of the last (4th) segment appear more or less cupreous or aeneous. On the middle of each of the 2nd, 3rd and 4th segments are two greyish white, narrow stripes, beginning in the centre of each segment, almost contiguous, and extending diagonally to the posterior corners. The whole surface of the abdomen is uniformly punctured, and is covered with short light yellow hairs, which are depressed, and which are much thicker on the last segment. Belly dull liver brown.

Legs.—Coxae blackish, with hoary bloom and greyish hairs. Anterior femora dull aeneous black, a little brownish yellow at base and tips; a fringe of pale yellow hairs on underside; hind femora considerably enlarged, distinctly aeneous, covered with yellowish grey hair; brownish yellow at base and tips.

Anterior tibiae with basal half brownish yellow, apical half or thereabouts, blackish; the tips brownish yellow, the whole tibiae with yellowish grey hair. Hind tibiae as aeneous as hind femora, considerably larger than the anterior ones, being covered with much more hair. Anterior tarsi moderately bright brown with yellowish grey hairs; hind tarsi blackish above with yellowish grey hairs, bright reddish brown below.

Wings very pale grey; stigma small, dark brown: halteres pale yellow.

Described from a perfect unique ♀ in the Indian Museum from Kurseong, 8-vii-08.

Allied to *splendens*, W., and *albifrons*, Walk. From the former it is distinguished by the black (not brilliant red) antennae, and its larger size; from the second species by the bright reddish orange hair of the scutellum and the black antennae; it is also rather larger and more robust than the specimen of *albifrons*, Walk., sent to the Museum by Herr Meijere. The differences, however, may be sexual and *pulcherrima* may prove to be the ♀ of Walker's species.

It is the most handsome eastern species of the genus known to me.

Eumerus aeneithorax, mihi, sp. nov.

♂ Simla. Long. 7 mm.

Head.—Eyes contiguous for a comparatively short space only. Frons and vertex brassy aeneous, shewing various tints when viewed from different directions; black hair on lower part of frons, yellow hairs on upper part and on vertex. Face dull blackish grey, with light tomentum which appears yellowish white viewed from above. Face clothed with white hairs. Antennae wholly black, 3rd joint with obtusely rounded tip. Occiput whitish grey with a narrow fringe of whitish hairs round the margins, some yellow hair on the brassy aeneous upper ocular margin, which is moderately puffed out.

Thorax and *scutellum*, shining brassy aeneous, both rather thickly clothed with brownish yellow pubescence; dorsum with a pair of widely separated whitish tomentose stripes and traces of a very narrow median line of the same colour. Pleura dull aeneous with a little greyish hair.

Abdomen dull aeneous black, 2nd, 3rd and 4th segments each with a pair of whitish dust lunule spots of the usual size and shape, placed diagonally; the upper ends approximate to one another above the centre of the segment, the posterior ends of 1st and 3rd pairs reaching the side margin near the posterior angles of the segment; the 2nd pair of spots not attaining the margin. All the spots bear a little yellowish white hair, which also occurs at the posterior angles of the segments and about the tips of the abdomen. The dark portions of the surface covered with almost microscopic black pubescence. Belly dull aeneous, with some pale yellowish hairs.

Legs.—Femora aeneous black, with rather thick yellowish pubescence on hinder and outer sides, and microscopic pubescence of the same colour on the remainder of the surface. Tibiae aeneous black, rather broadly pale reddish brown at base. Tarsi blackish, emarginations slightly reddish brown; hind metatarsi blackish on disc, reddish brown towards sides and on underside.

Wings grey, stigma blackish, inconspicuous; halteres pale yellowish.

Described from a single perfect ♂ taken by Capt. Evans, R.E., at Simla in August 1914, and generously presented by him, with other diptera, to the Indian Museum.

Eumerus sexvittatus, mihi, sp. nov.

♀ Western Himalayas. Long. 8 mm.

Head.—Black, rather dull; vertex and upper part of frons with short black hairs; ocelli small, dull, dark reddish; lower part of frons with yellowish grey hair. Face, seen from below, whitish grey, with whitish grey hair. Back of head black, with a little short whitish hair on the eye orbits. Proboscis black, reddish brown towards tip. Antennae black, rather large, lower part of 3rd joint white-dusted.

Thorax.—Black, rather dull, with very short yellowish brown hair covering all the dorsum and scutellum, and extending over the sides below the shoulders. Sides blackish.

Abdomen.—Black, dull, 1st segment only with a little aeneous tinge, 2nd, 3rd and 4th segments with a pair of diagonally placed whitish elongate spots, each beginning near the centre of the segment, but well separated from one another, and reaching towards but not attaining the hind corners. The abdominal pubescence is black on the black parts and yellowish on the spots; also towards the upper corners of the abdomen and at the sides.

Legs.—Black, with yellowish grey or whitish grey pubescence. Basal half of anterior tibiae (and, apparently occasionally, the extreme tips of the femora), reddish brown, the colour on the hind pair of legs much restricted; middle tarsi reddish brown, except towards tips. Hind femora greatly incrassated as usual, hind metatarsi considerably incrassate.

Wings.—Pale grey, stigma brownish; signs of a very slight brownish suffusion across the middle of the wing. Halteres yellow.

Described from one ♀ from Bhowali, Kumaon District, 5700 ft., October 1909 [*Imms*]. In the Indian Museum.

SERICOMYIA, Mg.

Sericomyia cristaloides, Brun.

Described from a ♀ (Rec Ind. Mus. viii, 167 ♀, 1913), from near Rotung, 2200 ft., 20-xi-11 [*Kemp*]. A unique specimen, in the Indian Museum.

Temnostoma nigrimana, mihi, sp. nov.

(Plate xiii, fig. 15).

♂ Western Himalayas. Long. 16 mm.

Head wholly bright yellow with concolorous tomentum and a little yellow hair along eye margins below antennae. Antennal prominence, facial bump and mouth opening a little more orange. Oral orifice, proboscis and a short black stripe from lower corner of eye reaching half way to end of snout, black. Antennae orange, 1st joint and basal half of 2nd black, arista dull orange. Vertex reddish brown with long black hairs in front and brownish yellow ones behind Occiput greyish, with a fringe of yellow hairs behind eyes, becoming longer on underside of head and hinder part of cheeks.

Thorax slightly shining black, a trace of a pair of narrow median grey stripes towards anterior margin; humeri conspicuously bright yellow, the anterior margin on inner side of them dull reddish orange. An elongate brownish orange spot on the side of the dorsum just above and in front of the wing, reaching to the similarly coloured posterior calli. A rather small oval bright lemon yellow spot on propleura. Pubescence of disc of dorsum rather thick, black; bright yellow on humeri and on pleura below the lemon-coloured spot; reddish on the marginal spot above the wings and on posterior calli, where there are black hairs intermixed. Scutellum reddish brown, the base black nearly to the middle, long yellow hairs on anterior half and brownish black hairs on posterior half. A large bunch of long reddish orange hair on mesopleura.

Abdomen black; 1st segment with bright brown hair at sides; 2nd with hind border reddish brown, the colour widest towards the sides, a bright chrome yellow, moderately narrow band in front of the middle; 3rd with a similar orange band in front of the middle and another on posterior margin; 4th similar to 3rd but the hinder band much wider; genitalia wholly reddish brown. Pubescence on dorsum of abdomen mainly bright yellow, becoming brown on the black parts of the surface; mainly black on 4th segment and genitalia. Belly black, with a rather narrow yellowish band on posterior margin of segments.

Legs principally orange; coxae, and a broad stripe on under side of hind femora, black; a black streak on front side of middle femora; apical half of fore tibiae and the fore tarsi wholly, black. Pubescence on legs mainly yellow, bright lemon yellow short pubescence on basal parts of tibiae.

Wings yellowish grey, anterior part brownish yellow as far inwards as to fill both basal cells. Halteres yellow.

Described from 2 ♂ ♂ in the Indian Museum from the Garhwal District, 11,000 ft., vi-14.

There is a considerable general resemblance at first sight between this species and my *Milesia ferruginosa*, which is not rare in the Kumaon District.

Arctophila simplicipes, mihi, sp. nov.

(Plate xiii, figs. 16—18).

♀ Western Himalayas. Long. 12—13 mm.

Head.—Frons blackish aeneous, with a transverse groove at base of antennal prominence, which is of the same colour; both frons and prominence covered with rather long yellowish hairs, intermixed on vertex with black hairs. The dull reddish ocelli placed flat on the vertex. Face blackish, with whitish tomentum and microscopic pubescence, and some long soft white hairs along inner orbit of eyes. A nearly bare irregular median stripe on face. Cheeks and underside of head blackish, with soft comparatively short yellowish hairs. Proboscis blackish. Antennae dull dark brown, 3rd joint with greyish tomentum, arista brownish yellow, with 16 or 17 long hairs along the entire upperside and about 12 shorter hairs on apical half of underside. Occiput blackish grey with a little minute yellow pubescence; some long brownish yellow hairs behind vertex and yellowish grey hairs on underside.

Thorax.—Black, barely shining, with a pair of median moderately narrow, barely perceptible greyish stripes and a narrower one between them The whole dorsum and the scutellum covered with thick long canary yellow pubescence, except narrowly on anterior margin. The pubescence extends thickly over the vicinity of the mesopleura.

Abdomen moderately shining black, with thick yellowish pubescence on anterior corners, and bright red pubescence on major (apical) part of last segment and on the concolorous red genitalia. On the rest of the dorsum the pubescence is black, short and very fine; a little longer on hind border of segments and obviously long and thick on the sides. Belly black, with short sparse yellowish hairs, hind margin of segments narrowly pale, last segment red.

Legs black, tarsi reddish; femora mostly covered with short black pubescence, except on upper side; rest of legs with minute black pubescence, some short yellow pubescence on outer side of middle tibiae.

Wings grey, a moderately wide dark brown band from middle of anterior margin to a little beyond the 4th longitudinal vein Halteres blackish; squamae brownish, with fringe of brown hair.

Described from several ♀♀ in the Indian Museum from the Garhwal District, Kumaon, 11,000 ft., 20-v-14 to 20-vii-14.

Arctophila, according to Schiner, its founder, should have considerably thickened hind femora and curved hind tibiae, but Verrall in describing *A. mussitans*, F., says, "hind femora rather thick, hind tibiae slightly curved", so, as the character is not so pronounced, the present species is referred to this genus though the hind femora and tibiae are but little thicker or more curved respectively than the others. The genus is, however, otherwise sufficiently characterized. Only three other species are known, two from Europe and one from North America.

MILESIA, Latr.

Meijere records *M. macularis*, W., from Sukabumi, Java, one ♂ [*Kramer*]: and I have noted a specimen from Sikkim which may be a variety of this species (Rec. Ind. Mus., ix, 268). Meijere also records *gigas*, Macq., from the environs of Semarang, 1000 metres [*Jacobson*]; and *variegata*, Brun., from Sikkim, one ♂. Among the diptera sent to the Indian Museum by Lord Carmichael were 3 *gigas* (♂ ♀) from Sikkim, v-12 and Singla, Darjiling, iv-13; and a good series of both *variegata*, ♂ ♀ and *balteata*, Kert. ♂ ♀ (with which my *himalayensis* is synonymous, as announced by Meijere), from both these localities. I have seen three ♀♀ from the same localities, in the same collection which may be *doriae*, Rond. *M. ferruginosa*, sp. nov., is described by me in Rec. Ind. Mus., ix, 268, ♀, pl. xiv, 12, from the Eastern and Western Himalayas.

Milesia sexmaculata, mihi, sp. nov.

♂ South India. Long. 23 mm.

Head.—The eyes touching for a distance equal to one-third of the height of the frons which is yellowish; in the form of an elongate isosceles triangle with yellowish hair; the ocelli red, inconspicuous. Eye facets in front for a short space just perceptibly larger than the others. Face moderately projecting with brownish yellow tomentose dusting, becoming paler yellowish about the mouth, the latter black, cheeks black. Occiput dark grey with pale yellowish grey margin, with a row of short grey hairs behind the eyes. Proboscis black, shining, projecting, two-thirds as long as the height of the head. Antennae dull ferruginous brown with concolorous style.

Thorax.—Dorsum dull black; shoulders and a lateral stripe extending above the wings from the shoulders to the scutellum, yellowish brown. Two dorsal median rather thin yellowish grey stripes, a little dilated on the anterior margin, and reaching nearly to the posterior border, on which latter is an indistinct yellow tomentose streak. Scutellum shining black, with a distinct yellowish brown posterior margin; metanotum shining black. Surface of thorax and scutellum covered with yellow hair. Sides of thorax blackish, apparently a yellowish spot on the mesopleurae. The stigmatic spots yellow.

Abdomen.—Black, shining, 1st segment wholly black; 2nd with a yellow transverse sublunate spot on each side near the base, and contiguous to the side margin; the two spots fairly widely separated from one another. On the 3rd segment a nearly similar pair of yellow spots which are more elongo-conical in shape and are similarly situated; on the 4th segment a pair of yellow nearly triangular spots similarly situated; all the six spots of about the same size and of the same colour. Abdomen with close black pubescence, except that over the spots, which is yellow. Belly black, yellowish at base of 2nd, 3rd and 4th segments, the colour forming two spots on the 2nd segment.

Legs.—Bright brownish yellow; anterior femora with a black streak above and below on basal half; hind femora considerably enlarged, with a conspicuous reddish tooth-like prolongation on underside towards the tip; black, except at tips, the reddish brown colour more extensive on underside.

All the legs with short yellow pubescence, but the hinder side of the middle and hind tibiae with a very thick long fringe of bright yellow hair; (hind tarsi missing).

Wings.—Yellowish grey, subcostal cell brownish yellow. Halteres very small, yellow.

Described from a single ♂ from Trivandrum, Travancore State: in the Indian Museum received from the Trivandrum Museum.

Subfamily *CHRYSOTOXINAE.*

Chrysotoxum convexum, mihi, sp. nov.

(Plate xiii, fig. 19).

♂ Western Himalayas. Long. 14 mm.

Head.—Frons with yellowish grey dust; antennal prominence shining black, with black hairs, a few of which extend to the adjacent parts of the frons; antennae all black, arista reddish brown on basal portion. Face bright yellow with a broad median black stripe: a black band from the corner of the eye to the mouth border, which latter is reddish and shining. Under side of head yellowish orange; proboscis dark brown with short yellowish hairs. Black hairs on vertex, and a fringe of yellow hairs along posterior orbit of eye.

Thorax moderately shining black with short sparse black pubescence, a few rather bright brown hairs in the middle of the disc. A pair of moderately narrow yellowish grey median stripes on anterior border, extending only for a short distance. Humeri, and posterior calli with a short lateral contiguous narrow stripe, bright yellow; a short stripe on the pleura just below but not touching the humeri, the base of the wings, a duller yellow. Scutellum bright yellowish orange on anterior margin, orange yellow on hind margin, the remainder, forming the bulk of the disc, moderately shining black.

Abdomen black, slightly shining: posterior border of 2nd, 3rd and 4th segments dull brownish red, the colour extending forwards in the centre of the 2nd and 3rd segments nearly to the middle of the disc. A pair of elongate triangular yellow spots on 2nd segment, placed at the middle of the side, their apices nearly reaching the middle of the disc. A pair of moderately wide, slightly curved, with the convex side placed anteriorly, extending from each hind corner of the 3rd segment to the anterior margin, where their ends nearly meet. The 4th segment similar, 5th mainly yellowish orange, a narrow median line from anterior margin, forking early, the ends not reaching the margin. Base of underside of abdomen yellowish white; a pair of oval yellowish spots placed transversely near anterior margin and near the sides of the 3rd and 4th segments; those on the 4th segment shorter, the hinder part of that segment more or less reddish orange. Dorsal side of abdomen with black hairs except on the yellow; markings, where the pubescence is concolorous. On the belly the whole pubescence is black.

Legs.—Coxae black with black hair; fore femora yellow, about the basal half black; anterior femora reddish brown, middle pair more broadly, hind pair very narrowly black at base. Tibiae and tarsi orange yellow, base of tibiae more lemon yellow. The femora bear short black pubescence, a little longer on the base, the hind pair with some very short yellow pubescence intermixed on lower side; tibiae and tarsi with yellow pubescence.

Wings grey, anterior margin narrowly brownish yellow: halteres yellow; squamae yellowish orange, with deeper edges and yellow fringe.

Described from a single ♂ in the Indian Museum from Andarban, Garhwal Distr., 11,000 ft., W. Himalayas, vi-14 (*Col. Tytler*).

This species has a considerable resemblance to the *C. intermedium* of Europe, differing in its larger size and the greater prominence of the buccal region.

It is just possible that it is a variety of the European species.

Subfamily *CERINAE*.

Dr. Meijere has described three new species, *C.* (he employs the name *Cerioides*, Rond., instead of *Ceria*) *flavipennis* (Tijd. v. Ent. li, 195, 1908), from Minahassa, Celebes, one ♂; *fruhstorferi* (*l.c.* 196, pl. vii, 1-2) one ♀ from Sikkim, and *himalayensis* (*l.c.* 198) one ♀ from Sikkim. He says his *fruhstorferi* is very near *obscura*, Brun., of which species he records a specimen, a ♂, from Sikkim.

He gives a useful table comprising 9 species. The types of his three new ones are in the Hungarian Museum.

Of *C. compacta*, Brun., described by me from a type ♀ in my collection from Mussoorie, I have found another specimen amongst my unnamed material, which is also a ♀ and from

Mussoorie, 4-iv-05. I have seen several specimens of *C. javana*, W., ♀, and *trinotata*, Meij., from Darjiling, v-1912, and have described a new species *triangulifera*, ♂ ♀ (Rec. Ind. Mus., ix, 273, pl. xiv, 10) from the same district and noted some specimens of further undescribed species.

Ceria fulvescens, mihi, sp. nov.

(Plate xiii, figs. 20—21).

♂ Western Himalayas. Long. 13 mm.

Head.—Hinder orbit of eyes lemon yellow. The whole front part of the head lemon yellow, except for a broad median brown stripe, extending to the mouth, and which is enlarged around the base of the antennae into a diamond-shaped patch which occupies all the upper part, except for the rather narrow lemon yellow border immediately contiguous to the eyes. The side corners of the diamond-shaped brown part just touch the eyes at about half their height, viewed from in front. The cheeks are wholly similarly brown coloured, leaving a broad lemon yellow space between them and the lower part of the median stripe. Ocellar triangle small, brown. Eyes closely contiguous for the short distance that they touch. Antennae with 1st joint reddish brown, or more nearly maroon; 3rd joint brownish yellow, lighter towards tip; style brownish yellow at base, the remainder yellowish white. Back of head more or less yellowish or brownish yellow.

Thorax.—Reddish brown or ferruginous. Humeral calli lemon yellow; a prealar lemon yellow callus at each end of the transverse suture, and lemon yellow coloured marks are placed as follows. Two faint short lines from the anterior margin which nearly meet, and short transverse similar marks placed longitudinally along the transverse suture, one on each side of the middle. An elongate triangular mark on hind margin of dorsum, the base of the triangle coinciding with the margin; a narrow sub-lateral streak towards each side near the wings; a rather large very clearly cut mark on each of the meso- sterno- and metapleurae, the first one approximately oval, the others roughly circular. Scutellum reddish brown, the base and hind margin rather broadly lemon yellow; metanotum reddish brown.

Abdomen.—Reddish brown or ferruginous; a large triangular lemon yellow spot on each side at the base of the very narrowed 2nd segment; posterior margins of 2nd and 3rd segments yellowish, that of the 4th also indistinctly so. Belly reddish brown, a small lemon yellow transverse spot towards the hind margin of 2nd segment.

Legs uniformly ferruginous brown.

Wings.—Pale yellowish; anterior half yellowish brown, the colour filling the marginal cell and extending partly into the 1st basal cell. Stigma a little darker brown. Halteres with yellowish white stems and reddish brown knobs.

Described from one ♂ in the Indian Museum from Bhowali (5,700 ft.), July 1909 [*A. D. Imms*].

Ceria ornatifrons, mihi, sp. nov.

(Plate xiii, fig. 22).

♀ Nepal. Long. 9 mm.

Head.—Occipital margin moderately wide directly behind frons and upper part of eyes, but disappearing at about the middle of the eyes. It is bright light reddish brown, with a small lemon yellow triangular spot at the inner corner of each eye. The space between the eyes across the middle of the head equal to nearly half that width. Upper part of frons light red. On each side of the frons, on a level with the antennae, is a semi-circular lemon yellow callus-like spot, its convexity contiguous to the eye margins. Barely separated from the lowermost part of this spot is, on each side, a nearly vertical lemon yellow stripe, contiguous to the eye margins for a short distance, and then, bending inwards, proceeding to the mouth, above which the two stripes meet. At the spots where the stripes quit the eye margin, there is (but on the inner side of each stripe) a finger-like projection (mark) running towards the centre of the face. The whole space around the base of the antennae and of the face comprised between these two pairs of yellow calli-like markings, is moderately dark brown, punctuated by a number of fine black spots. The sides of the head below the eyes (cheeks) are lemon yellow, a broad reddish brown stripe between the cheeks and the yellow vertical facial stripes. Antennal 1st and 2nd joints brownish yellow (3rd joint missing). The head is placed very broadly and squarely on the thorax, no vestige of neck being apparent.

Thorax.—Broad; reddish brown, with a little hoary bloom, viewed from certain directions. Humeral calli lemon yellow; a small oval lemon yellow spot on the mesopleura. Transverse suture very narrowly yellowish. Scutellum wholly dull lemon yellow; metanotum reddish brown.

Abdomen.—The 1st and 2nd segments reddish brown, a conspicuous lemon yellow callus on each side at the base of the 1st segment. An indistinct though obvious circular black spot in the middle of the dorsum of the 2nd segment; 3rd and 4th segments dark reddish brown or brown; posterior margin of each with a thick lemon yellow band, whole abdomen with a slight greyish bloom. Belly concolorous, with an indistinct yellow band on hind margins of 2nd and 3rd segments.

Legs (fore pair missing) light reddish brown with a hoary bloom; knees and base of tibiae a little yellowish in certain lights.

Wings clear; anterior part yellowish brown, the colour reaching to the spurious vein. A subapical blackish spot of some size from the costa, extending posteriorly just below the 3rd vein and reaching basally to about in a line with the anterior cross vein. The wing tip below this subapical spot lightly blackish. No obvious stigma. Halteres reddish brown.

Described from one ♀, Kumdhik, base of Nepal Himalayas, 22-iii-09. In the Indian Museum collection. This should be near but quite distinct from *eumenoides*, Saunds., described from North India, the latter is, however, double the length of the present species.

Ceria crux, mihi, sp. nov.

♀ Western Himalayas. Long. 10 mm.

Head black. A bright yellow, moderate-sized round spot on frons between base of antenna and eye, contiguous to latter but not to former. A broad yellow stripe on each side of face, beginning in a point just below the circular spot, broadening rapidly, thence gradually narrowing to a point at the mouth border. These four yellow spots leave a black cross, viewed from in front of the head, extending from vertex to mouth opening. Antennae black, 1st joint, which is nearly as long as 2nd and 3rd together (these two being subequal), reddish brown, especially on underside. Apical style of 3rd joint conical, with short narrow elongate tip; a little yellowish or greyish pubescence, almost tomentum, behind vertex, some slight grey pubescence on lower ocular orbit.

Thorax black. A bright yellow spot on each humerus, a triangular one at each end of the transverse suture, which itself bears a thin greyish line. A bright yellow vertical stripe on mesopleura and a round similarly coloured spot on sternopleura, both stripe and spot nearly in a line with the spot at the end of the transverse suture. Scutellum bright yellow, black at base.

Abdomen black, anterior corners of 1st segment with a round bright yellow spot; hind borders of 2nd, 3rd and 4th segments with a moderately wide well defined band of same colour. 1st segment contracted distinctly but not greatly towards tip, and 2nd segment equally contracted at base; the contracted part at its narrowest point being one-third as wide as the abdomen at its broadest part.

Legs.—Coxae blackish; anterior legs ferruginous brown, traces of an indistinct blackish ring on all tibiae beyond the middle; hind femora blackish, except at base, tip and underside; tarsi a little darker.

Wings grey; anterior half from base to tip, and as far hindwards as just beyond 3rd longitudinal vein, blackish brown, the colour darker here and there; basal half of 1st basal and whole of 2nd basal cell also dark brown, costal cell clearer. Halteres bright yellow.

Described from a perfect unique specimen in the Indian Museum from Kousanie, 6075 ft., Kumaon, vii-14 [*Col. Tytler*].

Ceria probably contains numerous as yet undiscovered species in the Himalayas. In the Indian Museum are five undescribed species with 8, 2, 2, 2 specimens, and 1 specimen, respectively, but all in bad condition.

Note on Ceria.

The name *Ceria* is of far too old standing to be changed now. Verrall (British Flies, *Syrphidae*, 665) enquires into the alleged synonymy and substantiates its retention, it having stood unchallenged since 1704. I cannot but agree with "continuity before priority" as did both Osten Sacken and Verrall, two of the greatest systematic dipterologists of recent times. The retrograde nature of the changes of the names of nearly all the old familiar genera (involving in many cases the change of the family name also!), as suggested in Kertesz's addenda to Vol. VII of his otherwise admirable catalogue of the world's diptera, consequent on the proposed adoption of the names of genera in Meigen's paper of 1800, is incalculable, and it is most unfortunate that some dipterologists have followed this lead.

The names in question were given up by Meigen himself in a further paper in 1803 and even this latter paper was regarded by him as wholly preparatory, since he hardly ever referred to either paper, as recorded by Verrall (British Flies, *Stratiomyidae*, 285); so that it is a poor compliment to him who has well been called the father of European diptera to ignore his wishes in the matter.

Moreover, as Williston, Aldrich, and others have pointed out no species were accorded to any of the generic names in Meigen's "1800 paper", so that on that score alone they are quite inadmissible. All the names of well-known genera in diptera which have stood unchallenged since the days of Meigen, Schiner, Zetterstedt, Macquart, Loew, Walker and their contemporaries, and more especially still, those which give their names to families or subfamilies must be regarded, in the best interests of zoology, to be *beyond the sphere of priority*, and exempt from change or modification through any cause whatever, and personally I shall most rigorously refuse to accept any such alterations.

The only way to obtain ultimate finality in nomenclature is rigidly to establish it *now* by upholding all time-honoured names and by ruthlessly ignoring the present fevered craze in some quarters for change.

Subfamily *CHRYSOTOXINAE*.

Chrysotoxum sexfasciatum, Brun.

Only the ♀ was described by me of this species (Rec. Ind. Mus., ii, 89). A ♂ has since been acquired by the Indian Museum taken by Dr. Annandale at Simla, 9-v-10. It agrees closely with the ♀ but is brighter and more lemon yellow in colour, the eyes are absolutely contiguous for the normal distance, the facial stripe is brownish; the hind femora have a pale brown broad band at the tip, the hind tibiae with a narrow brown apical ring, and all the tarsi are pale brown. A further ♂ was taken near Rotung (N.E. Front. India) 20-xi-11 [*Kemp*].

Subfamily *MICRODONTINAE*.

Microdon, Mg.

In the Tijd. v. Ent. li (1908), Dr. Meijere describes the following new species:—

fulvipes, p. 203, ♀, Tandjong Morawa Serdang (Sumatra) [*Hagen*]. Type in Leyden Museum.
fuscus, p. 204, ♀, Medan, Sumatra [*Bussy*].
simplicicornis, p. 205, pl. vii, 6, ♂, Buitenzorg, Java [*Jacobson*].
novae-guineae, p. 206, pl. vii, 5, ♀, Papua, several localities. Type in Hungarian National Museum.
grageti, p. 207, pl. vii, 10, ♂, Graget Is., Papua. Type in Hungarian National Museum, one ♂.
limbinervis, p. 208, pl. vii, 8, 9, Sattelberg, Huen Gulf, Papua [*Biro*]. Type in Hungarian National Museum, one ♂.
tricinctus, p. 208, pl. vii, 7, ♂ ♀, Batavia [*Jacobson*].
vespiformis, p. 210, pl. vii, 7, Batavia [*Jacobson*].
odyneroides, p. 213, Simbang, Huon Gulf, Papua [*Biro*]. Type in Hungarian National Museum.

The types of *fuscus*, *simplicicornis*, *tricinctus* and *vespiformis* are in the Amsterdam Museum.

Dr. Meijere records *M. stilboides*, Walk., from Sukabumi (Java) one ♂, and *indicus*, Dol., from Bali.

M. annandalei, Brun.

I described only the ♂ of this. Since then I have seen a ♀ from Bhowali, Kumaon, 2-vii-12 [*Imms*]

M. indicus, Dol.

Meijere records a pair *in cop* from Semarang taken in April [*Jacobson*].

Microdon unicolor, mihi, sp. nov.

♂ Orissa. Long. 10-11 mm.

Head dark violet, a dark bluish tint behind upper part of eyes; frons and face with rather long yellowish grey hair, leaving the centre of the latter bare. It is also sparser on the vertex and around the ocelli. A few stiff black hairs behind vertex. Middle and lower ocular orbits with short yellowish grey hair. Proboscis brown. Antennal 1st joint distinctly longer than 3rd, nearly as long as 2nd and 3rd together; 3rd three times as long as 2nd, 1st and 2nd joints black, 3rd black with dirty brownish grey dust.

Thorax and scutellum deep violet, only a little shining, with rather thick short black pubescence, which also occurs on the pleura; mesopleura with a little short grey hair.

Abdomen deep violet, a little shining; dorsum with very short black pubescence; longer all grey pubescence at sides, also sparsely on hind margins of segments. Belly deep violet, nearly bare.

Legs blackish violet with minute black pubescence; tibiae with grey pubescence except on inner sides; tarsi with a little grey pubescence above, with which at least on hind metatarsi, some black pubescence at the sides is intermixed.

Wings rather dark brown, a little paler on posterior half; halteres brownish yellow.

Described from a perfect ♂ from near Puri, Orissa, 6-xi-12 [*Gravely*]. In the Indian Museum.

The only other violet black species from the East is *sumatranus*, Wulp, which is punctuated freely on the body and legs with white hair spots.

Mixogaster vespiformis, Brun.

Described by me (Rec. Ind. Mus. viii, 169, ♀, pl. vi, 8—10, wing, head, abdomen, 1913), from a unique ♀ in the Indian Museum taken by Mr. Kemp on the Abor Expedition at Dibrugarh, Assam 17—19-xi-11.

ADDENDUM.

Whilst this paper was passing through the press a long one by Dr. Meijere on Javan diptera has appeared (Tijd. v. Entom. lvii, 1914), in which the following new species have been described.

Xylota decora,	p. 142,	one ♀.
,, *strigata*,	142,	♀.
Milesia simulans,	144,	♂ ♀.
Eristalis nebulipennis	145,	one ♀.
,, *simpliciceps*,	146,	♀.
Graptomyza cornuta,	149,	one ♂.
Chilosia javanensis,	150,	♀.
Syrphus konigsbergeri,	152,	♂.
,, *latistrigatus*,	153,	one ♂.
,, *depressus*,	153,	one ♂.
,, *torvoides*,	155,	one ♀.
,, *gedehanus*,	156,	one ♂.
,, *ichthops*,	157,	one ♂; pl. v. 3. head.
,, *cinctellus*, Zett. var. nov. *strigifrons*,	158,	♂ ♀.
,, *monticola*,	159,	♀.
Chamaesyrphus nigripes,	162,	♀.
Melanostoma 4 notatum,	163,	♂.
,, ,, var. nov. *gedehensis*,	163,	♂.
Sphaerophoria obscuricornis,	165,	♂.
,, *javana* var. nov. *medanensis*,	166,	♂ ♀ (Sumatra)

EXPLANATION OF PLATE XIII.

FIG.	1.—*Psilota cyanea*,	Brun. sp. nov.			head in profile.
,,	2.—*Melanostoma orientale*, W.				,,
,,	3.—*Melanostoma* undescribed form				,,
,,	4.—*Melanostoma univittatum*, W.				,,
,,	5.— ,,				abdomen ♂.
,,	6.— ,,				,, ♀.
,,	7.—*Syrphus distinctus*,	Brun.	sp.	nov.	abdomen.
,,	8.—*Baccha apicenotata*.		,,	,,	part of wing.
,,	9.—*Sphegina tricoloripes*,		,,	,,	wing.
,,	10.—*Graptomyza tinctovittata*,		,,	,,	,,
,,	11.—*Merodon ornatus*,		,,	,,	abdomen.
,,	12.—*Myiolepta himalayana*		,,	,,	head in profile.
,,	13.— ,,		,,	,,	wing.
,,	14.—*Criorhina imitator*,		,,	,,	full insect.
,,	15.—*Temnostoma nigrimana*,		,,	,,	abdomen.
,,	16.—*Arctophila simplicipes*.		,,	,,	full insect.
,,	17.— ,,				hind leg.
,,	18.— ,,				arista.
,,	19.—*Chrysotoxum convexum*,		,,	,,	abdomen.
	20.—*Ceria fulvescens*,		,,	,,	thorax
,,	21.— ,,		,,	,,	abdomen.
,,	22.—*Ceria ornatifrons*,		,,	,,	front view of head.

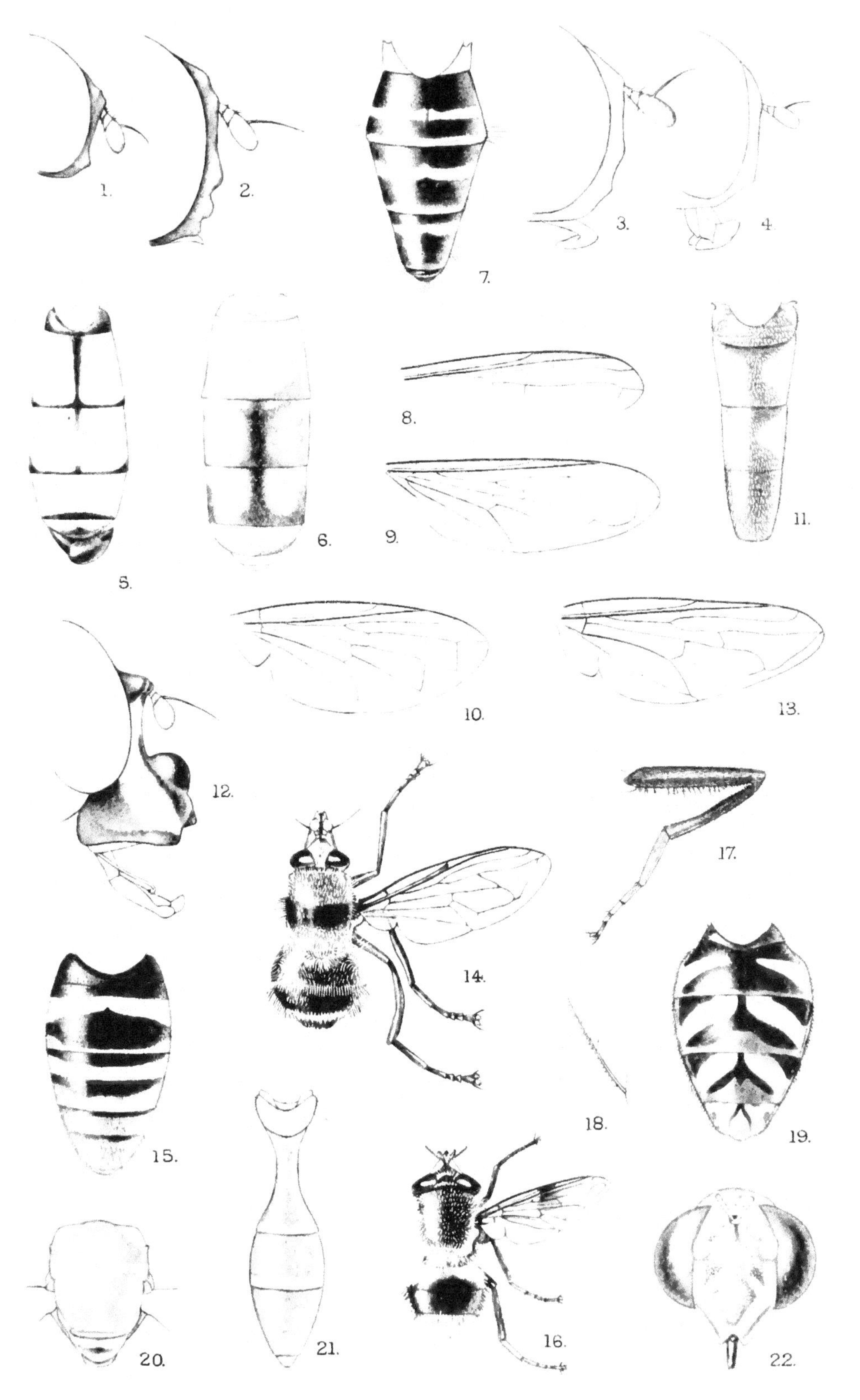

D Bagchi, del.

Bemrose, Collo, Derby

ORIENTAL SYRPHIDAE.

XIV. NOTES ON INDIAN MYGALOMORPH SPIDERS.

By F. H. GRAVELY, *M.Sc., Asst. Superintendent, Indian Museum.*

(Plate XV).

The present is intended to be the first of a series of papers on Indian spiders, based on the collections in the Indian Museum.

The earliest descriptions of species in this collection were published by Stoliczka, in the *Journal of the Asiatic Society* for 1869. He pointed out in a most forcible manner the extraordinary neglect with which the study of so important and fascinating a group as the Indian Arachnida had met, a neglect which he set himself to remedy. The variety of other groups with which he was occupied can have left him little time for such work, and he only published two papers [1] in connection with it. But he collected specimens vigorously right up to the time of his early death in 1874. The whole of his private collection was bequeathed to the Indian Museum, where most of it still remains in good condition.

Since Stoliczka's death several Orders of Indian Arachnids have been investigated by Kraepelin, Pocock, Thorell, Roewer, Nuttall, Warburton and others; but our knowledge of Indian spiders is still woefully incomplete.

In the years 1887-9 the spiders preserved in the Indian Museum formed the subject of a series of short papers contributed by Simon to the *Journal of the Asiatic Society of Bengal.* And a short paper on our Mygalomorphae was published by Hirst in the *Records of the Indian Museum* for 1909.

In 1895 the British Museum published an account of the spiders of Burma by Thorell, who in 1896 and 1898 respectively contributed two lengthy papers on the spiders collected in Burma by Fea, to the Annals of the Civic Natural History Museum of Genoa.

In 1899 the Bombay Natural History Society published a paper by Pocock on Indian spiders with which they had supplied him. This was followed in 1900 by a paper in the same Journal containing descriptions which "were drawn up for publication in a volume upon the Arachnida of India, forming part of the Fauna of India Series" but which "together with the diagnoses of many

[1] "A Contribution towards the Knowledge of Indian Arachnoidea" (*J.A.S.B.* xxxviii [II], pp. 201-251, pls. xviii-xx); and "Notes on the Indian species of *Thelyphonus*" (*J.A.S.B.* xlii [II], pp. 126-141, pl. xli).

previously established species" were omitted on account of "exigencies of space." Why any volume of a series of books, whose chief value lies in their completeness, should have been thus curtailed, it is difficult to understand, especially as the volume in question is one of the shortest of the series and attempts to deal with four comparatively small Orders as well as with the immense Order Araneae. It is particularly unfortunate that spiders should have been treated in this way, for there is probably no other group in the whole of the animal kingdom which is so universally distributed in India, and at the same time so striking and varied both in structure and in habit. New and interesting facts about spiders force themselves upon one's attention wherever one goes; but a satisfactory record of them is commonly rendered almost impossible by the difficulty of indicating with sufficient precision the different kinds of spider to which the various facts refer.

The extension in 1912 of the space available for the research collections of the Indian Museum allowed of a much needed expansion of our collection of spiders. Previous to this extension the space allotted to spiders was so crowded by bottles of mixtures from different localities that no attempt at organization could be made. Since then I have devoted such time as I could periodically spare to sorting out the contents of these bottles, and getting both the named and the far larger unnamed collections systematically arranged.

The present paper, and those with which I hope to follow it, are the outcome of this work, which is now approaching its provisional conclusion. These papers will not aim at an extensive revision of the Indian spiders, but will discuss, in the light of the specimens in our collection, the classification adopted by Simon in his "Histoire Naturelle des Araignées" (Paris, 1892 and 1897), and record the localities from which the specimens dealt with have been obtained.

The extremely scattered literature relating to species of spiders already described, often all too briefly and usually without figures, together with the means which a large proportion of these species possess in early life of travelling long distances through the air, render it hopeless for anyone who cannot work on spiders during the greater part of his time to determine with certainty whether a species he has been unable to name is new to science or not. There are, however, many indications that a large proportion of such species are actually new. For instance, some common Himalayan spiders were described as new by no less an authority than Simon as recently as 1906. It is highly desirable, I think, that as many as possible of our more distinctive species should be described and named without delay, even at the risk of the creation of a few synonyms. The final revision of each family of spiders will have to be made by a specialist in a position to deal with members of that family from all parts of the world, and the richer the published material at his disposal, provided that the descriptions and illustrations it contains are adequate and that reasonable care has been taken to avoid repetition, the more complete is his work likely to be. I propose therefore to describe a certain number of species as new, even though I may not be able to trace every possible description that may refer to them.

The very small number of extra-Oriental forms in our collection makes it impossible for me to criticize the relation which these bear to Oriental forms in Simon's system. Where, however, as in the case of the Aviculariinae dealt with in the present paper, the Oriental forms appear to exhibit definite structural zoogeographical relationships to one another, I have not hesitated to suggest the advisability of trying to alter his system in order to bring these into prominence. Such relationships have been found in all of the few groups in which I have looked for them. In the case of one of these groups—the Passalidae—in which such relationships recently led me to separate the Indo-Australian forms from those of the rest of the world, none of which I had seen, I have already obtained proof that the separation was justified; though some of the latter resemble certain Indo-Australian forms so closely that I, like previous authors, should probably have been misled by striking superficial characters, had not my earliest work on the family been confined to Indo-Australian species.

The characters on which the classification of spiders is at present based are to a great extent admittedly unsatisfactory; and it is quite possible that by dealing separately with the faunas of different zoogeographical areas—the extent of the areas that will have to be taken may be found to differ in different groups—local relationships may be brought to light which will lead to the discovery of new characters of deeper significance where we least expect them, especially among the more sedentary families.

References to Simon's "Histoire Naturelle des Araignées" and to Pocock's "Fauna" volume are so numerous that I have omitted the titles of these works throughout. Where not otherwise stated all references to these authors imply references to these works. Where no references to descriptions of species are given, these will be found in the "Fauna".

It has been convenient to put this paper into the form of a catalogue of the specimens in our named collection, a form which will probably be convenient for the rest of the series also. Our collection of spiders has recently been increased to a considerable extent by the generosity of collectors in different parts of India. This has made it more representative of India generally than would otherwise have been the case, and has greatly facilitated my work. Our thanks are due to all who help us in this way, and especially to H. E. the Governor of Bengal who, with the assistance of Mr. Möller, has been making large collections of the Invertebrata of the Darjeeling District; to Dr. Sutherland who has collected spiders extensively round Kalimpong in the same district; to Mr. M. Mackenzie who has sent numerous specimens from Siripur in Bihar; to Mr. G. Henry who has submitted to me the specimens he has been able to collect in Ceylon during his tours on behalf of the Colombo Museum; to Mr. T Bainbrigge Fletcher of Pusa; and to Mrs. Drake of Serampore.

Family LIPHISTIIDAE.

Genus **Liphistius**, Schiödte.

This interesting genus is represented in our collection by a single damaged specimen from Moulmein in Lower Burma.

Family AVICULARIIDAE.

Subfamily *CTENIZINAE.*

Group PACHYLOMEREAE.

Genus **Conothele**, Thorell.

Two female or immature specimens were collected by Theobald in the Nicobars. These differ from *C. birmanica*, Thorell, in having the posterior series of eyes procurved, and in having more teeth on the labium; but they may perhaps belong to some Malaysian species.

Group IDIOPEAE.

I am unable to follow Simon's final revision of this group (Vol. II, pp. 888-890) except as regards the union of *Acanthodon* with *Idiops*, a union the necessity of which is supported by the occurrence in our collection of the male of an Indian species with the eyes of the second group closely crowded and strongly unequal.

Simon separates the American genera of Idiopeae from those of the Old World on the grounds that in the former the eyes of the posterior line, seen from above, are lightly procurved whereas in the latter they are lightly recurved, the area occupied by the four median eyes being moreover parallel-sided in the former and broader behind than before in the latter.

In all our specimens, however, and apparently also in those described in the "Fauna," the posterior line of eyes is distinctly procurved and never recurved, the posterior margins of the large laterals never being behind, and the anterior margins of these eyes always being in front, of the corresponding margins of the smaller posterior median eyes. And the area occupied by the four median eyes is not always even slightly wider behind than before.

Further, when these characters are disregarded, and an attempt is made to put our three specimens of the group into the Old World genera which would otherwise receive them, only one of the three (*Heligmomerus* sp.) is found to fit. The other two resemble *Gorgyrella* in the structure of the chelicerae, and *Pachyidiops* and *Titanidiops* in the shape of the labium, differing markedly from all of these and from one another in their combinations of the other characters used by Simon in his generic definitions.

I have therefore fallen back on Simon's earlier revision of the group (Vol. I, pp. 90-92), which, when *Acanthodon* has been merged in *Idiops*, takes in all these forms conveniently.

Genus **Heligmomerus**, Simon.

Represented by one female, caught in the Royal Botanical Gardens at Sibpur near Calcutta, where it may easily have been introduced among plants from some other place. Its burrow, a short silk-lined tube, closed externally by a trap-door, is also in our collection.

Genus **Idiops**, Perty.

Represented by a female from Bellary in South India, and by a male whose characters seem sufficiently well defined to permit of its description here as a new species.

Idiops biharicus, n. sp. ♂.

(Pl. xv, figs. 1 *a-b*).

Locality.—Sahibgunge in Bihar.

Dimensions.—Carapace 6·0 × 5·2 mm.; sternum 3·0 × 2·8 mm.; legs in the order 1, 4, 2, 3. In the first legs the femur is fully, and the combined tarsus and metatarsus are scarcely, as long as the carapace. The patella and tibia combined are a little longer than either. The tibia and metatarsus of the second legs are about equal to the patella and tibia of the first in length, but are much slenderer. The tibia of the third leg on each side is nearly three times as long as wide; the femur and patella of these legs are together scarcely as long as the carapace, and are about equal to the femur alone of the fourth legs.

Colour.—Carapace plum-coloured; appendages dark reddish above, paler beneath especially basally; sternum and lower surface of abdomen also pale, almost ochraceous; upper surface of abdomen dull brown.

Structure.—The *carapace* is ovate, slightly narrower behind than in front, with the posterior margin short and faintly concave in the middle line. The anterior lateral eyes are situated close together on a prominent tubercle close to the anterior margin. The remaining eyes are situated in a compact group: of these the anterior medians are almost in contact, and are the largest; the posterior medians are separated by a distance about equal to a diameter of one of the anterior medians, and are the smallest; both are almost in contact with the posterior laterals, whose long diameter is about equal to that of an anterior median, and whose other diameter is about equal to that of a posterior median. The fovea is large and very deeply impressed in the form of a procurved semicircle.

In front of it the cephalic part of the carapace is strongly elevated, and bears a pair of broad longitudinal bands of sparse coarse tubercles, which become faint on either side of the posterior group of eyes and disappear before they reach the anterior margin of the carapace, this being quite smooth. The rest of the margin is granular except in the median concavity behind, and broadening bands of coarse tubercles radiate towards it from the fovea.

The *labium* is about as broad behind as it is long, and is slightly narrower in front. It is unarmed.

The *sternum* appears to have been spiney.

The *chelicerae* are provided each with a rastellum set on an apophysis overhanging the base of the fang. The chelicerae are armed each with 5 outer and 7 or 8 inner teeth.

There is no *stridulating organ.*

The tibia of the *palp* is excavate beneath in its distal third, the outer side of the hollow being armed with stout spines, of which those at the two ends are long and those in the middle short. The distal end of the tarsus bears a bluntly conical process on the outer side.

The bulb of the palpal organ (fig. 1*b*) is helicoid. The style consists of two parts, a basal lamina which is triangular in shape and somewhat narrower at the base than it is long, and a very slender, slightly curved, distal duct of about the same length.

The *legs* are spiney. The extremity of the tibia of the first legs (fig. 1*a*) is armed on the inner side with two stout conical apophyses, of which the proximal has a simple apex turned slightly downwards when viewed laterally, while the distal is strongly indented on the lower side below the somewhat upwardly directed apex. The metatarus is somewhat bent outwards and swollen on the inner side below the middle; it lacks the submedian conical spur found in *I. constructor* (Pocock), but bears numerous stout spines on the lower side, as does the tibia also.

The tibiae of the third legs are faintly excavate above, though not definitely so as in *Heligmomerus.*

This species seems to be most closely related to *I. constructor* (Pocock), from the male of which it differs chiefly in the large size of the anterior median eyes—assuming that Pocock's description of the eyes of the female applies also to the male, except as regards their proximity where he notes a difference between the sexes. The unarmed labium appears to be another distinguishing character. In any case the present species differs from *I. constructor* in the absence of the metatarsal spur of the first leg of the male.

Group CYRTAUCHENIEAE.

Genus **Atmetochilus**, Simon.

Represented by the type of *A. fossor*, Simon (*genotype*), and by an immature male from Upper Tenasserim.

Group AMBLYOCARENEAE.

Genus **Damarchus**, Thorell.

Represented by the type specimen (male) of *Damarchus assamensis*, Hirst, and by three females associated with it which "do not differ in structure from the female of *D. oatesii* (1909, p. 384)." Also by a small specimen from Gmatia in the Birbhum District of Bengal.

Group ARBANITEAE.

Genus **Scalidognathus**, Karsch.

Represented by specimens of *S. radialis* (Cambr.), from Kandy, Galagedara and Newara in Ceylon.

Genus **Nemesiellus**, Pocock.

Represented by specimens from Barkuda Island, Chilka Lake (north-eastern end of Madras Presidency), and from S. India. The lateral spacing of the eyes is distinctly less in both than it is in our specimens of the preceding genus, which makes the anterior line appear more procurved, and makes the anterior and posterior lateral eyes on each side appear relatively further apart.

Subfamily *BARYCHELINAE.*

Group DIPLOTHELEAE.

Genus **Diplothele**, Cambr.

Represented by one specimen of *D. walshi*, Cambr., from Waltair on the eastern side of the Madras Presidency.

Neither Simon nor Pocock appear to have been aware that Walsh described this species under the name *Adelonychia nigrostriata*, n. gen. and sp. (*J.A.S.B.*, LIX, [II], pp. 269-270) at about the same time that Cambridge described it from specimens which Walsh had sent him. Walsh's description was received on Oct. 27, read on Nov. 5, and published on Dec. 10, 1890. Cambridge's description was received on Oct. 23 and read on Nov. 18 of the same year; the date of publication is not recorded and cannot have been much if at all before Dec. 10. Cambridge's name has, however, been universally adopted, and it seems in any case undesirable to change it.

Group BARYCHELEAE.

Genus **Sasonichus**, Pocock.

The description of this genus, and of the single species on which it is based, are very imperfect. So far as I can tell the new species which I am referring to the genus differs from the original species in one only of the characters to which generic value has been attached. To avoid establishing a new monospecific genus this character, the presence or absence of apical apophyses on the tibia of the first leg of the male, may be given specific value.

Sasonichus arthrapophysis, n. sp. ♂.

(Pl. xv, figs. 2 *a-b*).

Locality.—Barkul in south-east Orissa.

Dimensions.—Carapace 7·5 × 6·0 mm.; sternum 2·7 × 2·5 mm.: legs in the order 4, 1, 2, 3.

The patella and tibia of the first legs are together equal to the length of the carapace; the tarsus and metatarsus are together slightly shorter, and the femur is shorter still, the femur and half the patella being about equal to the length of the carapace, as are also the femur of the second legs with the whole of the patella, the tibia and metatarsus together of the third legs, and the patella and tibia together and the metatarsus alone of the fourth legs.

Colour.—Dark brownish above, paler below, the ends of the tibiae of the legs silvery above—least so on the hind legs.

Structure.—The *carapace* is ovate, slightly broader behind than in front. The ocular area is very compact and is situated on a clearly defined tubercle approximately circular in outline. The anterior lateral eyes are oval, and are situated obliquely in front of the rest about a short diameter away from the anterior medians and fully a long diameter from one another. The anterior medians are round, their diameter fully as great as the long diameter of the anterior laterals; they are separated by a distance about equal to a diameter of the small posterior medians. The posterior medians and anterior laterals form a square; and the centres of the former are directly behind the outer margins of the anterior medians. The posterior laterals are quite as long as the anterior laterals, but much narrower. A line of low tubercles extends medially from the ocular tubercle to the fovea, which is linear as a whole, but distinctly recurved just at its extremities. Lines of tubercles radiate from the fovea. The whole carapace has probably been covered with long golden brown hair and scattered black spines, but most of these have disappeared. The spines are very long and thick posteriorly, where they project outwards and curve forwards.

The *labium* is very imperfectly separated from the sternum. It is armed behind the anterior margin with a line of four more or less distinct erect teeth, among long spiniform hairs.

The *sternum* is covered with erect spiniform hairs, and is bordered laterally and behind by a single row of long black slender spines The *coxae, trochanters* and *femora* are similarly armed; but long white silky hair surrounds the mouth, both on the labium and on the coxae of the palps. On the latter it hides a group of denticles like those which form a line on the labium, but much more numerous.

The *chelicerae* are armed with about ten inner but no outer teeth, and are provided with a rastellum whose spines are somewhat long and slender.

There is no *stridulating organ.*

The tarsus of the *palp* is lobed on the inner side below. The style of the palpal organ (fig. 2*b*) is more or less lamelliform and parallel-sided throughout the greater part of its length, and is twisted on its own axis through about 90°; distally it is sharply pointed.

The *legs* are spiney, with a series of very stout spines on the tibia and metatarsus. The tibia of the first legs is armed on the inner side near the end with two stout apophyses, of which the distal is ventral to the other (fig. 2*a*). They curve towards one another as a whole, but the extreme apices are slightly turned in the opposite direction. The distal part of each, which is greater in the proximal than in the distal, appears to be jointed on to the basal part. From this it seems probable that the former is moveable in life. I do not remember to have heard of any other Arachnids with jointed apophyses: but the jointed setae of Nereidiform Polychaet worms and the jointed tooth found on the mandibles of most Passalid beetles, afford instances of similar jointing of chitinous structures in other groups.

This species differs from *S. sulivani* chiefly in the presence of apophyses on the tibiae of the first legs.

Group SASONEAE.

Genus **Sason**.

Represented by specimens of *S. cinctipes*, Pocock, from Peradeniya in Ceylon, and by one undetermined specimen from the Nicobars. *S. cinctipes* lives on moss-covered rocks or walls where it constructs a curious flat, more or less **8**-shaped nest. The upper part of this nest consists of two rounded flaps hinged together along their contiguous borders, these borders forming the cross-piece of the eight. The double trap-door is attached to the basal part of the nest on either side of the cross-piece.

Subfamily *AVICULARIINAE*.

Five of the groups of this sub-family recognized in Simon's "Supplement" occur in the Indian Empire, and of these four are only known from the Oriental and Australian Regions. The fifth is the most primitive of them all, and has a much wider distribution; it may be looked upon as the ancestor of the other four.

This group, the Ischnocoleae, is almost confined in the Oriental Region to the Indian Peninsula and Ceylon. The genera which occur there are found nowhere else, except perhaps in the Eastern Himalayas and Burma. In Simon's arrangement they are scattered among genera from other parts of the world; but when taken by themselves they are found to fall into line, not only with one another, but also with the Thrigmopoeeae, each genus of the two groups (except perhaps *Annandaliella*, see below, p. 271) representing one stage in an evolutionary series culminating in the genus *Thrigmopoeus*.

Simon's final revision of the Ischnocoleae brings all the Oriental species of the group into three genera, *Phlogiodes*, *Heterophrictus* and *Plesiophrictus*, and to these Hirst has since added the genus *Annandaliella*. Of these the first appears to have been known to Simon only from Pocock's imperfect description of two forms which probably, as pointed out below (p. 269), are opposite sexes of a single species. Of the second he appears to have seen a female (the only sex known) of the single species as yet referred to it. Of the third the male was evidently known to him from Pocock's description only. It is, therefore, scarcely to be wondered at, that his definitions of these genera are somewhat unsatisfactory, and that several of the species described below differ from the genera in which I have placed them in one or more of the characters used in his keys; but as they differ at least equally widely from all extra Oriental genera and appear to be closely related to one another, I have thought it best to place them in these Oriental ones.

An account of the genera of Indian Ischnocoleae and of Thrigmopoeeae will be found below (pp. 269-280). It is designed to bring out the evolutionary sequence which the genera appear to illustrate. This sequence seems to me to indicate that the two groups should ultimately be united; and that if any characters can be found to separate both of them from the extra-Oriental Ischnocoleae, a new group should be instituted for them. But as I have no extra-Oriental forms for comparison I am not able to attempt this at present.

None of the genera of Indian Ischnocoleae and Thrigmopoeeae have attained so high a degree of specialization as have the genera *Poecilotheria* and *Chilobrachys*, which also live in the Indian Peninsula and Ceylon. The former lives in trees and in the thatch of houses, so can scarcely be regarded as entering into competition with ground-dwellers like the Indian Ischnocoleae[1]. I have elsewhere (1915, pp. 417-418) given reasons, largely zoogeographical, for supposing that it originated from a primitive stock—presumably of the Ischnocoleae or Thrigmopoeae—in the Indian Peninsula or Ceylon. It will be sufficient here to point out that it differs from the Selenocosmieae not only in important details of the stridulating organ, but also in the structure of the labium—for which reasons, among others, I prefer to follow Simon who established a special group, Poecilotherieae, for its reception, rather than Pocock who united it with the Selenocosmieae.

[1] Nothing appears to be known of the habits of the Thrigmopoeeae, which probably resemble those of the Ischnocoleae. The specimen I obtained in Cochin was not recognized when captured. If I caught it myself it must have been on the ground, like all the other Mygolomorphae I found. But it may have been brought to me by someone else.

The only remaining genus of Aviculariinae found in the Indian Peninsula or Ceylon is *Chilobrachys*, the most highly specialized genus of the group Selenocosmieae. The whole history of the evolution of this ground-dwelling genus can be read in the forms inhabiting the countries north and east of the Ganges to-day; and there seems no reason to doubt that its evolution took place there. The primitive forms left there are extremely rare, having no doubt suffered in the struggle for existence with their more highly specialized relatives. The most highly specialized genus of these has spread into the Indian Peninsula and Ceylon, a fact which probably accounts for the concentration southwards and westwards of the Indian Ischnocoleae and the Thrigmopoeeae.

The evolution of the Selenocosmieae has already been dealt with from a primarily zoogeographical point of view (Gravely, 1915), with the results indicated in the above summary. The morphological point of view must now be more fully considered.

Reference has been made above to the existence in parts of the Oriental Region north and east of the Ganges of a few primitive species of Aviculariinae. These appear to be extremely rare, and those hitherto described are known to me from descriptions only. There is, however, in the Indian Museum collection, a single immature specimen from the Darjeeling District which must be associated with them. The species already described are two in number: both were collected by Fea in Burma, and referred by Thorell to the genus *Ischnocolus* (1896, pp 170-175). More recently Simon (Vol. II, p. 925) has shown that this genus *Ischnocolus* must be restricted to species from the Mediterranean and Ethiopian regions; but he makes no mention of the position to be assigned to the Burmese forms. In describing the labium of one of these, "*Ischnocolus*" *brevipes*, Thorell says, "apice fascia transversa sat lata granulorm densissimorum praeditum." With regard to the labium of the other, "*Ischnocolus*" *ornatus*, which he described from two immature specimens, he says, "quod ... apice minus dense granulosum est—an ita etiam in adultis?" Now the presence of a densely granular transverse band on the apex of the labium is characteristic of the Selenocosmiae. In all other Oriental groups of Aviculariinae the anterior part of the labium is more sparsely armed. The distinction, although quantitative, is very marked; and except perhaps in very young and imperfectly hardened specimens such as no one could think of naming, a glance at the labium is sufficient to show whether a specimen belongs to the Selenocosmieae or not.

The only Burmese species in which the labium is sparsely armed, other than those referred by Thorell to the genus *Ischnocolus*, are those comprising the group Ornithoctoneae, which are separated from all other Oriental species by the densely hairy outer surfaces of their chelicerae. That the occurrence of a sparsely armed labium in a Burmese species without externally hairy chelicerae struck Thorell as very remarkable, seems to be indicated by his suggestion that its presence was due to the immaturity of his specimens, a suggestion which is not supported by the immature specimens of Selenocosmieae in our collection. The dense armature of the labium of the Selenocosmieae has been recorded as a group character by Simon (Vol. II, p. 953), though he does not appear to have attached much importance to it. In view of the fact, however, that it supplies a clearly defined character which, unlike the stridulating organ, appears unchanged in all genera of the group, and thus enables us to recognize as a primitive ally of the group "*Ischnocolus*" *brevipes* in which no stridulating organ occurs at all, its importance as a group character should, in my opinion, be ranked even higher than that of the stridulating organ itself. "*Ischnocolus*" *brevipes* may therefore be transferred to the Selenocosmieae, the evolution of the higher forms of which is discussed below (pp. 282-287).

"*Ischnocolus*" *ornatus* must now be considered. It differs from the Selenocosmieae not only in the structure of the labium, but also in the greater number of spines on its legs. In the former character it resembles all, and in the latter the more primitive, of the Indian Ischnocoleae. For the present then it will be best to associate it with this group and especially with the primitive genus *Plesiophrictus.* But its genus cannot be definitely determined in the absence of mature specimens of either sex. This applies also to the immature specimen referred to above, which was collected in the Darjeeling District, and is preserved in our collection. These two forms are presumably remnants of a primitive Himalayo-Malaysian fauna from which both the Selenocosmieae and Ornithoctoneae have originated: and their rarity is probably accounted for by their inability to compete successfully with these more highly specialized groups.

The Ornithoctoneae are the only Oriental Aviculariinae that have not been dealt with above. They form so compact and isolated a group that little or no direct morphological evidence of their affinities with other groups is to be found (see Gravely, 1915, p. 417).

The five Oriental groups of Aviculariinae as described above may now be defined.

1.	Anterior part of labium armed with denticles somewhat sparsely distributed ..	2.
	Anterior part of labium covered with closely crowded granules ..	*Selenocosmieae*, p. 282.
2.	Outer surface of chelicerae bare ..	3.
	Outer surface of chelicerae densely hairy	*Ornithoctoneae*, p. 280.
3.	No bacilli present on anterior surface of coxae of palps, this surface bearing at most small spines	4.
	A cluster of more or less claviform bacilli, accompianed by one or more stout denticles, present on anterior surface of coxae of palps	*Poecilotherieae*, p. 280.

4. { No stridulating organ present between chelicerae and coxae of palps .. *Ischnocoleae*, p. 269.
{ A stridulating organ present in this position .. .. .. *Thrigmopoeceae*, p. 278.

If any character can be found by means of which the Indian Ischnocoleae can be separated from the Ischnocoleae of other parts of the world it will be advantageous, as pointed out above (p. 266), to bring about this separation, at the same time uniting the former with the Thrigmopoeceae

Group ISCHNOCOLEAE.

Among Indian genera of this group there appears to be a marked sexual dimorphism. All known males are distinguished by the more or less extensive and conspicuous development of white hair on the feet, especially the anterior ones

In the two species of which males are known to me, the anterior tarsal scopulae, which, except in the genus *Phlogiodes*, are always more or less clearly divided in females,[1] are either undivided in the male or less clearly divided in the male than in the female; also the anterior median eyes tend to be enlarged in the male and the anterior laterals in the female.[2] As yet all species of this group appear to have been described from one sex only; but there can be little doubt, I think, that *Phlogiodes robustus*, Poc. (♀) = *P. validus, Poc.* (♂), since both are found at Matheran. In the former, according to Pocock (1899, pp. 748-9), the tarsal scopulae are broadly divided on legs 2-4, in the latter they are undivided except on the fourth leg where the division is narrow.

I have found it impossible to separate the genus *Heterophrictus* from *Plesiophrictus*. Pocock's distinction, based on slight differences in the shape of the fovea, is very unsatisfactory.

Simon separates them primarily on characters presented by the vestiture of the anterior surfaces of the coxae of the first legs. But these vary even in mature examples of one sex of a single species, and they are clearly correllated with size, the *Plesiophrictus* characters being found in the young of large forms whose adults have well-marked *Heterophrictus* characters, as well as in adults of species of small size similar to that of the species grouped together by Pocock in the former genus.

The genus *Annandaliella* ought also, perhaps, to be merged in *Plesiophrictus*; but as the spines on the inner surfaces of the chelicerae, by which it is characterized, are sharply distinctive I retain the genus provisionally. These spines are considered by Hirst to

[1] They are said to be undivided in *Annandaliella travancorica*, but fresh specimens show a median line of fine hairs such as accompany the spines by which the scopulae of the other feet are divided.

[2] In females of *Plesiophrictus sericeus, collinus* and *fabrei*, according to Pocock, the anterior laterals are not larger than the medians. Males do not appear to be known in any of these species.

be stridulatory structures; but so far as I know there is no direct evidence on this point. It is difficult to find any other explanation for them; in view, however, of the fact brought out by material recently added to our collection, that they do not occur in specimens less than half grown, or in mature males, their physiological homology with the stridulating organs of other Oriental Aviculariinae is open to question. But for the importance that has been attached to these spines the only species yet referred to the genus would find its natural place somewhere near the middle of the series of species composing the genus *Plesiophrictus*.

This series shows a gradual change from small forms with small marginal posterior sigilla and more distinctively Plesiophricticid anterior coxae, to larger forms with larger posterior sigilla more widely separated from the margin of the sternum and more distinctively Heterophricticid coxae, characters all of which are intensified in the genus *Phlogiodes*, which affords a transition to the Thrigmopoeeae.

If the genus *Phlogiodes* were only distinguished by the size and position of its sigilla, and by the shape of its fovea—the characters used by Pocock in his key—its distinctness from *Plesiophrictus* could hardly be maintained. Probably the most important character separating the two genera is the absence in *Phlogiodes* of the tibial apophysis of the first leg of the male—a character which separates it alike from *Plesiophrictus* and *Annandaliella*.[1] But this character does not help in the case of species (unfortunately the majority) known from females only. It appears, however, that *Phlogiodes* approaches the Thrigmopoeeae in the characters of its feet, as in so many other features. The feet of the Thrigmopoeae are very different from those of *Plesiophrictus*; and it is likely, I think, that the character will prove to be a valid one for the separation of *Phlogiodes* from *Plesiophrictus*, in spite of a certain amount of variation which it exhibits in the latter and perhaps in both genera.

The genera of the Indian Ischnocoleae may now be redefined thus:—

1. A row of stout spines present on the inner surfaces of the chelicerae of mature females; feet of first legs slender, the division of their tarsal scopulae more or less obsolete especially in male; male with tibial apophysis of first leg .. *Annandaliella*, p. 271.
   No spines on the inner surfaces of the chelicerae .. .. .. 2.

[1] The possession of this apophysis, and of somewhat numerous spines on the legs generally, suggests a possible relationship between the more primitive Indian Ischnocoleae and the Indian Barychelinae. In the Indian Barychelinae, however, the spines thickly cover all joints of the legs, and no definite arrangement of them can be recognized. In the Indian Ischnocoleae such an arrangement is recognizable among the few spines that may be present on the anterior legs, and is repeated on the posterior legs in all species in which their spines have been reduced to a small enough number (see below, p. 274).

2. Male with tibial apophysis of first leg: feet of first legs slender, their tarsal scopulae (? always) clearly divided .. *Plesiophrictus*, p. 273.
Male without tibial apophysis of first leg; feet of first legs stout, their tarsal scopulae (? always) undivided .. *Phlogiodes*, p. 278.

Genus **Annandaliella**, Hirst.

It will be convenient to deal first with this genus, which appears to form a lateral offshoot from the main trend of evolution, leading up towards the Thrigmopoeeae. It appears to have originated from some species near the middle of the evolutionary series of the genus *Plesiophrictus*, and to differ therefrom only in the presence of the characteristic spines on the inner surfaces of the chelicerae of the female, and perhaps in the absence of spines from among the fine hairs by which the anterior tarsal scopulae are divided in the female, hairs which are not sufficiently numerous in the male even to form a definite line. The absence of the characteristic spines from the mandibles of the male (and young) is very remarkable, if, as has hitherto been supposed, they constitute a stridulating organ comparable to that found between the chelicerae and palps of the more highly specialized Oriental genera of Aviculariinae.

The genus is represented in our collection by a number of specimens of *A. travancorica*, Hirst (1909). It is also represented by a specimen from Chalakudi in the cultivated low country of Cochin which may perhaps belong to the same species; by a specimen said to come from Hung in Persian Baluchistan—a locality which I have reason to think was at some time attributed to at least one bottle of mixed spiders from Southern India or Burma; by a mutilated specimen from Ootacamund; and by a young one from Coimbatore.

Annandaliella travancorica, Hirst.

(Pl. xv, figs. 4*a-b*).

This species is represented in our collection by the type from Travancore; by a female from Kulattupuzha in the same State, at the base of the western slopes of the Western Ghats: and by numerous specimens, including three males, from under stones and logs of wood, in the rich evergreen jungle at the base of the same range near Trichur (Cochin) and near the rubber estate between the tenth and fourteenth miles of the Cochin State Forest Tramway. It is very sluggish, at least by day, crouching down when discovered, and remaining quiet with its legs drawn up against the body when seized.

This species has hitherto been known from the type only. Now that more extensive material is available it may be redescribed as follows:—

♂. *Dimensions.*—Carapace 7·2 × 6·0 mm.–9·2 × 8·3 mm. Sternum 3·4 × 3·0 mm.–4·4 × 3·9 mm. The fourth leg longer than the first Tarsus and metatarsus together of first and third legs, and metatarsus alone of fourth legs, about equal to carapace in length; tarsus and metatarsus of second legs slightly shorter, of fourth legs longer by about half the length of the metatarsus, this joint being slightly longer than the femora of the first and fourth legs which are about equal to one another and to the femur together with half the patella of the second and third legs. The proportions all somewhat variable.

Colour.—Dark olivaceous brown, the tarsi and metatarsi of the two front pairs of legs, and the tarsi and distal halves of the metatarsi of the two hind pairs, white. The tarsi of the palps whitish.

Structure.—The *carapace* resembles in shape that of *Plesiophrictus satarensis* described below, but the fovea is lightly procurved, and the anterior median eyes vary from slightly smaller than, to distinctly larger than, the anterior laterals, the diameter of the former being in the latter case about equal to the long diameter of the latter.

The posterior sigilla of the *sternum* vary in position from being almost close to the margin to being separated from it by somewhat more than the diameter of one of them.

The *labium* and its teeth are normal.

The inner surfaces of the *chelicerae* lack the row of spines characteristic of females of this genus.

The *palps* are slender, their tarsi bilobed, with the outer lobe itself obscurely divided into two parts, one anterior to the palpal organ and the other on its outer side. The palpal organ is shown on pl. xv, fig. 4*b*: the spiral curvature of its gracefully bowed, slender, tapering style is very slight.

The first *legs* are unarmed except for the usual apical spine on the metatarsus and apophysis (fig. 4*a*) and spine (the latter sometimes absent) on the tibia. The metatarsus of the second legs is armed with three apical spines and one (rarely absent) about in the middle of the ventral side. The tibia of the same leg has two apical spines and often one mid-ventral one. The tibiae and metatarsi of the third and fourth legs are each armed with a number of spines in the distal two-thirds of their length.

Of the tarsal scopulae only the fourth is divided. The metatarsal scopulae are all apical only; those of the third and fourth legs are sometimes obsolete.

♀. *Dimensions.*—Carapace up to 11·0 × 8·4 mm. Sternum up to 4·5 × 3·8 mm. Legs in the order 4, 1, 2, 3, but relatively much shorter than in the male. Carapace of about the same length as sum of tibia and patella or metatarsus of first leg, to sum of femur and patella of second leg, to patella and tibia with half metatarsus of third, to tibia with patella or half metatarsus of fourth; metatarsus of fourth about equal to tarsus and metatarsus combined of first and second legs, slightly shorter than those of third legs. As in the male these proportions are not altogether constant: the fourth metatarsus is, for instance, sometimes relatively longer as compared with the other joints.

Colour.—Brown, much paler than in the male and not olivaceous.

Structure.—The *carapace* differs from that of the male in having the fovea transversely linear, and the anterior median eyes smaller than the anterior laterals. *Sternum* and *labium* as in male; posterior sigilla often obscure.

The tibia of the *palp* is armed with two apical spines as in *Plesiophrictus satarensis.* The tibia of the first *leg* has one or two apical spines and no apophysis; otherwise the armature of the legs resembles that of the male, except that the metatarsus of the second leg usually has one instead of three spines. The metatarsal scopulae are denser and more extensive than in the male, those of the first legs extending practically to the base of the joint. The first tarsal scopula is often somewhat indistinctly divided by a row of long hairs, rather than by a definite band of spines; the second is divided by a line of spines, the third and fourth also by bands of spines.

Genus **Plesiophrictus**, Poc.

Incl. *Heterophrictus*, Poc.

This genus appears to have given rise to both the other genera of Indian Ischnocoleae, and through one of them to the Thrigmopoeeae also. It is much larger than any of the four derived genera; and the following description, based mainly on the species by which it is represented in our collection, may serve as a standard by comparison with which these genera can be more briefly described. In *Plesiophrictus satarensis*, of which alone the male is known to me, the characters mentioned are found in both sexes unless otherwise stated.

The ocular area is rectangular, nearly or quite three times as broad as long. The eyes of the anterior line, which is lightly procurved, are about equally spaced, somewhat variable in relative size but together larger than the eyes of the posterior line together. The median eyes of the posterior line, which is very lightly recurved, are smaller than the posterior laterals, with which they are practically contiguous being widely separated from one another. The anterior medians are circular, the rest are more or less oval.

The position of the posterior sigilla of the *sternum* varies. In small species they are (? always) marginal; in larger ones they tend to be separated from the margin by a distance not (? ever) exceeding their own width.

The *labium* is about as long as broad, with slightly concave anterior margin, immediately behind which it is armed with a transverse band of somewhat sparsely scattered denticles, rather coarse in the female but sometimes very fine in the male. Similar denticles occupy a roughly equilaterally triangular patch on the lower surface of the coxa of the palp, a patch of which one side is formed by the anterior half of the basal margin.

The *chelicerae* are armed with a row of denticles on the inner side only.

The trochanters of the *palps* are not scopulate; their vestiture resembles that of the trochanters of the legs. The tarsal scopulae of the palps (♀) resemble those of the first legs. The penultimate joints are not scopulate.

The first *legs* are almost always shorter than the fourth[1], the second than the first, and the third than the second. The tarsal scopulae of the first legs are (? always) divided (? sometimes imperfectly especially in the male). The tarsal scopulae of the fourth legs are always divided in both sexes, and in the female at least the division is sometimes so broad that the scopula appears only as a pair of narrow lateral bands. The spiney armature of the legs does not reach its full development in all forms; and it is noteworthy that this is especially the case in relatively large forms whose posterior sigilla are situated away from the margin of the sternum. Such forms resemble *Phlogiodes* and the Thrigmopoeeae in these respects.

The spines develop only after the specimen has attained a moderate size: they appear in a definite order, and those which are normally developed last are the first to be lost in the larger and more highly specialized species. The complete armature may now be described. The spines are confined to the lower surface and sides of the tibiae and metatarsi. On the third and fourth legs they are relatively numerous in well-grown specimens of all species. On the first and second legs, however, they are less numerous and occupy very definite positions. The metatarsi of these legs may bear the following spines—one midapical, a pair of lateral apicals, and one median, of which the midapical always appears first, the order of appearance of the others being less constant; but I do not know of any species in which any of these except the first is developed on the front leg. The complete armature of the tibiae consists of the following spines—inner apical, outer apical, and median, developed in that order. The tibia of the palp is similarly armed, except that so far as I know the median spine is never developed.

The species of *Plesiophrictus* in our collection are as follows:—

Plesiophrictus satarensis, n. sp.

(Pl. xv, figs. 3*a*-*b*).

Localities.—Medha, 2200 ft., in the Yenna valley (♂ ♂); Umbri, 3500 ft., Taloshi, 2000 ft., Helvak, 2000 ft., and Kembsa, 2650 ft., in the Koyna valley (♀ ♀ and immature). All these localities are in the Satara district of the Bombay Presidency. The upper parts of the valleys of the Yenna and Koyna, rivers which flow into the Krishna, are only separated by one ridge of

---

[1] *P. tenuipes*, Poc., from Ceylon, is an exception.

hills, and I have no hesitation in regarding the males found in the one as belonging to the same species as the females found in the other. I have selected the largest male as type.

♂. *Dimensions.*—Carapace 5·0 × 3·4 mm.-7·2 × 4·9 mm. Sternum 3·0 × 2·4 mm.-2·1 × 1·6 mm. Fourth leg longer than first. Carapace about equal in length to patella and tibia of first and fourth legs, to tibia and metatarsus with patella or tarsus of second and third.[1] Legs relatively a little longer in small than in large specimens.

Colour.—Brown, sternum and coxae slightly brighter than the rest because less obscured by hair. The anterior metatarsi whitish.

Structure.—The *carapace* is ovate, broader behind than in front; it is smoothly rounded, free from tubercles, but clothed with hair. The anterior median eyes are as large as the anterior laterals. The fovea is transversely linear.

The posterior sigilla of the *sternum* are marginal.

The *labium* is armed with teeth so small as to be distinct only under a much higher magnification than is usually necessary.

The patella of the *palp* is swollen distally and the tibia proximally. The tarsus is bilobed. The palpal organ is shown in pl. xv, fig. 3*b*; its style is slender, tapering and spirally curved.

The first two pairs of *legs* are unarmed except for the usual apical spine on the metatarsi,[2] the apophysis and its accompanying stout spine on the tibia of the first leg (see pl. xv, fig. 3*a*) and one or two apical spines (not always found) on the tibia of the second leg. The metatarsus of the first leg is lobed on the outer side at the base (fig. 3*a*). The tibia and metatarsus of the third and fourth legs are armed ventrally with 2-3 transverse series of 2-4 spines each. The first tarsal scopula is undivided, the second very narrowly, the third and fourth more (but not very) widely divided.

The metatarsal scopula of the first legs is a little less dense than the tarsal, it is broad distally and narrow proximally, but extends over rather more than the distal half of the joint. On the second legs it is similar, but less obscured by long hair; on the third and fourth it is much smaller and confined to the sides of the distal part of the joint.

The male of this species appears to differ from *P. millardi*, Pocock (the only male hitherto described in the genus) in the denser metatarsal scopula of the first legs, and in the presence of a small apical metatarsal scopula on the fourth legs.

♀. *Dimensions.*—Carapace up to 6·0 × 4·5 mm., sternum up to 2·7 × 2·5 mm. The fourth leg longer than the first as in the male: the pieces which are about equal in length to the carapace in the male seem to be a little shorter in the female.

---

[1] These joints are a trifle longer in the second than in the third leg—slightly so in the type specimen, decidedly so in the other two, which are much smaller.

[2] Occasionally another near it in the second legs.

Colour.—Distinctly yellower than the male; no white hairs on any of the legs.

Structure.—The *carapace* resembles that of the male, but the anterior lateral eyes are somewhat larger than the anterior medians.

The *sternum* is somewhat broader in proportion to its length than in the male.

The teeth on the *labium* are stouter than in the male, normal.

The tibia of the *palp* is armed with two apical spines.

The first *legs* are armed only with the usual apical spine of the metatarsus, and sometimes with a small apical spine on the inner side of the tibia; the metatarsus is not lobed at the base. The tibia is similarly armed in the second legs, but the metatarsus of this pair has three apical spines. The metatarsi of the third and fourth legs are armed as in the male, but the tibiae of these legs appear to be unarmed in their basal halves. All the tarsal scopulae are divided, those of the anterior legs normally, those of the posterior legs very widely. The metatarsal scopulae resemble those of the male, but are perhaps a trifle less pronounced.

The female of this species differs from *P. tenuipes*, the only species previously described in which the anterior median eyes are smaller than the anterior laterals, in having the anterior legs distinctly shorter than the posterior.

Plesiophrictus raja, n. sp.

This handsome species resembles *Annandaliella travancorica* in habits. Its name is given in recognition of the facilities for collecting kindly afforded me by H. H. the Raja (now the ex-Raja) of Cochin, and of the interest which he took in my work.

Localities.—Kavalai, 1300-3000 ft. on the Cochin State Forest Tramway, and near the rubber estate on the lowest slopes of the Ghats between the tenth and fourteenth miles of that tramway. Only one specimen, however, was obtained from the latter place. I have selected the largest of the Kavalai specimens as type.

♂. Unknown.

♀. *Dimensions.*—Carapace up to 9·0 × 6·5 mm. Sternum up to 3·2 × 3·2. The fourth legs longer than the first. Carapace equal in length to femur and patella and to tibia metatarsus and tarsus of first legs, to patella tibia and metatarsus of second legs, to femur patella and tibia of third, and to femur and patella and to metatarsus and tarsus of fourth.

Colour.—Carapace and abdomen covered with hair, occasionally (in one faded-looking specimen from Kavalai) dull greenish brown throughout, usually deep blue above, giving the whole upper surface of the body a rich dark, steel-blue lustre. Legs and lower surface of body olivaceous, sternum and coxae more reddish; anterior tarsi and apical half of anterior metatarsi pale.

Structure.—The *carapace* is ovate, broader behind than in front. The fovea is lightly procurved. The anterior lateral eyes are at least as large as the anterior medians.

The *sternum* is no longer than it is broad. The posterior sigilla are fully a diameter distant from the margin.

The *labium* is normal.

On the tibia of the *palps* only the inner apical spine is developed.

The first *legs* are unarmed except for the usual apical spine on the metatarsus. The second legs have three apical and one median spines on the metatarsus; their other joints are unarmed. The third and fourth legs bear spines on the distal two-thirds of the length of each. All the tarsal scopulae are completely divided, but the spines between the two halves are stouter and more widely spaced on the two posterior pairs of legs than on the two anterior pairs. All protarsal scopulae are more or less obsolete.

This species differs from all that have hitherto been described in its deep steel-blue colour.

Plesiophrictus bhori, n. sp.

This species resembles *Annandaliella travancorica* and *Plesiophrictus raja* in its general habits. The jungle in which it lives is, however, largely of the deciduous type, instead of the evergreen type that predominates at the base of the hills and at Kavalai. A large proportion of the specimens were found under pieces of wood in open jungle consisting largely of bamboo, a type of jungle of which neither insects nor arachnids seem usually to be fond. The species is named after Mr. J. Bhore, the Dewan of Cochin, whose constant help enabled me to make interesting collections in places that I could not otherwise have reached during my short visit to the State.

Locality.—Parambikulam in the Western Ghats, Cochin State, at altitudes varying from 1700-3200 ft.

♂. Unknown.

♀. *Dimensions.*—Carapace up to 12·0 × 10·0 mm. Sternum up to 5·3 × 5·3. The fourth leg longer than the first. Carapace slightly shorter than femur and patella or tibia metatarsus and tarsus of first leg, about equal to (perhaps slightly longer than) patella and tibia of same leg, to femur and patella and to tibia metatarsus and tarsus of second legs, to trochanter femur and patella of third legs, and to tarsus and metatarsus of fourth, scarcely as long as femur and patella of fourth.

Colour.—Almost uniformly brown.

Structure.—The *carapace, sternum* and *labium* resemble those of the preceding species. The sternum is, however, somewhat more densely hairy. The tibia of the *palps* is armed with two apical spines only in the largest specimen seen (the type), in other large specimens only the inner one is present, the palps being as usual unarmed in the very young.

The first *legs* are armed as in the preceding species, except that in full grown specimens there is (? always) a small apical spine on the inner side of the tibia. The same applies to the second legs except that this inner apical spine of the tibia appears at an earlier stage, and is followed by a mid-ventral spine and an outer apical one. The third and fourth legs are armed as in that species. The tarsal scopulae resemble those of the preceding species. The metatarsal scopulae are dense on the first pair of legs, slightly thinner on the others; they occupy the distal half of the metatarsi of the first two pairs of legs, but are more restricted on the last two.

This species seems to be very closely allied to *P. milleti.* It agrees with Pocock's short description of that species in all structural characters, but differs in the colour of its pile which is distinctly brown, not red, being almost olivaceous on the abdomen; it also differs in the absence of white hairs from the extremities of the legs. The localities from which the two species come are very widely separated; and a fuller description of *H. milleti* will probably reveal structural differences between the two.

Genus **Phlogiodes**, Pocock.

This genus is not represented in our collection, unless it be by two immature specimens from the Bombay Presidency. I can add nothing to what I have already said about it above (pp. 269-270).

Group THRIGMOPOEEAE.

Pocock's key to the two genera recognized in this group seems quite satisfactory.

Genus **Haploclastus**, Simon.

The stridulating organ of the new species of this genus described below is of a very simple, almost rudimentary type. It has been figured elsewhere (Gravely, 1915, pl. xxxi, fig 1). The bacilli on the chelicerae are situated on the lower margin, into the general hairiness of which they merge, and the minute scattered bristles on the anterior surface of the coxa of the palp are scarcely if at all different from the more numerous bristles which cover this surface in the first legs. In other characters, the genus closely resembles the preceding [1] which has no stridulating organ, and the following in which the stridulating organ is of a somewhat more advanced type It may therefore be regarded as transitional between the two.

Haploclastus kayi, n. sp.

Locality.—Parambikulam, 1700-3200 ft., Cochin State, where the wide knowledge of the country and its jungles possessed by

[1] This refers to the female. No male Thrigmopoeeae yet appear to be known.

Mr. P. B. Kay enabled me to find without delay the most promising collecting grounds of the neighbourhood. Only one specimen of the present species was obtained.

♂. Unknown.

♀. *Dimensions.*—Carapace 13·0 × 9·8 mm. Sternum 6·1 × 5·5 mm. First legs fully as long as fourth which are slenderer than any of the others; second shorter than first and fourth, longer than third. Carapace not quite as long as patella and tibia of first legs, much longer than femur and than tarsus and metatarsus of same, and than femur or patella and tibia or tarsus and metatarsus of second legs; fully as long as femur and patella of third legs, scarcely as long as tibia metatarsus and tarsus of same; about equal to femur and half patella, to patella and tibia, and to tarsus and metatarsus of fourth legs.

Colour.—Reddish brown, except the upper sides of the patella and tibia of the second legs which are paler, almost golden; and of the patella, tibia, and base of metatarsus of the first legs which are paler still and greyish.

Structure.—The shape of the *carapace* resembles that of the preceding species. The anterior median eyes are larger than the anterior laterals, the diameter of the former being about equal to the long diameter of the latter. The fovea is lightly procurved and very deeply impressed. The *labium* is normal; the *sternum* is very hairy, with large sigilla which are rounded in front and pointed behind. The *chelicerae* resemble those of *Plesiophrictus* apart from the presence of stridulatory spines upon them. There is the usual mid-apical spine on the metatarsi of the first pair of *legs*, but it is much hidden by the dense scopula. I have not been able to detect any spine on the second metatarsus, but here too the scopula is very dense. The metatarsi of the two hind legs each have three apical spines. The tibiae of the palps and of all the legs are unarmed. The tarsal and metatarsal scopulae are divided in the fourth leg only. The metatarsal scopulae of the first two legs are very dense, and extend to the base of the segment. Those of the third legs, though dense, only cover the distal half of the segment. Those of the fourth legs are weaker and apical.

This species differs from *H. nilgirinus* in that the fourth leg is longer than the second, and from *H. cervinus* in that the patella and tibia of the first are together longer than those of the fourth. From both it appears to differ in colour, but this difference may be less real than it seems as its most striking feature—the light grey of the upper surface of the patellae and tibiae of the anterior legs—is not apparent as long as the specimen remains superficially wet.

Genus **Thrigmopoeus**, Pocock.

A single immature specimen from South Arcot is the only representative of this genus which we possess. Its stridulating organ differs from that of *Haploclastus kayi* in the more definite arrangement and slightly greater size of the spines on the palps, and also in the greater distinctness of the group of bacilli on the chelicerae from the hairs which clothe the lower sides of these appendages. The organ has been figured elsewhere (Gravely, 1915, pl. xxxi, fig. 2).

Group POECILOTHERIEAE.

This group, which contains only one genus, appears to have originated in the Indian Peninsula or Ceylon, from some form presumably allied to the foregoing genera of Aviculariinae, as a result of adaptation to a new mode of life (see Gravely, 1915, pp. 417-418).

Genus **Poecilotheria**, Simon.

Poecilotheria miranda, Pocock.

One female specimen from "Kharagpur Hills" [1], and another from near Chaibassa in the Singbhum District of Chota Nagpur.

Poecilotheria regalis, Pocock.

One male from Bangalore, and one female from the Annamalai Hills. The latter record extends the known range of this species to the hills south of the Palghat Gap, an extensive low-lying plain which cuts right across the hills of South India. The specimen is one determined by Mr. Hirst of the British Museum, who presumably had the type available for comparison. The discovery of a male in the Annamalais is greatly to be desired, as it is possible that its palpal organ may prove to differ from that of the male found on the opposite side of the Gap.

Poecilotheria striata, Pocock.

One female from South India, and one somewhat smaller specimen from Pamben on Rameswarem Island.

Group ORNITHOCTONEAE.

Only one species of each of three genera of this group are recorded from the Indian Empire. Of these *Melopoeus minax* is much the commonest, and is represented in our collection by females from "Burma", "Upper Tenasserim", Myawadi on the Burmo-Siamese frontier (Thoungyin valley, Amherst District of Tenasserim), the hills between the Thoungyin and Me-Ping in Siam, and from Pitsanuloke in Siam. It spends the day in silk-lined burrows devoid of a trap-door, but comes out in the evening. The only specimen I saw outside seemed very sluggish.

The road between Thingannyinaung and the base of the Dawna Hills, on the extreme west of the Thoungyin Valley,

[1] Kharagpur is situated in the Midnapur District of Bengal, in the western part of the flat country bordering on the Gangetic Delta. The hills referred to are probably those of Singbhum, a district of Chota Nagpur immediately to the west of Midnapur.

was lined with these burrows; but having little time to spare when I noticed them, and no proper digging implements, I only got one spider from them. This was a male which I had no hesitation in associating with the similar-looking female common in the district, *i.e.* with *Melopoeus minax*. Its characters were, however, those of a *Cyriopagus* rather than of a *Melopoeus*. This led me to consider whether *Cyriopagus* might not be simply the male of *Melopoeus*. The type of the former genus is recorded as a female; but it is in our collection, and there can, I think, be no doubt at all about its immaturity. It may therefore be a male. *Omothymus schioedtei*, Thorell, which Simon refers to the genus *Cyriopagus*, is described from a male only. The male of *Selenocosmia albostriata*, the species for which Pocock established the genus *Melopoeus*, is described by Simon (1886, p. 162) as "feminae subsimilis sed cephalothorace humiliore." The low cephalothorax is one of the two chief characters in which *Cyriopagus* differs from *Melopoeus*; and nothing is said as to the distance of the eyes from the margin of the carapace in either sex of the species in question. I conclude, therefore, that *Cyriopagus* and *Melopoeus* represent opposite sexes of one genus.

Of these two names the former has priority. This is unfortunate, inasmuch as the genus *Ornithoctonus*, which is known from the female sex only, differs from *Cyriopagus* in the same characters as does "*Melopoeus*", and may also very possibly have a male with *Cyriopagus* characters. The characters by which Pocock separates *Cyriopagus* (= *Melopoeus*) from *Ornithoctonus* are unsatisfactory even for females; and the two genera will very likely have to be united.

The material before me is not, however, sufficient to justify this course at present, so the probable relation of the genotype of the former, *Cyriopagus paganus*, to other members of these genera must be considered. The characters by which their females are separated are found in practice to be so unsatisfactory even in that sex, that it would be hopeless to try to apply them to the other. Pocock's figure of the stridulating organ of *Ornithoctonus* suggests, however, another means of separating that genus from *Cyriopagus*. For the stridulatory processes of the palp are shown as long spiniform structures, whereas in *Cyriopagus minax* they are short and denticuliform. And it may be mentioned that a specimen in our collection which seems to approach the genus *Ornithoctonus* rather than "*Melopoeus*" in the characters of its legs and fovea has spiniform, not denticuliform, stridulatory processes on the palp. Unfortunately the locality of the specimen is not known.

The stridulatory processes on the palp are denticuliform in the genotype of *Cyriopagus*; so it is *Melopoeus* rather than *Ornithoctonus* that must now be sunk as a synonym. Whether *Ornithoctonus* is to be sunk as well requires further investigation.

Apart from the immature type of *Cyriopagus paganus*, *C. minax* is the only named species of this group in our collection.

The male specimen is at present on loan in America so I cannot give the description of it which ought to appear here.

Group SELENOCOSMIEAE.

This group, as defined above (p. 268), contains all the Indo-Australian Aviculariinae in which the anterior part of the labium is covered with densely packed granules, no matter whether a stridulating organ is present between the chelicerae and palps or not.

The only known species which lacks the stridulating organ is "*Ischnocolus*" *brevipes*, Thorell, but in "*Ischnocolus*" *subarmatus*, Thorell, this organ is quite rudimentary. The latter species was removed by Simon (Vol. II, p. 925) to the genus *Phlogiellus*, Poc., a genus which has since been shown by Hirst (1909, p. 384) to be indistinguishable from *Selenocosmia* and *Chilobrachys*. For the species *subarmatus*, however, he instituted a new subgenus *Neochilobrachys*, on account of the rudimentary nature of the stridulating organ (*loc. cit.*, p. 389).

Neochilobrachys subarmatus differs from species belonging to the genus *Chilobrachys* in having a much smaller number of stridulating rods on the coxa of the palp, and *Chilobrachys* differs from *Selenocosmia* in the same way. It was presumably for this reason that Hirst decided to regard *Neochilobrachys* as a subgenus of the former rather than of the latter. The change from the *Selenocosmia* to the *Chilobrachys* type of stridulating organ—of which many stages can be illustrated from species found at the present day—has, however, been accompanied by a marked increase in the specialization of the stridulating rods. The whole organ is clearly of a more advanced type in *Chilobrachys* than in *Selenocosmia*, and the reduction in the number of the rods cannot be regarded as in any way indicating a tendency towards degeneration—the only process which could bring them to the rudimentary condition of the "rods" found in *Neochilobrachys subarmatus*. The "rods" of *N. subarmatus* are, indeed, mere spines, comparable to those composing the dorsal and lateral parts of the groups of "rods" found in *Selenocosmia*, in which genus only the middle and ventral elements of these groups are really bacilliform.

In my opinion, therefore, *Neochilobrachys subarmatus* should be regarded as a primitive form transitional between "*Ischnocolus*" *brevipes* with no stridulating organ, and the genus *Selenocosmia* which possesses a stridulating organ of some complexity. In this case *Neochilobrachys* cannot remain as a subgenus of *Chilobrachys*; and as it differs from *Selenocosmia* more widely than does that genus from certain species of *Chilobrachys*, it may be regarded as a distinct genus. For the present it will be best, I think, to define this genus somewhat loosly, so that "*Ischnocolus*" *brevipes* may be included in it. Otherwise yet another monospecific genus would be required.

The three genera *Neochilobrachys*, *Selenocosmia* and *Chilobrachys* represent three stages in the evolution of the type of stridulating organ found in the group. The *Selenocosmia* stage is found in the localized genera *Lyrognathus* (3 species), *Coremiocnemis* (2 species), *Selenostholus* (1 species) and *Selenotypus* (1 species) as well as in the large and widely distributed genus *Selenocosmia*. Of these the last two are Australian, and I am not in a position to say anything about them. Of the desirability of keeping the first two distinct from *Selenocosmia* I am very doubtful. The Indian species of *Selenocosmia* appear to be transitional between this genus and *Coremiocnemis*, a genus which should certainly, I think, be abandoned. And the value of the single character by which the genus *Lyrognathus* is distinguished is probably small. I have, however, provisionally retained *Lyrognathus* as a subgenus. Similarly, I am inclined to doubt the advisability of keeping distinct from the larger and more widely distributed genus *Chilobrachys* the mono-specific genus *Orphnoecus* from the Philippines.

The genera of Selenocosmieae found in the Indian Empire may be distinguished as follows:—

1. { Stridulating organ between chelicerae and palps rudimentary or absent .. *Neochilobrachys*.
   Stridulating organ well developed .. 2.

2. { Stridulating organ consisting of a dorsal crescent of fine spines, merging into and partially surrounding a ventral group of more or less claviform, but always somewhat slender, bacilli .. *Selenocosmia*, p. 284.
   A few of the bacilli in the ventral row very large and strongly claviform; the number of rows, both of bacilli and of spines, often greatly reduced; the ventral row of bacilli usually extending beyond the spines in one or other direction .. .. .. *Chilobrachys*, p. 285.

Genus **Neochilobrachys**, Hirst.

So far as I know, only two species have yet been described which can be placed in this genus. They are *N. brevipes* (Thorell, 1896, pp. 170-173) and *N. subarmatus* (Thorell, 1891, p. 13). In the former there is no stridulating organ between the chelicerae and palps. In the latter, which is represented in our collection by a number of specimens from the Nicobars, there is a row of 2-6 (see Hirst, 1909, p. 388) stout spinules on the palp, and a group of somewhat similar but scattered and smaller spinules on the sides of the chelicerae close to the hair on the proximal part of the lower margin. This stridulating organ has been figured elsewhere (Gravely, 1915, pl. xxxi, fig. 3). The part on the chelicerae has also been figured by Hirst (1909, pl. xxiv, fig. 2).

Genus **Selenocosmia**, Ausserer.

The palpal parts of the stridulating organs of this genus have already been described (p. 283). The parts situated on the sides of the chelicerae normally consist of a number of long and slender spines mixed with, and not always sharply differentiated from, scattered hairs which are continuous with the thicker covering of the lower margins. In *S. himalayana*, however, these spines are shorter, and not mixed with hair, resembling those found in *Chilobrachys assamensis* and *fumosus*, rather than those found in other members of its own genus. In this species, too, the group of claviform bacilli on the palp is elongated at the expense of the downwardly curved ends of the group of simpler bristles, the two groups being almost equal in extent. It approaches *Chilobrachys* in these characters of the palpal part of the stridulating organ, to a greater extent than does any other species of *Selenocosmia* known to me. I have elsewhere figured a typical stridulating organ of the genus *Selenocosmia* (1915, pl. xxxi, fig. 4).

Two Indian subgenera may be recognized, though their value is uncertain. They may be distinguished thus:—

Fourth legs much thicker than first, their metatarsal scopulae entire and extending to the base of the segment *Lyrognathus.*
Fourth legs not thicker than first, their metatarsal scopulae weaker and apical .. .. .. *Selenocosmia.*

Lyrognathus is represented in our collection by two specimens. One, from the Khasi Hills, has been determined by Hirst as *L. crotalis*; the other, from the Garo Hills, has been determined by myself as *L. pugnax.* I am very doubtful whether they are really distinct.

Selenocosmia is represented by one specimen from the Andamans doubtfully referred by Hirst to *S. javanus*; by two immature specimens from the same group of islands; by several specimens with slenderer legs from Sibsagar in Assam; and by several specimens (mostly males) of *S. himalayana.* The last named species was described by Pocock from a specimen from Dehra Dun, said to be a female, the length of whose carapace was 15 mm. Hirst (1907, pp. 523-4, text-fig. 2) has since described a male from Kasauli, 6,600 ft., with a carapace length of 20 mm., and a female from Dalhousie, 6000 ft., with a carapace length of 18 mm.; he therefore concludes that the type was immature. This was not necessarily the case, however, for we have adult males whose carapace lengths range from 8·5-12·0 mm. Two of these are from Dehra Dun, two from Almora, 5500 ft., and two from Naini Tal. We also have one female from Dehra Dun. The species is evidently very variable in size. In one of the males from Dehra Dun the characteristic projection on the outer side of the palpal organ, though present on that of the left side, is absent on that of the right.

Genus **Chilobrachys**, Karsch.

Simon's definition of this genus applies only to those species in which the stridulating organs approach or attain their highest degree of specialization. To the simpler forms it is inapplicable. This is especially so in the case of *C. assamensis* and *C. fumosus*, species which resemble *Selenocosmia himalayana* in the structure of the parts of the stridulating organs situated on the chelicerae, and approach it more closely than do any other species of *Chilobrachys* known to me, in the structure of the parts of these organs situated on the palps. The stridulating organ of *C. assamensis* has already been figured (Gravely, 1915, pl. xxxi, fig. 5).

C. assamensis and *C. fumosus* are closely allied to one another. *C. fumosus* appears invariably to attain a much greater size than *C. assamensis*; but in view of the great range in size shown by *S. himalayana* (see previous page) and by *C. hardwickii* (see following page) this, the only difference known to me in females, cannot be regarded as an altogether satisfactory character.

The palpal organs of males of the two species are remarkably alike in their general features, but the style is longer and more abruptly spatulate at the end in *C. fumosus* than in *C. assamensis*. The latter species is represented in our collection by cotypes of both sexes from Sibsagar in Assam. The former is represented by two males from Kurseong, on one of which (that collected by Dr. Annandale) Hirst's description of this sex (1909, pp. 386-7—the only one yet published) was based. Females and young, which must provisionally be referred to this species, are represented in our collection by specimens from Chitlong in Nepal; Singla, 1500 ft., Darjeeling, Sureil and Kalimpong in the Darjeeling District; the Assam-Bhutan Frontier of Mangaldai District; and Burroi at the base of the Dafla Hills. It remains to be seen, however, whether the males from all these localities belong to a single species. If not, as the type is a female labelled "North India", the name *C. fumosus* should be kept for the Kurseong (Darjeeling District) form. The arrangement of the spines on the chelicerae in parallel rows is more or less clearly marked in certain specimens of this and other species; it cannot be regarded as a good specific character.

In *Chilobrachys assamensis* and *fumosus* the largest bacilli on the palps are situated in the distal half of the ventral row, and the stridulating processes of the chelicerae are slender and spiniform as in *Selenocosmia*. In all other species which I have seen the largest bacilli are proximal, and the projections against which they work are short and denticuliform. In a species from the Malay Peninsula, however, the former are practically median, though the dorsal spines are concentrated a little on the distal side of them. It is possible that *Chilobrachys assamensis* and *fumosus* have originated independently of the rest of the genus, in which case the former might be made the type of a new genus containing the latter and perhaps also *Selenocosmia himalayana*. But the evidence is not yet conclusive; and in any case these species furnish an interesting indication of the manner in which the more typical forms of the genus *Chilobrachys* must have arisen. I have already figured elsewhere the stridulating organ of the type specimen of *C. stridulans* (1915, pl. xxxi, fig. 6).

Of the three remaining Indian species of *Chilobrachys*, which I am able to identify in our collection, *C. fimbriatus* appears to be the most primitive, *i.e.* the least removed from *Selenocosmia*, in the structure of its stridulating organs; for the rows of small bacilli are more numerous than is usual in either of the others. In *Chilobrachys hardwickii* the extent of these small bacilli appears to be somewhat variable, but it is usually less than is the case in our single specimen of *C. fimbriatus*: the shape of the whole group of bacilli in *C. hardwickii* is, moreover, longer and narrower, and so more like that of *C. stridulans*, in which the rows of small bacilli left exposed by the dorsal fringe of hair are still fewer. Another variable feature of *C. hardwickii*, and one in no way correlated with the variations found in the stridulating organ, is its size. The mature males in our collection have carapaces varying in length from barely 10 to over 16 mm. in length. The smallest males are associated with specimens of similar dimensions which are presumably mature females. Pocock's suggestion that females of this species may always be distinguished from those of *C. nitellinus* by their larger size can no longer, therefore, be maintained.

C. fimbriatus is represented in our collection by a single male from Hoshali in the Shimoga District of Mysore. *C. hardwickii* is represented by specimens from Dharhara (Monghyr District) and Sahibgunge in Bihar; from Chakardharpur (Singbhum District) in Chota Nagpur; and from Gmatia (Birbhum District) and Murshidabad in Bengal. *C. stridulans* is represented by specimens from Punkabari at the foot of the Darjeeling Hills, and from Goalpara, Shamshernager (Sylhet), Silcuri (Cachar), Aideo[1] and Sibsagar in Assam.

Subfamily *DIPLURINAE*.

Group MACROTHELEAE.

Genus **Macrothele**, Ausserer.

Macrothele vidua, Simon.

(Pl. xv, fig. 5).

I have little hesitation in referring to this species specimens sent me by Dr. Sutherland from Kalimpong. The species was described by Simon (1906, p. 306) from the "bas plateaux de l'Himalaya"; and the only way in which our specimens appear to differ from it is in the armature of the anterior tarsi, which is present on the outer as well as on the inner side.

[1] I do not know in which district Aideo is situated.

The Kalimpong series includes one male, and we have a male of the same species from Kurseong. This sex differs from the female in having the anterior median eyes more distinctly larger than the anterior laterals, and the posterior medians much smaller than the posterior laterals. The lower surface is inclined to be somewhat reddish throughout—more so in our Kurseong specimen than in the other. The legs and spinerettes are much slenderer in the male than in the female; and the abdomen is shorter in proportion to the length of the spinerettes. The palpal organ is lightly constricted below the stout conical base of the remarkably long slender and almost straight style (see pl. xv, fig. 5).

LIST OF LITERATURE.

1886. Simon, E. "Arachnides Recueilles par M. A. Pavie dans le royaume de Siam, au Cambodge et en Cochinchine." *Actes Soc. Linn. Bordaux* (4) X, 1886, pp. 136-187.

1891. Thorell, T. "Spindlar från Nikobarerna och andra delar af Södra Asien." *K. Sv. Vet. Akad. Handl.* XXIV (2), 1891, 149 pp.

1892. Simon, E. "Histoire Naturelle des Araignées." Vol. I, Paris, 1892.

1895. Pocock, R. I. "On a New and Natural Grouping of some of the Oriental genera of Mygalomorphae, with descriptions of new genera and species." *Ann. Mag. Nat. Hist.* (6) XV, 1895, pp. 165-184, pl. x.

1896. Thorell, T. "Secondo Saggio sùi Ragni Birmanie I, Parallelodontes—Tubitelariae." *Ann. Civ. Mus. Genova* (2*a*), XVII (XXXVII), 1896-7, pp. 161-267.

1897. Simon, E. "Histoire Naturelle des Araignées." Vol. II, Paris, 1897.

1899. Pocock, R. I. "Diagnoses of some new Indian Arachnida." *J. Bombay Nat. Hist. Soc.* XII, 1898-1900, pp. 744-753.

1900. Pocock, R. I. "The Fauna of British India.—Arachnida." London, 1900.

1906. Simon, E. "Descriptions de quelques Arachnides des bas plateaux de l'Himalaya." *Ann. Soc. Ent. France* LXXV, 1906, pp. 306-314.

1907. Hirst, A. S. "On Two Spiders of the genus *Selenocosmia.*" *Ann. Mag. Nat. Hist.* (7) XIX, pp. 522-524, 2 text-figs.

1909. Hirst, A. S. "On some new or little-known Mygalomorph Spiders from the Oriental Region and Australia." *Rec. Ind. Mus.*, III, 1909, pp. 383-390, pl. xxiv.

1915. Gravely, F. H. "The Evolution and Distribution of Oriental Spiders belonging to the sub-family Aviculariinae." *J.A.S.B.* (in the press).

EXPLANATION OF PLATE XV.

FIG. 1.—*Idiops biharicus*, type (♂).

a. Junction of tibia and metatarsus of first leg of right side, from above and in front.

b. Outer side of right palpal organ.

,, 2.—*Sasonichus arthrapophysis*, type (♂).

a. Junction of tibia and metatarsus of first right leg from the inner side.

b. Outer side of right palpal organ.

,, 3.—*Plesiophrictus satarensis*, type (♂).

a. Junction of tibia and metatarsus of first right leg from below.

b. Outer side of right palpal organ.

,, 4.—*Annandaliella travancorica* (♂).

a. Tibial apophysis of first right leg from below.

b. Outer side of right palpal organ.

,, 5.—*Macrothele vidua* (♂). Outer side of right palpal organ.

5

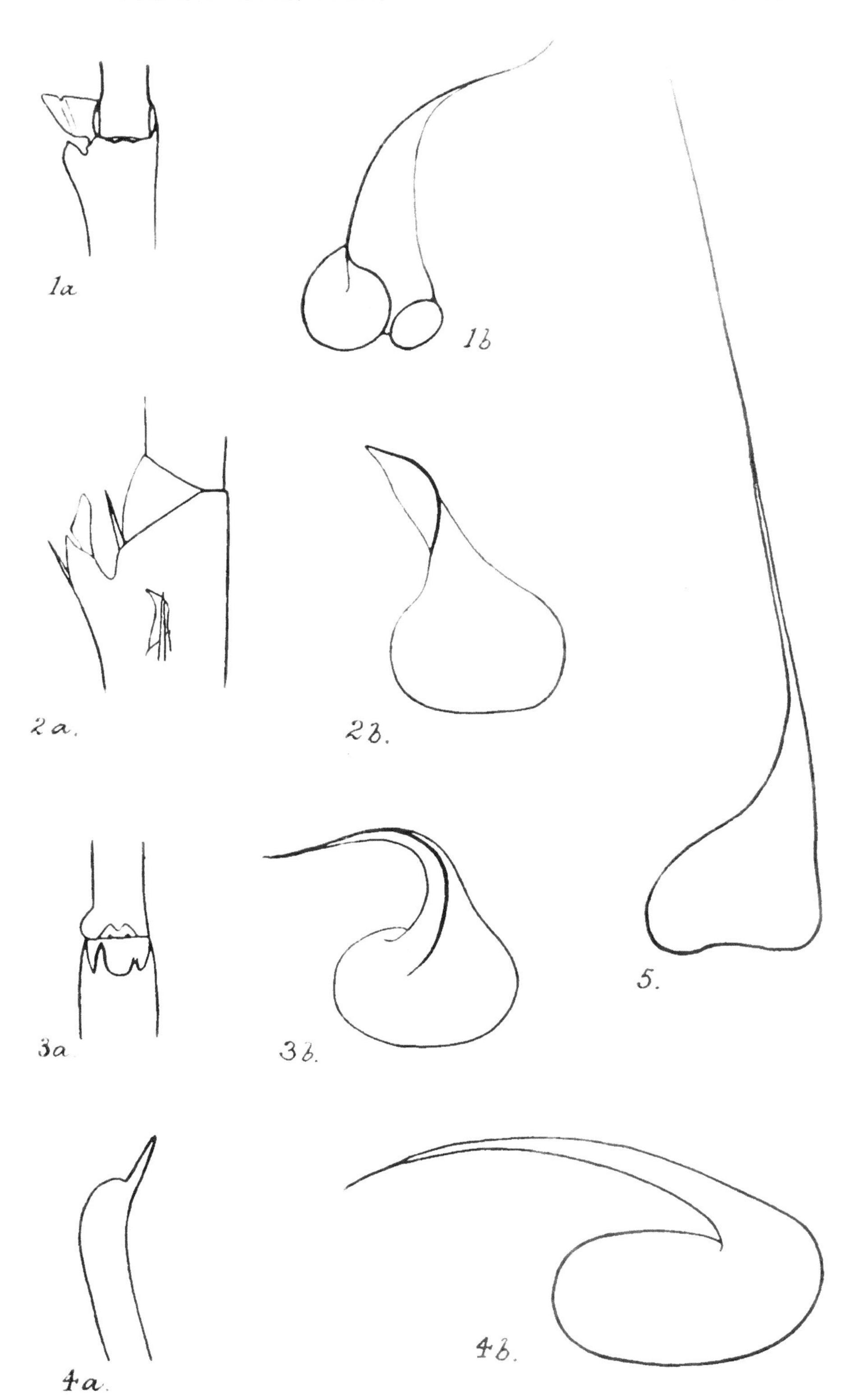

Indian Mygalomorph Spiders.

XV. A FURTHER REPORT ON MOLLUSCA FROM LAKE CHILKA ON THE EAST COAST OF INDIA.

By H. B. PRESTON, *F.Z.S.*

At the request of Dr. Annandale of the Indian Museum the author has examined a second collection of Mollusca from Lake Chilka, a report on which is given below. The first collection was made by Dr. Annandale and Mr. Kemp in 1913 and included thirty-four species of which twenty-one were described as new,[1] though two of these (*Velorita satparaënsis* and *Tornatina soror*) have from the examination of considerable further material proved to be unworthy of retention, as will be seen later. The present collection was made by the same collectors during September and December of 1913 and January, February, March, July and September of 1914, and contains sixty-seven species of which twenty-five appear to have hitherto escaped notice and are described and figured in the present report, the type specimens in all cases being returned to the Indian Museum. This large number of new forms in both collections may be accounted for by the fact that practically no systematic collecting has ever been previously done in the lake.

As was to be expected a large proportion of the species originally recorded are again included in this second collection, some however very sparingly while others are conspicuous by their total absence. This is largely because examples of some easily recognized species were not sent.

In conclusion the author would take this opportunity to express his thanks to Mr. E. A. Smith, I.S.O., for much help ungrudgingly given in the generic determination of the smaller Pelecypods of which both collections contain a large number.

Class GASTROPODA.

Order *PROSOBRANCHIA*.

Family TEREBRIDAE.

Terebra rambhaënsis, Preston.

Rec. Ind. Mus., X, p. 297.

Nalbano and channel S.E. of Nalbano, 4-8 ft.

[1] *Rec. Ind. Mus.*, X, pp. 297-310 (1914).

Family NASSIDAE.

Nassa sistroidea, G. and H. Nevill.

J. As. Soc. Bengal, XLIII, pt. 2, pl. i, fig. 6.

Serua Nadi, 5-9 ft. (two immature examples).

Nassa marratii, Smith.

J. Linn. Soc., XII, p. 543; *Proc. Zool. Soc. London*, 1878, p. 809.

Channel off Satpara Point, 8-12 ft.

Nassa orissaënsis, Preston.

Rec. Ind. Mus., X, p. 299.

Channel off Barhampur Id., 6-9 ft.; 1 mile N.E. by E. of Chiriya Id., 5½-6¾ ft.; opposite Barkul bungalow; 2-8 miles N.E. ½ E. of Kalidai, 5-6 ft.; S.E. of Barkuda and Samal Id., ¼ mile off shore, 6 ft.; 2-3 miles S.E. by E. ½ E. of Patsahanipur, 4¾-5½ ft; Serua Nadi, 5-9 ft.; channel between Barnikuda and Satpara, 6½ ft.; about 4 miles E.N.E. of Kalupara Ghat; 1-1¼ miles off Kalupara Ghat. channel from Satpara towards Barnikuda, 9-12 ft.

Nassa denegabilis, Preston.

T. c., pp. 297-299.

Channel between Satpara and Barhampur, 8-20 ft.; Nalbano and channel S.E. of Nalbano, 4-8 ft.; Sand-dunes opposite Manikpatna; Serua Nadi, 5-9 ft.

Bullia vittata, Linn.

Syst. Nat., Edit. 12, p. 1206.

Outer bar close to mouth (a single young and dead specimen).

Family STROMBIDAE.

Strombus isabella, Lk.

Anim. s. vert., IX, p. 700.

Chilka Lake (a very young specimen).

Family CERITHIIDAE.

Potamides (Tympanotonos) fluviatilis, Pot. and Mich.

Cat. Moll. de Douai, p. 363.

Channel between Satpara and Barhampur, 8-20 ft.; channel from Satpara towards Barnikuda, 9-12 ft.; off Barnikuda, 5-12 ft.

Potamides (Telescopium) fuscum, Schumacher.

Essai Nouv. Syst., p. 233.

Outer bar opposite Manikpatna temple.

Family FOSSARIDAE.

Chilkaia, gen. n.

Shell minute, ovate, rimate, with large body whorl and aperture, spirally lirate and transversely plicate; operculum?

Hab.—Lake Chilka, E. coast of India.

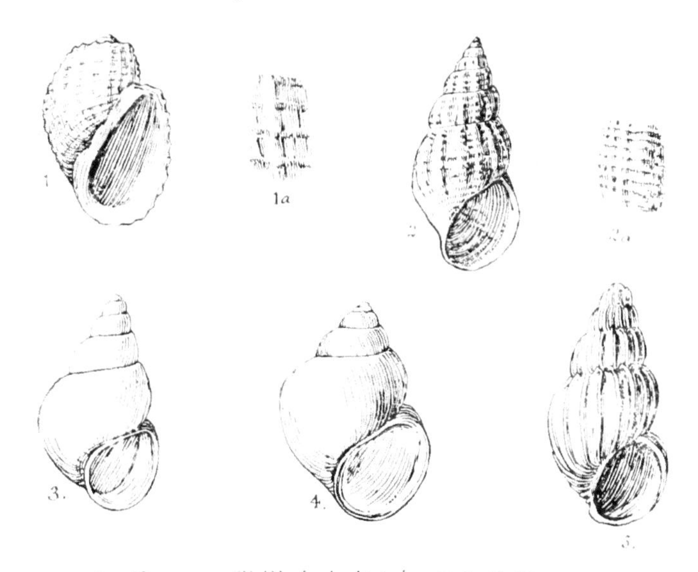

FIG. 1.—*Chilkaia imitatrix*, sp. n. × 10.
,, 1*a*.— ,, ,, (sculpture) × 30.
,, 2.—*Litiopa* (*Alaba*) *copiosa*, sp. n. × 8.
,, 2*a*.— ,, ,, ,, (sculpture) × 12.
,, 3.—*Stenothyra trigona*, sp. n. × 14.
,, 4.—*Stenothyra obesula*, sp. n. × 8.
,, 5.—*Epitonium hamatulae*, sp. n. × 4.

Chilkaia imitatrix, sp. n.

(Figs. 1, 1*a*.)

Shell small, oblong ovate, covered with a light reddish periostracum; whorls 3, finely and wavily spirally lirate and slightly distantly obliquely transversely plicate, the last whorl shouldered in the infra-sutural region; suture impressed; perforation very narrow; columella whitish, descending in a curve, extending above into a thick, white, restricted, parietal callus which unites it with the labrum above; labrum acute, a very little dilated below; aperture oblique and rather elongately ovate.

Alt. 2·5, diam. maj. 1·75 mm.

Aperture: alt. 1·25, diam. ·75 mm.

Hab.—Mahosa, southwards towards sandhills, 4-8 ft. (*Type*); Serua Nadi, 5-9 ft.

Bearing, in miniature, an extraordinary resemblance to the more ovate forms of *Paramelania* from Lake Tanganyika.

Family LITIOPIDAE.

Litiopa (Alaba) kempi, Preston.

Rec. Ind. Mus., X, p. 300.

8 miles W. by S. of Breakfast Id., 5-5¾ ft.; E. side of Rambha Bay, 1-4¾ ft.; 1 mile N.E. by E. of Chiriya Id., 5½-6¾ ft.; Breakfast Id., midway between Ganta Sila and Chiriya Id. South Pt., 5-6 ft.; Nalbano and channel N.E. of Nalbano, 4-8 ft.; southwards from Mahosa, 5-9 ft.; Mahosa, Barhampur Id.; off Sankuda Id., Ganjam District, Madras; Rambha Bay: among grass-like weeds on sandy bottom in 3-3½ ft. of water close to shore of Barkuda Id.

Litiopa (Alaba) copiosa, sp. n.

(Figs. 2, 2*a*, p. 291.)

Shell small, fusiformly turrite, whitish, showing traces of having been covered with a thin greenish periostracum, the last whorl narrowly transversely banded with reddish brown; whorls 6, the last two moderately convex, the upper whorls flattish, sculptured with fine, closely-set, spiral lirae and slightly oblique, rounded, rather distant, transverse costulae; suture impressed, crennellated by the terminations of the transverse costulae; columella obliquely descending, scarcely reflexed, diffused above into a restricted, well defined, parietal callus which reaches to the upper margin of the labrum: labrum simple; aperture oblique, ovate.

Alt. 3·75, diam. maj. 1·75 mm.

Hab.—Serua Nadi, 5-9 ft. (*Type*); channel between Barnikuda and Satpara, 6½ ft.; Mahosa, southwards towards sandhills, 4-8 ft.; channel from Satpara towards Barnikuda, 9-12 ft.; Nalbano and channel S.E. of Nalbano, 4-8 ft.

Family HYDROBIIDAE.

Stenothyra minima, Sow.

Ann. Mag. Nat. Hist. London (Charlesworth's series), I, 1837, p. 217 (as *Nematura*).

Opposite Barkul bungalow; 1-5 miles N. by E. of Kalidai, 7-7½ ft.; 1 mile N.E. by E. of Chiriya Id., 5½-6¾ ft.; southernmost island of Manikpatna series; 1 mile S. of Kalidai, 4-8 ft.; Serua Nadi, 5-9 ft.; between Barkuda and mainland, 6-8 ft.

Stenothyra chilkaënsis, Preston.

Rec. Ind. Mus., X, p. 300.

Serua Nadi, 5-9 ft.; 4-9 miles E., ½ S. of Barkul bungalow, 5¼-5¾ ft.; channel off Barhampur Id., 6-9 ft.; 1 mile N.E. by E. of Chiriya Id., 5½-6¾ ft.; channel between Barnikuda and Satpara, 6½ ft.; off Barnikuda, inside lake, 6 ft.; 2-8 miles N.E. ½ E. of Kalidai, 5-6 ft.; 1 mile S. of Kalidai, 4-8 ft.; between Barkuda and mainland, 6-8 ft.; Nalbano and channel S.E. of Nalbano, 4-8 ft.; southwards from Mahosa, 5-9 ft.; S.E. of Barkuda and Samel Id., ¼ mile off shore, 6 ft.; channel from Satpara towards Barnikuda, 9-12 ft.

Stenothyra orissaënsis, Preston.

T.c., pp. 300-301.

Serua Nadi, 5-9 ft.; Rambha Bay, about 6 ft. of water, among weeds: about 4 miles E.N.E. of Kalupara Ghat, 6-7 ft.; 1 mile N.E. by E. of Chiriya Id., 5½-6¾ ft.; Nalbano and channel S.E. of Nalbano, 4-8 ft.; 2 miles N.E. by N. ½ N. of Kalidai, 7 ft.

Stenothyra trigona, sp. n.

(Fig. 3, p. 192.)

Shell rimate, thin, turrite, semitransparent, greyish white; whorls 5, smooth, rather rapidly increasing, the last inflated, ascending a little in front; suture impressed, margined below; perforation appearing as a narrow and not very deep chink; columellar lip descending in a curve; labrum continuous: aperture oblique, ovate.

Alt. 2·5 (nearly), diam. maj. 1·5, diam. min. 1 mm.

Hab.—Lake Chilka, opposite Barkul bungalow (*Type*); Rambha Bay, among weeds; Serua Nadi, 5-9 ft.; 1 mile N.E. by E. of Chiriya Id., 5½-6¾ ft.; Nalbano and channel S.E. of Nalbano, 4-8 ft.

Stenothyra obesula, sp. n.

(Fig. 4, p. 291.)

Shell rimate, ovately fusiform, of an olive colour; whorls 4, smooth, the first three small and regularly increasing, the last large, inflated, descending in front; suture impressed, very narrowly margined below; perforation reduced to a mere chink; labrum continuous, the margin dark brown; aperture slightly oblique, ovate; operculum normal.

Alt. 3·25, diam. maj. 2·25 mm.

Hab.—Southernmost island of Manikpatna series.

This species stands out, owing to the obese form of the last whorl, from any other yet described from the Indian region.

Family NATICIDAE.

Natica marochiensis, Gmel.

Syst. Nat., p. 3673, No. 15.

Satpara, Chilka Lake.

Dead shells inhabited by *Coenobita.*

Natica maculosa, Lamarck.

Anim. s. vert. (Desh. ed.), VIII, p. 641.

Satpara, Chilka Lake.

Dead shells inhabited by *Diogenes.*

Family SCALARIIDAE.

Epitonium hamatulae, sp. n.

(Fig. 5, p. 291.)

Shell imperforate, turritely fusiform, whitish flesh colour; remaining whorls 5, sculptured with rather fine, erect and closely-set, transverse costulae, the terminations of which are bent forward in a hook-like manner in the immediate super-sutural region and of which there are seventeen on the last whorl, the interstices being quite smooth; suture impressed, crenellated by the hook-like terminations of the transverse costulae; columella descending in a slightly angular curve; aperture oblique, oval.

Alt. 7·75, diam. maj. 4·5 mm.

Aperture: alt. 3, diam. 2 mm.

Hab.—Channel off Barhampur Id., 6-9 ft.

Family PYRAMIDELLIDAE.

Chrysallida (Mormula) humilis, Preston.

Journal of Malacology, XII, 1905, p. 6 [as *Pyramidella (Mormula)*].

Channel off Barhampur Id., 6-9 ft.; Nalbano and channel S.E. of Nalbano, 4-8 ft. (a single deformed specimen): E. side of Rambha Bay, 1 4¾ ft.; southwards from Mahosa, 5-9 ft.; main channel W. of Satpara Id., 3-8 ft.: Serua Nadi, 5-9 ft.; Mahosa, southwards towards sandhills, 4-8 ft.

This appears to be a very variable species in sculpture, convexity of the whorls and size. Specimens differing in all these characters merge into one another but perhaps a single tapering specimen, which the author was at first inclined to regard as distinct, from "Breakfast Island midway between Ganta Sila and Chiriya Island South Point (5-6 ft.)", may be especially mentioned, the dimensions of this individual being:

Alt. 8·75, diam. maj. 2·75, diam. min. 2·5 mm.

Aperture: alt. 2, diam. 1 mm.

While those of the type specimen from Ceylon with which a number of Lake Chilka examples fairly agree are—

Alt. 4·75, diam. maj. 1·5 mm.

Aperture: alt. 5 mm.

Nevertheless one race would seem to stand out from all others and to be worthy of subspecific rank, while two other forms are so distinct as to warrant the accordance of full specific status.

Chrysallida (Mormula) humilis chilkaënsis, subsp. n.

(Figs. 6, 6*a*.)

Shell differing from the typical *P.* (*M.*) *humilis*, Preston, in its less tapering, shorter and proportionately broader form, finer transverse costulae, coarser spiral striae and more oblique aperture.

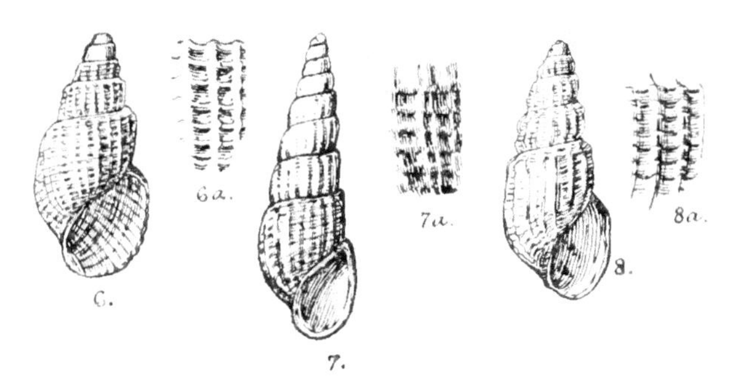

FIG. 6.—*Chrysallida* (*Mormula*) *humilis chilkaensis*, sub-sp. n. × 8.
,, 6*a*.— ,, ,, ,, ,, (sculpture) × 12.
,, 7.—*Chrysallida* (*Mormula*) *ecclesia*, sp. n. × 4.
,, 7*a*.— ,, ,, ,, (sculpture) × 8.
,, 8.—*Chrysallida* (*Mormula*) *nadiensis*, sp. n. × 8.
,, 8*a*.— ,, ,, ,, (sculpture) × 12.

Alt. 3·25, diam. maj. 1·75 mm.

Hab.—Serua Nadi, 5-9 ft. (*Type*); main channel W. of Satpara Id., 3-8 ft.; southwards from Mahosa, 5-9 ft.; channel between Barnikuda and Satpara, 6½ ft.; Nalbano and channel S.E. of Nalbano, 4-8 ft.; Mahosa, southwards towards sandhills, 4-8 ft.; channel from Satpara towards Barnikuda, 9-12 ft.; off Barnikuda, inside lake, 6 ft.

Chrysallida (Mormula) ecclesia, sp. n.

(Figs. 7, 7*a*.)

Shell subulately fusiform, tapering, reddish brown; whorls 8½, regularly increasing, not very convex, sculptured with rather closely-set, perpendicular, transverse plicae, crossed by fine, spiral lirae; suture impressed; columella descending in a curve, porcellenous, slightly reflexed, extending above into a thin, well defined, restricted, parietal callus which reaches to the upper margin of the labrum; labrum acute, a little dilated at the base; aperture ovate, very slightly oblique.

Alt. 8·75, diam. maj. 2·75, diam. min. 2·25 mm.

Aperture: alt. 2, diam. 1 mm.

Hab.—Breakfast Id., midway between Ganta Sila and Chiriya Id S. Point, 5-6 ft.

Chrysallida (Mormula) nadiensis, sp. n.

(Figs. 8, 8*a*, p. 295.)

Shell fusiform, white; whorls 6, regularly but rather rapidly increasing, shouldered above and below, transversely costulate and finely spirally striate, the costulae becoming obsolete and the striae coarser on the base of the shell; suture well impressed; columellar margin somewhat obliquely descending and a little inwardly bulging above, curved below, extending above into a thickish, well-defined, parietal callus which unites it with the upper margin of the labrum; labrum acute, rather dilated at the base; aperture slightly oblique, ovate.

Alt. 3·25, diam. maj. 1·25 mm.

Hab.—Serua Nadi, 5-9 ft. (*Type*); Mahosa, southwards towards sandhills, 4-8 ft.; Nalbano and channel S.E. of Nalbano, 4-8 ft.; 1 mile S. of Kalidai, 4-8 ft.

Odostomia chilkaënsis, Preston.

Rec. Ind. Mus., X, pp. 301-302.

Mahosa, southwards towards sandhills, 4-8 ft.

Family NERITIDAE.

Neritina souverbiana, Montrouzier.

J. Conchyliol., Paris, XI, 1863, pp. 75, 175, pl. v, fig. 5.

Mahosa, Barhampur Id.; Serua Nadi, 5-9 ft.; 1 mile E. by N. of Patsahanipur, 4½ ft. (dead shells only).

Family CYCLOSTREMATIDAE

Cyclostrema (Tubiola) innocens, sp. n.

(Figs. 9, 9*a-b*, p. 299.)

Shell small, discoidal, almost planulate, milk white, smooth throughout; whorls 3, rather rapidly increasing, the last convex, marked only with growth striae; suture well impressed; umbilicus moderately wide; labrum continuous, simple; aperture rather large for the size of the shell, subcircular.

Alt. ·5, diam. maj. 2, diam. min. 1·75 mm.

Hab.—Serua Nadi, 5-9 ft.

Family TROCHIDAE.

Umbonium vestiarum, Lin.

Syst. Nat., X, p. 758.

Outer bar close to mouth; outer bar opposite Manikpatna temple.

Solariella satparaënsis, Preston.

Rec. Ind. Mus., X, pp. 302-303.

Mahosa, southwards towards sandhills, 4-8 ft. (a single dead specimen).

Order *OPISTHOBRANCHIA*.

Family BULLIDAE.

Bulla (Haminea) crocata, Pease.

Proc. Zool. Soc. London, 1860, p. 19.

Serua Nadi, 5-9 ft. (a single small, though apparently, fully grown specimen).

Family TORNATINIDAE.

Tornatina estriata, Preston.

Rec. Ind. Mus., X, p. 303

Channel from Satpara towards Barnikuda, 9-12 ft.; one mile N.E. by E. of Chiriya Id., $5\frac{1}{2}$-$6\frac{3}{4}$ ft.; Serua Nadi, 5-9 ft.; off Barnikuda, inside lake, 6 ft.; channel off Barhampur Id., 6-9 ft.; main channel W. of Satpara Id., 3-8 ft.; Nalbano and channel S.E. of Nalbano, 4-8 ft.; southwards from Mahosa, 5-9 ft.; 2 miles N.E. by N., $\frac{1}{2}$ N. of Kalidai, 7 ft.

In view of the plasticity of the members of this genus, the author considers it necessary to unite *T. soror* [1] with the present species, this conclusion having been come to as a consequence of the examination of a very large further series of examples from the lake in which more or less connecting links between the two originally described forms occur.

Class **PELECYPODA**

Order *TETRABRANCHIA*.

Sub-order *MYTILACEA*.

Family MYTILIDAE.

Mytilus smaragdinus, Chemnitz.

Conch. Cab., VIII, pl. lxxxiii, fig. 745.

Manikpatna, oyster-beds (a single, very immature specimen).

[1] *Rec. Ind. Mus.*, X, p. 303.

Modiola undulata, (Dkr.).

Proc. Zool. Soc. London, 1856, p. 363.

Channel off Barhampur Id., 10-20 ft.; channel between Satpara and Barhampur, 8-20 ft.; Nalbano and channel S.E. of Nalbano, 4-8 ft. (young examples only); 1 mile N.E. by E. of Chiriya Id., $5\frac{1}{2}$-$6\frac{3}{4}$ ft. (young only).

Var. **crassicostata**, Preston.

Rec. Ind. Mus., X, p. 304.

Serua Nadi, 5-9 ft. (young specimens only); main channel W. of Satpara Id., 3 8 ft. (a young specimen).

Sub-order *ARCACEA*.

Family ARCIDAE.

Arca (Fossularca) lactea, Lin.

Syst. Nat. p. 1141.

Channel between Satpara and Barhampur Id., 6-8 ft., and 8-20 ft.; main channel W. of Satpara Id., 3-8 ft.; Satpara.

After very careful comparison the author is unable to separate the present shells from the common European form, which has already been recorded from Bombay and Mergui as well as from S. Africa, the Red Sea, Aden, Ascension Id. and (somewhat doubtfully) from the Philippines.

Sub-order *ERYCINACEA*.

Family ERYCINIDAE.

Kellya chilkaënsis, sp. n.

(Figs. 10, 10*a*, p. 299.)

Shell very thin, flattened, oblong-ovate, transparent, pale brownish, except towards the margins where it is covered with a thin membranaceous reddish brown periostracum, concentrically striate; umbones very small; dorsal margin arched; ventral margin gently rounded; anterior side obtusely rounded; posterior side very slightly produced, rounded; hinge-teeth normal.

Long. 4·25, lat. 5·75 mm.

Hab.—Channel between Satpara and Barhampur Id., 6-8 ft. (*Type*); Mahosa, southwards towards sandhills, 4-8 ft.; 1 mile S. of Kalidai, 4-8 ft.; channel off Barhampur Id., 6-9 ft.; 4-7 miles E. $\frac{1}{2}$ S. of Patsahanipur, 4-$4\frac{1}{4}$ ft.

Kellya mahosaënsis, sp. n.

(Fig. 11, p. 299.)

Shell minute, inequilateral, oblong-ovate, transparent, pale yellowish horn colour, reddish at the margins, concentrically striate; umbones comparatively large and rather prominent; dorsal margin strongly arched; ventral margin contracted in the median part; anterior side very obtusely rounded; posterior side rounded; hinge-teeth normal.

Long. 1·5, lat. 1·25 mm.

Hab.—Mahosa, southwards towards sandhills, 4-8 ft.

Family GALEOMMIDAE.

Scintilla chilkaënsis, sp. n.

(Figs. 12, 12*a*.)

Shell oblong ovate, considerably flattened, very thin, transparent, pale yellowish, polished, shining, irregularly concentrically striate; umbones very small; dorsal margin arched at either side, slightly sloping in the umbonal region; ventral margin

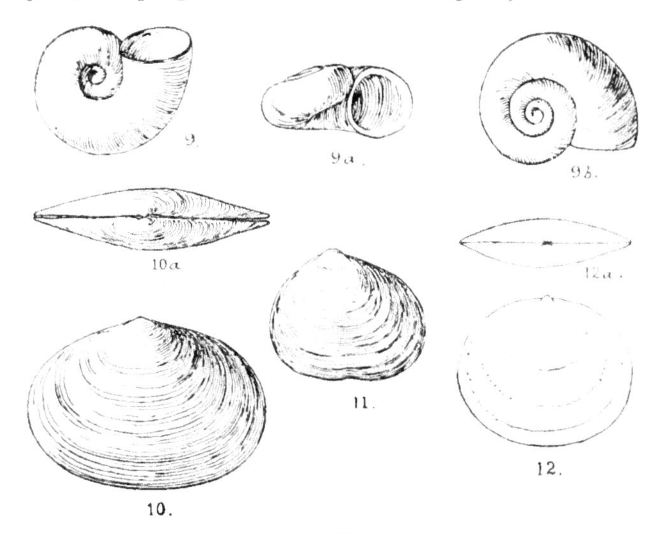

FIGS. 9, 9*a*, 9*b*.—*Cyclostrema* (*Tubiola*) *innocens*, sp. n. × 10.
„ 10, 10*a*.—*Kellya chilkaensis*, sp. n. × 6.
„ 11.—*Kellya mahosaensis*, sp. n. × 14.
„ 12, 12*a*.—*Scintilla chilkaensis*, sp. n. × 4.

gently rounded; anterior side bluntly rounded: posterior side sloping above, rounded below; hinge-teeth normal.

Long. 4·5, lat. 5·75 mm.

Hab.—Mahosa, southwards towards sandhills, 2-8 ft. (*Type*); channel south of Satpara Point, 8-12 ft.; channel between Satpara and Barhampur, 8-20 ft.

Sub-order *CARDIACEA*.

Family CARDIIDAE.

Cardium (Fulvia) rugatum, Gronov.

Gronovius, *Zoophylaceum*, pl. xviii, fig. 5.

Outer bar close to mouth (juvenile examples only).

Sub-order *CONCHACEA*.

Family VENERIDAE.

Meretrix casta, Chem.

Anim. s. vert., VI, p. 299.

Satpara Bay: Mahosa, Barhampur Id.; outer bar, opposite Manikpatna temple (young only): swamp inside bar, N. of Barhampur Id.; Manikpatna Id.

A very large example was secured, but has reached the writer with no other locality attached than "Chilka Survey"; the dimensions of this specimen are as below.

Long. 67, lat. 73 mm

Having now had the opportunity to examine fresh specimens of this species from the outer channels of the lake, the author is of opinion that the shell recently described by him as *Corbicula (Velorita) satparaënsis*[1] cannot stand, the worn subfossil remains upon which the species was based clearly proving it to be identical with individuals of *M. casta* now to hand; this conclusion is borne out by Blanford's record of the occurrence of *M. casta* in the Rambha Island Beds.[2]

Meretrix ovum, Hanley.

Proc. Zool Soc. London, 1846, p. 21.

Satpara Bay: Mahosa, Barhampur Id.

Meretrix morphina, Lk.

Anim. s. vert., VI, p. 300.

Channel off Barhampur Id., 10-20 ft. (a single valve).

Tivela dillwyni (Deshayes).

Cat. Brit. Mus., Conchif., 1853, p. 49; *Cytherea mactroides*, Sowerby, *Thes. Conch.*, II, p. 615, pl. 128, fig. 56, *non* Born, *nec* Chemnitz, *nec* Lamarck.

Serua Nadi, 5-9 feet (young specimens only).

Meroë scripta, Gray.

Rumphius, *Mus Amb.*, pl. xliii, figs. L, M.

Outer channel, Lake Chilka (a single much worn valve).

Tapes pinguis, Chem.

Conch. Cab., VI, p. 355, pl. xxxiv, figs. 355-357 (as *Venus*).

Manikpatna Id. (a somewhat inflated and rounded variety); Manikpatna Id. (a normal specimen); swamp inside bar N. of Barhampur Id.; S. side of Satpara Id., opposite bungalow.

---

[1] *Rec. Ind. Mus.*, X, p. 306. [2] *Rec. Geol. Surv. India*, V, p. 61.

Tapes ceylonensis, Sow

Thes. Conch., I, p. 683, pl. cxlvi, figs. 24-25.

Sand-dunes opposite Manikpatna (juvenile examples); channel near Mirzapur, 8-12 ft.: Mahosa, Barhampur Id; Serua Nadi, 5-9 ft. (young specimens only).

Clementia annandalei, Preston.

Rec. Ind. Mus., X, p. 306.

Serua Nadi, 5-9 ft. (young examples): Nalbano and channel S.E. of Nalbano, 4-8 ft (also young).

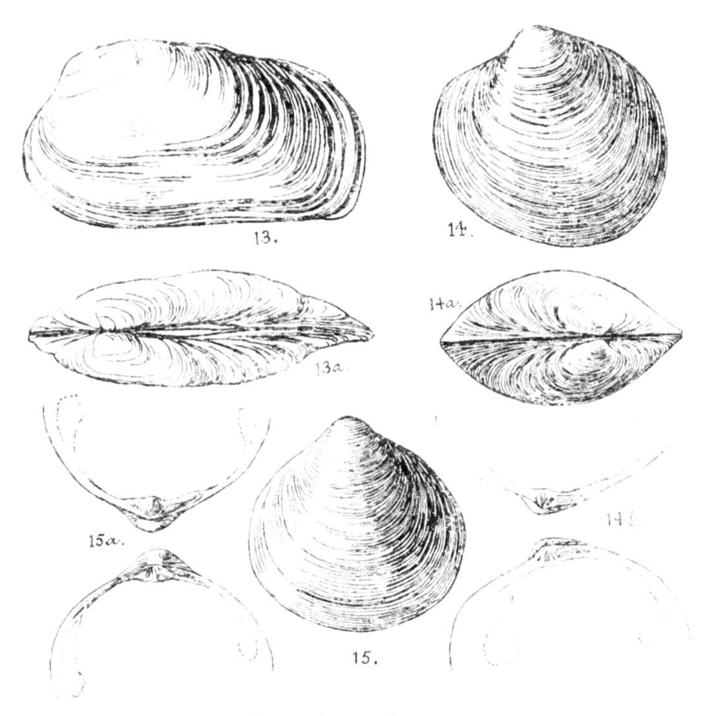

FIGS. 13, 13*a*.—*Petricola esculpturata*, sp. n. × 3.
,, 14, 14*a*.—*Diplodonta satparaensis*, sp. n. × $1\frac{1}{2}$.
,, 14*b*.— ,, ,, (hinge) × $1\frac{1}{2}$.
,, 15.—*Diplodonta barhampurensis*, sp. n. × 3.
,, 15*a*. ,, ,, (hinge) × 3.

Family PETRICOLIDAE.

Petricola esculpturata, sp. n.

(Figs. 13, 13*a*.)

Shell oblong, rather solid, white, showing traces of having been covered with a thin pale greenish brown periostracum, concentrically striate with lines of growth, but without other sculpture; umbones small, not very prominent, very anteriorly situated; dorsal margin sloping towards the posterior side; ventral margin nearly straight; anterior side obliquely and rapidly sloping above, somewhat rounded below; posterior side produced, very obtusely rounded, angled below; hinge normal.

The dimensions of the type specimen are as follows:—

Long. 13·25, lat. 23, diam. 7·5 mm.

Those of a larger, though imperfect individual are—

Long. 17, lat. 31·5, diam. 11·5 mm.

Hab.—Manikpatna, oyster-beds.

Since the above was written the author has seen more specimens of this species from the Chilka Lake, the largest of which yields the following measurements:—

Long. 27·5, lat. 47, diam. 17·25 mm.

Family UNGULINIDAE.

Diplodonta satparaënsis, sp. n.

(Figs. 14, 14*a-b*, p. 301.)

Shell ovately rhomboidal, fragile, slightly glossy, whitish, covered with a very thin, pale greenish yellow periostracum, irregularly finely and closely concentrically striate; umbones small, but slightly prominent; dorsal margin arched; ventral margin anteriorly sloping, posteriorly rounded; anterior side angled above, then descending in a rather oblique curve; posterior side sloping, rather abruptly rounded; right valve bearing two small cardinal teeth, of which the anterior is short and sloping and the posterior narrowly bifid above, broadly so below; left valve also bearing two cardinals of which the anterior is rather narrowly bifid and the posterior very fine and oblique

Long. 19, lat. 20·5 mm.

Hab.—Channel between Satpara and Barhampur, 8-20 ft. (*Type*); swamp inside bar N. of Barhampur Id.; channel off Satpara, 16-20 ft.; Kalidai Id.; between Mahosa and Satpara, 6 ft.; Satpara Bay; channel off Barhampur Id., 10-20 ft.; channel off Barhampur Id., 6-9 ft.; channel between Barnikuda and Satpara, 6½ ft.; southwards from Mahosa, 5-9 ft.; Mahosa, Barhampur Id.

Diplodonta barhampurensis, sp. n.

(Figs. 15, 15*a*, p. 301.)

Shell inflated, subequilateral, roundly trigonal, covered with a dark brown periostracum, finely concentrically striate; umbones rather large, prominent; dorsal margin somewhat angularly arched; ventral margin rounded; anterior side rounded; posterior side obtusely rounded; hinge normal.

Long. 13·75, lat. 14·75, diam. 10 mm.

Hab.—Channel off Barhampur Id., 10-20 ft.

Diplodonta (Felania) annandalei, Preston.

Rec. Ind. Mus., X, p. 307.

Channel between Satpara and Barhampur Id., 6-8 ft.; outer bar close to mouth; main channel W. of Satpara Id., 3-8 ft.; channel off Barhampur Id., 6-9 ft.; 3-4 miles E. by S. ½ S. of Patsahanipur, 5-5¼ ft; 4-9 miles E. by S. ½ S. of Patsahanipur, 4-5 ft.; near Barnikuda, inside lake, 5½ ft.; Satpara Bay; outer bar opposite Manikpatna temple; Maludaikuda Id.; 2-8 miles N.E. ½ E. of Kalidai, 5-6 ft.

Diplodonta (Felania) chilkaënsis, Preston.

T.c., p. 307.

Manikpatna Id., sand-dunes opposite Manikpatna; outer bar 1 mile S.W. of mouth, 6 ft.; swamp inside bar N. of Barhampur Id.; outer bar opposite Manikpatna temple; S. side of Satpara Id., opposite bungalow.

Diplodonta (Felania) ovalis, Preston.

T.c., pp. 308-309.

Outer bar, 1 mile S.W. of mouth, 6 ft.; channel between Satpara and Barhampur Id., 6-8 ft.

Family DONACIDAE.

Donax pulchella Hanley.

Proc. Zool. Soc. London, 1843, p. 6.

Outer bar close to mouth (a single valve).

Family PSAMMOBIIDAE.

Psammobia mahosaënsis, sp. n.

(Figs. 16, 16*a-b*, p. 304.)

Shell small, very inequilateral, ovately cuneiform, concentrically striate, whitish, covered with a thin brown periostracum which is chiefly noticeable round the margins, both valves angled posteriorly; umbones small; dorsal margin angularly arched; ventral margin gently rounded; anterior side rounded; posterior side rather abruptly descending; right valve furnished with two minute cardinal teeth, the anterior being placed at a very obtuse angle to the posterior; the left valve also bearing two minute cardinals of which the anterior fits between those of the right valve.

Long. 6·25, lat. 9·25 mm.

Hab.—Southwards from Mahosa, 5-9 ft. (*Type*); channel between Satpara and Barhampur, 8-20 ft.; channel between Barnikuda and Satpara, 6½ ft.; on swamp inside bar N. of Barhampur Id.; Mahosa, southwards towards sandhills, 4-8 ft.

Family SOLENIDAE.

In the writer's former paper, on the Mollusca of the Chilka Lake, he recorded what he supposed to be juvenile examples of *S. truncatus*, Wood [1] this determination however cannot stand: Dr. Annandale has since gone thoroughly into the matter and reports that there are, in his opinion, no less than three forms in the lake, one, a form only found on a bottom of fine mud inside the lake, a second, which only differs in its larger size and thicker shell and which according to Dr Annandale is found "at sandy localities in the outer parts of the lake", and a third, which is found with the second, but has a much narrower shell.

The first (*i.e.* the smallest) of the three forms would appear,

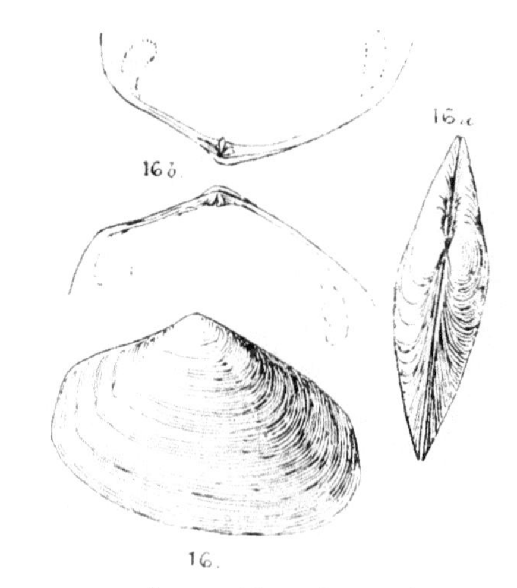

FIGS. 16, 16*a*.—*Psammobia mahosaensis*, sp. n. × 4.
" 16*b*.— " " (hinge) × 4.

from Dr. Annandale's investigations, to be sexually adult and, as no such small forms are known from the Indian region, the shells of the last two are described in the present paper, no material being at the time of writing on hand from which to draw up a diagnosis of the first, the specimens having been returned to the Indian Museum with the author's previous report.

Solen annandalei, sp. n.

(Figs. 17, 17*a*, p. 305.)

Shell small, covered with a yellowish brown, glossy, shining, polished periostracum, and plainly marked with concentric growth lines; dorsal and ventral margins quite straight; anterior side obliquely sloping in an anterior direction; posterior side obtuse,

[1] Sowerby, *Genera of Shells*; Reeve, *Con. Icon.*, *Solen*, XIX, 1874, pl. i, fig. 1.

slopingly rounded at the ventral corner and sharply so at the dorsal corner; right valve bearing a single cardinal tooth which is grooved below and posteriorly erect above; left valve furnished with a very anteriorly erect, somewhat claviform, cardinal tooth.

Long. 49·25, lat. 9·5 mm.

Hab.—Satpara Bay.

Solen kempi, sp. n

(Figs. 18, 18*a*.)

Shell differing from *S. annandalei*, in its smaller and proportionately much narrower form, it is of a thinner texture and the concentric growth lines are not so clearly marked, the anterior side is more obliquely truncate, the cardinal tooth in the right

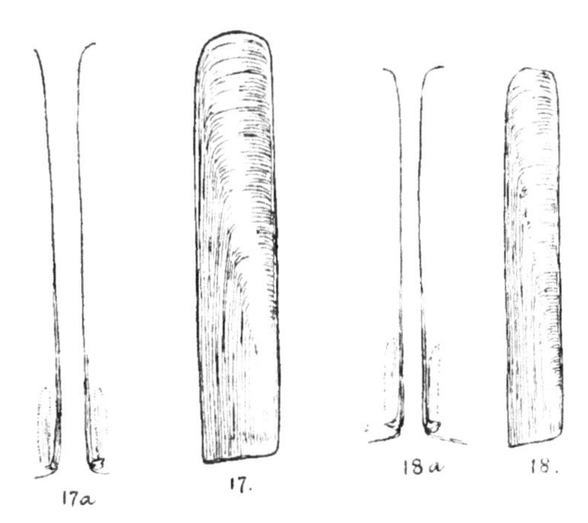

FIG. 17.—*Solen annandalei*, sp. n. × 1½.
,, 17*a*.— ,, ,, (hinge) × 1½.
,, 18.—*Solen kempi*, sp. n. × 1½.
,, 18*a*.— ,, ,, (hinge) × 1½.

valve is grooved, though only shallowly, throughout its whole breadth, while that in the left valve is more rigidly erect, even than is the case in *S. annandalei*.

Long. 43, lat. 6·5 mm

Hab.—20 miles S.E. by S. of Patsahanipur, 5½ feet.

Sub-order *MYACEA*.

Family MACTRIDAE.

Standella annandalei, sp. n.

(Figs. 19, 19*a-b*, p. 306.)

Shell thin, fragile, gaping posteriorly, broadly cuneiform, whitish, covered with a thin brownish yellow periostracum, concentrically striate; umbones small; dorsal margin arched anteriorly, sloping posteriorly; ventral margin very gently rounded; anterior side rounded; posterior side produced, rather sharply rounded; right valve furnished with a **V**-shaped, somewhat anteriorly erect and jagged cardinal; left valve bearing an even more erect, but not jagged, cardinal tooth; lateral teeth in both valves short and sloping.

Long. 15·5, lat. 23·5 mm.

Hab.—N.E. side of Nalbano (*Type*); Satpara Bay.

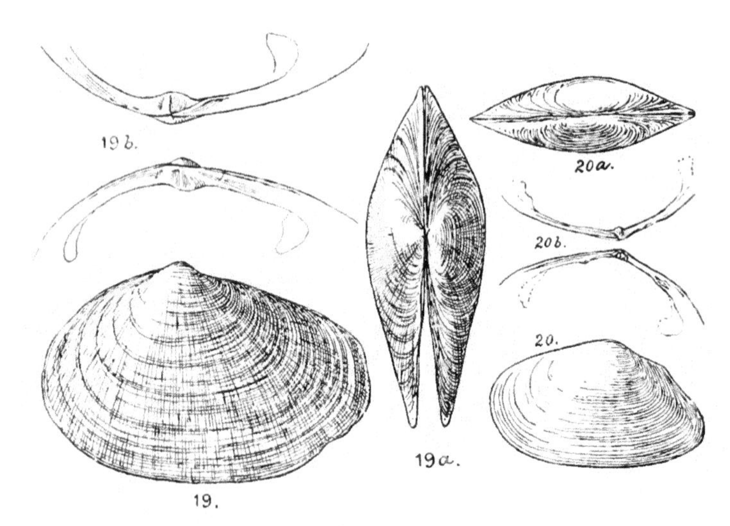

FIGS. 19, 19*a*.—*Standella annandalei*, sp. n. × 2.
" 19*b*.— " " (hinge) × 3.
" 20, 20*a*.—*Tellina chilkaensis*, sp. n. × 3.
" 20*b*.— " " (hinge) × 3.

Sub-order *ADESMACEA*.

Family TEREDINIDAE.

Xylotrya stutchburyi, Sow.

Con. Icon., XX, pl. ii, fig. 5, *a*, *b*, *c*.

Post in channel off Satpara Point, 3-8 ft.

Order *DIBRANCHIA*.

Sub-order *TELLINACEA*.

Family TELLINIDAE.

Tellina chilkaënsis, sp. n.

(Figs. 20, 20*a*-*b*.)

Shell small, elongately ovate, yellowish flesh coloured, polished, shining, somewhat iridescent, concentrically striate; umbones small, rather flattened; dorsal margin anteriorly sloping and slightly arched, posteriorly shortly excavated; ventral margin very gently rounded; anterior side sharply rounded above, slopingly so below; posterior side bluntly rostrate; right valve bearing two divergent, short, grooved, cardinal teeth, a strong, anteriorly overhanging, anterior lateral and a very weak posterior lateral; left valve bearing a rather strong bifid anterior and a very weak posterior cardinal only.

Long. 6·25, lat. 9·75, diam. 3·25 mm.

Hab.—Channel off Barhampur Id., 6-9 ft.

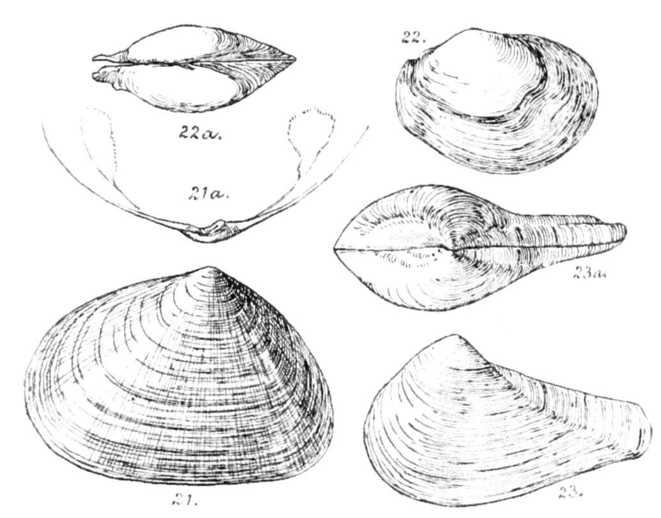

FIG 21.—*Tellina barhampurensis*, sp. n. × 3.
" 21a.— " " (hinge) × 3.
" 22, 22a.—*Cumingia hinduorum*, sp. n. × 4½.
" 23, 23a.—*Cuspidaria annandalei*, sp. n. × 6.

Tellina barhampurensis, sp. n.

(Figs. 21, 21*a*.)

Shell oblong-trigonal, whitish, concentrically and faintly radiately striate; umbone small; dorsal margin strongly arched; ventral margin almost straight; anterior side obtusely rounded; posterior side broad, rounded; anterior cardinal tooth very oblique, rather broadly bifid; posterior cardinal tooth angularly bent in a posterior direction; anterior lateral tooth very weak; posterior lateral short, erect.

Long. 14·5, lat. 20·5 mm.

Hab.—Channel between Satpara and Barhampur, 8-20 ft.

Unfortunately only a single valve (the right) has been available for description; but this is so characteristic that the author considers that no further apology is necessary for the founding of a species upon such scanty material.

Family SCROBICULARIIDAE.

Theora opalina, Hinds.

Proc. Zool. Soc. London, 1843, p. 78 (as *Neaera*).

Main channel W. of Satpara Id., 3-8 ft.; main channel between Satpara and Barnikuda, 6 ft.; 2-8 miles N.E ½ E. of Kalidai, 5-6 ft.; 1 mile N.E. by E. of Chiriya Id., 5½-6¾ ft.; Serua Nadi, 5-9 ft.; 1-1½ miles off Kalupara Ghat, 6-7 ft.: channel between Satpara and Barhampur Id., 6-8 ft.

Cumingia hinduorum, sp. n.

(Figs. 22, 22*a*, p. 307.)

Shell ovately rhomboidal, very thin, greyish, concentrically striate with growth lines only; umbones small; not very prominent; dorsal margin arched; ventral margin anteriorly rounded, posteriorly sloping; anterior side produced, rather sharply rounded; posterior side very obtusely rostrate; hinge normal.

Long. 6·5, lat. 9 mm.

Hab.—Main channel W. of Satpara Id., 3-8 ft. (*Type*): channel off Barhampur Id., 6-9 ft.; Mahosa, southwards towards sandhills, 4-8 ft; 3-4 miles E. by S. ½ S. of Patsahanipur, 5-5¼ ft.; channel between Satpara and Barnikuda, 6-10 ft.; southwards from Mahosa, 5-9 ft.; Mahosa, Barhampur Id.

Sub-order *ANATINACEA.*

Family CUSPIDARIIDAE.

Cuspidaria annandalei, sp. n.

(Figs. 23, 23*a*, p. 307.)

Shell small, irregularly triangulate, thin, yellowish white, except for the posterior prolongation and ventral margin which are covered with a light brown, membranaceous periostracum, concentrically striate, both valves being obliquely angled from the umbones downward, the angle being more accentuated in the right valve and extending to the ventral margin on the posterior side; umbones small, somewhat prominent; dorsal margin anteriorly arched, posteriorly gently sloping; ventral margin gently rounded especially anteriorly; anterior side rounded; posterior side rostrately produced, obliquely truncate, bearing a depression in both valves for about three-fourths of its length.

Long. 4, lat. 6·5 mm.

Hab.—4-9 miles E. by S. ½ S. of Patsahanipur, 4-5 ft. (*Type*); Serua Nadi, 5-9 ft.: 4-7 miles E. ½ S. of Patsahanipur, 4-4½ ft.; about 4 miles E.N.E. of Kalupara Ghat, 6-7 ft.; 4-9 miles E. ½ S. of Barkul bungalow, 5¼-5¾ ft.; 1-1½ miles off Kalupara Ghat, 6-7 ft.; 1 mile S. of Kalidai, 4-8 ft.; Nalbano and channel S.E. of Nalbano, 4-8 ft.; main channel W. of Satpara Id., 3-8 ft.; main channel between Satpara and Barnikuda, 6 ft.; Mahosa, Barhampur Id.

Family LYONSIIDAE.

Lyonsia samalinsulae, Preston.

Rec. Ind. Mus., X, p. 310.

Main channel W. of Satpara Id., 3-8 ft.; Serua Nadi, 5-9 ft.; Mahosa, southwards towards sandhills, 4-8 ft.

Family ANATINIDAE.

Anatina barkudaënsis, sp. n.

(Figs. 25, 25*a*.)

Shell rather small, oblong, gaping at both ends, but especially posteriorly, concentrically striate and minutely pustulate except at the posterior side, where the pustules cease abruptly and the shell is only very coarsely and irregularly concentrically striate;

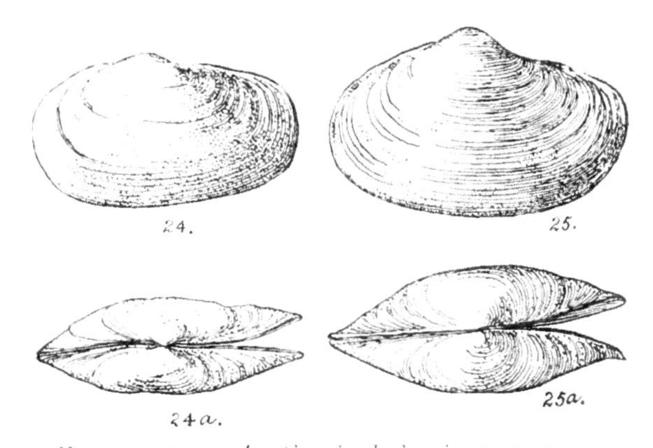

FIGS. 24, 24*a*.—*Anatina barkulensis*, sp. n. × 3.
,, 25, 25*a*.—*Anatina barkudaensis*, sp. n. × 3.

umbones of moderate size, slightly prominent; dorsal and ventral margins nearly straight; anterior side rounded; posterior side very abruptly rounded; hinge quite normal.

Long. 9·25, lat. 17 mm.

Hab.—Barkuda Id. (*Type*); Chiriya Id.; Manikpatna Id.; in mud at edge of Lake Rambha, Ganjam District, Madras; swamp inside bar, N. of Barhampur Id.: Satpara Bay; channel between Satpara and Barnikuda, 6-10 ft.; E. side of Rambha Bay, 1-4¾ ft.

Anatina barkulensis, sp. n.

(Figs. 24, 24*a*.)

Shell oblong ovate, thin, gaping at both sides, white, minutely pustulous, the margins showing traces of a reddish laminiferous periostracum, concentrically striate; umbones rather large, moderately prominent; anterior margin somewhat straight;

ventral margin gently rounded: anterior side rounded above, receding below; posterior side rounded; hinge normal.

Long. 11·25, lat. 20 mm.

Hab.—Barkul Point (*Type*); Mahosa, southwards towards sandhills, 4-8 ft.

XVI. NOTES FROM THE BENGAL FISHERIES LABORATORY, INDIAN MUSEUM.

No. 2. On some Indian Parasites of Fish, with a note on Carcinoma in Trout.

By T. Southwell, *A.R.C.S., F.L.S., F.Z.S., Dy. Director of Fisheries, Bengal, Bihar and Orissa, Honorary Assistant, Indian Museum.*

(Plates xxvi–xxviii).

CONTENTS. *Page*

The following paper deals with a variety of fish diseases, all of which are—with two exceptions—caused by parasites.

The "lice" which live on the skin of Bhekti are not more harmful than other lice which live on other animals. The "lice" which are described from Rohu are, however, much more dangerous than those found on Bhekti. In tanks and confined water-areas these parasites may cause great mortality amongst Rohu, and every fish in the tank may die.

The larval Trematodes which live in the skin and flesh of a number of fish are of some importance. Heavy infection most probably interferes somewhat with normal growth, and it is not impossible that certain of these parasites may mature in the human intestine.

The "pox" recorded from *Rasbora daniconius* (Bengali "dankona") may become a very fatal disease when occurring in limited water-areas such as tanks.

The parasite from the abdomen of *Diagramma crassispinum* (a fish somewhat like the one known in Bengal as Khora Bhekti) is not of any commercial importance.

An epidemic of goitre amongst the trout at Naini Tal is of a serious nature, and if not kept in check will interfere greatly with the successful cultivation of this species.

A case of Glaucoma of the eye in a marine fish recorded in this paper is of pathological interest only.

(1) A skin disease found on Rasbora daniconius.

(Plate xxvi, fig. 3).

Four specimens were collected by Capt. R. B. Seymour Sewell, B.A., I.M.S., from a stream near Katiwan, Mirzapore (U. P.), India.

In all, six cysts were found on the four fish in question. They were situated immediately below the scales, in the epidermis, and were milky-white, soft, flattened, and roughly oval in shape. The largest measured 1·1 mm. No pigment was present.

Preparations of the contents of these cysts showed that they contained Myxosporidia, or parasitic protozoa. The order Myxosporidia, Butschli, contains a series of parasites which occur in both fresh water and marine fish. They are usually found beneath the skin, as small wart-like nodules near the fins and on the gills. The parasite causing the well-known silkworm disease (*Glugea bombycis*) is closely related to the parasite recorded in the present paper. Representatives of the Myxosporidia have also been found in the urinary bladder and gall bladder of fishes, and they are also recorded as occurring in Crustacea, frogs, and crocodiles. At the present time over 60 species of fish are known to harbour parasites included in the order Myxosporidia and about 50 distinct species of parasites are recognized. In Europe epidemics amongst fish have frequently been traced to the presence of such parasites, although it appears that the mortality is not directly due to their presence, but to the presence of bacilli which develop within the cysts and give rise to ulcerations, which, discharging, not only kill the fish, but spread the disease.

Our parasites belong to the family Myxobolidae, Thelohan. The characters of this family are as follows:—

Spores with one or two polar capsules and with a peculiar iodinophilous vacuole in the sporoplasm (Minchin).

The genus *Myxobolus*, Butschli, 1882, to which our specimens belong, are characterized by the presence of one or two polar capsules and by the absence from the spore-membrane of a tail-like process. Minchin (Lankester's Zoology, Part I, London 1903,

p. 296) states that the genus is divisible into three sections. One section possess pear-shaped spores each with a single polar capsule. In the second section the two polar capsules are of unequal size. In the third section are numerous forms characterized by two polar capsules of equal size. *M. pfeifferi*, Thel., gives rise in Europe to the deadly "barbel disease." *M. cyprini*, Dofl. has been recorded from carp. As far as I have been able to ascertain, the only papers available in Calcutta which deal with the genus *Myxobolus* are the following :—

(1) Minchin, *vide ante*.

(2) Gurley. On the classification of the Myxoporidia, a group of Protozoan parasites infesting fishes. *Bulletin United States Fish Commission*, vol. XI for 1891, Washington, 1893, pp. 407-420.

(3) Ludwig Cohn. Über die Myxosporidien von *Esox lucius* und *Perca fluviatilis*. *Zool. Jahrb.*, *Anat. Abth.*, vol. IX, text and plates, Jena, 1896, pp. 228-272.

(4) (*a*) Linton. On certain wart-like excresences occurring on the short Minnow, *Cyprinodon variegatus*, due to Psorosperms.

(*b*) Notice of the occurrence of Protozoan parasites (Psorosperms) on Cyprinoid fishes in Ohio. *Bulletin United States Fish Commission*, vol. IX for 1889, Washington, 1891, pp. 99-102 and 359-361.

Our specimens apparently do not belong to any of the species described in the above papers, but they are very closely related to the Psorosperms obtained by Linton from *Cyprinodon variegatus*. Owing to lack of literature I have been unable to determine whether our parasites represent a new species or not. Further, as our material was scanty, I have not been able to work out all the details regarding the shape of the spores. I have therefore deemed it advisable to leave our specimens unnamed, at least for the present. *Myxobolus cyprini* has been recorded from *Cyprinus carpio*, and it is quite possible that our parasites may belong to *M. cyprini*. The following details were ascertained.

Cyst.—Lenticular. Greatest length 1·1 mm.

Spore.—Length 13μ ; breadth 13μ.

Capsules; 2, equal, 4μ in length, 4μ in breadth, with a very short anterior tail-like process.

Vacuole present.

As all my specimens were at once stained and mounted in balsam, I was unable to conduct re-actions with iodine, and sulphuric acid.

Habitat.—Sub-cutaneous intermuscular tissue of *Rasbora daniconius*, Day (=*Cyprinus daniconius*, Ham. Buch.).

(2) A parasite encysted in the skin of Cirrhina latia.

(Plate xxvii, fig. 10).

Three specimens from Mr. Mitchell, Srinagar, Kashmir, September, 1914. Mr. Mitchell stated that such diseased fish were fairly plentiful. Other genera or species affected were not defined.

One fish measured 77 mm. long and 28 mm. broad.

The size of the largest cyst found on this fish was 4 mm. × 2·5 mm., and that of the smallest was 1·25 mm. × 1 mm. There were 7 cysts on one specimen distributed as follows:—

A. (1) Behind junction of upper and lower left lips.
(2) One mm. behind left eye.
(3) Ten mm. behind base of left pectoral fin.
(4) Three mm. behind base of right pectoral fin.
(5) Near posterior extremity of right pectoral fin.
(6) Anterior and a little to left of anus.
(7) Four mm. in front of anus.

B. This fish measured 89 mm. long and 18 mm. broad. The positions of the cysts were as follows:—

(1) Under left eye.
(2) Above right eye.
(3) Between the branchial apertures.
(4) Near posterior extremity of left pectoral fin.
(5) One and a half mm. posterior to base of left pectoral fin.
(6) Two mm. behind left eye, near middle line.
(7) On the right side of the dorsal fin.
(8) At the base of the caudal fin on the right side.
(9) On anal fin.
(10) Anterior to anus.
(11) Near middle of right side of body-wall.
(12) Eighteen mm. anterior to base of caudal fin, on the left side.
(13) Twenty-nine mm. anterior to base of caudal fin, on the left side.
(14) Ten and a half mm. posterior to right eye.
(15) Mid-ventral line, between mouth and anus.

In every case the cysts were situated in the epidermis and were covered by scales. No cysts were found in the muscles. The wall of one cyst was 1·1 mm. thick and was densely pigmented with black. To the naked eye the cysts appeared of a dark steel-grey colour, due to the unpigmented covering of scales and epidermis. The wall of the cyst was tough and fibrous, and, as we have already noted, densely black.

The cysts contained Cercaria of a milky-white colour. They measured ·7 mm. long and were bent upon themselves. The fish were not well preserved and the Cercaria were of a pasty consistency which did not allow of a careful examination of their anatomy. One sucker, however, appeared quite distinct. It is, of course, impossible to state the probable identity of the adult species represented by these immature forms. Similar cysts and parasites have been recorded by Linton from a "cunner", *Tautogolabrus adspersus* (*Bull. U.S. Fish Comm.*, Washington

1889, page 296, pt. 40, figs. 76-81), and by Ryder (*Bull. U.S. Fish Comm.*, 1884, pages 37-42).

The 'cunner' however is a marine fish of the Wrasse family, and it is unlikely that our parasites are identical either with those obtained by Linton or Ryder.

Hofer (*Handbuch der Fischkrankheiten*, München, 1904) describes a number of similar cysts from European fresh-water fishes and classifies them as *Holostomum cuticula*. It is quite possible that our larvae belong to the same species.

If these encysted immature forms of Trematodes are actually the young of *Holostomum cuticula*, it is almost certain that the adult forms will eventually be found in the intestine of fish-eating birds. The two species of fish infected are both very small and would be easily available and readily eaten by such birds.

(3) Cercaria in skin of Nuria danrica.

(Plate xxvii, fig. 9).

Eight specimens presented by Mr. J. Taylor, Angul, Orissa, 4-v-14. All these fish (which are known in Bengal as "danrika") were very heavily infected. In the largest fish, which measured 36 mm. in length, 27 cysts were counted scattered all over the body. The smallest fish measured 21 mm. and 18 cysts were counted on this specimen. The older cysts were black pigmented, but the pigmentation was not nearly so dense as in the preceding form. The capsule was very delicate and easily ruptured. The amount of black pigment that was present varied, but was never very great. In two very young larval forms, which were removed from the gills, no pigment was present. In slightly older stages only a little pigment was observed, whilst in the largest and oldest forms obtained the pigment was never sufficiently abundant to obscure the larval cyst.

The largest cyst observed was oval and measured 1·1 mm. by ·9 mm. The larva only occupied about one-half the interior space of the capsule. It measured ·4 mm. and was folded upon itself. What appeared to be a sucker was discernible in the older forms.

It will be obvious that these larval forms differ from those obtained from *Cirrhina latia* in being younger and much smaller. Whether or not they are identical with those found on that species remains to be determined.

(4) Cysts from the skin of Nuria danrica var. grahami.

(Plate xxvi, figs. 5, 5*a*, 5*b* and 5*c*).

Champadanga, R. Damodar near Calcutta, July, 1913.

This fish, which only measured 17 mm. long, was caught along with the young of a number of carp, *Ambassis* spp., and *Barbus* spp.

Three specimens were obtained having black cysts in the skin. In one specimen there were 7 cysts, in another 13 or 14, and in the third there were over 20. The cysts were distributed generally over the body and were situated in the epidermis. They were surrounded with dense black pigment. On opening the cyst a milky-white larva was obtained which measured ·7 mm. long and ·5 mm. broad. This larva was enclosed in a thin, but pigmented sac, which was unattached and easily removed. Figures of the parasite are given on plate xxvi, figs. 5*a*, *b* and *c*. There was an outer, somewhat egg-shaped membrane, which was tough and transparent. The contents of this membrane were disposed towards one pole. A few cells in an active state of division were observed, towards the pole. The larva is evidently too young to admit of certain identification.

It is probable that the adult of this parasite will be found in fish-eating birds as its host is commonly eaten by them.

(5) Encysted Cercaria in the superficial muscles of Labeo rohita and Catla buchanani.

(Plate xxvi, fig. 4).

Locality.—(I) *Labeo rohita* and *Catla buchanani* from Rajmehal, Bihar, India, October, 1913.
(II) *Labeo rohita* (other specimens). No history.

The cysts were smaller but similar in outward appearance to those found in *Cirrhina latia*. They were, however, situated in the superficial muscular layer. The two suckers were prominent. At present, the identification of these larval forms is impossible.

Their occurrence in the muscular tissue of these fish is a fact of considerable importance. These two species of carp are the two most important food-fishes in Bengal and they frequently attain a weight of over 25 lbs. The fact that the larvae occur in the muscular tissue and not in the skin, means that they are not removed during the ordinary process of cleaning, prior to the fish being placed on the table. It is true that if well cooked, the larvae are destroyed. Even if not destroyed by cooking we have no information at present as to whether these larval forms mature in the human intestine, or not.

In a previous paper (*Rec. Ind. Mus.* vol. IX, part II, June, 1913) I called attention to the fact that the rare Trematode, *Gasterodiscus hominis*, Lewis and McConnel, has been recorded twice from man in Calcutta. It is quite possible that the larval form of this Trematode may occur in the skin of fish. In this connection it is to be remembered that fish is one of the staple articles of food in Bengal.

Although I have examined several thousands of marine fish during an experience of roughly six years on the Ceylon Pearl Banks, I have never found either Trematodes or Sporozoa in the skin.

(6) Carcinoma of the Thyroid in rainbow trout (Salmo irideus) from Naini Tal.

(Plate xxvi, figs. 1 and 2).

During the early part of 1914, the Dy. Conservator of Forests, Naini Tal, United Provinces, India, in a letter to me, stated that numbers of rainbow trout (*Salmo irideus*) were dying in the hill-waters in the vicinity of Naini Tal. I requested him to forward to me specimens of the dead fish, preserved in spirit. In all, I received 13 specimens. The largest measured 15½ inches and the smallest 10 inches. There were 4 or 5 females with ripe eggs. No external or internal parasites were discovered. Three of the fish had a small abrasion on the body. These wounds, however, were occasioned during packing and transit, and were in no way connected with the death of the fish. Out of the 13 fish sent, tumours were found on the gills of three. Excepting the tumours just mentioned, the fish appeared normal and well fed. The location of the tumours was as follows :—

(1) *A small trout* 10 *inches long.*—The tumour was situated in the gills, on the convex (postero-ventral) edge of the gill-arches on the right side. Only one tumour was present (plate xxvi, fig. 2). It measured 17 mm. long, 9 mm high and 4 mm. thick. The outer surface of the gills in the vicinity of the tumour was slightly pigmented with black. The tumour did not involve the bony branchial arches, but only the gill-filaments in the vicinity, *i.e.* the tumour replaced the gill-filaments. The gill on the last branchial arch was the only one not involved.

(2) *A large trout* 15½ *inches.*—Two tumours, one in each branchial cavity, visible ventrally as coarsely nodulated or lobated masses in the anterior extremity of each branchial chamber. The one on the right side was larger than that on the left. The measurements were as follows :—

Large tumour—long 20 mm., high 13 mm., thick 11 mm.
Small tumour— ,, 14 ,, ,, 7 ,, ,, 5 ,,

The external gill-filaments on the right side were only slightly affected. The corresponding gill-filament on the left side was not involved.

(3) *A medium-sized fish.*—Two tumours situated as in (2), both of the same size and measuring 20 mm. long, 14 mm. high and 15 mm. thick.

The latter tumours were situated at the anterior extremity of the branchial chamber just below the eye, and were sufficiently bulky to project into the buccal cavity. The tumours consisted for the most part of a yellow-white cheesy substance enclosed in a thin, but slightly tough, fibrous capsule. The tumours were not pedunculated, but were supported by the capsule, which was attached at various places to the anterior wall of the branchial chamber. Posteriorly the cyst was free. Anteriorly the two tumours were in contact in the centre line. The tumours consisted entirely of the caseous substance referred to above.

Unfortunately only spirit specimens were available and hence I have been unable to make observations on fresh material. Sections were made from one of the larger tumours. These tumours are undoubtedly due to a disease of the thyroid variously known as gill-disease, thyroid tumour, endemic goitre and carcinoma of the thyroid. The disease was first noted in 1883 by R. Bonnet (1)[1]—in *Trutta lacustris* obtained from a hatchery in Torbole on the Gardasee. The disease accounted for the death of 3000 fish.

The first investigator to define the tumour as Carcinoma was Scott (9) who found it in *Salmo frontinalis* from ponds at Opoho belonging to the Dunedin Acclimatisation Society, New Zealand.

In 1902, Marianne Plehn (8) recognized that the tumours were due to a disease of the thyroid gland. In 1903, L. Pick (7*b*) described the disease fully. Gilruth (3) in the reports of the New Zealand Department of Agriculture (Veterinary Division, 1901 and 1902) described a similar disease in *Salmo salar* as "Epithelioma affecting the branchial arches of Salmon and Trout." Later, this author recorded the disease from *Salmo irideus*.

L. F. Ayson, Chief Inspector of Fisheries, New Zealand, stated in a letter to Gilruth that he had noticed the disease in *Salvelinus frontinalis* in 1890. In 1908, Jaboulay (5) reported the disease in six trout from Thonon. Up to the present, five species of fish from Europe and 21 species from America have been recorded suffering from the disease. In America the disease appears to have been known for a long time although not described until 1909, when Dr. Gaylord read a paper on "an epidemic of cancer of the thyroid in brook trout" before the American Association for Cancer Research. The initial investigations into diseased thyroids in the Salmonidae by the American Association for Cancer Research were due to the papers on the subject which had previously appeared by Plehn and Pick.

The Association continued to make extensive observations on the disease, and an excellent and exhaustive report on the subject was published in the Bulletin of the Bureau of Fisheries vol. XXXII, 1912 ("Carcinoma of the Thyroid in the Salmonoid fishes," by Harvey R. Gaylord, etc., Doc. No. 790. Issued April 22, 1914). To these authors I am indebted for most of the details set forth in the present paper. In Salmonoid fishes the thyroid is a more or less diffuse, unencapsuled organ, distributed along the course of the ventral aorta. The gland, however, appears to be much more diffuse in domesticated trout than in wild trout. Gudernatch states that in wild species the gland may extend into the gill-arches and even into the muscle bundles of the isthmus. This circumstance serves to explain the fact that tumours may be found in places as far removed as the jugular pit and the rectum. The disease is universal where trout are artificially cultivated, and in certain hatcheries it may become endemic. Artificial cultivation is obviously a predisposing factor since the disease is rare in nature.

[1] These numbers refer to the literature cited on p. 320.

Healthy wild fish placed in hatcheries where the disease is endemic soon become affected. The first symptoms of disease of the thyroid is simple hyperplasia usually marked in living specimens by a redness of the throat ("red floor") near the second gill arches, and caused by an increase in the blood supply to the thyroid, and to hyperaemia of the adjacent tissues. This condition of simple hyperplasia passes gradually into the stage of visible tumour. Various structural types of infiltrating tumour are known amongst Salmonoid fishes, including the alveolar, solid, tubular, papillar and mixed, and the investigations of workers in America show that the tumour is of a true malignant nature.

The cause of the disease has not been definitely determined, but it is known that trout fed on animal proteid food, in an *uncooked* condition, are more heavily infected than those fed on *cooked* animal proteid food. The crowding together of fish in confined spaces, and other generally unsatisfactory hygienic conditions also favours the spread of the disease.

General insanitary conditions alone are, however, insufficient to account for the phenomenon. A specific living organism is suspected, although no such organism has been isolated up to the present. In America it was found that scrapings from the inner surface of the wooden tanks in which diseased fish were kept, if suspended in water and administered to certain mammals, produced in such animals a definite condition of goitre, and it was accordingly believed that this agent was the cause of the disease amongst the fish in the tank.

It was further shown that by boiling the water the effective agent was destroyed. The disease does not appear to be directly transmissible from one individual to another.

Some species of the Salmonidae are practically immune from the disease, and in other species spontaneous recovery frequently occurs. Especially is this the case if the diseased fish are removed to natural conditions and allowed to feed on natural food. Moreover the disease is directly susceptible to treatment. At all times the normal thyroid contains traces of iodine. During hyperplasia the proportion of iodine appears to be reduced. Occasionally human goitre reacts favourably to treatment with iodine. In all stages of its growth the tumour in fish is favourably affected by solutions of iodine as well as by those of mercury and arsenic. It is thus possible, in a limited way, to treat these diseases in hatcheries and in limited water-areas. At the same time it will be obvious that the object of fish culturalists should be to prevent the disease rather than to effect its cure.

A tumour, apparently of a similar nature, is recorded by Williamson (Fisheries Scotland, Scientific Investigations for 1911, Glasgow, 1913, page 23) in the following words.

"*Tumour in the pharynx of a Salmon caught in the sea.*"

"It was found loose in the gill cavity after the fish had been killed by a blow on the head. Two of the gills were found to be damaged. There was a sinus in the free edge of the gill. It is not clear how the tumour was attached, but the connection was apparently a slender one. The tumour is lobulated, fibrous in structure, without any distinct lamina."

Simple hyperplasia of the thyroid has also been recorded in pike, bass, and occasionally in herring, by Marine and Lenhart.

During an experience of about five years on the Ceylon Pearl Banks, although I had occasion to examine very many thousands of fish, in no instance did I see a case of visible tumour in any examined.

LIST OF LITERATURE CITED.

1. Bonnet, R.—Studien zur Physiologie und Pathologie der Fische. *Bayerische Fischerei-Zeitung, München*, nr. 6, p. 79, 1883.

2. Gaylord, H. R. and Marsh, M. C.—Carcinoma of the Thyroid in the Salmonoid Fishes. *Bulletin of the Bureau of Fisheries*, vol. XXXII, 1912. Document No. 790. Issued April 22, 1914.

3. Gilruth, I. A.—Epithelioma affecting the branchial arches of salmon and trout. *Report of the New Zealand Department of Agriculture, Division of Veterinary Science*. 1902.

4. Gudernatsch, J. F.—The structure, distribution and variation of the thyroid gland in fish. *Journal of the American Medical Association*, vol. 54, no. 3, Jan. 15, 1910, p. 227. (American Association for Cancer Research, meeting held November 27, 1909).

5. Jaboulay.—Poissons atteints de goitres malins hereditaires et contagieux. *Journal de Médécine et de Chirurgie pratiques*, t. 79, p. 239, 1908.

6. Marine, David and Lenhart, C. H.—On the occurrence of goitre (active thyroid hyperplasia) in fish. *John Hopkins Hospital Bulletin*, vol. XXI, no. 229, p. 95, April 1910.

7(*a*). Pick, L.—Der Schilddrusenkrebs der Salmoniden. Aus dem Laboratorium der L. und Th. Landauschen Frauenklinik, Berlin. *Berliner klinische Wochenschrift*, 1905, nos. 46-49, p. 1435-1542.

7(*b*). Pick, L.—Ueber einige bemerkenswerte Tumorbildungen aus der Tierpathologie, insbesondere uber gutartige und krebsige Neubildungen bei Kaltblutern. *Berliner klinische Wochenschrift*, 1903, nos. 23-25. Abstract in *Journal American Medical Association*, August 8, 1903, p. 401.

8. Plehn, Marianne.—Bosartiger Kropf (Adeno-Carcinom der Thyreoidea) bei Salmoniden. *Allgemeine Fischerei-Zeitung, München*, no. 7, p. 117-118, April 1, 1902.

9. Scott.—Note on the occurrence of cancer in fish. *Transactions and Proceedings of the New Zealand Institute, Wellington, N.Z.*, vol. 24, 1891 (issued May, 1892), pp. 201, 1 plate.

(7) Description of a new species of Isopod Crustacean parasitic on the Bhekti (Lates calcarifer).

Rocinella latis, n. sp.

(Plate xxviii, figs. 12–15).

No. $\frac{9056}{10}$	From skin of *Lates calcarifer*, 15-ii-1915.	Diamond Harbour (R. Hughli), near Calcutta.	T. Southwell.

All the specimens caught were males. The head projects well beyond the basal joints of the first and second antennae as a broadly rounded plate, convex from above. The first peraeon segment is longer than the rest. The first five segments of the pleon are a little narrower than those of the peraeon. The telson is slightly narrower than the preceding segments of the pleon.

The eyes are large, well separated, and situated laterally. They extend on to the ventral surface.

The bases of the first and second antennae are hidden in dorsal view by the frontal lamina which extends well beyond them. The first antennae are much stouter and much shorter than the second antennae. The flagellum is 8-jointed and terminates abruptly. The last joint extends to the middle of the first peraeon segment. The second antennae are, as noted, slender.

The flagellum consists of 10 (possibly 11) joints and it extends to the posterior extremity of the first peraeon segment. In length it exceeds that of the first antennae by its terminal 3 joints. The basal joints of the antennae are not distended.

The upper lip is crescentic, thin and membranous.

The mandibles have the palp somewhat elongated.

The first three pairs of legs are prehensile and have long and evenly curved dactyli, the extremities of which are of a dark brown colour in most specimens. The propodus is broad and crested and is armed with about 8 long, pointed, comb-like spines. There is a single elongated spine at the distal and exterior angle of the merus. The spines on the propodus of the first pair of legs are not quite as well defined as those of the second and third pairs.

The four gressorial legs are very similar to the first three pairs of legs, but a little more slender. The dactylus bears two spines near its base. The propodus bears 8 spines, but the spines are not borne on a crest, *i.e.* the propodus is not nearly so broad as is the case in the first three pairs of legs. The carpus bears four spines and the merus two. All these spines are situated on the internal surface. The last pair of legs is slightly smaller than the rest.

The largest specimen measured 14 mm. long and the greatest breadth was 4·5 mm.

In young specimens the whole surface is marked with minute pigment spots, hardly visible to the naked eye. In adults, however, the pigment consists of three very narrow longitudinal bands, one on each side and one running along the centre of the carapace.

This latter band spreads out between the eyes. Each band of colour consists of very numerous pigment spots.

The Bhekti on which these parasites occurred was caught in the vicinity of Diamond Harbour in the Hughli river near Calcutta on February 15, 1915. When placed on deck alive, some three dozen parasites were found to be moving over the skin of the fish. A few left the host and were picked up from the deck of the ship. I have never seen these parasites on any Bhekti in the markets and I believe that after the fish are removed from water the parasites quickly drop off.

The parasites are Isopods of the tribe Flabellifera. They are included in the family Aegidae and fall in the genus *Rocinella*, Leach, 1818. The characters of this genus are as follows:—

"Form of body resembling that of *Aega*, though being somewhat less compact and more depressed. Metasome generally less broad, with the terminal segment rounded off at the end and finely ciliated. Eyes well developed with very large and conspicuous cornea. Antennae slender, the superior ones much shorter than the inferior, and with the basal joints not expanded. Epistomal plate very small and narrow. Mandibles considerably produced, with the cutting edge expanded inside to a linguiform lamella (molar expansion); palp well developed with the basal joints much elongated. Maxillae nearly as in *Aega*. Maxillipeds with the palp composed of only two joints, the terminal one armed with strong recurved teeth. The three anterior pairs of legs having the propodus more or less expanded and armed inside with strong spines, dactylus forming a very large and evenly curved hook. The four posterior pairs slender, resembling in structure those in *Aega*. Pleopoda and uropoda normal" (Sars).

The conspicuous cornea, the non-expanded base of the antennae, the nature of the propodus in the three anterior pairs of legs, the evenly armed dactylus on the three anterior pairs of legs, and the four-jointed maxillipeds, distinguish the parasite as belonging to the genus *Rocinella*. An outstanding feature of the species is the broad elongated head-shield which extends well beyond the bases of the two pairs of antennae.

The specimens have been deposited in the Indian Museum.

LITERATURE.

Stebbing.—History of Crustacea, London, 1893.

Stebbing.—Ceylon Pearl Oyster Reports, vol. V, London, 1905.

Sars.—Crustacea of Norway, vol. II, Isopods, Bergen, 1899.

Hansen.—Bulletin Museum Comparative Zoology, Harvard College, vol. XXXI, no. 5, Cambridge, Mass., 1897.

Richardson.—Proc. U.S. Nat. Museum, vol. 21, 1899.

Richardson.— do. vol. 23, 1901.

Richardson.— do. vol. 27, 1904.

Richardson.—Bull. U.S. Fish Comm., 1903.

Moore.— do. do. vol. 20, part 2, 1900.

(8) Argulus foliaceus, Linnaeus, from the skin of Labeo rohita (Rohu).

(Plate xxviii, figs. 16–18).

Agricultural Farm tank, Siripur, Bihar, India. August 20, 1913.

During the year 1913 I was engaged making observations with reference to the breeding habits of Indian Carp. For this purpose three tanks were dug in the Agricultural Farm at Siripur. The measurements of the three tanks were the same, namely, 50 ft. in length, 37 ft. in breadth and 7 ft. in depth.

These tanks were situated in a line at right-angles with, and very close to, a neighbouring stream, from which they received water.

In addition, it was found that a small spring of water existed in the middle tank. The tanks were only separated from each other by a narrow bund and they were in connection with each other by means of a pipe running through each bund near the surface. The tank which we will consider as No. 1 was situated nearest to the stream, from which it was distant only about three or four yards.

About eighteen mature specimens of *Labeo rohita*, both males and females, were placed in these three tanks during the latter part of July. It was found that large numbers of frogs entered the tank nearest to the river. In order to exclude frogs from this tank a matted fence was erected all round it. About the middle of August, Mr. Mackenzie, the Superintendent of the Agricultural Farm, noticed that the fish became sluggish and floated on the top of the water. On examining one or two it was found that they were covered with external parasites. These were preserved in spirit and forwarded to me. The steps taken by the Farm Superintendent to remedy the disease were as follows:—First, all the fish were captured and scraped as clean as possible. The fence matting was then removed giving free access to frogs, etc. Lastly, an upright bamboo was erected in the centre of the pond. The Farm Superintendent, whose observations and statements are thoroughly reliable, states that the fish proceeded forthwith to rub themselves against this bamboo. There were no deaths.

About the end of September all the fish were captured and killed and were then found to be perfectly clean; not a single parasite was found. The fish present in the second and third tank were not affected. I have since ascertained that extensive deaths amongst carp in tanks due to "external parasites" have occurred, within the last four years, in the districts of Mymensingh and Murshidabad, and I have no doubt that the parasite causing these diseases was identical with the one obtained from Siripur. The forms examined by me are undoubtedly *Argulus foliaceus*, Linn., and have been recorded as external skin and gill parasites from a large number of European freshwater fishes, such as *Tinca vulgaris*, *Gasterosteus* spp., *Cyprinus carpio*, *Esox lucius*, *Perca fluviatilis*, *Salmo trutta*, etc., and even from the tadpoles of frogs.

The parasites attach themselves to their host by means of two strong suckers which are the modified anterior maxillipeds. Like other parasitic Copepoda they suck the blood of their host. This is effected by means of a proboscis or dart which is evertable, and which is formed by a modification of certain of the mouth parts. The posterior maxillipeds are also modified for the purpose of clasping, and thus enable the parasite to cling to its host. In addition, the basal joints of the anterior antennae are modified for a similar purpose. The parasites lie inserted between the scales of the fish, with their long axis parallel to that of the host. They are, however, by no means stationary and fixed, but may be seen to skip about over the fish's body as if in search of a better position. During the breeding season they voluntarily leave the body of the fish and swim about actively in the water by means of four pairs of swimming legs. Unlike other Copepoda, the eggs, which are laid in gelatinous strings of two rows, are usually shed into the water and not carried about by the female. On being shed, the gelatinous covering hardens and thereby firmly attaches the eggs to the object on which they were deposited. Observations made in Europe show that the parasite breeds three times a year. Under these circumstances it is clear that there are three occasions each year when infected fish may free themselves from their parasites. The development of the egg occupies about a month.

Wilson states that the newly-hatched larvae have the general characters of the adults and on hatching begin to swim at once. The nauplius, metanauplius and early cyclops stages are passed inside the egg. After a few moults they become adult. Certain species of *Argulus* appear to be capable of living on both fresh and salt water fishes. This circumstance, together with the fact that the parasites can swim freely and frequently leave their host, accounts for the fact that the same species of parasite is often found on different species of fish.

As far as I am aware this is the first definite record of this parasite in India. In April, 1910 Mr. S. W. Kemp, Senior Assistant Superintendent, Indian Museum, inspected a tank near the palace of the Maharaja of Cossim Bazar in which diseased fish (Rohu, *Labeo rohita*) were living. He found that the disease was associated with scanty food, and the presence of large numbers of leeches and parasitic Copepoda, the latter belonging to the genus *Argulus*.

On April 17th, 1911, Dr. Annandale obtained a free-living *Argulus* from the Atrai river, near Siliguri, at the base of the Himalayas (Jalpaiguri District, Bengal). I have examined the latter specimen and found it to be a very small male ***Argulus foliaceus***, Linn. Unfortunately I have been unable to obtain the specimens collected by Mr. Kemp, but it is very probable that they also belong to this species.

It is significant that the parasites, both at Cossim Bazar and Siripur, only attacked *Labeo rohita*, although other species of fish, such as *Catla buchanani* and *Cirrhina mirgala*, were living in the same tank at Cossim Bazar, and in the next tanks at Siripur. The mortality amongst tank fish in particular, due to the presence of this parasite, is, in all probability, fairly extensive in Bengal. In nature, however, the parasites are rarely dangerous. The practice of stocking tanks with fish—so prevalent in Bengal—undoubtedly favours conditions under which the parasite thrives. It is known that in Europe the parasites themselves are eagerly devoured by roach, dace and bream. The presence of such fish therefore tends to check the distribution of the parasite and thus to protect other fish from their attacks. It seems possible that in our tanks at Siripur, the frogs, when allowed to enter, also devoured the parasites, but no direct observations were made in this connection, and hence we have no certainty that such was really the case.

The family Argulidae, Muller, belongs to the sub-order Branchiura, Thorell, and to the order Copepoda, Muller. It contains three genera *viz. Argulus*, *Chonopeltis* and *Dolops.* In the former two genera the first maxillipeds are modified into sucking discs and in the latter genus sucking discs are absent. The genus *Argulus* contains about thirty-two species of which only three are European, the rest being found in American waters. The majority of the forms are marine. The various species exhibit but little trace of degeneration, a circumstance one would expect considering the alternation which exists between temporary parasitism and a free life. The males differ but slightly from the females but are considerably smaller.

Our specimens, of which we have over 200, are small, the extreme length of the largest female was 3·2 mm. and the greatest breadth was 2·2 mm. The carapace is elliptical. The posterior sinus extends nearly half-way up the length of the carapace. The abdomen is almost square and about $\frac{1}{4}$ the length of the body. The suckers are small, placed quite anteriorly and well separated from each other. Their diameter is almost ·4 mm. The basal plate of the posterior maxillipeds is triangular in shape with three well defined, sharp, rectangular teeth. All the swimming legs extend well beyond the edge of the carapace. In the male the abdomen is much longer and narrower than in the female, and the sinus is narrow, sharply cut, and deep.

The specimens have been deposited in the Indian Museum and are numbered $\frac{9054}{10}$.

LITERATURE

Wilson, C. B.—North American Parasitic Copepoda of the family Argulidae, with a Bibliography of the group and a systematic review of all known species. *Proc. U.S. Nat. Mus.*, vol. XXV, pp. 635–742, pls. viii–xxvi (no. 1302). Washington, 1902.

(9) Amphilina magna, n. sp. from the coelom of Diagramma crassispinum.

Group **CESTODARIA**, Mont., 1892.

= *Cestoidea monozoa*, Lang.
= *Cestodes monogeneses*, V. Ben.
= *Atomiosoma*, Monticelli, 1892 ?

Cestodes in which the animal consists of a single segment, containing a single set of reproductive organs. In addition to the male pore and female (vaginal) pore, there is a third aperture, that of the uterus (birth-pore). The apparatus by which fixation is effected, consists usually of a single sucker, but presents considerable variation in form, as well as in disposition, with regard to the genital pores.

Family AMPHILINIDAE, Braun, 1883.

Oval or leaf-shaped, without a distinct " head ", but with a single small acetabulate sucker at one end.

Genus **Amphilina**, Wagener, 1858.

" Body flat. Long egg-shape to leaf-like. Anterior and posterior ends pointed. Dorsal surface more arched than ventral surface. Skin with a net pattern caused by regular pit-like depressions. Anterior extremity usually with a pit, deep, or otherwise, according to the degree of contraction. This extremity may also present the appearance of a papilla or glandiform snout. On this papilla numerous one-celled glands occur, with long excretory ducts. The excretory system consists of anastomosing vessels with pore posterior. Testes numerous. Cirrus-sac absent. Ovary and reproductive aperture posterior. Opening of vagina a little way from posterior extremity, marginal or on surface. Uterus a long N-shaped canal, first running forward, then turning round and running posteriorly, then again curving round and running forward " (Wagener).

Apmhilina magna, n. sp.

(Plate xxvii, figs. 6-7).

Z.E.V. $\frac{8946}{7}$ Coelom of *Diagramma crassispinum.* Pearl Banks, Ceylon. T. Southwell.

Three specimens. Two damaged, one perfect.

A description of the superficial characters of this worm was given by me in the Ceylon Marine Biological Reports, Vol. VI, Jan. 1912, page 273, and this I reproduce here.

" During the examination of a number of specimens of *Diagramma crassispinum*, three specimens were found to (each) contain a most remarkable free living parasite in the coelom. Unfortunately I have not had time to make a careful examination of this parasite, and I am at present uncertain of its strict zoological position........... In the living condition it measured 15 inches long and $1\frac{1}{8}$ inches broad. It was quite flat, and had a thickness of $\frac{1}{16}$th inch. The preserved specimens, of which I have 3, measure $9\frac{1}{2}$ inches long, $\frac{3}{4}$ inch broad, and are about $\frac{1}{8}$ inch thick. The extremities are rounded and terminate in a minute, acute point. At one extremity there is a minute sucker-like aperture situated centrally, whilst at the other extremity there is a similar but slightly larger aperture situated laterally. This latter aperture appears to open to the interior of the worm. The edges of the worm are straight and parallel. A pair of narrow blackish tubes run along the lateral margins, one on each side. Down the centre of the worm, and stretching from one extremity of the worm to the other, is an opaque milky-white mass $\frac{1}{4}$ inch broad. On each side of this mass there are a series of black coiled tubes, $\frac{1}{16}$th inch in diameter disposed in bunches, also running the entire length of the worm, but situated for the most part on one side. No other apertures could be detected. In consistency the worm is that of a stiff (milky-white) jelly, (in formalin).''

For assistance in working out the anatomical details of this worm I am indebted to Dr. Ekendranath Ghosh, L.M.S., M.Sc., Assistant Professor of Zoology in the Medical College, Calcutta.

The following measurements of the specimens (preserved in formalin) have been taken recently :—

	Length.	Breadth.
Specimen I	250 mm.	20 mm.
,, II	180 ,,	14 ,,
,, III	160 ,,	15 ,,

The testes lie scattered about through the parenchyma. At a point about 15 mm. from the posterior extremity, the paired vas deferens unite in the middle line, and open by a minute pore 1 mm. from the posterior extremity.

The germarium is situated in the middle line 15 mm. from the posterior extremity. It is 7 mm. long and 3·5 mm. broad. Immediately posterior to it are the paired follicular shell-glands. These are each 3 mm. long and together are 3 mm. broad. The vaginal pore is 2 mm. from the posterior extremity. The vitelline glands are paired, linear, and cylindrical, one on each side, near the lateral margins. They extend the whole length of the worm. At 4 mm. from the posterior extremity of the worm their ducts curve towards the vagina and open close to the shell-gland. The uterus is a long convoluted tube having exactly the same form as that figured for *A. foliacea*. It opens by a minute pore, which is situated at the base of the small anterior end of the worm. In this respect it agrees with *A. liguloidea*, Diesing, and differs from *A. foliacea* (Rud., 1819). It differs, however, from *A. liguloidea* in the absence of a vagina, anterior to the germarium. The situation of the male and the vaginal pores also resemble that of *A. liguloidea*, and differs from that of *A. foliacea*.

No hooklets were observed on the penis of our specimens. The eggs are large, measuring almost ·1 mm. In shape they resemble half a sphere, the flat surface of which has became concave. No filament was observed. Compared with other known species of *Amphilina*, our specimens are enormous, as will be seen from the following table:—

	Length.	Host.
A. foliacea	.. 60 mm	.. *Acipenser* sp.
A. liguloidea	.. 80 mm.	.. ?
A. neritina	.. 18 mm.	.. *Acipenser* sp.
A. magna	.. 250 mm.	.. *Diagramma crassispinum*.

The nearly related genus *Wageneria* contains the following species:—

W. proglottis (Wag., 1854), Mont., 1892.
W. aculeata, Cohn, 1902.
W. porrecta (Lühe), Cohn, 1902.
W. impudens (Crep., 1846), Cohn, 1902.
W. sp. Lühe, 1902.
W. sp. (Mont.)?

Unfortunately I have been unable to compare my types with descriptions of any of the above species of *Wageneria*, as literature was not available in India. The type of *Amphilina magna*, n. sp. is deposited in the Indian Museum.

LITERATURE.

1. Wagener, G.—Enthelminthica No. V. Ueber *Amphilina foliacea* mihi (*Monostomum foliaceum* Rud.), *Gyrocotyle* Diesing und *Amphiptyches* Gr. W. (*Arch. f. Naturges.* 24 Jahrg. I Bd., pp. 244-249, Taf. viii, Berlin, 1858).
2. Salensky, W.—Ueber den Bau und die Entwickelungsgeschichte der *Amphilina*. (*Zeits. f. wiss. Zool.* Bd. XXIV, pp. 291-342, Taf. xxviii-xxxii, Leipzig, 1874).
3. Monticelli, F. S.—Appunti sui Cestodaria. (*Atti Accad. Napoli.* Ser. 2, Vol. V, No. 6, pp. 67-78, 4 figs., 1892).
4. Braun, M.—Vermes in Braun's Thierreichs. Bd. IV, Abt. I*a*, 1893.
5. Cohn, L.—Zur Kenntnis des genus *Wageneria*, Monticelli, und anderer Cestoden. (*Centrbl. Bakter.*, Bd. XXXIII, pp. 53-60, 7 figs., Jena, 1902).
6. Pintner, Th.—Ueber *Amphilina*. (*Verhandl. Ges. deuts. Natf. und Aerzte.* 1905.)
7. Janicki, C. V.—Uber den Bau von *Amphilina liguloidea*, Diesing. (*Zeits. f. wiss. Zool.* Bd. LXXXIX, pp. 568-599, Taf. xxxiv-xxxv, Leipzig, 1908.)
8. Southwell, T.—Ceylon Marine Biological Reports. Pt. VI, Jany. 1912, Colombo.

(10) Syndesmobothrium filicolle, Linton, parasitic in the flesh of Harpodon nehereus ("Bombay duck") from Diamond Harbour.

(Plate xxvii, fig. 8).

Z.E.V. $\frac{6869}{7}$	Flesh of *Harpodon nehereus.*	Diamond Harbour. 17-ii-1915.	T. Southwell

The original description of this parasite was given by Linton in the Report of the Commissioner of Fish and Fisheries for 1887 (published 1891, Washington), page 861, plate xv, figs. 2 and 4. The description was from a single specimen and the details given are so meagre that the identification of the parasite is attended with a little uncertainty. The adult form was obtained by Linton from the spiral valve of *Trygon centrura* at Woods Hole, Mass. *Encysted* forms were also obtained by Linton from various species of Teleosts such as *Pomatomus saltatrix*, *Cybium regale*, etc.

The present writer also obtained the same parasite on the Ceylon Pearl Banks from the intestines of *Cybium guttatum* (Seer fish) and *Chorinemus lysan* (Southwell, Ceylon Marine Biological Reports, Part VI, page 269, plate ii, figs. 16 and 17. Colombo, 1912).

The flesh of *Harpodon nehereus* (when alive) is transparent, resembling that of a jelly-fish. The cysts being milky-white were easily discernible with the naked eye. Many cysts were 16 mm. long and the smallest obtained was 4 mm. They were all roughly tadpole-shaped, but the "tail" portion varied greatly in length and thickness. When removed, the cyst moved about actively. The parasite itself could be seen under a low power as a more densely milk-white spot in the head of the cyst. Over 80 cysts were taken from the flesh of a single fish which measured 3½ inches long.

Harpodon nehereus was very plentiful in the river between the sea and Diamond Harbour; in the vicinity of the latter place it was scarce and ten miles further north entirely absent. Every fish caught was infected. As far as I am aware, this is the first record of a cestode parasite occurring in the flesh of any fish, east of Suez. The parasite becomes adult in the larger species of *Trygon* and *Hypolophus*. In other words when rays and skates eat the infected Bombay duck, the larval parasites in the latter become adult tapeworms in the intestines of the rays. The parasites do not inhabit the human intestine and hence there is not the slightest danger of human beings becoming infected by eating the infected fish.

Diamond Harbour is on the river Hughli and is situated about 40 miles from the sea.

LITERATURE.

Linton.—Notes on entozoa of Marine fishes. *Report U.S. Fish Comm.* for 1887, pp. 862-866, plate xv, figs. 5-9. Washington, 1891.

Southwell.—Ceylon Marine Biological Reports, Part VI, page 273, plate iii, fig. 40. Colombo, 1912.

(11) Disease in the eye of Holocentrum rubrum.

(Plate xxviii, fig. 11).

The opportunity is taken of recording in this paper the occurrence of a disease of the eye in *Holocentrum rubrum*. This fish is a marine species.

I noticed the disease in question in January, 1915 during a visit to the Marine Aquarium, Madras, where the fish was then living in one of the tanks.

I am indebted to Dr. J. R. Henderson, Superintendent of the Madras Museum, for kindly presenting the specimen. Dr. Henderson suggested that the disease was a glaucoma, probably caused by an accident whilst the fish was being captured at sea or during its transference to the aquarium.

Since the above paper was written I have obtained very large cysts containing young Trematodes from the flesh of *Ophiocephalus striatus*, *Ophiocephalus marulius* and *Ophiocephalus gachua*. Smaller cysts, apparently containing Cercaria, have also been obtained from the flesh of *Saccobranchus fossilus* and *Trichogaster fasciatus*. In *Saccobranchus fossilus* over 140 cysts were counted in the flesh of a small specimen.

The infected fish were obtained from *beels* in the Khulna district. I hope to describe the parasites in a future paper.

EXPLANATION OF PLATE XXVI.

FIG. 1. Carcinoma in *Salmo irideus*. Ventral view, × $\frac{2}{3}$.

,, 2. ,, ,, ,, Latero-ventral view, × $\frac{2}{3}$.

,, 3. *Rasbora daniconius*, showing white spot near dorsal fin containing *Myxobolus* sp., × $1\frac{1}{2}$.

,, 4. *Labeo rohita*, showing encysted Cercaria, × $1\frac{1}{2}$.

,, 5, 5*a*, 5*b*, 5*c*. *Nuria danrica* var. *grahami*, showing cysts removed.

Explanation of Lettering.

(*a*) anterior; (*b*) vas deferens; (*c*) germarium; (*d*) shell glands; (*e*) vaginal pore; (*f*) gills; (*g*) vitteline glands; (*h*) carcinoma; (*p*) posterior; (*t*) testis; (*u*) uterus.

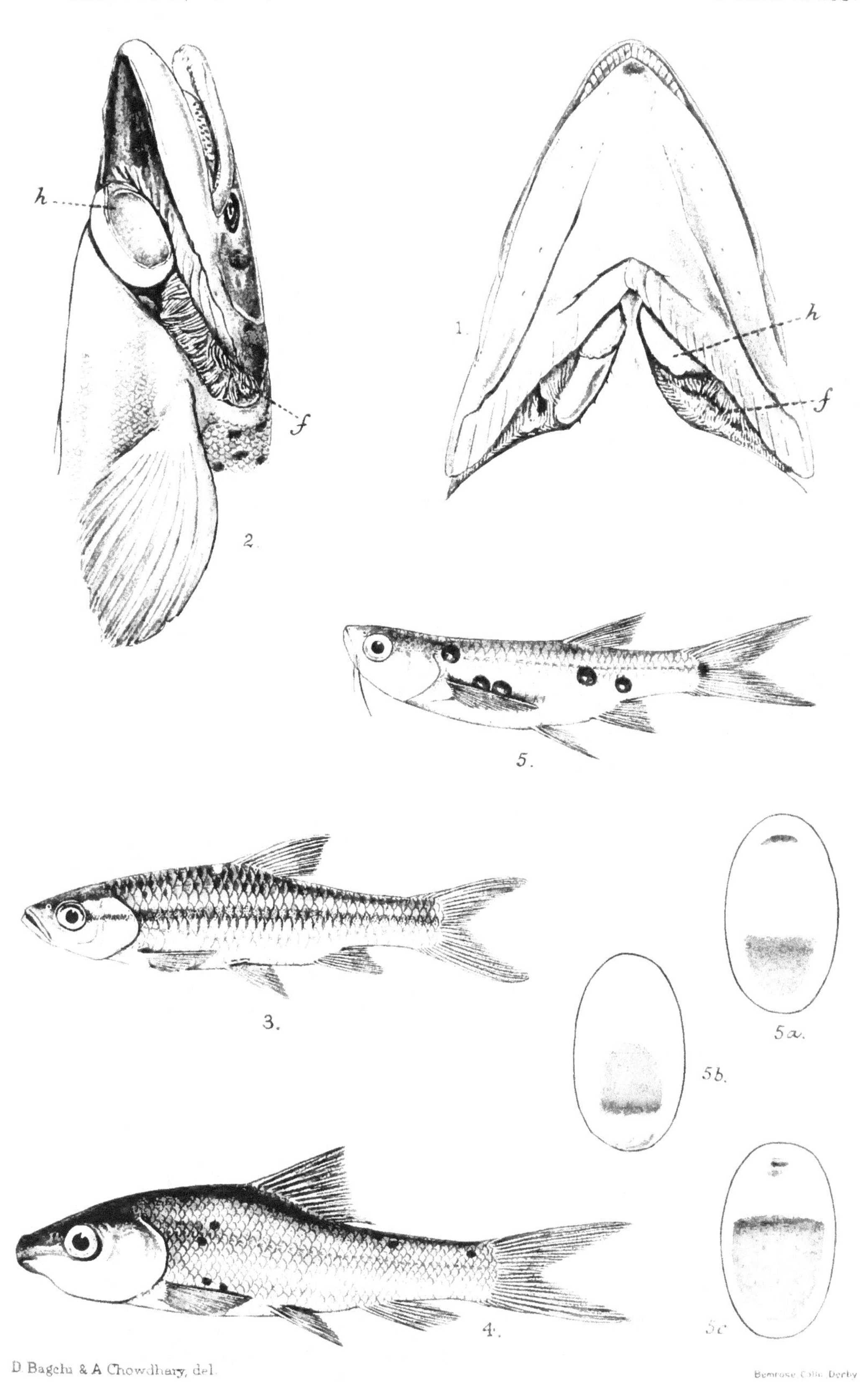

D. Bagchi & A. Chowdhary, del. Bemrose Collo. Derby

PARASITES OF INDIAN FISH.

EXPLANATION OF PLATE XXVII.

FIG. 6. *Amphilina magna*, n. sp., × ⅓.
,, 7. ,, ,, posterior extremity, × 2.
,, 8. *Harpodon nehereus,* showing encysted larvae of *Syndesmobothrium filicolle*, Linton, × 1½.
,, 9 *Nuria danrica*, showing Cercaria in skin, × 1½.
,, 10. *Cirrhina latia*, showing parasites encysted in skin, × 1½.

Explanation of Lettering.

(*a*) anterior; (*b*) vas deferens; (*c*) germarium; (*d*) shell glands; (*e*) vaginal pore; (*f*) gills; (*g*) vitteline glands; (*h*) carcinoma; (*p*) posterior; (*t*) testis; (*u*) uterus.

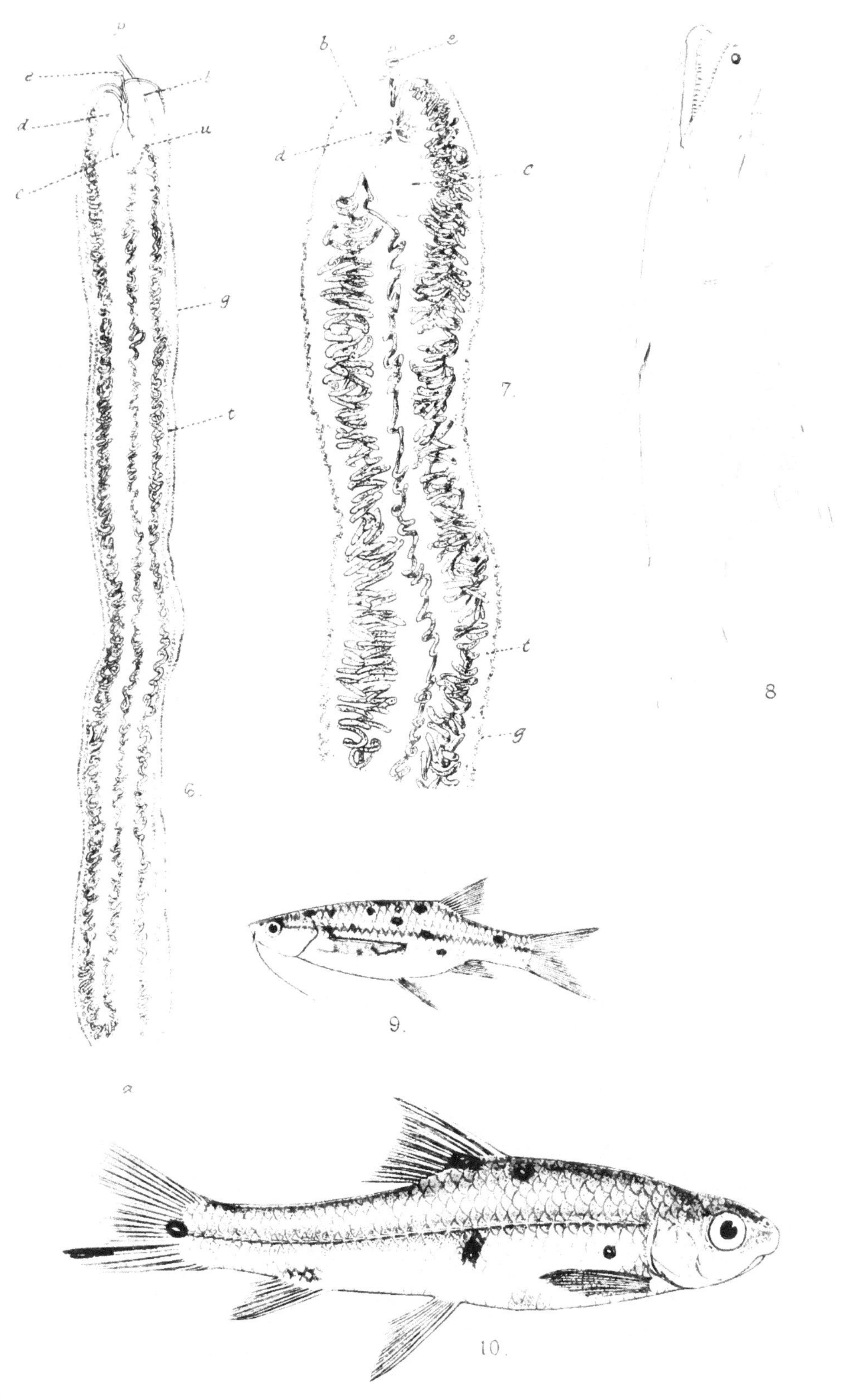

D. Bagchi & A. Chowdhary, del.

Bemrose Collo. Derby.

PARASITES OF INDIAN FISH.

EXPLANATION OF PLATE XXVIII.

FIG. 11. Glaucoma in eye of *Holocentrum rubrum*, natural size.
,, 12. *Rocinella latis*, n. sp., male, ventral view, × 6¼.
,, 13. ,, ,, dorsal view, × 6¼.
,, 14. ,, ,, first gnathopod, × 30.
,, 15. ,, ,, second peraeopod, × 30.
,, 16. *Argulus foliaceus*, Linn., female, ventral view, × 30.
,, 17. ,, ,, dorsal view, × 30.
,, 18. ,, ,, male, dorsal view, × 30.

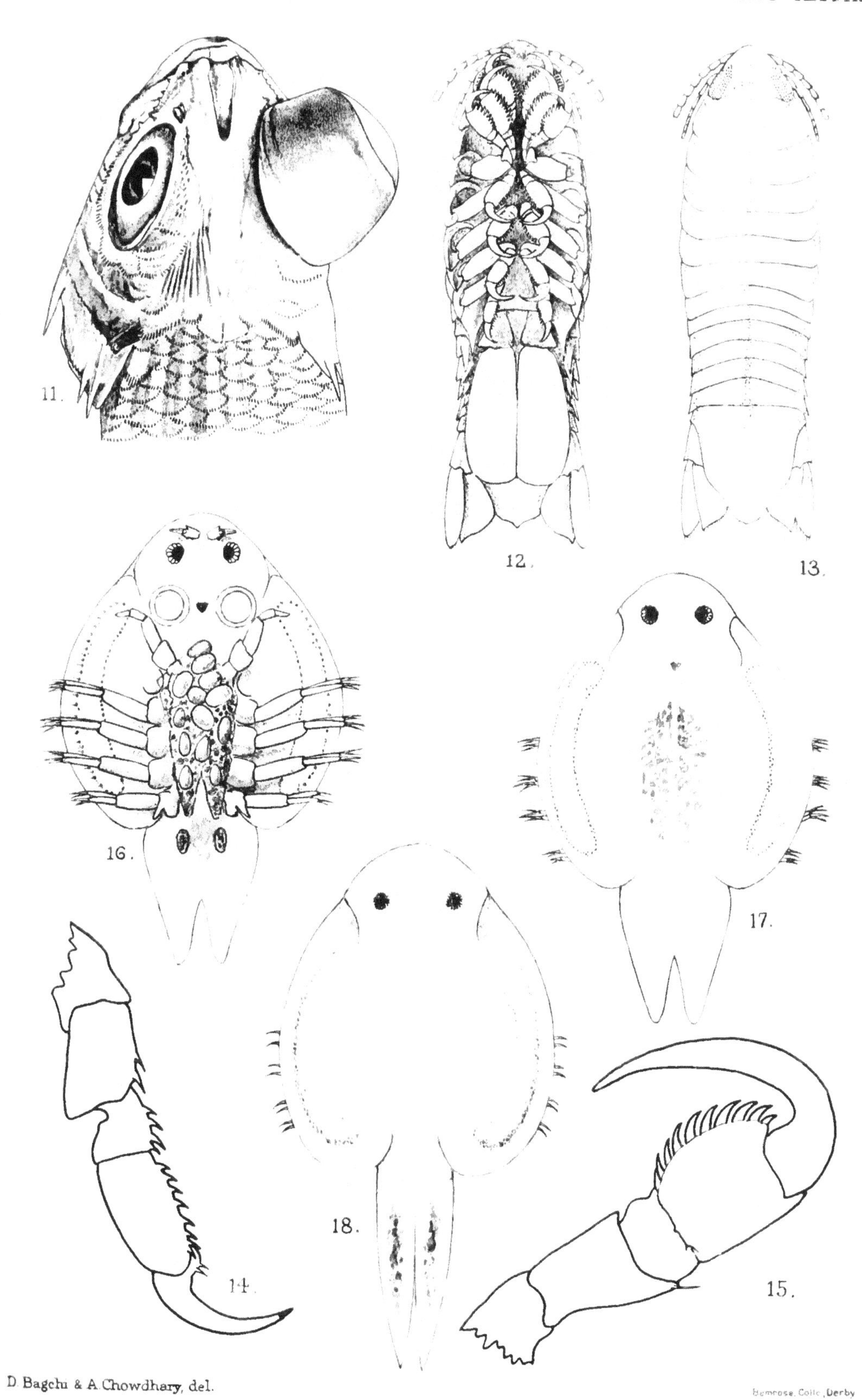

D. Bagchi & A. Chowdhary, del.

Bemrose, Collo, Derby

PARASITES OF INDIAN FISH.

XVII. NOTES FROM THE BENGAL FISHERIES LABORATORY, INDIAN MUSEUM.

No. 3.—On Helminths from Fish and Aquatic Birds in the Chilka Lake.

By T. Southwell, *A.R.C.S., Lond., F.L.S., F.Z.S.,*
Dy. Director of Fisheries, Bengal, Bihar and Orissa,
Honorary Assistant, Indian Museum.

Dr. Annandale visited the Chilka Lake during the last two weeks of November 1914, in order to continue his enquires into the fauna of this area I had the pleasure of accompanying him, and the parasites described in the following paper were collected during the investigations.

The fish from which the parasites were taken were caught in an otter-trawl about two miles east of Rambha. They appear to feed principally on prawns and crabs.

A number of small rays (*Trygon imbricata*) were also caught and examined, but no parasites were found. This species appears to feed on small crustaceans and on thin-shelled molluscs.

The little cormorant was shot on the shore of the lake near Rambha, and the pochard (*Nyroca ferina*) was shot in a swamp near Nalbano Island.

Other birds and fish were examined but no parasites were found.

The collection is an interesting one, but possesses no outstanding or distinctive features.

Family TETRABOTHRIDAE, Linton, 1891.

Tribe *TETRABOTHRIINAE*, Perrier, 1897.

Genus I. **Phyllobothrium,** Van Beneden, 1849.

Body articulate, taeniaeform; head separated from the body by a neck, with four opposite sessile bothria, each bothrium lacinio-cristate on the margin and provided with a single ampulla-like supplementary disc. Genital apertures marginal.

Phyllobothrium pammicrum, Shipley and Hornell.

Z.E.V. 6570/7	*Hypolophus sephen.*	Main area Chilka Lake, Dec. 1914.	T. Southwell.
Thirteen specimens.			

Extreme length, 4·2 mm. to 5·5 mm.
Breadth of head, ·4 mm.

Breadth of last segment, ·25 mm.
Length of last segment, 1 mm.
Length of neck, ·25 mm.

The head consists of four sessile, crimpled bothridia, which have their edges slightly thickened. There is no myzorhynchus, and accessory suckers are absent.

There is a short neck. The worms consist of 5, or at most 6 segments. In many, what appeared to be the terminal vesicle was still intact. As in Shipley's specimens (*Ceylon Pearl Oyster Reports*, V, London, 1906, p. 53) the genital organs were developed in the very first segment, and no short, shallow, young proglottides were observed in any of our specimens. The genital aperture was only obvious in the last segment. This segment had the sides slightly curved, and the greatest breadth was across the middle, through the genital aperture.

The testes are very numerous and large, and were disposed on each side of the longitudinal axis of the proglottid. The cirrus pouch is not conspicuous. No spines were observed on the penis. The deferent canal runs transversely to the genital pore. The vitteline glands were disposed parallel, and external to the testes. The ovary and shell gland were situated posteriorly. The ducts from the vitteline glands also unite in the centre line posteriorly. The oviduct occupies a central position and runs anteriorly in a loosely coiled manner.

The genus *Phyllobothrium*, Van Beneden, is closely related to the genus *Crossobothrium*, Linton. The latter differs from the former only in having the bothria pedicelled and in possessing no neck. It will be noted that our specimens do not possess accessory suckers. No mention is made of accessory suckers in *Phyllobothrium blakei*, Shipley and Hornell (*Ceylon Pearl Oyster Reports*, V, London, 1906, p. 70, figs. 72 and 73), although suckers are shown in *P. pammicrum*, Shipley and Hornell. Johnstone was unable to find accessory suckers in specimens of *P. lactuca*, Van Beneden (*Trans Biol. Soc. Liverpool*, XX, 1906, pp. 159-160), and he refers to the absence of a myzorhynchus in both *P. lactuca*, Van Beneden, and *P. thridax*, Van Beneden.

No myzorhynchus was observed in our specimens and no myzorhynchus is described or figured for the following species:—

P. minutum, Shipley and Hornell,
P. pammicrum, Shipley and Hornell,
P. blakei, Shipley and Hornell,
P. lactuca, Van Beneden,
P. thysanocephalum, Linton.

A neck is absent in *P. blakei*, Shipley and Hornell, long in *P. lactuca*, Van Beneden, *P. minutum*, Shipley and Hornell, and *P. thysanocephalum*, Linton.

The characters of the genus *Spongiobothrium*, Linton, are as follows :—

Body articulate, taeniaeform. Head separated from the body by a neck. Bothria 4, opposite, pedicelled, broken up into laciniocristate folds which are transversely costate. Unarmed. Auxiliary acetabulum none, terminal papilla none. Genital apertures marginal.

The genus *Phyllobothrium* would thus also appear to be closely related to the genus *Spongiobothrium*, from which it differs only in the absence of the cristate folds on the rostellum. It will be observed that the characters of the genera *Crossobothrium* and *Spongiobothrium* relate almost entirely to external features, and the anatomical details are few and unsatisfactory. It is highly desirable that such details should be worked out so that the true relationships of the genera could be determined. It is not impossible that subsequent research may suggest the desirability of regarding external features such as the presence or absence of a neck, or supplementary suckers, or of a myzorhynchus, as specific, rather than generic, characters.

LITERATURE.

Shipley and Hornell, *Ceylon Pearl Oyster Reports*, Part V, London, 1906.

Genus II. **Parataenia**, Linton, 1889.

Parataenia medusia, Linton, 1889.

Z.E.V. 6571/7	Intestine of *Hypolophus sephen.*	Main area, Chilka Lake, Nov. 29, 1914.	T. Southwell.

Only two species of *Parataenia* are known, viz. *P. medusia*, Linton, and *P. elongatus*, Southwell. The latter differs from the former in being ten times longer, in possessing a neck, and in the ripe segments being broader than long.

Linton's specimens of *P. medusia* measured 6 mm. long, but he observed that "they must grow somewhat longer than this." Our specimen (we obtained only one) measured 15 mm. In other respects it agreed with Linton's description.

LITERATURE.

Linton, Notes on Entozoa of Marine Fishes. *U.S. Fish Comm. Report for* 1887, pp. 862-866, plate xv, figs. 5-9 (Washington, 1891).

Southwell, *Ceylon Marine Biological Reports*, Part VI, p. 273, pl. iii, fig. 40 (Colombo, 1912).

Tribe *CALLIOBOTHRIINAE*, Perrier, 1897.

Genus **Calliobothrium**, Van Beneden, 1850.

Calliobothrium eschrichtii, Van Ben.

Z.E.V. 6572/7	Spiral valve of *Hypolophus sephen.*	Main area, Chilka Lake, Dec. 1914.	T. Southwell.

A single specimen is referred here with some hesitation. A definite identification was impossible because the spines were somewhat damaged. In other details it agreed with Van Beneden's species *eschrichtii*. The length of the worm is 9 mm. There were only about 13 segments the last of which measured 1 mm. long.

LITERATURE.

Linton, *U.S. Fish Comm. Report for* 1887, pp. 812-1816, figs. 5-12 (Washington, 1891).

Van Beneden, *Acanthobothrium eschrichtii*. *Bull. Acad. Belg.*, Ser. 2. Vol. XVI, p. 280.

Van Beneden, *Calliobothrium eschrichtii*. *Mem. Acad. Belg.*, Vol. XXV, 1850, pp. 145 and 195.

Family HYMENOLEPIDIDAE, Railliet and Henry, 1909.

Genus **Hymenolepis**, Weinland, 1858.

Hymenolepis breviannulata, Führmann, 1906.

Z.E.V. 6875/7	*Phalocrocorax javanicus* (the little Cormorant).	Chilka Lake, Dec. 1914.	T. Southwell.

Two specimens, 57 mm. long and 5 mm. broad. Rostellum with 20 hooks.

Führmann in his original description states "leider fehlt der scolex."

The genital openings are unilateral and are situated near the anterior extremity of the proglottid.

LITERATURE.

Führmann, *Centrbl. Bakter.*, I. Abt. Bd. XLI, Heft 4, pp. 445-446, fig 25, 1906.

Family TAENIIDAE, Ludwig, 1886.

Genus **Diploposthe** Jacobi, 1896.

Diploposthe laevis, Jacobi, 1896.

Z.E.V. 6871/7	*Nyroca ferina*. (the Pochard).	Chilka Lake, Nov. 1914.	T. Southwell.

A single specimen is referred here with some hesitation. No spines could be detected on the rostellum when the head was cleared in clove oil. Three testes were observed near the posterior border of the proglottides. The vesiculae seminalis is large. The cirrus is large, tubular, and armed with strong spines.

The female genital organs lie in the centre of the proglottid. No other details could be observed save that masses of eggs lay in what appeared to be a lateral extension of the uterus, close to the cirrus sac, and of about the same size as the latter.

The synonymy of this species is extensive. For a full account the reader is referred to Johnstone (i).

LITERATURE.

(1) Johnstone, T. H. H., On a re-examination of the types of Krefft's species of Cestoda in the Australian Museum, Sydney, Part I.
(2) Führmann, *Centrbl. Bakter.*, Vol. XL, 1906.
(3) Führmann, *Zool. Jahrb.*, Supplt. Bd. X, Heft I, 1908.

Anaporrhutum largum, Lühe.

Z.E.V. $\frac{6782}{7}$	*Hypolophus sephen.*	Chilka Lake, Stations 140-141, Nov. 1915.	T. Southwell.

This Trematode was found on the liver and in the coelom of the above ray. It is a transparent leaf-like form, and occasionally occurs in such numbers as to completely cover the outer surface of the liver. The fish was caught in the main area of the lake.

LITERATURE.

Lühe, in Herdman's *Ceylon Pearl Oyster Reports*, Pt. V, London, 1906.
Southwell, *Rec. Ind. Mus.*, Vol. IX, Pt. II, Calcutta, June, 1913.

XVIII. NOTES ON ORIENTAL DRAGON-FLIES IN THE INDIAN MUSEUM, No. 2.

By F. F. LAIDLAW.

LIBELLULINAE.

Genus **Amphithemis.**

See Ris, *Monogr. Libell.*, pp. 88-91 in *Coll. Zoolog. Selys*, Fasc. IX, 1909.

Anal loop feebly developed, containing few cells. Arculus lying between the second and third antenodal nerve. Costal side of discoidal triangle of fore-wing relatively long, broken. Radial supplementary sector moderately developed, median supplement not at all. Proximal side of discoidal triangle of hind-wing a little distal to the arculus. Three or four median nerves on the hindwing. Discoidal triangle *followed by one or two rows of cells*, increasing.

Colouration black with yellow or brick-red markings; (adult male of *A. vacillans*, Selys, with pruinescence on basal segments of abdomen). Abdomen rather long, slender and cylindrical; segment 8 of female without dilatation. Legs long, the hairs on femora rudimentary.

The discovery of *A. mariae* necessitates a slight modification in the definition of the genus shown in the character italicized.

This interesting genus is confined to S. India and to the Indo-Chinese Peninsula so far as is at present known; the S. Indian species *A mariae*, here described as new, being very distinct from its congeners. This distribution of a genus is somewhat unusual. Its nearest ally is *Pornothemis* from Sumatra and Borneo.

A. mariae is readily distinguished from the two remaining species by wing characters, the discoidal triangle in each wing being followed by a single row of cells, in *A. vacillans* and *A. curvistyla* by two rows. The Deccan species is moreover more brilliantly coloured, with bright yellow bands on the dorsum of the thorax.

The Burmese species are not so readily distinguished from one another. *A. curvistyla* is distinctly the smaller (hind-wing 18·22 mm. as opposed to 21·24 mm. in *A. vacillans*).

It has also much red on the abdomen, whilst *A. vacillans* has yellowish markings in the young male and female, its adult males having the body entirely black.

Males of these two species are well characterized by the anal appendages. In *A. vacillans* the upper pair are twice as long as the lower appendage and end in a fine upturned point, in *A. curvistyla* the upper pair are about equal to the lower appendage, rather stout, curved downwards and divaricate from one another.

Amphithemis vacillans, Selys.

2 ♂ ♂ 1 ♀ $\frac{1551}{20}$ $\frac{1552}{20}$ $\frac{1553}{20}$ Sibsagar, Assam.

These specimens were named by de Selys in whose handwriting are the labels. They are unfortunately in poor preservation, the best preserved specimen, an adult male, has the body entirely black. A male from the Abor Expedition collection, still more adult, has the second and third segments of the abdomen covered with bluish-white bloom.

Distribution: Burma, Assam.

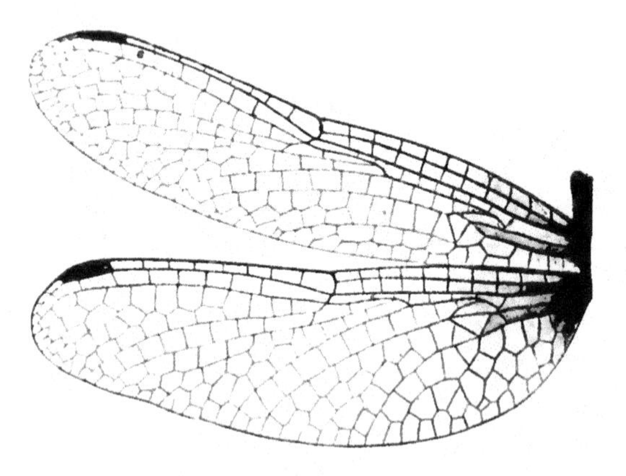

FIG. 1.—Wings of *Amphithemis mariae*, sp. n.

Amphithemis curvistyla, Selys.

1 ♂ 1 ♀ $\frac{1554}{20}$ $\frac{246}{6}$ Sibsagar, Assam.[1]

As with the last species the specimens were named by de Selys. Their condition is too bad to admit of a satisfactory examination.

Distribution: Burma, Tonkin, Assam.

Amphithemis mariae, sp. n.

4 ♂ 4 ♀ (in spirit.) $\frac{8260}{20}$ $\frac{8259}{20}$ $\frac{8280}{20}$ $\frac{8279}{20}$ $\frac{8281}{20}$ Forest tramway, mile 29-30, 1600 ft.; Parambikulam, 1790-3200 ft., Cochin State, 16—24-ix-14 (*F. H. Gravely*).

(For the photograph reproduced in text-figure 1, I am much indebted to Messrs. H. and F. E. Campion).

Types ♂ ♀ in the Indian Museum.

[1] On several labels of specimens from this locality de Selys wrote "Palone." It is not clear what he means.

Abdomen ♂ 19 mm., ♀ 19 mm. Hind-wing ♂ 20 mm., ♀ 21 mm.

Venation characters (see text-fig.).

Wings relatively shorter and broader than in the other species of the genus. Triangles and supratriangles normally uncrossed. Cubito-anal (median) space of hind-wing with three cross-nerves. Discoidal triangles in both wings followed by a single row of cells: supplementary radial sector feebly developed. Base of wings tinged with yellow to level of discoidal triangles.

♂ Lower lip cream-colour edged with black. Upper lip, post- and anteclypeus and vertical part of frons also creamy-yellow; upper surface of frons, vertex, and occiput metallic green.

Prothorax black.

Thorax dorsal surface black with a broad greenish yellow humeral stripe on either side; inter-alar space brick-red. Lateral surface brownish yellow with two well-defined dark bands on either side, ventral surface yellow.

Abdomen. Segments 1, 2, 3 brick-red, the last with a narrow black terminal ring. The remaining segments black, 4-7 with a basal yellow ring, most marked laterally, and progressively smaller from before backward.

Legs black, first pair of femora with a yellow line on their posterior side.

Anal appendages black, rather short, upper pair regularly curved downwards, moderately stout. Lower appendage a little shorter than the upper pair. Resemble in general the appendages of *A. curvistyla*, Selys.

Genital structures on segment 2. Anterior lobe very small, hamulus on either side with a fine backwardly directed spur. Lobe of segment 2 small, triangular, curved a little forward.

♀ Head coloured as in the male.

Prothorax pale yellow.

Thorax brownish black anteriorly, with a pair of very wide pale yellow ante-humeral bands, much larger than those of the male and united above at the base of the wings. The rest of the thorax is pale yellow in colour.

Abdomen. Segments 1, 2, 3 pale yellow, 3 with a fine black terminal ring, 4-8 yellow at the base, the apical half of the segment (two-thirds in 8) black; the yellow ring divided dorsally by a fine black line along the mid-dorsal carina. Segments 9, 10 entirely black.

Legs as in the male, anal appendages black. Lateral margins of segment 8 not widened. Valvulae vulvae very small.

Distribution: Southern Peninsular India.

XIX. HERPETOLOGICAL NOTES AND DESCRIPTIONS.

By N. ANNANDALE, *D.Sc., F.A.S.B., Superintendent, Indian Museum.*

(Plate xxxiii.)

Trionyx sulcifrons, sp. nov.

(Plate xxxiii, figs. 1, 1*a*, 2.)

The head is relatively small, triangular in shape and somewhat flattened; the tubular nostrils are relatively long and have a well-developed median longitudinal groove on the dorsal surface; the interorbital space is narrow and in fresh specimens the supercillary regions are raised.

A number of small longitudinal grooves originate between the eyes and, proceeding forwards, diverge on the forehead. These give an excuse for the specific name.

The disk is sub-circular, coarsely tuberculate in front of and behind the bony carapace; there is no dorsal ridge or groove, but a large prominence occurs on the bony carapace in front.

In the young (pl. xxxiii, fig. 2) the head is olivaceous, with a smallish yellow spot beneath each eye, and another rather larger one at the junction of the jaws; the following black linear marks occur on the dorsal surface—a sinuous line originating behind the lower part of each eye, proceeding upwards and then bending downwards and running along each side of the head to disappear on the nape, and a large Y-shaped mark situated in the middle of the dorsal surface some distance behind the eyes and connected somewhat indefinitely with diverging lines on the nape. In the adult these dark marks break up as shown in fig. 1 (p. 342) and perhaps disappear finally. The yellow spot at the junction of the jaws persists but its limits become somewhat indefinite. In the adult living animal the eyelids are reddish-brown and such dark marks as persist are bordered with a brighter shade of the same colour. The disk of the young bears (? 4 or) 5 relatively small ocelli, the ground-colour being dark olivaceous obscurely reticulated; there is a narrow yellow margin. In the adult the ocelli disappear and the whole disk becomes dark olivaceous green obscurely marbled with a paler shade.

The pupil of the eye (in the only living individual, an adult female, examined) was black and the iris dark olivaceous with a yellow ring internally.

The skull (pl. xxxiii, figs. 1, 1*a*) resembles that of *T. gangeticus* in general appearance, but is considerably smaller and narrower. The interorbital space is slightly concave and considerably narrower than either the nasal cavity or the orbit; the post-orbital arch is rather more than half as broad as the orbit and the post-orbital foramen remarkably small. The snout is longer than the orbit and distinctly declivous; it is more pointed than in *T. gangeticus*, but less so than in *T. hurum* and *T. leithii*. The symphysis of the lower jaw is long, equalling the orbit in length: the jaw itself is bluntly pointed; there are no ridges either longitudinal or transverse in this region; the two rami are more convergent than in *T. gangeticus*.

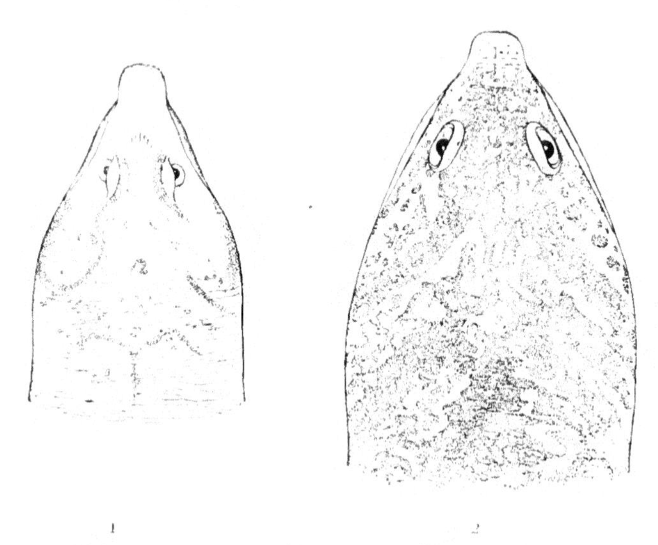

FIG. 1.—Head of *Trionyx sulcifrons* (from life), ½ nat. size.
,, 2.—Head of *Trionyx gangeticus mahanaddicus* (from type specimen), ½ nat. size.

The branchial skeleton resembles that of *T. gangeticus*,[1] but the greater cornua are more slender and the ceratobranchials stouter and shorter; the hypobranchials are distinct and show traces of segmentation into 2 or 3 pieces; this is, however, less marked than in *T. gangeticus*.

The margin of the bony carapace is concave in front and almost straight behind; the sculpturing of the posterior region is coarser than that of the anterior and near the posterior border there are small scattered bony tubercles. There are two or three neural plates between the first pair of costals. The plastron closely resembles that of *T. gangeticus*, but the hypoplastra and hyoplastra of the two sides apparently remain widely separated in the adult and all the bones are smaller.

[1] Annandale, *Rec. Ind. Mus.* VII, p. 159, fig. 1 (1912).

Type.—An adult female, which was examined alive and is now preserved as a skeleton in the Indian Museum (No. 17973: *Rept. Ind. Mus.*). The skin of the head is preserved in spirit.

I have also examined a slightly larger female (stuffed) and a young individual (in spirit) (pl. xxxiii, fig. 2), both the property of the Nagpur Museum.

The following are the measurements of the type:—

Disk.		Skull.	
Total length ...	407 mm.	Length	84 mm.
,, breadth	427 ,,	Breadth	54 ,,
Bony carapace (length) ...	335 ,,	Orbit	15 ,,
,, ,, (breadth) ...	331 ,,	Snout ...	24 ,,
		Interorbital width	11 ,,
		Nasal aperture (width) ...	13 ,,
		Postorbital arch...	8 ,,
		Mandibular symphysis	15 ,,

Distribution.—The type is from a tank in the town of Nagpur, the capital of the Central Provinces of India, as is also the adult specimen in the Nagpur Museum; while the young example in that museum is from a canal or stream at the same place.

This species is related to *T. gangeticus*, Cuv., the chief differences being (1) the presence of ocelli on the disk of the young and the absence of forwardly directed V-shaped markings on the head, (2) the more pointed snout, (3) the smaller postorbital foramen, (4) the longer symphysis of the lower jaw and the absence of a transverse ridge on its inner margin.

The eggs are small, the diameter being only 31 mm. in examples found ready for deposition in the type-specimen, which was killed in June. Another female killed at Nagpur was found to contain fully formed eggs in January.

I have to thank Mr. E. A. D'Abreu for the opportunity of making this very noteworthy addition to the herpetological fauna of India. He has also sent me for examination two specimens of the form I recently described as *T. gangeticus* subsp. *mahanaddicus*.[1] One is a skeleton of an adult slightly larger than the type (fig. 2, p. 342), while the other is a much smaller stuffed example. The localities are (?) Jubbulpore and Seonath R., Bilaspur district; both places being in the Central Provinces.

Trionyx leithii, Gray.

1915. Annandale, *Rec. Ind. Mus.* XI, p. 189, fig. 1.

In a recent paper I cast doubt on the occurrence of *T. leithii* in the Gangetic river-system, but I now take the earliest opportu-

[1] *Rec. Ind. Mus.* VII, p. 202 (1912).

nity to note that I have found in the old collection of the Indian Museum a number of young specimens from Allahabad and the River Hughli. There is no specimen of this species in the Nagpur Museum, though it occurs in the Central Provinces

Gonatodes bireticulatus, sp. nov.

(Plate xxxiii, figs. 3, 3*a*.)

Head small, ovate, moderately convex above; snout obtusely pointed, declivous, a little longer than the distance between the eye and ear-opening and more than twice as long as the eye; forehead grooved; ear-opening moderate, oval; 7 upper and 8 lower labials.

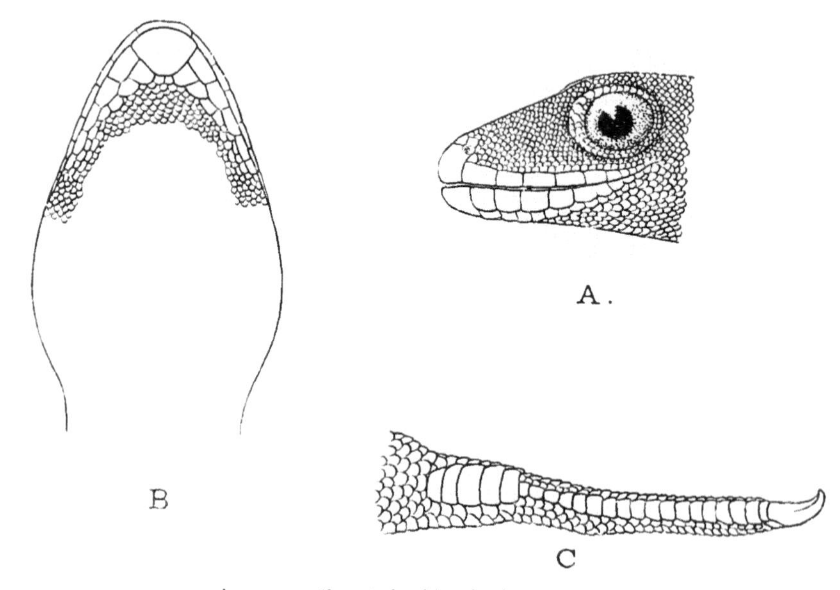

FIG. 3.—*Gonatodes bireticulatus*, sp. nov.
A. Snout in lateral view, × 3. B. Lower surface of head, × 3.
C. Lower surface of fifth toe, × 8.

Body and limbs moderate, the hind limb reaching the axilla; digits slender, basal joints not dilated and without transverse plates; five relatively large plates below the first articulation. Dorsal surface covered with conical keeled tubercles, which vary considerably in size, and are much smaller on the head than on the body; throat covered with similar tubercles; mental moderate in size, subtruncate posteriorly and followed by two small flattened scales placed transversely; several enlarged scales, which decrease in size from before backwards, on either side below the labials. Ventral scales small, leaf-shaped imbricate. Male with 7 femoral pores on each side. *Tail* cylindrical, tapering, covered above with small, oval, sub-imbricate, almost smooth scales, and below with flattened scales, the central row of which is distinctly enlarged.

Colouration.—Brown above, with a coarse black reticulation and, superimposed upon it, a much finer one of dotted white lines; two parallel white lines running backwards from the eye to above the ear; throat brownish, with a coarse, irregular white reticulation and with a white line running along each side: chest and abdomen brownish grey speckled with white. Ventral surface of tail greyish brown speckled with white; about 12 pale transverse bars on the dorsal surface. Fingers and toes with alternate brown and white bands.

Measurements.

Total length	..	.. 95 mm.
Length of tail	..	.. 50 ,,
Length of head	..	.. 15 ,,
Breadth of head	..	.. 9 ,,

Type.—No. 17970: *Rept. Ind. Mus.*

Locality.—In jungle at Kavalai, 1300-3000 feet, Cochin State (*F. H. Gravely*).

Numerous specimens of *G. wynadensis* (Beddome) and a few of *G. gracilis* (Beddome) were taken with the type, which is an unique specimen.

The species is closely related to *G. wynadensis*, but is distinguished from it (among other characters) by its colouration, by the larger number of femoral pores and by differences in the scaling of the feet.

Tropidonotus sancti-johannis, Boulenger.

1893. Boulenger, *Cat. Brit. Mus.* I, p. 230, pl. xv, fig. 1.

This snake, which seems to me to be a distinct species, has been recorded from several widely separated localities in Kashmir, the Himalayas and central India. I am not aware, however, that it has been found hitherto in the Malabar Zone. A typical specimen was obtained by Mr. F. H. Gravely at Chalakudi in the State of Cochin in September last.

Chirixalus simus, sp. nov.

Head large, broader than long; snout truncated, considerably shorter than the diameter of the orbit; canthus rostralis barely distinguishable; loreal region vertical, slightly concave; nostrils much nearer tip of snout than eye; interorbital region broader than upper eyelid, flat; tympanum about one-third the diameter of the eye.

Limbs.—Inner fingers with a very slight rudiment of a web; toes about two-thirds webbed; disks of fingers smaller than tympanum, slightly larger than those of toes; subarticular tubercles well-developed; a small and rather indistinct inner metatarsal tubercle. The tibio-tarsal articulation reaches the tip of the snout.

Skin.—Skin of head with small round scattered warts, of back nearly smooth: sides and throat with similar warts, abdomen and inner surface of thighs coarsely granular; a glandular fold extending from the supercilliary region to above the shoulder, and another, somewhat interrupted, from the gape to the same point.

Colouration.—Dorsal surface pale buff with several indistinct longitudinal dark lines and numerous scattered black specks. Ventral surface yellowish; throat and chest with minute black and white specks. Limbs without definite markings, inner surface of thighs reddish.

Measurements.

Total length of head and body ..	..	22 mm.
Length of head ..	..	.. 8 ,,
Breadth of head ..	..	.. 9 ,,
Length of hind limb	..	.. 32 ,,

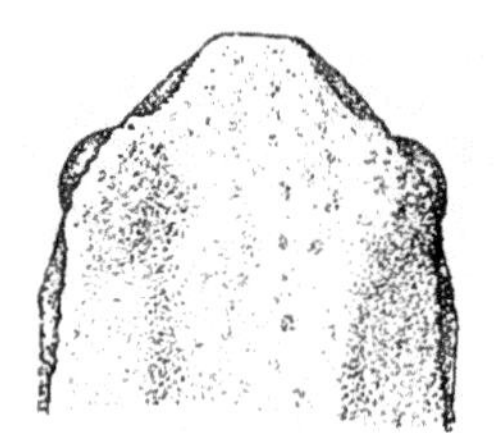

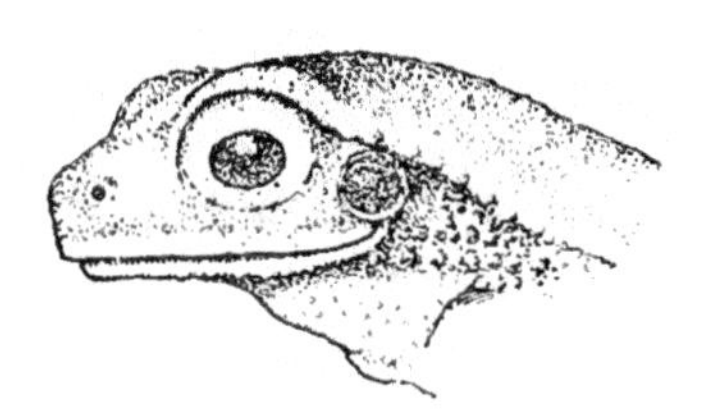

FIG. 4.—Head of *Chirixalus simus*, sp. nov., × 3.

Type.—No. 17971: *Rept. Ind. Mus.*: an unique specimen.

Locality.—Mangaldai, Assam north of the Brahmaputra (*S. W. Kemp*, 6-i-11).

This species differs from *C. doriae*,[1] the only other as yet known, in its larger head, truncated snout, smaller tympanum and rather longer hind legs, and in possessing a glandular fold between the eye and the shoulder.

C. doriae, which was described from Upper Burma, has recently been recorded from the Himalayan foot-hills[2] immediately to the north of Assam. The discovery of a second species near the base of the same hills is therefore interesting.

Ichthyophis glutinosus var. tricolor, Annandale.

1909. Annandale, *Rec. Ind. Mus.* III, p. 186.

Two specimens of this variety or local race were found by Mr. F. H. Gravely on the eastern slopes of the Western Ghats in Cochin in September last, the exact locality being Parambikulam

---

[1] Boulenger, *Ann. Mus. Genova* (2) XIII, p. 341, pl. x, figs. 5, 5*a* (1893).

[2] Annandale, *Rec. Ind. Mus.* VIII, p. 18 (1912).

(alt. 1700-3200 ft.). The specimens are considerably larger than the type, one of them being 280 mm. long. The yellow lateral band on each side is separated from the white median ventral band by a dark one, which is greyish in spirit. This dark band varies considerably in breadth.

Ichthyophis monochrous (Bleeker).

1912. Boulenger, *Faun. Malay Penin., Rept.*, p. 286.

Boulenger notes (*op. cit.*) that this species has not been recorded from Ceylon, but there are two specimens from that island in the collection of the Indian Museum. They were taken some years ago at Pattipola in the hills of the Central Province (alt. *ca.* 6000 feet) by Mr. F. H. Gravely.

ADDENDA.

Since these notes went to the press I have received specimens of three species of Chelonia from Mr. W. Lancelot Travers, who obtained them near Baradighi in the Jalpaiguri district of Bengal. The same gentleman had already sent me examples of two others from the same locality. As our knowledge of the Chelonia of northern Bengal is still far from complete, this little collection is of considerable importance. It includes the following species:—*Chitra indica* (several young specimens), *Emyda granosa* (a half-grown specimen of the typical form), *Testudo elongata* (one young specimen), *Geoemyda tricarinata* (one adult), *Geoemyda indopeninsularis* (a large male).

It is of particular interest to find that the range of *T. elongata* actually extends, as Anderson thought probable, along the sub-Himalayan tract to the west of Assam, and that *G. indopeninsularis* occurs north of the Ganges.

The specimen of the latter species agrees well with the male type.[1] The shell is actually deeper as a whole than in *G. trijuga* var. *edeniana*, but the bridge has relatively a much smaller vertical depth. The specimen from Assam referred doubtfully to *edeniana* (*op. cit.*, pp. 69, 70) should probably be assigned to *G. indopeninsularis*, in spite of its broad second vertebral shield; it is much smaller than the other three in the collection.

[1] *Rec. Ind. Mus.* IX, p. 71, pl. v, fig. 2.

EXPLANATION OF PLATE XXXIII.

All the figures on this plate are from untouched photographs of natural size.

FIGS. 1, 1*a*, 2.—*Trionyx sulcifrons*, sp. nov.

1.—Skull of type-specimen as seen from above: 1*a*.—Lower jaw of the same specimen.

2.—Young specimen preserved in spirit.

FIGS. 3, 3*a*.—*Gonatodes birceticulatus*, sp. nov.

3.—Type-specimen from above: 3*a*.—Same specimen from below.

Photo by S. C. Mondul.

1, 1a, 2. TRIONYX SULCIFRONS. 3, 3a. GONATODES BIRETICULATUS.

MISCELLANEA.

HYDROZOA.

A Short Note on *Hydra oligactis*, Pallas.

On the occasion of a recent visit to Ludhiana (Punjab) I found a few specimens of *Hydra oligactis*, Pallas, in a small pond full of the pond-weed *Potamogeton pectinatus*, Linn. One of these specimens of *Hydra* was rather peculiar in having seven tentacles. Dr. Annandale in his account (*Fauna of British India, Freshwater Sponges, Hydroids and Polyzoa*, p 159) says that he has not seen any Indian specimen with more than six tentacles, while quite a large number of specimens that I have examined from Lahore and Ferozpore had usually four, and in exceptional cases five tentacles. The manner of capturing food was also observed, it exactly corresponds to Dr. Annandale's account (*Fauna*, p. 152) of *Hydra vulgaris* phase *orientalis* Annandale. The food consisted of very young individuals of the Aphis *Siphocoryne nymphae* which was infesting the plant in large numbers.

BAINI PARSHAD, *B.Sc.*,
Alfred Patiala Research Student, Zoological Laboratory.
Government College, Lahore.

BATRACHIA.

The larva of *Rhacophorus pleurostictus*, Boul. (*Fauna*, p. 479.)

The tadpoles, which were collected in Coorg, presented some difficulty in the matter of identification. This was due to the absence of any four-legged forms in the collection; but Dr. N. Annandale, who had received a fine collection of tadpoles from Cochin, has by a process of exclusion identified them as the larvae of *Rhacophorus pleurostictus*; as he has pointed out, the character of the feet at once excludes these larvae from the genus *Rana*.

The head and body are moderately flattened above and broadly oval, ventrally convex. The snout is rounded. The length of the body is to the breadth as 7 : 5. The body is finely pitted above, perfectly smooth below; but in specimens in which the hind limbs have not sprouted it is smooth above as well as below. Two conspicuous oval parotoids are present

The eye and nostril are both small, dorsally placed, by no means prominent. The nostril is very small, directed almost anteriorly, equidistant between the eyes and the tip of the snout. The internasal space is twice the interorbital

The mouth is subterminal, small. Its greatest width is only slightly greater than the interorbital space (as 7 : 5). The lower lip is strongly developed, directed backwards; the upper is not prominent. The distribution of tubercles varies; they are generally absent or only sparsely present on the upper lip, while stronger ones fringe the corners and the lower lip. The dental formula is liable to vary. It may probably be expressed thus:—

$$2-3:5-7+5-7 \mid 5-7+5-7:4-6.$$

The uninterrupted tooth rows on both the lips are longest.[1] The lower jaw is V-shaped with granulate or dentate edge. The upper beak is broadly crescentic. Almost every specimen in the

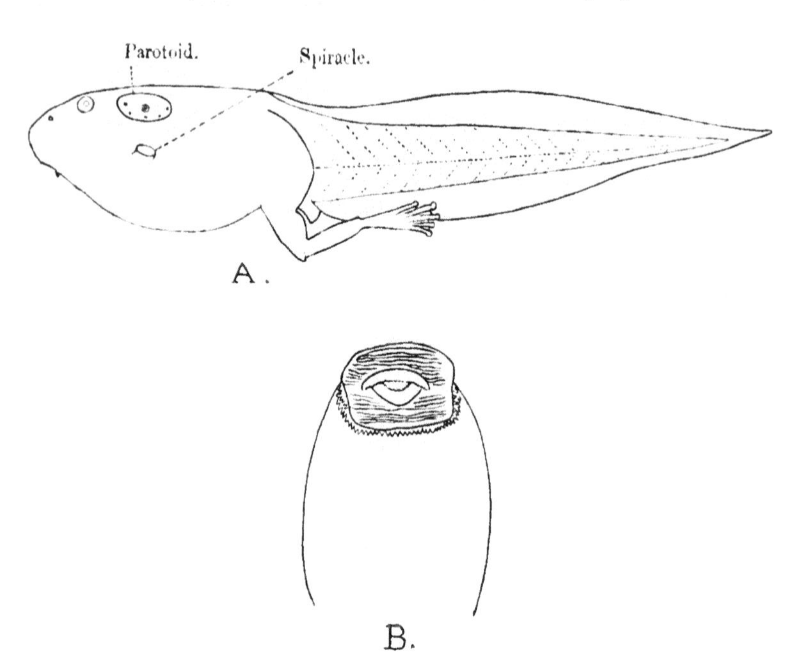

FIG. 1.—Tadpoles of *Rhacophorus pleurostictus*.
A. Lateral view. B. Mouth, showing the beak, tooth-rows and tubercles.

collection shows the lower jaw to be cornified, but the upper beak is very fully developed. No glandular swellings are present at the corners of the mouth.

Skin and glands.—In specimens in which the hind limbs have not developed, the skin is smooth and no whitish glandular pits occur. When the hind limbs have grown, the dorsal surface is beset with numerous cutaneous glands, which are distributed over the head as well. Two most conspicuous large oval parotoids, twice the diameter of the eye, are present; they may be

[1] In the specimens in my collection I could discover no horny teeth, only bare ridges; but Dr. Annandale, to whom examples have been sent, states that he has found patches of small teeth in a few.

yellow or darker in colour. One or two irregular rows of white round glandular swellings exist on both the caudal crests.

Spiracle.—Sinistral, large, very slightly tubular; opens just below the parotoid; nearer to it than to the eye. The opening is directed slightly upwards, not visible from above or below, quite as large as the eye.

Vent.—Tubular, median or slightly dextral, quite as large as the spiracular opening.

Tail.—Gradually pointed or only slightly rounded at the tip. The muscular portion is very strongly developed, the membraneous crests are thin and transparent. The total length of the tail is nearly 1½ times the length of the body and head, and the muscular part is only ½ the total width. Both crests are strongly convex and of equal depth.

Colouration.—The young tadpoles are nearly transparent and the parotoids are bright yellow with a dark central spot. A few dots occur on the tail. In older specimens, the body is dull grey (slightly bluish in spirit specimens) with more numerous blotches on the back and the muscular part of the tail. Generally there is a ring of small dots with a bigger one in the centre on the parotoids. The central surface is dirty white, in most specimens immaculate.

Dimensions.—The measurements of (A) an individual in which the hind limbs have not sprouted, and of (B) an individual in which they have fully grown are as follows:—

	A.	B.
Total length	52 mm.	92 mm.
Length of head and body ..	24 mm.	36 mm.
Length of tail	28 mm.	56 mm.
Maximum breadth of body ..	13 mm.	25 mm.
Maximum depth of body ..	11 mm.	20 mm.
Maximum depth of tail ..	10 mm.	16 mm.

Biological.—These tadpoles occur in abundance in tanks in the vicinity of houses where fish are reared. The bottom of the tanks being more are less clayey, the tadpoles can hardly be made out in the water. Water snakes destroy them in large numbers.

C. R. NARAYAN RAO.

BIRDS.

An Albino Bulbul.

A very fine specimen of an Albino Bulbul, *Molpastes burmanicus*, has recently been sent to the Museum from Mr. A. H. Ricketts.

The bird was captured when only just commencing to learn to fly, and was at that time wholly white, with the typical white bill, feet and claws and bright red eyes of a true albino.

After being about six months in captivity this bird acquired the normal brilliant crimson colouring on the under tail coverts and (in the skin) a faint shade of reddish or buffy brown may be noticed both on the head and the rectrices. At the same time the bill and claws are still entirely colourless showing that, as a whole, there was no probability of increase in the pigmentation. Mr. Ricketts records that when about a year old the bird developed epileptic fits and died.

This specimen is very interesting not only in that it possesses one patch of most brilliant normal colouration in spite of the rest of the plumage remaining that of a true albino, but also in the fact that the faint tinge of colouration elsewhere discernible is buff or reddish.

Reds and yellows are the most volatile of all colours and in skins of birds exposed to sun and weather the first colours to evaporate are the yellows and then the reds, yet we find in this bulbul, as in many albino snipes, etc., the buff persisting to some extent to the exclusion of the far more permanent browns whilst the one vivid colour retained is crimson.

The conjunction of epileptic characteristics with albinism is also worthy of note as the same is known to obtain in human beings and other animals.

E. C. Stuart Baker.

XX. THE LARVAE AND PUPAE OF SOME BEETLES FROM COCHIN.

By F. H. GRAVELY, *M.Sc., Assistant Superintendent, Indian Museum.*

(Plates xx-xxi).

I Cucujidae—*Uleiota indica*, Arrow.

(Plate xxi, figs. 13-19).

The specimens on which the following descriptions are based were found by Mr. B. Sundara Raj under bark at Parambikulam, 1700-3000 ft.

The adult agrees with the description [1] of the species to which I have referred it in every detail, except that the third joint of the antenna is slightly shorter instead of longer than the succeeding ones. In this respect, however, I find it to be in agreement with cotypes from Kanara, presented by Mr. H. E. Andrewes to our collection and to that of the Agricultural Research Institute at Pusa, and with others which Mr. Andrewes very kindly sent me for examination.

LARVA.

The larva of *U. indica* is whitish in colour, and closely resembles larvae of other species of the genus in general appearance.

The antennae arise from collar-like sockets which Perris (see "Larves de Coléoptères," p. 61) has supposed to represent a distinct segment, making four in all. The first segment beyond this is about half as long as the second, which bears a minute conical process on the inner side of its distal end and is slightly longer and much stouter than the third.

Immediately behind the base of each antenna are five ocelli. Normally four of these appear to be arranged in a row, with the remaining ocellus immediately behind the middle of the space between the upper two. But on one side of one specimen the solitary ocellus is in front of the space between the lower two members of the row.

The apex of the mandibles is strongly bidentate, and is followed by a row of about four small teeth on the inner edge.

The blade of the maxilla is strongly fringed at the apex. The three joints of the maxillary palps are of about equal length, but the

[1] *Trans. Ent. Soc.* 1901, pp. 599-600.

third is much slenderer than the other two. The terminal joint of the two-jointed labial palps is slightly longer and slenderer than the basal.

The anterior margin of the first tergite is convex, overlapping the back of the head. The posterior margins of all segments are straight, both above and below. All segments are distinctly broader than long. The two joints of the appendages of the eighth abdominal segment are distinct as in *U. planatus*, the basal joint being stout and the distal spiniform. The three joints of the appendages of the ninth abdominal segment are more or less completely fused as in *U. serricollis.*

U. serricollis is a Ceylonese species and its larva appears to resemble that of *U. indica* more closely than does any other larva yet described.

The larvae at present referred to the genus *Ulciota* may be distinguished from one another as follows:—

1.	Appendages of ninth abdominal segment two-jointed, not spiniform; only one ocellus on each side ..	*Gernet's undetermined larva.*
	Appendages of ninth abdominal segment spiniform, joints three in number when not all fused together; several ocelli on each side ..	2.
2.	Apex of mandibles bidentate; all three joints of maxillary palps equally distinct	3.
	Apex of mandibles tridentate; basal joint of maxillary palps very short and obscure (ocelli 3 + 2 on each side)	*U. crenatus.*
3.	Ocelli 4 + 2 on each side; appendages of ninth abdominal segment distinctly jointed	*U. planatus.*
	Ocelli 4 + 1 on each side; appendages of ninth abdominal segment rigid throughout	4.
4.	Appendages of eighth abdominal segment two-jointed (*i.e.* the terminal spine articulated, not fused, to the basal part); seventh abdominal segment wider than long	*U. indica.*
	Appendages of eighth abdominal segment rigid; seventh abdominal segment longer than wide ..	*U. serricollis.*

PUPA.

The pupa is white in life, and is very like that of *U. serricollis.* The antennae are much shorter than in the pupa of *U. serricollis*

(? in both sexes). They are ornamented with fleshy processes, of which the larger are placed in circlets round the ends of the developing segments of the antennae of the adult, and the smaller round the middle of each of these segments except the long basal one on which they are more numerous. The abdomen is armed on either side with a series of long, fleshy, more or less forwardly-directed processes, on to the end of each of which a large and more or less backwardly-directed spine is articulated.

A considerable number of Cucujid life-histories have already been worked out wholly or in part, and the following is a list of the descriptions known to me.

Key for the determination of genera of Cucujid larvae.

P. de Peyerimhoff, *Ann. Soc. Ent. Fr.*, LXXI, 1902 (1902-3), pp. 717-8.

Catogenus rufus, Fabr.

* G. Dimcock, *Psyche*, III, pp. 341-2.
* W. F. Fiske, *Proc. Ent. Soc. Washington*, VII, p. 90.

Prostomis mandibularis, Fabr.

W. F. Erichson, *Arch. Naturg.*, 1847, pp. 285-6.

Chapuis and Candèze, "Catalogue des Larves des Coléoptères", *Mem. Soc. R. Sci. Liége*, VIII, 1853, p. 425.[1]

J. Curtis, *Trans. Ent. Soc. London* (n.s.) III, 1854-6, pp. 37-39, pl. v, figs. 23-24.

E. Perris, "Larves de Coléoptères", Paris, 1877, p. 56.

Cucujus clavipes, Fabr.

* Wilson, *Bull. Brooklyn Soc.*, I, p. 56.

Cucujus coccinatus, Lewis.

A. S. Olliff, *Cist. Ent.*, III, 1882-5, pp. 59-60, pl. iii, fig. 7.

Cucujus haematodes, Erichs.

* W. F. Erichson, *Naturg. Ins. Deutschl.*, III, 1845, p. 310.

H. Assmann, *Stett. Ent. Zeit.*, XII, 1851, p. 352, pl. ii, figs. C-D.

Chapuis and Candèze, *Mem. Soc. R. Sci. Liége*, VIII, p. 426, pl. ii, fig. 8 (figure reproduced in Lefroy's "Indian Insect Life," p. 301).

---

* Papers marked thus are not available in Calcutta.

[1] Apparently = p. 85 of reprint (see Perris, "Larves de Coléoptères", p. 56).

Platisus integricollis, Reitter.

A. M. Lea, *Proc. Linn. Soc. N.S. Wales*, XXIX, 1904, pp. 88-9, pl. iv, fig. 6.

Inopeplus praeustus, Chevol.

P. de Peyerimhoff, *Ann. Soc. Ent. Fr.*, LXXI, 1902-3, pp. 715-8, 3 text-figs.

Uleiota[1] **crenata**, Payk.

F. B. White, *Ent. Mo. Mag.*, VIII, 1871-2, pp. 196-8.
E. Perris, "Larves de Coléoptères", pp. 60-62.

Uleiota[1] **planata**, Linn.

* W. F. Erichson, *Naturg. Ins. Deutschl.*, 1846, p. 332.
Chapuis and Candèze, *Mem. Soc. R. Sci. Liége*, VIII, 1853, pp. 428-9.
E. Perris, *Ann. Soc. Ent. Fr.* (3) I, 1853, pp. 621-626, pl. xix, figs. 127-137 (2 figs. reproduced by Sharp, Camb. Nat. Hist., Insects, pt. ii, p. 234, fig. 115), and "Larves de Coléoptères", pp. 57-59.

Uleiota[1] **serricollis**, Candèze.

M. E. Candèze, *Mem. Soc. R. Sci. Liége*, XVI, 1861, pp. 341-343, pl. ii, figs. 1-1*c*.

? **Uleiota**[1] sp.[2]

C. v. Gernet, *Horae Soc. Ent. Ross.* VI, 1869, pp. 3-6, pl. i, figs. 7-7*g*.

Laemophloeus ater, Oliv.

J. O. Westwood ("*Cucujus spartii*": see Perris, "Larves de Coléoptères", p. 60, concerning this synonymy), "Introduction to the Classification of Insects" I, pp. 149-150, fig. 12 (19).
E. Perris, "Larves de Coléoptères", p. 62.

Laemophloeus bimaculatus, Payk.

E. Perris, "Larves de Coléoptères", p. 62.

Laemophloeus clematidis, Erichson.

E. Perris, "Larves de Coléoptères", p. 62.

---

* References marked thus are not available in Calcutta.

[1] Or *Hyliota* = *Brontes*, incl. *Dendrophagus*; see Arrow, *Trans. Ent. Soc.* 1901, p. 593.

[2] Not *U. crenata*; see White, *Ent. Mo. Mag.*, VIII, 1871-2, p. 198. The larva was not reared, and White thought it could not belong to the genus *Uleiota* at all. But it has all the distinctive characters of the larvae of this genus given in Peyerimhoff's key.

Laemophloeus dufouri, Laboulbène.

E. Perris, *Ann. Soc. Ent. Fr.* (3) I, 1853, pp. 618-621, pl. xix, figs. 122-6.

Laemophloeus ferugineus, Stephens.

Carpentier, *Bull. Soc. Linn. nord France*, 1877, pp. 239-241.
H. S. Olliff, *Entomologist*, XV, 1882, pp 214-5.

Laemophloeus hypobori, Perris.

E. Perris, "Larves de Coléoptères", p. 62.

Laemophloeus juniperi, Grouvelle.

F. Decaux, *Bull. Soc. Ent. Fr.*, 1890, pp. cxxv-cxxvi.

Laemophloeus monilis, Fabr.[1]

* Bellevoye. *Bull. Soc. Metz* (2) XIV, 1876, pp. 183-9.

Laemophloeus testaceus, Fabr.

E. Perris, "Larves de Coléoptères", pp. 59-60, pl. ii, figs. 43-45.

Lathropus sepicola, Müller.

* E. Perris in Gobert's Cat. Col. Landes, fasc. 3, p. 122, and "Larves de Coléoptères", pp. 62-65, pl. ii, figs. 46-53.

Pediacus dermestoides, Fabr.

E. Perris, *Ann. Soc. Ent. Fr.* (4) II, 1862, pp. 190-2, pl. v, figs. 535-543.

Prostominia convexiuscula, Grouvelle.

P. de Peyerimhoff, *Tran. Linn. Soc. London* (2 Zool.) XVII, 1914, pp. 156-159, figs. A.-F.

Silvanus advena, Waltl.

E. Perris, "Larves de Coléoptères", pp. 65-68.

Silvanus surinamensis, Linnaeus.[2]

J. O. Westwood, "Introduction to the Classification of Insects" I, p. 154, fig. 13 (10-12).

---

* References marked thus are not available in Calcutta.
[1] = *denticulatus*, Preyssl. (Munich Catalogue).
[2] The larvae figured by different authors are not all alike, and it scarcely seems possible that all of them can belong to one species.

J. F. J. Blisson (*S. sexdentatus*), *Ann. Soc. Ent. Fr.* (2) VII, 1849, pp. 163-172, pl. vi, fig 1.

C. Coquerel (*S. sexdentatus*), *Ann. Soc. Ent. Fr.* VII, 1849, p. 172.

F. H. Chittenden, *U. S. Agric. Ent. Bull.* (n.s.) 4, 1896, pp. 121-2, figs. 59 *a-d* (figure of larva reproduced with new figure of adult in Fletcher's "South Indian Insects", p. 290).

* Jablonouski, *Termes. Kosl.*, 1899, pp. 126-130, text-figs.

* J. Curtis, "Farm Insects", Lond., 1883 (figure reproduced in *Ind. Mus. Notes* III [3] p. 120).

Lefroy, "Indian Insect Life", pp. 300-301, text-figs. 179-180.

Silvanus unidentatus, Fabr.

E. Perris, *Ann. Soc. Ent. Fr.* (3) I, 1853, pp. 627-633, pl. xix, figs. 138-143.

E. Perris, "Larves de Coléoptères", p. 65.

? Nausibius dentatus, Marsh.

J. O. Westwood, "Introduction to the Classification of Insects" I, pp. 153-4.

II. Lycidae—*Lyropaeus biguttatus*, Westwood,[1] and some "Trilobite Larvae."

(Plate xx, figs. 1-12).

Larvae, pupae and an adult of this species were found clustered together on the under side of a large slab of stone, which was resting on other stones in such a manner as to leave a clear space above the ground beneath it. The pupae hung head downwards from the mid-dorsal fissure of the cast larval skins, which remained unshrivelled on the stone in the positions taken up by the larvae prior to pupation.

Adults were obtained in Cochin at altitudes varying from the level of the base of the hills to two or three thousand feet above the sea, and there is one specimen in our collection from the Nilgiris. The distribution of black pigment is very variable, and the black spots on the elytra are often absent. A specimen from Nedumangad in Travancore, determined by Bourgeois himself as *L. aurantiacus*, Bourgeois,[2] evidently belongs to the same species; and *L. aurantiacus* may therefore be regarded as a synonym of *L. biguttatus*.

LARVA.

The larva is flattened as a whole, and is of a blackish brown colour.

* References marked thus are not available in Calcutta.

[1] *Ann. Mag. Nat. Hist.* (5) V, 1880, p. 213.

[2] *Ann. Soc. Ent. Fr.* LXXVII, 1908-9, pp. 503-4.

The head can be retracted into a tubular pouch opening below the anterior margin of the prothorax, and the short thick antennae can be retracted into the head. The almost globular termination of each antenna is ornamented with more or less labyrinthine markings. The mandibles are small and are inserted in the middle line as in other Lycid larvae. They are very slender and project almost vertically downwards as a whole, but are directed slightly backwards basally and forwards distally, being lightly curved throughout. Their extremities rest in grooves on the upper surfaces of the somewhat fleshy blades of the maxillae, and as the mandibles are rather long they press the maxillae downwards till they too project almost vertically. The maxillary palps are three-jointed (excluding the basal support), and the labial palps two-jointed; both have the form of a slender cone.

The pronotum is roughly triangular, nearly as long as wide, truncate in front, and slightly rounded at the two posterior angles. The mesonotum and metanotum are roughly rectangular, slightly more than twice as wide as long, with the anterior angles somewhat obtuse and the posterior somewhat acute, especially those of the metanotum. Equally well developed spiracles are present on the mesothorax and metathorax.

The first eight abdominal tergites are much alike. The anterior ones are somewhat, and the posterior ones much, narrower than the thoracic segments, and all are very much shorter. Each is produced laterally into a simple stout backwardly-curved process. The terminal abdominal segment is somewhat longer than the segments immediately in front of it, being little more than twice as wide as long.

The abdominal sterna bear a pair of small conical processes on their posterior margins. These processes are more distinct on the posterior than on the anterior segments, and bear a tuft of bristles on the last two. The sternum of the terminal segment is without these processes, and bears the sucker-like anus.

PUPA.

The pupa is white in life, but the preserved specimens have become brownish.

The pronotum is quadrangular with almost straight sides; it is broader behind than in front, and even in front is broader than long. It does not overlap the head, which is bent downwards.

Each of the first three abdominal segments bears on either side above the stigma an elongate simple process with conical base, and below it a similar but moniliform (? jointed) process. The five following segments bear only a pair of conical processes above the stigmata, those of the first of these segments being the smallest. The terminal segment bears a pair of much slenderer processes.

The appendages are smooth, and not distinctly segmented.

"TRILOBITE LARVAE."

The *Lyropaeus* larva described above belongs to the group known as "Trilobite Larvae." The "Trilobite Larvae," which have hitherto attracted most attention, have been of extraordinarily large size, and the group has been a puzzle to entomologists ever since Perty described his *Larva singularis* in 1831. The following references to "Trilobite Larvae" are known to me:—

*1831. Perty, M. "Observationes Nonnullae in Coleoptera Indiae Orientalis", p. 33, pl. i, figs. 8-9.

1839. Westwood, J. O. "Introduction to the Classification of Insects" I, p. 254, figs. 27 (1) and 28 (1).

1841. Erichson, W. F. "Zur systematischen Kenntniss der Insectenlarven." *Arch. Naturg.*, VII, pp. 91-92.

1861. Candèze, M. E. "Histoire des Metamorphoses de quelques Coléoptères exotique." *Mem. Soc. R. Sci. Liége*, XVI, 1861, pp. 358 (apparently p. 34 in reprint) and 403-4, pl. vi, fig. 12.

1887. Kolbe, H. J. "Ueber einige exotische Lepidopteren- und Coleopteren-Larven, (6) Perty's '*Larva singularis*'." *Ent. Nachr.*, III, pp. 37-39.

1887. Lucas, M. H. *Bull. Soc. Ent Fr.*, 1887, pp. xxxv-xxxvii, reprinted in "Mission Pavie Indo-Chine 1879-1895", 1904, pp. 104-5.

1898. Gahan, C. J. "Dipeltis a Fossil Insect?" *Nat. Sci.* XII, pp. 42-44, 2 text-figs.

*1899. Bolivar, I. "Anomalous Larvae from the Philippines." *Act. Soc. Espan.* 1899, pp. 130-133, text-figs.

1899. Bourgeois, J. "Description de deux larves remarkables appartenant probablement au genre *Lycus*." *Bull. Soc. Ent. Fr.*, 1899, pp. 58-63, 2 text-figs.

1899. Sharp, D. "On the Insects from New Brittain," Willey's Zool. Results, p. 383, pl. xxxv, figs. 4-4*b*.

1899. Sharp, D. Cambridge Natural History, Insects, pt. II, p. 251.

1900. Hanitsch, R. "An Expedition to Mount Kina Balu, British North Borneo." *J. Straits R. Asiatic Soc.* No. 34, pp. 77-79.

1901. Shelford, R. "Notes on Some Bornean Insects." *Rep. Brit. Ass.*, 1901, pp. 690-691.

1908. Gahan, C. J. "Lampyridae from Ceylon." *Proc. Ent. Soc. London*, 1908, p. xlviii.

1913. Gahan, C. J. "On some Singular Larval Forms of Beetle to be found in Borneo." *J. Sarawak Mus.* I, pp. 61-65, 3 text-figs.

Perty thought his *Larva singularis* was to be ascribed to a Necrophagous rather than to a Malacodermatous insect; but Westwood disagreed with him, and suggested that it belonged

* Papers marked thus are not available in Calcutta.

rather to some species of *Lycus*. To this genus—which has since been subjected to extensive subdivision—he was also inclined to refer the slender parallel-sided insect of the "Trilobite" group, which he was the first to notice and figure.

Erichson accepts these insects as Malacoderms, but in spite of their weak mandibles regards them, because of their shape and because the head is completely retractile, as Lampyrids rather than Lycids. Candèze agrees with Erichson; but Kolbe returns to Westwood's view, and even goes so far as to suggest that the specimens which were sent to him were probably the larvae of *Lycus* (*Lycostomus*) *melanurus*, Blanchard.[1] The opinions of other authors are similarly divided.

Gahan (1913) favours Lycidae, but does not think the insects can belong to the genus *Lycus*, as they are very unlike the authenticated larvae of that genus. He thinks it more probable that they belong to some genus in which only the male—perhaps not even the male[2]—is winged. Further, he points out that the known distribution of "Trilobite Larvae" corresponds to that of the genus *Lyropaeus*, of which only males are known to him; and he suggests an association with this genus. His conclusion is in a measure confirmed by the above observations on the development of *Lyropaeus biguttatus*, and it is noteworthy that all the winged specimens that I have seen are males.

The larvae which give rise to these winged insects are, however, not particularly large, and throw no certain light on the status of the much larger insects with which the name "Trilobite Larvae" is more particularly associated. Two large insects of the "Trilobite" type were also, however, found in the Cochin forests. These are figured on pl. xx, figs. 9-12.

One of them (pl. xx, figs. 9-10) is very like the larvae found to develop into males of *Lyropaeus biguttatus*. The principal differences are the presence of more definite tubercles at the angles of the thoracic terga in the former than in the latter; the paler colour of the upper surface; and the yellow colour of the legs and sterna and of the lower surface of the lateral extensions of the terga, which contrast strongly with the black pleural structures. These, however, are features which may well be acquired only as maturity is approached. The specimen is not nearly so large as many species are known to become, and dissection has shown it to be immature; but it may perhaps represent a stage in the development of the female of *Lyropaeus biguttatus*, a female which in that case will almost certainly prove to be larviform.

The other specimen of "Trilobite Larva" found in Cochin (pl. xx, figs. 11-12) is slightly smaller, is black in colour, and is ornamented with more numerous and more elaborate tubercles and

[1] Authenticated larvae of this species have since been briefly described by Shelford (*Rep. Brit. Ass.*, 1901, p. 690). They do not appear to be of the "Trilobite" type, and are only 25 mm. long when full grown.

[2] See also Shelford's comment on a previous note by Gahan (*loc. cit.* 1908)

papillae, and appears to have shorter mandibles as these do not press the maxillae downwards and so are completely hidden. It differs greatly in this way from the larvae of *Lyropaeus biguttatus*, and need not be further discussed here.

Another South Indian species is represented in our collection by a dried specimen whose head, prothorax and legs are missing. It is transitional in character between the two preceding, resembling the former in colour, but having a double row of rudimentary tubercles down the back, and rudimentary tubercles on the abdominal epimera and episterna. It may represent a further stage in the development of that species; or it may be more nearly allied to a series of smaller larvae from Naduvotam (Nilgiris, 7000 ft.) which are preserved in the collection of the Agricultural Research Institute, Pusa, whence two specimens have been presented to our collection. It closely resembles these larvae in structure, but in them the yellow on the lower surface is confined to the anterior part and lateral angles of the prothorax, the anterior parts of the mesosterum and metasternum near the middle line, the abdominal sterna, and the bases of the legs.

The occurrence in the Pusa collection of a male insect from Naduvotam, belonging to the *Lyropaeus*-like genus *Calochromus*, suggested the possibility that this might be an adult of the species to which the "Trilobite Larvae" from that locality belonged. *Calochromus* is placed by Bourgeois (*Ann. Soc. Ent. Fr.* XI, 1891, p. 348) in the *Lygistopterus* group of genera, which immediately precedes in his system the *Dilophotes* group containing *Lyropaeus*[1]; and the larva of *C. melanurus* which has been briefly described by Shelford (*Rep. Brit. Ass.*, 1901, p. 690) appears to be of the "Trilobite" type. Males of *Calochromus* are much more numerous than females among the few specimens I have examined; but this may be due to their being more active, and females undoubtedly occur in some species. It is, however, possible, that some species of the genus may have large larviform females, or even that winged and larviform females may occur together in some or all species.

Our collection contains, in addition to the above South Indian specimens of the *Lyropaeus* or broad type of "Trilobite Larva", specimens of this type from the following localities:—

Ceylon: Peradeniya (? two species[2]).
Bengal: Chittagong—Rangamatti.
Burma: Sadon (Myitkyina Dist.); Pegu.
Malay Peninsula: Lankawi; Singapore.
Philippines.

---

[1] The genera *Calochromus* and *Lyropaeus* are, however, placed almost at opposite ends of the family by Westwood (*Trans. Ent. Soc. London*, 1878, pp. 96 and 104-5, and "Illustrations of Typical Specimens of Coleoptera in the collection of the British Museum, Pt. I, Lycidae", London, 1879, pp. 2-8 and 78).

[2] In one of these, represented by a single small specimen, the metathoracic stigmata are absent, and the prolongations of the angles of the abdominal terga and of other plates are very feebly developed.

Specimens of the slender type are represented from the following localities:—

Malay Peninsula: Johore.
Sinkep Island (near Sumatra).

I have examined the mouthparts of one specimen of the latter type from Johore, and of one of the specimens of the former type from Lankawi and of those from Ceylon. They are all constructed on the same plan, but are apt to be less slender than in the larva of *Lyropaeus biguttatus.*[1]

It is difficult to see how these creatures can feed. The mandibles are presumably used to pump juices along the grooved maxillae in much the same way as the maxillae are used to pump juices along the grooved mandibles of Hemerobiid larvae. But "Trilobite Larvae" seem to have no means of grasping prey. Presumably therefore they must eat something which they need not grasp securely, such as snails or planarians. Dr. Annandale tells me that he found these "larvae" in great abundance in the Malay Peninsula. He noticed that the broad and slender types always occurred together, which led him to think that the difference might conceivably be sexual[2]; and that they were only found where planarians were plentiful and snails scarce. It seems not unlikely, therefore, that they feed on planarians. It is also possible that they may feed on the juices of decaying wood, etc., which might account for the long periods of time during which they have been known to live without being known to feed (Gahan, 1913, p. 62).

Trilobite larvae are known in some instances at least to be luminous. This was first recorded by Kolbe (*loc. cit.*) on very uncertain authority, but Shelford (*loc. cit.*) has since noticed that one species has a pair of phosphoresecent organs on the penultimate segment of the abdomen.

III. Tenebrionidae—*Catapiestus indicus*, Fairmaire.

(Plate xxi, figs. 20-25).

Fairmaire described this species (*Ann. Soc. Ent. Belge.* XL, 1896, p. 28) from specimens collected in Kanara, and noted that it occurred in "Sikkim" also. It appears to have a wide distribution extending from the Western Ghats of Southern India to the Abor country and Lower Burma (for details see Tenebrionidae of the Abor Expedition, *Rec. Ind. Mus.* VIII).

The specimens described below were taken with adults from under the bark of a fallen log. A cast larval skin was found close behind the pupa.

[1] Other authors refer to the maxillary and labial palps as four and three-jointed respectively, instead of as three and two-jointed as they appear to me to be both in cleared cast-skins and potashed specimens.

[2] The slender type does not seem to occur in the Indian Peninsula or Ceylon; but this may mean that it is only in the Malay Region, where "Trilobite Larvae" appear to reach their highest development, that larviform males occur.

LARVA.

The larva of *Catapiestus indicus* is a parallel-sided, elongate, flattened insect, brownish in colour, and terminated behind by a pair of long spiniform processes (see pl. xxi, figs. 20-21).

The head is almost semicircular, with a well-defined and somewhat prominent clypeus which bends downwards, so that the semicircular labrum is almost vertical and only partly visible from above. The suture limiting the frons behind is (? always) very distinct; it extends on either side from a point in the middle line immediately in front of the anterior margin of the pronotum, almost in a straight line towards a point on the margin of the head immediately behind the base of the antenna; but after traversing nearly half this distance, it turns abruptly forwards to run a short distance parallel to the sagittal plane and then bends straight outwards till it regains its former line, which it resumes and follows to the margin of the head.

The ocelli are four in number on each side, three in a line situated immediately behind the base of the antenna, and one a little behind them on the dorsal surface.

The antennae are four-jointed. The basal joint is scarcely as long as broad; the second joint is somewhat longer than broad; the third joint is fully twice as long as the second and scarcely as thick; the fourth joint is minute, being only about as long as the third joint is broad, and about one-third as broad as long.

The mandibles are stout and are tridentate distally, the middle tooth being the largest and most prominent, the lowest the smallest and more or less fused with it. There is a very large molar tooth.

The lobe of the maxilla is about twice as long as broad, simply rounded distally. The maxillary palps have three joints, of which the middle one is a little the longest and the third is slenderer than the other two, which latter are of uniform width throughout and are together about as long as the lobe. The labial palps have two joints of about equal length; the basal is stouter than the distal.

The terga are traversed, except in the terminal segment, by a median longitudinal groove or suture which does not, however, extend across the slightly darkened transverse band by which each is bordered behind. Each segment except the last bears laterally a few long erect hairs.

The last segment bears on each side two stout backwardly-curved spines, of which the posterior is followed dorsally by three similar spines. The last four form a straight line lying obliquely across the base of the long terminal spine. The terminal spine bears two long erect hairs rather more than half way along the ventral surface. One such hair is associated with each of the smaller spines, except the middle one of the three above the base of each terminal spine; and six are arranged in a semicircle on the ventral surface of the body of the segment, between the anal papilla and the margin. The anal papilla is semicircular, and bears one pair of blunt conical spinules in the angles, and four smaller spinules arranged in a square medially. Of these four the two anterior are distinctly smaller than the two posterior.

PUPA.

The pupa is white in colour. Its form is shown on pl. xxi, figs. 22-23. Each of the marginal denticulations of the prothorax is continued into a papilla which is empty and transparent in the preserved specimen and so does not show in the photograph, and these papillae are tipped with long erect hairs. Similar hairs are present one on either side of the labrum, three on either side of the clypeus, two immediately in front of each eye, two between and behind the eyes, one in the middle of the anterior margin of the pronotum, two on either side mounted on papillae a little behind the anterior margin of the pronotum, one on either side a little in front of the posterior margin of the pronotum, two on either side of the meso- and metanotum,[1] one on either side of the third and two on either side of the fourth to eighth abdominal sterna.

The first six abdominal sterna are quadrangular, the seventh and eighth more nearly triangular. There is a pair of short divergent styles in the position of the anal papilla of the larva. The terminal segment is very like that of the larva; the anterior pair of marginal spines and the semicircle of hairs behind the anal papilla have, however, disappeared ; and the two hairs on each of the terminal spines are now mounted on strong spinules.

The most important works on Tenebrionid larvae appear to be [2] :—

1839. Westwood, J. O. "Introduction to the Classification of Insects" I (London, 1839), pp. 316-324, text-figs.

1853. Chapuis and Candèze. "Catalogue des Larves des Coléoptères." *Mem. Soc. R. Sci. Liége*, VIII, pp. 513-517, pl. vi, figs. 5-6*a*.

1877. Perris, E. "Larves de Coléoptères" (Paris, 1877), pp. 252-294, pl. viii, fig. 277, pl. ix, fig. 310.

1877. Schiodte, J. C. "De Metamorphosi Eleutheratorum Observationes" *Naturhist. Tidsskr.* XI, pp. 479-598, pls. v-xii.

All known larvae of the subfamily Tenebrioninae, in which Gebien places the genus *Catapiestus* (Junk's "Coleopterorum Catalogus", Tenebrionidaem-Trictenotomidae), appear to be described or referred to in these works, except that of *Menephilus*

[1] Three on the left side of the mesonotum in our only specimen.

[2] A useful list of Tenebrionid larvae, with a key to generic characters, is given by Kiesenwetter and Seidlitz, *Naturg. Ins. Deutschl.*—Coleoptera V (1) Tenebrionidae (Berlin, 1898), pp. 207-217.

cylindricus (=*curvipes*).[1] This larva, and two others belonging to the same subfamily, seem to resemble the larva of *Catapiestus indicus* more closely than do any other Tenebrionid larvae of which I have seen descriptions. The other two are *Iphthimus italicus*[2], and the South American species of *Upis* referred to on p. 319 of the first volume of Westwood's "Introduction to the Classification of Insects."[3] The larva of the last named insect is, however, known only from fragments of its cast-skin, and many of its characters are consequently somewhat uncertain.

---

[1] Described by Perris, *Ann. Soc. Ent. France*, (3) V, 1857, pp. 361-7, pl. viii, figs. 444-457.

[2] Described by Mulsant and Revelière, *Opusc. Ent.* XI, 1859, pp. 63-66.

[3] Described by Westwood, *Trans. Ent. Soc. London*, II, 1837-40, pp. 157-162, pl. xiv, figs. 11-18.

EXPLANATION OF PLATE XX.

FIG. 1.—*Lyropaeus biguttatus*, Westwood. Larva from below. × 2.

,, 2.— ,, ,, ,, ,, ,, above. × 2.

,, 3.— ,, ,, ,, Part of ventral surface of abdomen more highly magnified.

,, 4.—*Lyropaeus biguttatus*, Westwood. Head of larva in prothoracic sheath, from in front.

,, 5.—*Lyropaeus biguttatus*, Westwood. Pupa with larval skin attached, from the side. × 2.

,, 6.—*Lyropaeus biguttatus*, Westwood. Pupa with larval skin removed, from above. × 2.

,, 7.—*Lyropaeus biguttatus*, Westwood. Male from below. × 2.

,, 8.— ,, ,, ,, ,, ,, above. × 2.

,, 9.—? Immature female of *Lyropaeus biguttatus*, Westwood, from below. × 2.

,, 10.—? Immature female of *Lyropaeus biguttatus*, Westwood, from above. × 2.

,, 11.—Another form of "Trilobite Larva" from Cochin, from above. × 2.

,, 12.—Part of ventral side of abdomen of same specimen more highly magnified.

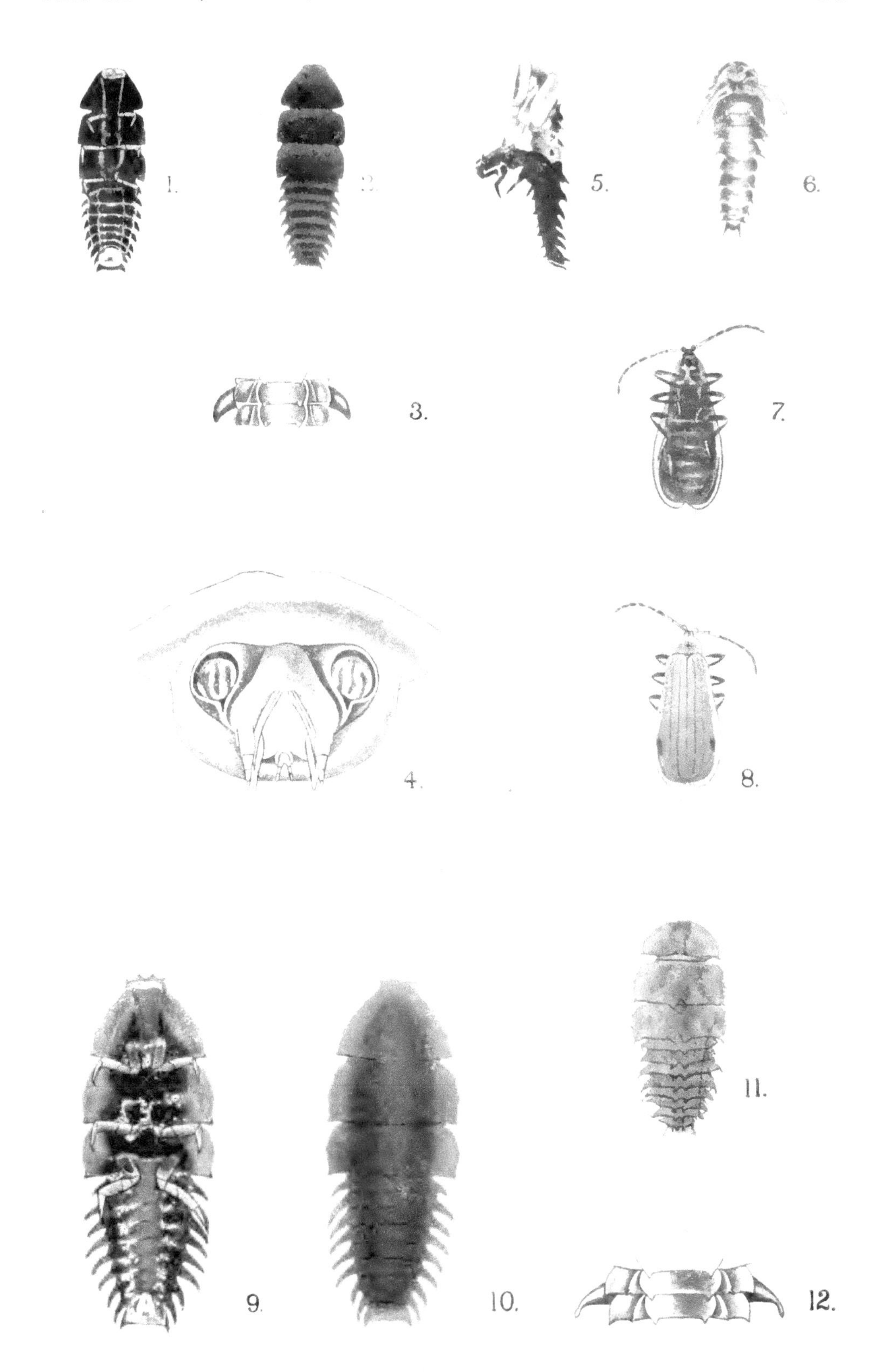

BEETLES FROM COCHIN.

EXPLANATION OF PLATE XXI.

FIG. 13.—*Uleiota indica*, Arrow. Larva from above. × $4\frac{1}{2}$.

,, 14.— ,, ,, ,, Posterior end of larva more highly magnified.

,, 15.—*Uleiota indica*, Arrow. Left spine of ninth abdominal segment of larva. × 30.

,, 16.—*Uleiota indica*, Arrow. Left spine of eighth abdominal segment of larva. × 40.

,, 17.—*Uleiota indica*, Arrow. Pupa from above. × $4\frac{1}{2}$.

,, 18.— ,, ,, ,, ,, ,, below. × $4\frac{1}{2}$.

,, 19.— ,, ,, ,, Adult from above. × $4\frac{1}{2}$.

,, 20.—*Catapiestus indicus*, Fairm. Larva from above. × 2.

,, 21.— ,, ,, ,, ,, ,, below. × 2.

,, 22.— ,, ,, ,, Pupa from above. × 2.

,, 23.— ,, ,, ,, ,, ,, below. × 2.

,, 24.— ,, ,, ,, Adult from above. × 2.

,, 25.— ,, ,, ,, ,, ,, below. × 2.

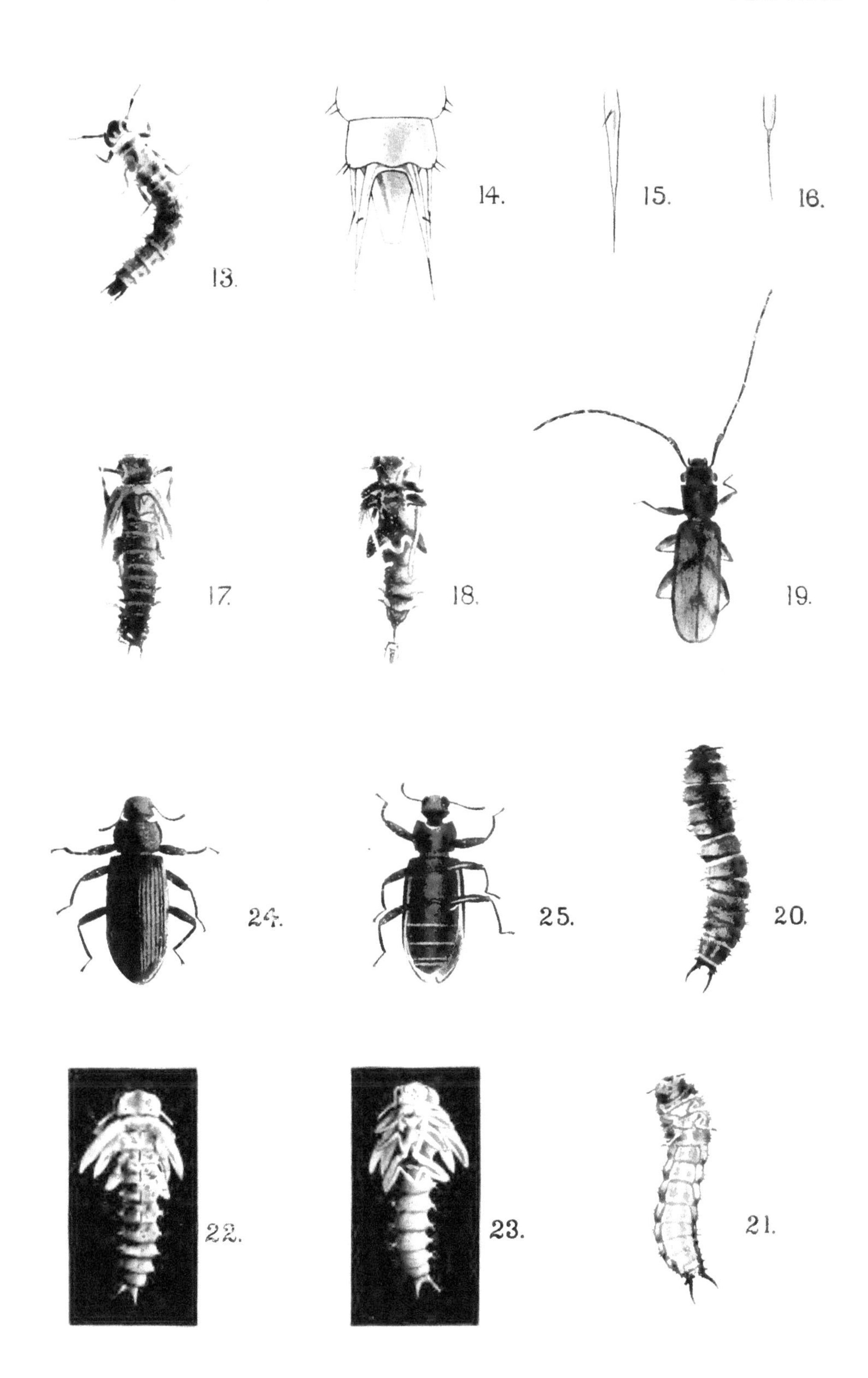

BEETLES FROM COCHIN.

XXI. CRYPTOSTOMES OF THE INDIAN MUSEUM.

PART II.

By S. MAULIK, *B.A.* (*Cantab*), *F.E.S.*, *Imperial College of Science and Technology, University of London.*

This paper is my second report on the Cryptostomes contained in the collections of the Indian Museum, my first having appeared in this journal (Vol. IX, part II, No. 7, 1913). In preparing it, I have followed the same method as before. The usual notes regarding distribution and variation have been added. Twenty-two species of the Hispinae are enumerated here, six of which are new to science, as is shown in the following list:—

1. *Botryonopa sheppardi*, Baly (var.)
2. *Macrispa krishnalohita*, n. sp.
3. *Anisodera guerini*, Baly.
4. ,, *excavata*, Baly.
5. *Prionispa himalayensis*, n. sp.
6. *Oncocephala quadrilobata*, Guér. (var.)
7. *Javeta pallida*, Baly.
8. *Agonia saundersi*, Baly.
9. *Gonophora bengalensis*, Ws.
10. ,, *haemorrhoidalis*, Weber.
11. *Monochirus sthulacundus*, n. sp.
12. *Hispella stygia*, Chap.
13. ,, *ramosa*, Gyll.
14. ,, *andrewesi*, Ws. (?)
15. *Rhadinosa girija*, n. sp.
16. ,, *laghu*, n. sp.
17. *Asamangulia cuspidata*, n. g., n. sp.
18. *Dactylispa spinosa*, Weber.
19. *Hispa armigera*, Oliv.
20. *Platypria echidna*, Guér.
21. ,, *hystrix*, F.
22. ,, *erinaceus*, F.

I have to thank the Indian Museum authorities for sending me their material here. To Mr. Andrewes my acknowledgments are due for his kindness in letting me see the types in his collection and also for letting me have three specimens of one new species described here. Dr. Gestro, of the Genoa Museum, has very kindly sent me some of his types, for which I wish to express my thanks. My obligations are also due to the British Museum authorities and Dr. Gahan for affording me all facilities in the Museum.

Family CHRYSOMELIDAE.

Group CRYPTOSTOMATA.

Subfamily *HISPINAE*.

Tribe *BOTRYONOPINI*.

Genus **Botryonopa**, Blanch.

Blanchard. *Hist. Nat. Ins.* II, 1845, p. 181.
Baly, *Cat. Hisp.* 1858, p. 91, t. 2, f. 6.
Chapuis, *Gen. Col.* XI, 1875, p. 291.
Hispopria, Baly, *Cat. Hisp.* 1858, p. 94, t. 2, f. 7.
,, Chapuis, *Gen. Col.* XI, 1875, p. 297.

Botryonopa sheppardi, Baly (var.).

Baly, *Cat. Hisp.* 1858, p. 92, t. 7, f. 4.
Weise, *Stett. Ent. Zeit*, LXIX, 1908, p. 214.

Locality.—Silchar, Cachar (*J. Wood-Mason*). One example.

It is a small specimen. The upper portion of the elytra and the prothorax are yellow and not of the usual red colour.

Genus **Macrispa**, Baly.

This genus was erected by Baly in 1858 (*Cat. Hisp.* 1858, p. 90) for the reception of *Macrispa saundersi*, Baly. The locality of this insect was not known at that time. Twenty-one years later, in working out the Phytophagous Coleoptera collected by Chennell in Assam, Baly found a very imperfect specimen of *Macrispa*. This localised the habitat of the genus (*Cist. Ent.* II, 1879, p. 405). The imperfect specimen has been indentified as *M. saundersi*, which, as I shall show, is not correct. In 1906 Gestro in a little note (*Ann. Mus. Civ. Gen.* 1906, p. 130) said that in the Oberthür collection he had found one example reported from British Bhutan. Thus there exist in the collections only three examples of the genus. I have before me three more examples (1 ♂ 2 ♀ ♀) which clearly belong to *Macrispa*. But it will be necessary to describe them as a new species.

In enumerating the generic characters, Baly states in reference to the antennae:—"Corporis dimidio longitudine, super tubercula duo inter oculos insertae, subfiliformes, ad apicem subincrassatae, articulo primo incrassato, secundo brevi, duobus proximis elongatis, gracilioribus, caeteris fere aequalibus, obconicis, perparum leniter incrassatis, subcompressis."

The following points in this description call for notice:—

(1) As the length of the antenna differs in the sexes (Baly had one ♀ specimen before him when he drew up the description) its relation to the length of the body cannot be made a generic character.

(2) In the specimens before me the third and fourth joints of the antenna are not slenderer than the rest.

(3) The antenna does not gradually increase in thickness towards the apex.

As these characters are not present in the specimens before me, they cannot be made generic characters.

One of the secondary sexual characters of this genus is a semilunate depression on the last abdominal sternite of the female. The depression varies in different species. Judging from this character, *M. saundersi*, Baly (one example in British Museum) is a female, and the imperfect specimen (British Museum) is also a female, but the depression being different, its identity as *M. saundersi* (*Cist. Ent.* II, 1879, p. 405) is doubtful. Besides, the elytra of the imperfect *Macrispa* is rufous and subnitid, whereas *M. saundersi* has opaque fulvous elytra.

Macrispa krishnalohita,[1] n. sp.

Macrispa krishnalohita, n. sp. is distinguished from *M. saundersi*, Baly, by the following characters :—

	M. krishnalohita.	*M. saundersi.*
1.	Smaller insect, 22 mm.	Larger insect, 25·5 mm.
2.	Apices of the joints of antennae not knobby.	Apices of the joints of antennae knobby.
3.	Thorax suddenly constricted in front.	Thorax less constricted in front.
4.	Colour of elytra subnitid, rufous.	Elytra opaque, fulvous.
5.	Semilunate depression on the last abdominal sternite (♀) broader.	The depression narrower.

Elongate; head, antennae, prothorax, abdomen, legs, shining black; elytra rufous, subnitid; the disc of the prothorax with a large finely punctate area in the middle, base transversely strigose.

Length: 22 mm.

Locality.—Dejoo, North Lakhimpur, base of hills, Upper Assam (*H. Stevens*, iv—viii-1911).

Described from three examples 2 ♀ ♀, 1 ♂.

Type in Mr. Andrewes' collection, London.

Co-type in the Indian Museum, Calcutta.

Fuller description.

Head.—Surface rugose, coarsely and deeply punctate, a deep groove from the vertex running along the middle line; 7 proximal joints of the antennae with coarse and elongated punctures and shining, 4 distal joints covered with a bloom, apical joint pointed, apices of all joints (except the last) impunctate and shining. Mouth parts covered with fulvous hairs.

Prothorax quadrate, abruptly narrowed in front, anterior angles obtuse and rounded, sides parallel, their margins slightly

[1] The specific name is derived from two Sanskrit words: *krishna* = black, *lohita* = red, thus indicating the two colours of the insect.

sinuate, subreflexed, posterior angles are sharp right angles; above shining black, anterior half of disc smooth, finely and sparsely punctate, this smooth shining surface narrows along the middle line and extends a little beyond the middle, one or two deep punctures on this smooth surface; on each side of the middle line a deep depression with punctures in it,—this character is not marked in *M. saundersi, Baly*; posterior half of disc coarsely and deeply punctate; at the base in front is a depression, base itself transversely strigose, the sides of the base sharply cut off, a character not present in *M. saundersi*, Baly.

Scutellum longer than broad at base, at a quarter of its length from the base it is bent, depressed in the middle, one or two transverse ridges on the surface near the apex, apex rounded.

Elytra broader than the prothorax, elongate, subparellel in front, slightly dilated behind, extending considerably beyond the sides and apex of abdomen, their apex rounded, sutural angles armed with an acute tooth; surface subnitid; nine costae on each elytron, 1st an abbreviated one anastomosing with the sutural ridge, 2nd-5th run parallel to each other down the whole length of the elytron, 6th a short one terminates by breaking up into deep punctures, 7th runs down the whole length of the elytron, meeting the 5th at the apex, 8th short and similar to 6th, 9th runs down the whole length of the elytron; deep punctures between the costae, between the 5th and the 7th and between the 7th and the 9th confusedly and deeply punctate; these costae are thicker at their bases than at the apices, where there is a tendency to their being obliterated by the deep punctures. Margins of the elytra subreflexed.

Underside shining, black; femora armed with a short flattened tooth, finely punctate.

♀ Antennae shorter, femora of fore legs not incrassate, last abdominal sternite with a semilunate depression.

♂ Antennae longer, femora of fore legs incrassate, last abdominal sternite without a semilunate depression.

Tribe *ANISODERINI*.

Genus **Anisodera**, Baly.

Baly, *Cat. Hisp.* 1858, p. 101, t. 2, f. 8.
Chapuis, *Gen. Col.* XI, 1875, p. 295.
Weise, *Deutsche Ent. Zeitschr.* 1897, p. 118.

Anisodera guerini, Baly.

Baly, *Cat. Hisp.* 1858, p. 101 (*ferruginea*), p. 168, t. 7, f. 8.
Gestro, *Ann. Mus. Civ. Gen.* 1885, p. 163.
,, *l.c.* 1860, p. 233, et 1897, p. 50.
ferruginea, Guer., *Rev. Zool.* 1840, p. 333.

Locality.—Sonapur, Assam (*L. W. Middelton*). One example.

It has a wide distribution, having been reported from Java, Burma, Mungphu Sikkim, Tenasserim.

Anisodera excavata, Baly.

Baly, *Cat. Hisp.* 1858, p. 105, t. 8, f. 1.

Locality.—Sadon, U. Burma, 5,000 ft., April 1911 (*E. Colenso*). One example.

It has been reported from the Himalayas, Tonkin, and Mungphu. The excavation on the disc of the prothorax is variable; it is not always deep, and in some specimens it has almost disappeared. The blackness of the prothorax also is not constant, for in some cases the prothorax is of the same chestnut colour as the body. These notes are taken from the numerous examples in the collection of the British Museum.

Tribe *CHOERIDIONINI.*

Genus **Prionispa**, Chap.

Chapuis, *Gen. Col.* XI, 1875, p. 337.
Gestro, *Ann. Mus. Civ. Gen.* 1899, p. 220.

Prionispa himalayensis, n. sp.

Cuneiform, rufo-testaceous, legs pale flavous, eyes, mandibles, labrum, and the apical four joints of the antennae black; external apical angles of the elytra are right angles, not produced into a spine; six large and small tubercles on each elytron. Length from head to apex of elytron 5 mm.

Described from one example.

Locality.—Kurseong, E. Himalayas, alt. 4,700-5,000 ft., 21-xi-10 (*Annandale*).

Type in the Indian Museum, Calcutta.

Fuller description.

Head rather projected, cylindrical, interantennal protuberance prominent, a few punctures on the vertex, underside smooth, shining; eyes oval, black; antennae, 1st joint small, 2nd joint longer than 1st, constricted at base, 3rd joint longest, 4th-7th gradually thickened towards the apex and each being shorter than the preceding. Joints 1-7 have got a peculiar transparency and a thin red ring at the apices; joints 8-11 opaque, black, 11th joint pointed.

Prothorax cylindrical, longer than broad, base bisinuate, sides with straight dark red margins, anterior angles toothed, disc coarsely and deeply punctate.

Scutellum longer than broad, narrowed at the apex, apex broadly rounded.

Elytra much broader at base than the prothorax, punctate-striate, shoulders elevated and projected; at about the middle of each elytron is a large shallow depression. There are two costae from the elevated humeral angle, one along the elevated surface up to the depression, the second below the elevated surface along the side to the apex of the elytron. There are six tubercles on each elytron, disposed as follows:—

A little distance posterior to the base of the elytron is a small tubercle, at about the middle of the elytron between the suture and the elytral depression is the largest tubercle, which is concave on its outer side; posterior to this tubercle are two small tubercles, one very close to the suture and the other beyond the line on which the largest tubercle is situated; external to this tubercle a little thickening of the second costa looks like a minute tubercle, but is not really so. Finally, there are two minute tubercles on the sloping apical portion of the elytron, one on the line of the preceding sutural tubercle, the other on the line of the largest tubercle. The tubercles are darker in colour. Suture raised, widely divergent at base for the reception of the scutellum.

Underside.—Legs pale flavous, transparent; underside of thorax, coxae and claws dark red.

Tribe *ONCOCEPHALINI.*

Genus **Oncocephala**, Chevr.

Chevrolat in Dorbigny, *Dict. Univ. Hist. Nat.* IX, 1847, p. 110.
Chapuis, *Gen Col.* XI, 1875, p. 308.
Weise, *Deut. Ent. Zeit.* 1897, p. 313.
Gestro, *Ann. Mus. Civ. Gen.* 1899, p. 313.
Nepius, Thomson, *Arch. Ent.* II, 1858, p. 225.

Oncocephala quadrilobata, Guér. (var.)

Locality.—Dawna Hills, 2000-3000 ft., L. Burma, 2—3-iii-08 (*Annandale*). Six examples.

This species has not been reported from this locality before.

Tribe *COELAENOMENODERINI.*

Genus **Javeta**, Baly.

Baly, *Cat. Hisp.* 1858, p. 108, t. 2, f. 10.

Javeta pallida, Baly.

There are four examples from Calcutta. Baly records it from Madras.

Tribe *GONOPHORINI.*

Genus **Agonia**, Ws.

Weise, *Deut. Ent. Zeit.* 1905, p. 116.
Gonophora, Baly, *Cat. Hisp.* 1858, p. 108 (pars.)
Chapuis, *Gen. Col.* XI, 1875, p. 303.
Distolaca, Baly, *l.c.*, p. 116 (pars.)
Chapuis, *l.c.*, p. 305.
Gestro, *Ann. Mus. Civ. Gen.* 1897, p. 67.

Agonia saundersi, Baly.

Baly, *l.c.*, p. 110, t. 8, f. 4.

Locality.—Mungphu. One example.

Genus Gonophora, Baly.

Baly, *Cat. Hisp.* 1858, p. 108, t. 2, f. 11.
Chapuis, *Gen. Col.* XI, 1875, p. 303.

Gonophora bengalensis, Ws.

Weise, *Stett. Ent. Zeit.* LXIX, 1908, p. 214.

Locality.—Rungpur, Bengal. Two examples.

Gonophora haemorrhoidalis, Weber.

Weber, *Obs. Ent.* 1801, p. 64.
Fabricius, *Syst. El.* II, 1801, p. 60.
Illiger, Mag. I, 1802, p. 183 (*Hispa*).
Baly, *Cat. Hisp.* 1858, p. 112.
Gestro, *Ann. Mus. Civ. Gen.* 1885, p. 167.
,, *l.c.*, 1897, p. 56, et 402.
,, *Notes Leyd. Mus.* XIX, 1897, p. 174.
,, *Bull. Soc. Ent. Ital.* 1902 (1903), p. 141.
Var. *niasensis*, Gest., *Ann. Mus. Civ. Gen.* 1897, p. 57.
Var. *undulata*, Ws., *Arch. f. Naturg.* 1905, p. 98.

Locality.—Johore, Malay Pen. (*Motiram*). One example.

Tribe *HISPINI*.

Genus Monochirus, Chap.

Chapuis, *Gen. Col.* XI, 1875, p. 330.
Hispellinus, Weise., *Deut. Ent. Zeit.* 1897, p. 144.
,, *l.c.*, 1905, p. 317.

There are six specimens which belong to this genus, but as they are not in perfect condition, I do not wish to pronounce any opinion as to their specific character, although they appear to be new to science. All of them were found at Calcutta, 12-viii-07, 4-ix-07, 21-x-11, Maidan; these dates show that they are obtainable in August, September and October. It is possible, therefore, to get some more specimens, so that they may be specifically determined.

Monochirus sthulacundus,[1] n. sp.

Black, shining, elytra spiny, basal six joints of the antennae bare, punctate, apical 5 joints formed into a very thick club which is covered with brown pubescens, 1st joint with a spine.

Length from head to apex of elytra 4 mm.

Described from one example.

---

[1] The specific name is derived from two Sanskrit words, viz., *sthula* = thick, *cundum* = antenna.

Locality.—Berhampur, Murshidabad district, Bengal, 1-1-08 (*R. E. Lloyd*).

Type in the Indian Museum, Calcutta.

Fuller description.

Head rugose, coarsely punctate, a fine groove from the vertex runs down the middle, an incomplete ridge enclosing a row of short brownish hairs round the eyes; basal 6 joints of the antennae black, bare, and punctate, apical 5 joints form a very dilated, round club which is covered with reddish brown pubescence, basal joint bearing a long spine on the dorsal side, 2-4 joints small, rounded, 5-6 joints subequal and together as long as 2, 3, and 4, apical joint pointed.

Prothorax more opaque than the elytra, as long as broad, narrowed in front, lateral margins rounded; surface coarsely punctate, covered with brown pubescence; a bare longitudinal area in the middle, the bare area is more or less elevated; two transverse shallow depressions; two pairs of bifid and erect spines on the front margin, one pair of similar bifid spines and a single one on each lateral margin; base bare, transversely channelled; each of the four lateral angles ends in a minute blunt tooth.

Elytra shining, sides parallel, rounded at the apex, deeply and coarsely punctate-striate, thinly covered with stout and erect spines, the marginal row of spines horizontal.

Legs short, stout, punctate, sparsely covered with brown pubescence; a pointed tooth on the underside of the fore femora, 3 in similar positions on each of the mid and hind ones, fore and hind tibiae straight, emarginate at the apices, mid tibiae curved.

Genus **Hispella**, Chap.

Chapuis, *Gen. Col.* XI, 1875, p. 334.
Weise, *Ins. Deutschl.* VI, 1893, p. 1061 and 1064.
Weise, *Deut. Ent. Zeit.* 1897, p. 143.

In erecting *Hispella* as a subgenus of *Hispa*, Chapuis stated the characters as follows:—

"Antennes de 11 articles, courtes, et ne dépassant pas la base du pronotum, comprimées et spinuleuses, 1 article assez gros, prolongé en dessus en une longue épine *arquée en avant*, 2 plus court, muni d'une spinule plus courte, 3-6 *légèrement dilatés de la base à l'extrémité, les angles de celle-ci assez saillants, les supérieurs plus que les inférieurs*, 7 en cône, 8-10 transversaux, trés-serrés, 11 aigu à l'éxtrémité, pattes courtes et robustes, *tibias droits, comprimées, dilatés au bord externe*, anguleux et souvent épineux avant l'extrémité.

"Cette division a pour type la *Hispa atra*, de Linné, qui habite les contrées tempérées et méridionales de l'Europe."

The italics are mine.

At present *Hispella* comprises six species, including the type *H. atra*, L., from which the above description is taken. The other five species are all from the Indian region. The Indian forms differ from the type in the following characters:—

(1) The long dorsal spine of the first joint of the antenna is not bent forward.

(2) 3-6 joints of the antennae are not dilated as in *H. atra*, L.

(3) The tibiae are not dilated as in *H. atra*, L.

The middle tibiae in the Indian forms are curved, which is not so in the case of *H. atra*, L.

The differences of the characters between the type and the Indian forms, the homogeneity of those of the Indian forms, and the fact that *H. atra*, L. is found in the temperate zone, all point to the conclusion that the Indian forms may be separated and formed into a new genus. On the other hand it may be pointed out that a slight gradation in the characters is noticeable in the Indian forms. I do not, therefore, propose to separate them at present, unless more material from the Indian region establishes this fact beyond doubt.

Instead, for the sake of convenience, I shall characterise the genus as follows:—

Antennae.—1-6 joints spiny, 3-6 may be dilated, apical 5 joints forming a club.

Claws.—Completely separate.

Tibiae.—Straight, dilated or not dilated, middle tibiae may be curved.

A table will distinguish the forms thus:—

1. 3-6 joints of antennae dilated (flattened)	*atra*, L.
2. 3-6 joints of antennae not dilated ..	3
3. Antennae short, stout, 1st joint with 5 dorsal spines	*brachycera*, Gestro.
4. Antennae long, slender, 1st joint with less than 5 dorsal spines ..	5.
5. 1st joint of antennae with 4 dorsal spines, 2nd joint with 2 dorsal spines	*stygia*, Chapuis.
6. 1st joint of antennae with less than 4 dorsal spines	7.
7. 1st joint of antennae with 3 dorsal spines, 2nd joint with 1 dorsal spine	*ramosa*, Gyll.
8. 1st joint of antennae with 2 dorsal spines, one very minute ..	*andrewesi*, Weise.

Owing to the reasons stated by Weise (*Deut. Ent. Zeit.* 1897, p. 127) I do not include Motschulsky's species *ceylonica* in this table.

Hispella stygia, Chap.

Chapuis, *Ann. Soc. Ent. Belg.* XX, 1877, p. 51.
Gestro, *Ann. Mus. Civ. Gen.* 1897, p. 124, f. 14.
Weise, *Deut. Ent. Zeit.* 1897, p. 126.

Locality.—This example has "Bombay" on its label. I have seen other specimens taken at Belgaum which is in the Bombay Presidency. This specimen may have been taken at the same place.

Hispella ramosa, Gyll.

Gyll. in Schonh., *Syn. Ins.* I, 3, App. 1817, p. 6.
Gestro, *Ann. Mus. Civ. Gen.* 1897, p. 124, f. 13.

Localities.—Paresnath, W. Bengal, 4,000-4,400 ft., 15-iv-09 (*Annandale*); Bangalore, S. India, 3,000 ft., 15-x-10 (*Annandale*); Dhikala, Naini Tal District, U.P., 26-iv-08 (*Mus. Collr.*). Three examples.

This species is apparently confined to the hills.

Hispella andrewesi, Ws.

Weise, *Deut. Ent. Zeit.* 1897, p. 126.

Locality.—Monda, Nepal, 12-v-08 (*Mus. Collr.*) One example.

The spines on the first and second joints of the antennae being broken, I doubtfully indentify this example. There is also a difference in the colour of the elytra, but no structural difference is observable. *H. andrewesi*, Ws. was taken at Kanara.

Genus **Rhadinosa**, Weise.

Weise, *Deut. Ent. Zeit.* 1905, p. 318.

Rhadinosa laghu,[1] n. sp.

Oblong, small, not thickset as the other members of this genus, black, with a faint metallic sheen, in some specimens the colour is a mixture of testaceous and black, subnitid, thoracic and elytral spines are long and slender as compared with the size of the insect, sparsely covered with white adpressed hairs; elytra deeply punctate-striate; besides these deep punctures, the surface is very minutely punctate. This character distinguishes this species from all others of the genus.

Length from head to apex of the elytra 3-5 mm.

Described from 15 examples.

Type in the Indian Museum, Calcutta.

Localities.—12 examples from Calcutta, 3—4-viii-07 (*N.A.*); Mangaldai, Assam, 16—18-x-10 (*Kemp*); Siliguri, base of E. Himalayas, 3—4-vi-1911 (*N.A.* and *S.K.*); Basanti, Forest Station, 24 Parganas, Sunderbuns, 16-xi-09 (*T. Jenkins*).

Fuller description.

Head coarsely punctate, not rugose, from the vertex to a point between the bases of the antennae deeply sulcate, a row of

[1] *Laghu* is a Sanskrit word meaning light. The name is applied to this species in reference to its light build.

white hairs round the eyes, a few similar hairs on other parts of the head: antennae long, slender, thickened towards the apex, apical 5 joints form a club, thickly covered with brownish pubescence, apical joint bluntly pointed, basal joint long and stout, with a long dorsal spine pointing forward, 2nd joint short and rounded, 3rd, 4th, 5th joints longer than 2nd, and almost equal to each other in length, 6th joint shorter than the preceding ones, 1st-6th joints with a few scattered white hairs.

Prothorax quardrate, as long as broad, lateral margins rounded, two pairs of bifid spines in front, on each lateral margin one pair of bifid spines, the space enclosed between these spines is rugose and coarsely covered with short white hairs, on the portion of the disc posterior to the single lateral spines is a shallow transverse depression, each of the 4 anterior and posterior angles of the prothorax ends in a blunt tooth.

Scutellum finely punctate, apex rounded, in the ♀ rather broader than long, slightly depressed in the middle, apex widely rounded.

Elytra sparsely covered with short white hairs, thinly covered with long spines, marginal row horizontal.

Underside.—Legs finely punctate, mid tibiae curved, all the femora with 3 small, pointed, curved teeth on the underside, the third tooth may be very minute.

Rhadinosa girija,[1] n. sp.

Oblong, black, shining, sparsely covered with long, erect, brownish hairs, as compared with the size of the insect, the prothoracic and elytral spines are short and stout. The structure of the disc of the prothorax distinguishes it from all others.

Length from head to apex of elytra 4 mm.

Locality.—Chutri Gouri, Nepal Terai, 26—27-iv-07 (*Mus. Collr.*). One example.

Type in the Indian Museum, Calcutta.

Fuller description.

Head rugose, forehead depressed in the middle, interantennal space elevated into a sharp ridge, spaces between the bases of the antennae and the eyes are also elevated; antennae thickest in the middle, *i.e.* the 7th joint is the thickest, gradually becomes thinner towards the apex, apical 5 joints form a club, covered with brownish pubescence, basal joint long, stout, with a dorsal stout spine, 2nd joint short, rounded, 3rd joint longest, 4-6 joints equal in length, basal 6 joints bare.

Prothorax quadrate, almost as long as broad, narrowed in front, lateral margins rounded, 2 frontal (bifid), 2 marginal (bifid), 2 marginal (single) spines, short and stout. The surface of the disc

[1] The specific name is derived from a Sanskrit word *giri*, meaning mountain, *girija* = originating in a mountain.

is broken up into many shallow hollows. In the centre there is a shining depressed elevation. Posterior to the single marginal spines the portion of the disc is a shallow and wide depression. Base smooth; each of the four anterior and posterior angles ends in a small blunt tooth.

Scutellum as long as broad, finely punctate, apex rounded.

Elytra punctate-striate, punctures large and shallow, the spines short and stout.

Underside black, shining, legs short, femora with a small tooth on the underside, mid tibiae curved.

There are two specimens of this genus from Shillong. They appear to be new to science. I do not describe the species because the examples are not perfect.

Asamangulia,[1] new genus.

Body elongate, antennae 11-jointed, 1st joint with a dorsal spine, claws completely separate, unequal, inner claw being smaller than the outer; frontal and marginal spines of the prothorax short, robust, and suberect. Elytra punctate-striate, tuberculate or spinose, with a row of horizontal marginal spines, at the apex the spines are longer.

This genus is distinguished from all the other genera of the Hispini by the *unequal claws and the single dorsal spine on the first joint of the antennae.* I attach generic importance to the inequality of the claws, because, since Chapuis laid stress on the character of the claws in founding the genus *Monochirus* in 1875, they have been found useful in separating the spiny Hispinae into genera. Except in the present case, however, the claws have not been found unequal, although they have afforded many other characters.

Asamangulia, n.g., is related to *Phidodonta*, Ws., by the form of the body, and to *Rhadinosa*, Ws., by the completely separated claws. I place the new genus *Asamangulia* after the genus *Brachispa*, Gestro.

Asamangulia cuspidata, n. sp.

Elongate, black, shining; prothorax sparsely covered with brownish adpressed hairs. Apical 5 joints of the antennae form a pointed club and are covered with reddish brown pubescence. Scutellum depressed in the middle. Elytra deeply punctate-striate, cuspidate; these cusp-like tubercles on the elytron are smaller at the base of the elytron, becoming larger (almost stout spines) towards its apex.

Length from head to apex of elytron 5-6 mm.

Locality.—Pusa, Bihar. Eleven examples.

Type in Mr. Andrewes' collection, London.

Co-types in Genoa Museum of Natural History, in the Indian Museum and in the British Museum.

---

[1] The generic name is derived from two Sanskrit words: *asama* = unequal, *anguli* = claw.

Fuller description.

Head rugose, prominently elevated round the bases of the antennae; antennae thickest in the middle, 1st joint large, dorsally produced into a long spine; 2nd joint small, rounded; 3rd joint longest; 4-6 joints subequal; 2-6 joints surface strigose.

Prothorax more opaque than the elytra, disc rugose, with two transverse depressions, a longitudinal deep furrow down the middle, sides rounded, front margin with two pairs of bifid spines, a few longer hairs between these spines, each lateral margin with one pair of bifid spines and a single one; the spines are short, stumpy and suberect.

Scutellum rounded, punctate, depressed in the middle.

Elytra deeply punctate-striate.

Mid tibiae curved.

Genus **Dactylispa**, Ws.

Weise, *Deut. Ent. Zeit.* 1897, p. 137.
Weise, *Arch. f. Naturg.* 1899, p. 265.
Podispa, Chap., *Gen. Col.* XI, 1875, p. 335 (pars.).
Hispa, Chap., *Gen. Col.* XI, 1875, p. 333 (pars.).
Monohispa, Ws., *Deut. Ent. Zeit.* 1897, p. 147.
Triplispa, Ws., *l.c.*, 1897, p. 147.
Gestro, *Bull. Soc. Ent. Ital.* 1902, p. 59.

Dactylispa spinosa, Weber.

Weber, *Obs. Ent.* 1801, p. 65.
Fabr., *Syst. El.* II, 1801, p. 58.
Gestro, *Ann. Mus. Civ. Gen.* 1897, p. 86 (*Hispa*).
,, *Bull. Soc. Ent. Ital.* 1902 (1903), p. 150.

Locality.--Sarawak, Borneo (*C. W. Beebe*). Two examples.

In the latest catalogue of the Hispinae by Weise, it is not mentioned that *H. saltatrix*, F. is a synonym of this species of Weber's.

Genus **Hispa**, L.

Linné, *Syst. Nat.* ed. XII, 1767, p. 603.
Chapuis, *Gen. Col.* XI, 1875, p. 334.
Weise, *Ins. Deutschl.* VI, 1893, p. 106.
Weise, *Deut. Ent. Zeit.* 1897, p. 137.
Dicladispa, *Gestro*, *Ann. Mus. Civ. Gen.* 1897, p. 81.
,, ,, *l.c.*, 1899, p. 329.

Hispa armigera, Oliv.

Oliver, *Ent.* VI, 1808, p. 763, t. 1, f. 8.
cyanipennis, Motsch., Schrenck's *Reise Amur.* II, 1861, p. 238.
aenescens, Baly, *Journ. Asiat. Soc. Beng.* 1887, p. 412.
aenescens, Cotes, *Ind. Mus. Notes*, 1889, p. 37.
Gestro, *Ann. Mus. Civ. Gen.* 1890, p. 248.
Gestro, *Ann. Mus. Civ. Gen.* 1897, p. 82.
Ws., *Deut. Ent. Zeit.* 1904, p. 457.

Localities.—Calcutta, 2-xi-07, 22-v-09, 28-viii-06, 14-viii-06, 12-ix-07; Howrah, near Calcutta; Midnapore and 24 Parganas,

Lower Bengal (*Cotton* and *Lyall*); Goalbathan, East Bengal, 10-vii-09 (*R. Hodgart*); Balighai, near Puri, Orissa coast, 16—20-viii-11; Malabar district, W. India (*E. Thurston*); Mandalay, U. Burma (*H.M.S. Matthews*); Khulna, E. Bengal (*Rainy*); Mungphu, near Darbhanga, N. Bengal (*H. S. Beadon*); Sibsagar, Assam; Backergunge, E. Bengal; Bilaspur, Darbhanga, N. Bengal (*G. W. Llewhelin*); Saraghat, N. Bengal; Katmundu, Nepal. Eighty-four examples and about 412 in alcohol.

Distribution.—This insect has a wide distribution. Dr. Modigliani reports it from Sumatra: Siboga, Baligha, Pangherang-pisang and Pedang (ref. 6). Nothing about the food-plant of this insect in these localities is mentioned. In India it is a pest of the Rice plant.

Weise has sunk Motschulsky's species *cyanipennis* as a synonym of *armigera*, Oliv. (ref. 7). Comparing Motschulsky's description (ref. 2) with Olivier's, and also Baly's, I find no reason why *cyanipennis*, Mots. should be considered as a synonym of *armigera*, Oliv. Motschulsky writes: "Corslet assez lisse, *sans epines dorsals*; elytra fortement ponctuees *avec quatre epines sur leur milieu*." Olivier in his description of *armigera* says: "Le corcelet est armee de cinq epines de chaque cote; la quatra anterieures ont une base commune; la cinquieme la plus courte de toutes, est places un peu au-dila. Les elytres sont d'un bleu fonce luisant; elles sont des points enfonces et un grand nombre d'epines." Baly's description of *aenescens* (ref. 4) runs as follows:—"Thorace rugoso-punctato lateribus anti medium spinis quatuor, basi connatis et pone medium spina unica armatis; elytris anguste oblongis, fortiter seriato-punctatis, spinis validis triseriatium dispositis instructis."

From the above it is evident that *cyanipennis*, Mots., cannot be a synonym of *armigera*, Oliv.; *cyanipennis* has no spines on the thorax and only four spines on the elytra. In his description I have italicised these portions. *Armigera*, Oliv., and *aenescens*, Baly, the descriptions of which agree well, both have five spines on the thorax and a great many on the elytra. In the absence of any reason from Weise for sinking *cyanipennis*, Mots., I consider it necessary to point out that Motschulsky's description does not warrant it. The type of *cyanipennis* is supposed to exist in the Museum of the University of Moscow.

Genus **Platypria**, Guér.

Guérin, *Revue Zool.* 1840, p. 139.
Chap., *Gen. Col.* XI, 1875, p. 336.
Gest., *Ann. Mus. Civ. Gen.* 1890, p. 229.
,, *l.c.*, 1897, p. 110; l.c., 1905, p. 515.

Platypria echidna, Guér.

Guér., *Rev. Zool.* 1840, p. 139.
Gest., *Ann. Mus. Civ. Gen.* 1890, p. 246, fig.; 1897, p. 112.

Localities.—The Nilgiris; Kanara Two examples.

Platypria hystrix, F.

Fabr., *Suppl. Ent. Syst.*, 1798, p. 116.
Fabr., *Syst. El.* II, 1801, p. 59 (*Hispa*).
Guérin, *Rev. Zool.* 1840, p. 140.
Gestro, *Ann. Mus. Civ. Gen.* 1897, p. 113.
erinacea, Oliv., *Ent.* VI, 1808, p. 762, t. 1, f. 6 (*Hispa*).
digitata, Gest., *l.c.*, 1888, p. 178.

Localities.—Sadon, U. Burma, alt. 5,000 ft., April, 1911 (*E. Colenso*); Katmundu, Soondrijal, Nepal; Calcutta, 4-vii-1907. Four examples.

Platypria erinaceus, F.

Fabr., *Syst. El.* II, 1801, p. 59 (*Hispa*).
Ill., *Mag.* III, 1804, p. 160.
Guér., *Rev. Zool.* 1840, p. 141.
Gest., *Ann. Mus. Civ. Gen.* 1897, p. 111.
Var. *bengalensis*, Gest., *l.c.*, 1897, p. 112.

Locality.—Jafna, Ceylon, June 1910. One example.

XXII. NOTES ON PEDIPALPI IN THE COLLECTION OF THE INDIAN MUSEUM.

V.—TARTARIDES COLLECTED BY MR. B. H. BUXTON IN CEYLON AND THE MALAY PENINSULA.

By F. H. GRAVELY, *M.Sc., Assistant Superintendent, Indian Museum.*

A valuable collection of Pedipalpi has recently been presented to the Indian Museum by Mr. B. H. Buxton, who obtained them in Ceylon and the Malay Peninsula when collecting further material for his work on Arachnid morphology. The Thelyphonidae and Tarantulidae will be dealt with in papers dealing with the Indo-Australian members of these groups as a whole. The time does not, however, appear to be ripe for the preparation of a general account of the Tartarides, of which group Mr. Buxton's specimens form the subject of this paper.

The chief points of interest brought out by Mr. Buxton's collection of Tartarides are (1) the unsatisfactory nature of the distinction between *Schizomus* and *Trithyreus*[1], a distinction involving the separation into different subgenera of such obviously allied species as *crassicaudatus* and *perplexus*; and (2) the increasing number of Oriental species whose females closely resemble the Papuan *modestus*, Hansen. It seems to me undesirable to go on describing these species in the absence of males on the basis of measurements alone.

Schizomus (Trithyreus) perplexus, n. sp.

Locality.—Polonuruwa, North-Central Province, Ceylon (under bricks 4 ♂ ♂, 1 ♀; under leaves 1 ♀ and several young).

♂. *Cephalothorax.*—Eye-spots absent. Cephalic sternum about as long as broad.

Arms.—Nearly as long as the body. Trochanter slender as in *S.* (*s. str.*) *crassicaudatus*[2]; lower margin lightly sinuous, convex basally, convex distally; anterior angle long and spiniform, directed slightly upwards, with a similar but somewhat smaller, lightly upturned process arising on the inner side at its base; anterior margin strongly convex. Femur with a ventral tubercle at the base as in *S. crassicaudatus*, but prolonged beyond this, the total length of the ventral margin in front of the trochanter being more

[1] See Hansen and Sörensen, *Arkiv för Zoologi* II (8), 1905, pp. 33-34.

[2] See Hansen and Sörensen, *Arkiv för Zoologi* II (8), 1905, pp. 40-42, pl. iii, figs. 1*a*—1*i*.

than half as great as the depth of the femur at its distal end. Patella also somewhat slenderer than in *S. crassicaudatus*, the median ventral tooth shorter and not directed forwards, the margin strongly concave behind it, more lightly concave in front. Tibia concave ventrally at base, then abruptly swollen and gradually tapered; the basal concavity hidden when the arm is not extended. Upper margin of tarsus two and a half times as long as claw.

First legs.—Nearly half as long again as body. Coxa terminating behind base of trochanter of arm. Femur about three quarters of length of patella, slightly longer than tibia. Tibia about one-fifth as long again as foot. Foot about ten times as long as deep, deepest at end of metatarsus. Second metatarsus about three-fifths as long as whole tarsus and about equal to five

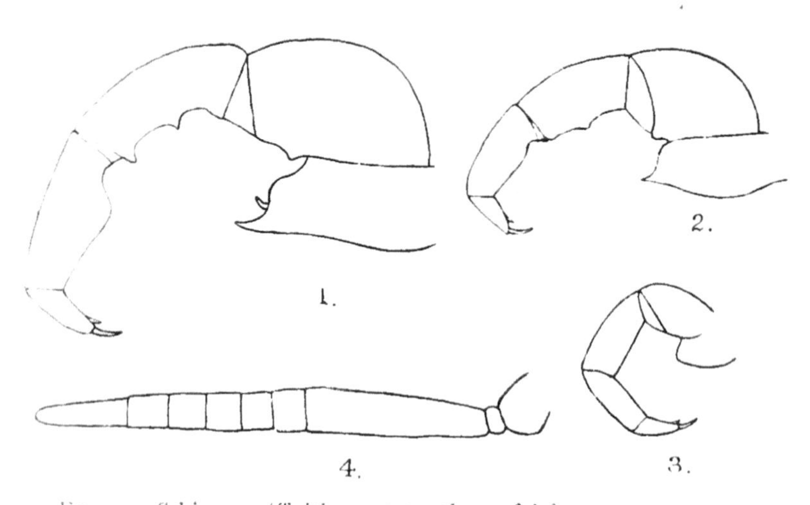

FIG. 1.—*Schizomus (Trithyreus) perplexus* ♂ left arm, × 30.
„ 2.— „ „ „ ♀ „ × 30.
„ 3.— „ „ *buxtoni* ♂ „ × 30.
„ 4.— „ „ „ ♂ foot of antenniform leg, × 65.

proximal tarsal joints which are subequal in length, the basal being perhaps somewhat shorter than the others.

Fourth legs.—Femur slenderer than in *S. crassicaudatus*, about two and a third times as long as deep.

Tail.—Resembles that of *S. crassicaudatus.*

♀. *Cephalothorax.*—As in the male.

Arms.—About three-quarters the length of the body. Trochanter with both margins lightly and evenly convex, practically straight; anterior angle less strongly produced than in male, a small spine present on inner surface some distance from it. Femur shorter than in male, free ventral margin not longer than basal tubercle. Lower margin of patella biconcave; ventral spine represented only by a tubercle between these concavities. Ventral margin of tibia concave basally, then lightly swollen. Upper margin of tarsus twice as long as claw.

First legs.—Nearly a quarter as long again as body. Coxa terminating behind base of trochanter of arm. Femur about three quarters length of patella, about as long as tibia. Tibia about a quarter as long again as foot. Foot about ten times as long as deep, deepest at end of metatarsus. Second metatarsus scarcely as long as sum of five proximal tarsal joints, about half as long again as terminal tarsal joint. First tarsal joint slightly shorter than any of the succeeding four.

Fourth legs.—Like those of male.

Tail.—Long and slender, about six times as long as deep. Basal joint nearly twice, second scarcely more than once as long as deep. Separation of third and fourth joints obscure.

Colour of both sexes.—Pale reddish brown, the abdomen and legs faintly greenish.

Length.—♂ about 3·5, ♀ about 3·0 mm. The arms of the male show this species to be closely related to *Schizomus* (*s. str.*) *crassicaudatus* from Ceylon; but its thoracic terga have the structure characteristic of the subgenus *Trithyreus*.

The distinctive features of the arms are fully developed in large specimens only. They are scarcely distinguishable in small ones, which are often most difficult to distinguish from immature specimens of the next species.

Schizomus (Trithyreus) buxtoni, n. sp.

Localities.—Polonuruwa, North-Central Province, Ceylon (several ♂ ♂, ♀ ♀; under bricks, many under leaves); Minneriya, North-Central Province (3 ♂ ♂); Sigiri, Central Province (many ♂ ♂, ♀ ♀).

♂. *Cephalothorax.*—Eye-spots absent. Cephalic sternum slightly longer than wide.

Arms.—Slender and of moderate length, without distinctive tubercles or spines. Trochanter with lower margin distinctly convex, anterior angle obtuse and more or less rounded with a small spine on the inner side behind it, anterior margin practically straight. Femur slender, with free ventral margin about equal to anterior margin of trochanter. Claw about half as long as upper margin of tarsus.

First legs.—Very slender, about one and a half times as long as body. Coxa terminating behind base of trochanter of arm. Femur much shorter than patella (7 : 9), slightly longer than tibia, much longer than foot (7 : 5). Second metatarsus about as long as five succeeding tarsal joints, which increase regularly in length from basal to distal.

Fourth legs.—Femora fully two and a half times as long as deep.

Tail.—Somewhat like that of *S. suboculatus*, but the disc broader and more evenly rounded behind, with the sides more convex distally—sometimes almost circular or even squarish. When seen from the side it lacks the profound dorsal excavation seen in Hansen and Sörensen's figure of that species.

♀. Closely resembles the male in general features, but the first legs are only about twice as long as the body. The tail is slender, being about five times as long as deep. The first joint is longer than the second, which is scarcely as long as broad. The first and second joints combined are scarcely as long as the third and fourth which are indistinctly separated.

Colour of both sexes.—Pale brown, sometimes with a greenish tinge in large specimens.

Length.—Up to about 3 mm.

This species seems to be allied to *S. vittatus*[1], but is paler and usually browner in colour, and lacks the eye-spots so conspicuous in that species. The tail of the female (the only sex known in *S. vittatus*) is, moreover, much slenderer, and lacks the swelling characteristic of that species.

Schizomus (Trithyreus) spp. *aff.* modestus, Hansen.

Localities.—Malay Peninsula (outside Kubang Tiga and Jerneh caves Perlis; Grik and Lengong, Perak).

The specimens, although fairly numerous, are all female or immature. The terminal joint of the tarsus of the antenniform legs is somewhat more than half as long as the metatarsus, as in *S. modestus*,[2] which the specimens appear to resemble in a general way, as do also the females of *S. vittatus*,[1] *greeni*,[3] *buxtoni*, etc. In the absence of any really definite characteristics, such as would doubtless be found in the tail of the male, it seems undesirable either definitely to record the Papuan species from the Malay Peninsula, or to provide the specimens before me with a new specific name. It is possible that more than one species may be represented.

---

[1] *Schizomus (Trithyreus) vittatus*, Gravely, *Spolia Zeylanica* VII, 1911, pp. 138-139, text-fig. 2c.

[2] *Trithyreus modestus*, Hansen and Sörensen, *Arkiv för Zoologi* II (8), 1905, pp. 63-65, pl. vi, figs. 3a-3f.

[3] *Schizomus (Trithyreus) greeni*, Gravely, *Rec. Ind. Mus.* VII, 1912, p. 109, text-fig. B.

XXIII. NOTES ON ORIENTAL DRAGON-FLIES IN THE INDIAN MUSEUM.

No. 3.—INDIAN SPECIES OF THE 'LEGION' PROTONEURA.

By F. F. LAIDLAW.

The distribution of the species belonging to this 'Legion' in British India and Burma is very interesting, although probably still inadequately known. The species of the group have as a rule a restricted range and are all to a great extent forest-haunting insects, at least they are not commonly found in areas which have been much affected by human industry.

The museum collection contains what are, I believe, the first examples of the Legion recorded from the Himalayas. From what is known of the group it appears probable that whilst Ceylon and the Deccan are inhabited by a rich and peculiar series of species, the great river valleys have no representatives of the group, whilst the great mountain ranges of the north possess few species, only one, namely that here described as a new species under the name of *Protosticta carmichaeli*, being recorded. Burma shows a distinct Malayan influence in the possession of three species, all with a range right down the Malay Peninsula. With the somewhat scanty material available it is impossible to dogmatize as to the distinctness of the Ceylon fauna from that of the Deccan. But it may be noted that whilst *Disparoneura quadrimaculata* (Ramb.) appears to be common in the Satara district, and was first recorded from 'Bombay', it does not occur amongst the material collected in Cochin State by Mr. Gravely, and so far as I know is not recorded from any locality so far south. Further, it is worth remark that none of the species from Ceylon have been recorded from the mainland, and also that no mainland species is known from Ceylon. The sole exception is *Platysticta maculata*, Selys, which has a distinct representative race in Cochin State readily distinguished from the typical Ceylon form. The following table shows the recorded species with their known distribution.

CEYLON.

Platysticta maculata, Selys.	*Platysticta tropica*, Selys.
,, *apicalis*, Kirby.	* ,, *hilaris* (Hagen).
,, *montana*, Selys.	,, *digna*, Selys.

Species marked thus * are represented in the museum collection.

Disparoneura caesia (Selys).
,, *centralis* (Selys).
,, *tenax* (Selys).
Disparoneura sita, Kirby.
,, *oculata*, Kirby.

S. INDIA.

(Nilgiri Hills; Cochin State.)

Disparoneura westermanni (Selys).
,, *gomphoides* (Ramb.).
**Platysticta maculata deccanensis*, subsp. n.
**Protosticta gravelyi*, sp. n.

BOMBAY PRESIDENCY.

**Disparoneura quadrimaculata* (Hagen).

HIMALAYAS

(Darjiling District).

**Protosticta carmichaeli*, sp. n.

BURMA-ASSAM.

Platysticta quadrata, Selys.
Disparoneura atkinsoni, Selys.
Disparoneura verticalis (Selys).
,, *interrupta* (Selys).

Platysticta maculata deccanensis, subsp. n.

(Text-fig. 1.)

$\frac{8830}{20}$ 5 ♂♂. Kavalai, Cochin State, 24—27-ix-14 (in spirit).

Length of abdomen 45 mm., of hind-wing 32·5 mm.

Differs from the typical race from Ceylon as follows :—

The prothorax is dark brown above. The thorax is brown without markings save for a fine black line along the mid-dorsal carina. The brown colouring becomes paler on the sides and ventrally.

Segments 8-9 of the abdomen vivid turquoise blue above. Segment 10, which is very short, is entirely black.

FIG. 1.—Anal appendages of one side of *Platysticta maculata deccanensis* ♂, seen rather obliquely from above.

I have figured the anal appendages of the male; they are evidently very similar to those figured by Kirby for his *Platysticta greeni*, which he subsequently regarded as a synonym of *P. maculata*, Selys (see Kirby, *Proc. Zool. Soc. Lond.*, 1891, p. 203, pl. xx, figs. 3, 3*a* and *J. Linn. Soc.* XXIV, p. 561, 1893).

Species marked thus * are represented in the museum collection.

Platysticta hilaris (Hagen).

Platysticta hilaris, Kirby, *Cat. Odonata*, p. 132 (1890).
" " *id.*, *J. Linn. Soc.* XXIV, p. 562 (1893).

$\frac{6515}{20}$ 1 ♂, Kandy, Ceylon, 21-i-10.

The prothorax in this specimen appears to be uniformly dark on the dorsal surface. The middle lobe of the prothorax carries a pair of small rounded bosses, one on either side of the middle line. The colouring of the abdominal segments is evidently much faded, but the specimen is, I believe, identical with that described in de Selys' synopsis under this name.

Protosticta gravelyi, sp. n.

(Text-fig. 2.)

$\frac{8231}{20}$ 1 ♂ 1 ♀, Kavalai, 1300-3000 ft., Cochin State, 24—27-ix-14 (spirit specimens).

♂. Length of body 44 mm., of hind-wing 21 mm.

Head, under surface brownish black, upper lip, genae and anteclypeus white, the upper lip with a fine black margin, the rest of the dorsal surface black.

Prothorax white; a black triangle occupies the posterior lobe and its apex extends forward on to the middle lobe. Thorax black, with a metallic lustre on the dorsum; laterally marked with two moderately broad bands of white, of which the anterior encloses the stigma; ventral surface black, but the infra-episternum is white.

Abdomen, segments 1-2 black above, sides and ventral surfaces white, but the genital appendages on 2 are tinged with dark brown. Segments 3-7 each with a white sub-basal ring, which laterally and ventrally is more extensive than it is dorsally. In the case of segment 7 the white mark is divided dorsally by a fine longitudinal line which is black, and it occupies about the first third of the dorsum of the segment; ventrally it extends for two-thirds of the length of the segment. On end of segments 3-6 the white mark is much less extensive occupying only a small fraction (one-tenth or less) of the dorsum of the segment. There are no markings on segments 8-10 which are entirely black.

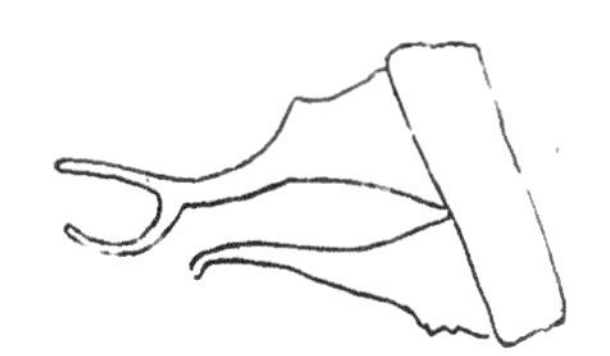

FIG. 2.—Anal appendages of *Protosticta gravelyi* ♂, seen from the side.

The relative length of the abdominal segments is as follows:

$$1 \cdot 2\tfrac{1}{2} \cdot 9 \cdot 9 \cdot 10 \cdot 10 \cdot 8 \cdot 3 \cdot 1 \cdot \tfrac{1}{2}.$$

Legs white, with black ridges and cilia.

Anal appendages about twice as long as segment 10. Upper pair stout at their bases with a small angular projection on their inner side; curved inwards and downwards, strongly chelate at their distal extremities. Lower pair rather slender, simple, curved a little upwards, shorter than the upper pair. Colour entirely black.

Venation, 13 postnodals. Pterostigma rather large, covering one and a half cells, its anal margin longer than the costal and its inner side oblique. M_3 rising from nerve descending from nodus, *Rs* one cell more distal. The rudiment of Cu_2 lies rather nearer the level of the second antenodal nerve than of the first.

The female specimen is very immature and too much macerated for measurement. It is possible to determine that the colouring is generally similar to that of the male, but that segments 8 and 9 of the abdomen have white lateral markings; also that the posterior lobe of the prothorax is simple and that segment 8 of the abdomen is about equal in length to segment 9.

This species is readily distinguished from other members of the genus by the remarkable chelate superior anal appendages of the male.

Protosticta carmichaeli, sp. n.

(Text-fig. 3.)

C.c. 1066-67 2 ♂ 2 ♀, Darjiling Dist.: Singla, 1500 ft. (spirit specimens, all in poor condition).

♂. Length of abdomen 35 mm, of hind-wing 22 mm.

Head, upper lip bluish-white, the whole of the rest of the dorsal and posterior surfaces bronze-black.

Prothorax and thorax also bronze-black dorsally; underneath dull black.

Abdomen, segment 1 dark brown. Segments 2-6 yellowish-brown, darker in the middle part of the segment. Each segment has a light apical ring and a dark terminal ring. Segment 7 is all dark brown save for a small apical ring which is light yellowish-brown. The three terminal segments are uniformly black, the tenth segment is very short.

FIG. 3.—Anal appendages of *Protosticta carmichaeli* ♂, seen a little obliquely from the left side.

The legs are yellowish-brown, with cilia of the same colour. Anal appendages black, upper pair more than twice as long as the tenth segment. They are Λ-shaped when seen in profile; towards its distal extremity each is flattened a little from side to side. The lower pair is shorter than the upper pair, cylindrical and nearly straight; each curves inwards a little at its free extremity.

Venation, 13 postnodals on the fore-wing. Pterostigma covering one cell, its anal margin a very little longer than its costal. M_3 rising from nerve descending from nodus, *Rs* about one cell distally. Rudiment of Cu_2 half-way between level of first and second antenodals.

The condition of the female specimens is such as to make description impossible. Generally speaking the colouring is similar to that of the male. The posterior margin of the prothorax is simple.

Note on the genus *Protosticta*.

Seven species of this genus have been named. It ranges from S. India, the Himalayas and the Malay Peninsula to Borneo and to Celebes. The genus appears to be a specialized form derived from *Platysticta*. It is even possible that the genus is polyphyletic, and in support of this view one might urge that *P. gravelyi* bears a very strong resemblance to the large species of *Platysticta* which occur in Ceylon, whilst the Bornean species resemble rather the small Malayan *Platystictas*. On the other hand all the species of *Protosticta* are alike in the great relative length of the very slender abdomen, and generally in venation; whilst the rather large Celebesian species resemble the large Ceylon *Platysticta* spp.

Disparoneura quadrimaculata (Ramb.).

Disparoneura quadrimaculata, Selys, *Bull. Acad. Belg.* (2) X, p. 440 (1860).
" " *id.*, *Mem. Cour.* XXXVIII, p. 163 (1836).
" " Kirby, *Cat. Odonata*, p. 133 (1890).

$\frac{6510}{20}$ 1 ♂ 1 ♀ (pinned), Medha, Yenna Valley, Satara district, *ca.* 2200 ft., 17-iv-12 (*F. H. Gravely*).
$\frac{6560}{20}$ 4 ♂ 1 ♀ (in spirit), from the same locality.

XXIV. NOTES ON ANT-LIKE SPIDERS OF THE FAMILY ATTIDAE IN THE COLLECTION OF THE INDIAN MUSEUM.

By KARM NARAYAN, *M.Sc., Professor of Biology, St. John's College, Agra.*

(Plate XXXII.)

The present paper describes the ant-like spiders of the family Attidae in the Indian Museum collection. Most of the specimens have been collected in Bengal, while a few from Ceylon, Madras and other places have also been described.

The work of identifying these spiders has been rather laborious, as the family Attidae has not been studied systematically in India so far. Mr. Gravely, in a recent paper (*Rec. Ind. Mus.* XI, p. 257, 1915), has called attention to the neglect which the study of spiders has met with in India. The remark applies much more forcibly to the Arachnomorph spiders than the Mygalomorphs. Pocock, in the "*Fauna of British India* (*Arachnida*)", omits the family Attidae altogether and says, "The group contains a vast number of species and is very imperfectly known—so imperfectly that no satisfactory account of it can at present be given." The most complete work on ant-like spiders is Peckham's "*Ant-like Spiders of the family Attidae*" published in 1892, but since then a good deal of work has been done and the literature added to. It is rather unfortunate that the literature relating to species of these spiders already described is extremely scattered and the descriptions are mostly brief and very often no diagrams are given. In certain cases immature specimens have been made the basis of new species. However, I have followed McCook who, in his book "*American Spiders and their Spinning Work,*" says that the epigynum and male palpus are essential structures on which specific characters can be based with certainty and that immature specimens are not worth keeping in a collection. Consequently, I have not referred to any of the immature specimens that I came across in working out the collection, except those accompanied by adults. At the end of the paper I have put together most of the literature so far published on the species from the Oriental region of the two genera dealt with in this paper.

I have to thank my Professor, Lt.-Col. J. Stephenson, I.M.S., who very kindly obtained permission for me to work in the Indian Museum and also got a number of books for me from the research grant of the Government College, Lahore. My thanks are also due to Dr. Annandale and Mr. Gravely for their valuable suggestions and kind help given while I was working at the Indian Museum.

Harmochirus lloydii, sp. nov.

(Plate xxxii, figs. 1*a*-*c*.)

The genus *Harmochirus* was first described by Simon (Faune Arachnologique de l'Asie Méridionale, *Bull. Soc. Zool. de France* X, 1885, p. 440), who named his species *Harmochirus malaccensis*. Peckham describes another species which he calls *H. albi-barbis* (Spiders of the *Homalattus* Group, Milwaukee, 1895). Still a third species has been described by Thorell as *H. brachiatus*.

It is a curious fact that in all these descriptions only ♂ spiders have been described. I have nowhere found any descriptions or diagrams of a female *Harmochirus*. The present description is based on a female specimen collected by Major R. E. Lloyd, I.M.S., from the Calcutta Medical College compound and preserved in the Indian Museum.

Measurements.

Total length 3·4 mm.

Cephalothorax: length 1·4 mm.; width at dorsal eyes 1·2 mm.; cephalic part 1 mm.

Legs 1423.

The cephalic part is moderately high, but a little lower than the abdomen. The thoracic part is very short and is on a sharp declivity behind the cephalic part. The eyes of the 2nd row are nearer the 3rd than the 1st row. The anterior eyes are directed forwards but the middle and dorsal eyes are situated on the sides. The interesting point about the chelicerae in this specimen (pl. xxxii, fig. 1*b*) is that, on the inferior margin from the ventral side, the right chelicera is *fissidentate* and the left is distinctly *unidentate* (*cf.* Simon, *Hist. Nat. Araign.*, vol. ii, p. 383), but Simon includes this genus in Salticidae fissidentati. The 'pièces buccales' are shown in pl. xxxii, fig. 1*c*, and the shape of the lower lip and the maxillary process of the palp are quite different from those of *H. brachiatus* (Simon, *Hist. Nat. Araign.*, vol. ii, p. 867).

The 1st leg has the characteristic shape shown in pl. xxxii, fig. 1*a*, with the femur compressed and much dilated, claviform, and the tibia disciform and subglobose. There are black stiff bristles on both edges of the tibia together with three special sharp spines dorsally as well as ventrally. The femur of the 2nd leg is compressed, while that of the 3rd leg, as also of the 4th leg, is cylindrical.

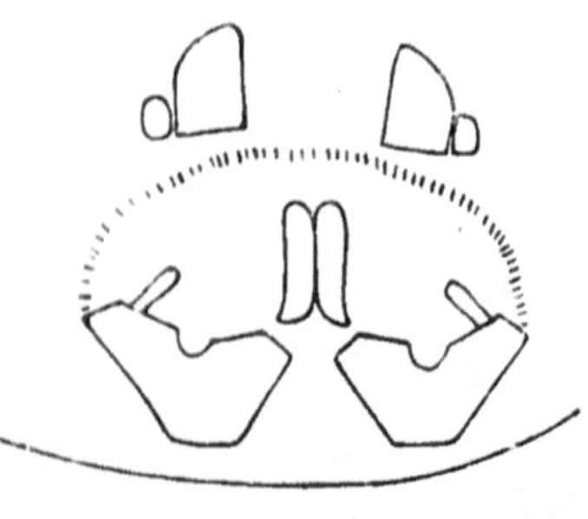

TEXT-FIG. 1.—Epigynum of *Harmochirus lloydii*, sp. nov.

The epigynum (text-fig. 1) consists of two dark-red tubercles which are produced both antero-laterally and internally into short processes. There is also a median elongated tubercle which seems double at its anterior end. There is a sort of a "halo" or crown of short black hairs extending from the outer extremity of one tubercle to that of the other. Rows of hairs are also seen projecting inwards from the tubercles internal to the lateral margins of the "crown." In front of the epigynum are two yellowish-white areas as shown in the diagram.

Colour.—The cephalothorax is dark brown, the cephalic part being covered with small white hairs which are longest towards the anterior eyes; there is a fine row of white hairs on the inferior lateral of the cephalic part. The thoracic part occupies a trapezoidal area dorsally and is devoid of hair; its posterior edge is emarginate. The falces are medium brown.

The 1st leg is medium brown except the tibia which has a dark tinge. The metatarsus is lined with black. The remaining legs are yellowish-white. The femur of the 2nd leg has a dark brown line on its anterior side, while the tibia has a black line anteriorly. The femur of the 3rd leg is black-lined anteriorly and posteriorly and the tibia only anteriorly. Also the posterior half of the femur is black-lined anteriorly and posteriorly, but the tibia only posteriorly. The sternum and lower lip are dark brown but the maxillary process of the palp and chelicerae are light brown.

The abdomen is dull brown with very few white hairs. There is, however, a group of white hairs just behind the top of the anterior end of the abdomen, where it forms a white spot. There are yellowish-white punctate spots all over the abdomen; they are arranged in regular rows and lines, running, for the most part, antero-posteriorly. There are also a few gold-coloured spots on the dorsal side of the abdomen.

KEY TO THE SPECIES OF *Harmochirus*.

I. Tibia thick but cylindrical, not flattened. No special spines besides those that are situated internally and externally on the tibia ... ... *H. albi-barbis* (♂).

II. Tibia flattened, disciform and subglobose; 3 special spines dorsally and ventrally on the tibia.

- A. Lower lip longer than broad; apex of the maxillary process of the palp directed outwards ... *H. brachiatus* (♂).
- B. Lower lip broader than long; apex of the maxillary process directed inwards ... ... *H. lloydii* (♀).

The following characters mentioned by Simon for *H. malaccensis* are not found in this species:

"Cephalothorax supra valde clathrato-rugosa et sat dense fulvo-squamulata. Clypeus fere glaber parcissime cinereo-setosus. Scuto nigerrimo et nitidissimo supra obtectum. Pedes I nigro-aenei metatarsis tarsisque paulo dilutioribus. Femora nigricantia supra albo-lineata, tibiae metatarsique obscure fulvi postici nigro-lineati."

Myrmarachne plataleoides, Camb. (♂)

Salticus plataleoides, Cambridge, *Ann. Mag. Nat. Hist.* (4) III, p. 68 (1869).
Salticus plataleoides, Peckham, *Ant-like Spiders*, 1892, p. 33.

Cambridge described this species from a single specimen in the Hope collection at Oxford, the habitat of which was unknown. He, however, confirmed his identification on receiving specimens from Ceylon. Peckham also describes the species from Ceylon. There are 5 specimens in the Indian Museum collection; their localities together with the names of collectors are given below:—

Peradeniya, Ceylon (*F. H. Gravely*).
Pusa, Bihar (*F. H. Gravely*).*
Sibpur, near Calcutta; 1894 (*T. H. T. Walsh*).*
Calcutta (*G. C. Chatterjee*).*
Calcutta (*F. H. Gravely*).*

Size.—These specimens vary from 6 to 7·5 mm. in total length; the falces are from 2 to 5 mm. long. In at least two specimens the falces exceed the length of the cephalothorax.

Simon says, "Anterior eyes are in a straight row", but in all the specimens, these eyes are a little recurved. The trochanter of the 4th leg is whitish and the posterior two-thirds of the abdomen ventrally and laterally is of a drab colour. There is a yellowish band in the mid-ventral line of the hinder two-thirds of the abdomen.

It is interesting to note that the tube containing the specimen collected by Mr. Gravely at Pusa also contains specimens of the ant *Oecophylla smaragdinca*, which the spider mimics. Dr. Annandale tells me that he has seen this or a very similar spider eating specimens of this ant.

Myrmarachne incertus, sp. nov. (♀)

(Plate xxxii, fig. 2.)

This species resembles in general shape and appearance *M. plataleoides* and was for some time mistaken for the latter by me, but there are important differences which justify its being placed in a different species. The following description, which embodies differences of this species from *M. plataleoides*, is based upon 3 specimens as given below:—

1. Calcutta (*N. Annandale*).
2. Pusa, Bihar (*F. H. Gravely*).
3. Pusa, Bihar.

Measurements.

	Total length.	Cephalothorax (length).	Cephalothorax (width).	Legs.
1	7·1 mm.	3·2 mm.	1·4 mm.	4132
2.	7 mm.	2·9 mm.	1 mm.	4132
3.	8 mm.	3 mm.	1·5 mm.	4132

* See note regarding these specimens under *M. incertus*, p. 397.

The thoracic part at its apex is almost as high as the cephalic, and not lower as in ♂ *M. plataleoides*. The cephalic part rounds off behind the dorsal eyes but not so abruptly as in *M. plataleoides*. In *M. plataleoides* the thoracic part is almost flat dorsally but in this species there is a hump just in front of the middle. There is a sharp declivity in front of the hump, but it slopes gradually behind.

The constriction in the abdomen is not so well-marked as in *M. plataleoides*; it may possibly be due to its being full of eggs.

The epigynum is characteristic (pl. xxxii, fig. 2) and serves to distinguish this species at once from the female of *M. plataleoides*, the vulva of which has an entirely different structure and shape (*cf.* Peckham, *Ant-like Spiders*, 1892, plate iii, fig. 1C). The vulva here consists of two circular white spots between which lies the genital armature. This is formed of two club-shaped masses which are fused just opposite the circular spots but diverge a good deal posteriorly; they diverge a little anteriorly but soon converge again. Posteriorly, at the meeting point of the diverging flanks, there are 2 spine-like processes, one on each side.

Colour.—The colours are mostly the same as in *M. plataleoides*, but the abdomen is yellowish-white and is covered all over with very small polygonal areas, flaky in appearance. In one of the specimens the abdomen is flat ventrally and is depressed in the middle line.

It is worthy of note that the specimen collected by Mr. Gravely at Pusa was found along with a ♂ *M. plataleoides* and a few of the ants of the species *Oecophylla smaragdinea*. It is possible that *M. plataleoides* and *M. incertus* are distinct in the female sex only, and that the males from Bihar and Bengal, which I have identified with the former species, belong in reality to the latter.

Myrmarachne tristis, E. Simon. (♀)

(Plate xxxii, fig. 3.)

This species was first described by Simon in *Ann. Soc. Ent. France*, 1889, p. 115, but the description is based on a ♂ specimen. Peckham also describes the species but gives no diagrams of the epigynum or other ♀ characters, although he gives measurements of the ♀ type. I have found 3 females in the Indian Museum collection which I have identified as belonging to this species.

Calcutta (*F. H. Gravely*).
Madras.
Madras (*Prof. Ramunni Menon*).

Measurements.

Calcutta specimen.

Total length 6·2 mm.
Cephalothorax: length 3 mm.; width 1·4 mm.
Legs 4312.

Peckham's description of this species in "*Ant-like Spiders*" holds for these specimens. A few additional observations may, however, be added. The eyes of the 2nd row are situated about midway between the first and the 3rd rows; there are 7 teeth on the inferior and 4 on the superior margin of the falces; there are 4 pairs of spines on the anterior tibia and 2 pairs on the anterior metatarsus, while there are 3 pairs of spines on the tibia of the 2nd leg.

The 1st tibia is black-lined anteriorly and the femur posteriorly; similarly, the 2nd femur has a black line on its anterior margin The abdomen is olivaceous with a dark band running across the middle of the posterior two thirds of the abdomen, which is depressed ventrally.

The epigynum has a characteristic shape (pl. xxxii, fig. 3). There are two obliquely elliptical white areas, between which lie the chitinous genitalia. The latter consist of two halves which meet about midway but are separated anteriorly and posteriorly.

Myrmarachne laetus, Thorell.

Ascalus laetus, Thorell, *Spiders of Burma*, 1895, p. 320.
Synemosyna laeta, Thorell, *Ann. Mus. Genova* XXV, p. 339 (1887).

This is the commonest ant-like spider in India. The Museum collection contains 6 specimens of the male of this species, of which 3 were collected by Prof. Ramunni Menon at Madras, one by Mr. Gravely and another by Mr. L. L. Fermor at Calcutta, and the last has been obtained from the Nicobars.[1]

One female specimen from Madras was collected by Prof. Ramunni Menon and another by Mr. Paiva from Katihar (Purnea district) in Bihar.

Measurements.

(Calcutta ♂ specimen).

Total length 7 mm.
Cephalothorax: length 3·1 mm.: width 1·5 mm.
Falces 2·1 mm.
Legs 4132.

The specimens agree in almost all essential features with the description given by Thorell; a few minor points brought out by the examination of the males may be noted here. It may be mentioned that I have compared these specimens carefully with an identified specimen of this species sent to the Indian Museum by A. S. Hirst from the Brit. Mus. collection.

The falces are divisible into two portions: a small basal portion from which the greater part of the falx is separated by a constriction. This basal portion is very prominent in some specimens, while in others it is sunk in the cephalothorax, but can be

[1] Since the above was written I have got three more ♂ specimens, one collected by Mr. Gravely at Calcutta, the other by Mr. Kemp at Port Blair (Andamans) and the third by Mr. Paiva at Katihar, Purnea (Bihar).

seen with a little difficulty. As regards the colour, the Indian specimens are darker than the Brit. Mus. specimen in the cephalic part, the falces and the abdomen.

Myrmarachne laetus var. flavus, n. var. (♂)

A specimen, collected by Mr. Paiva at Katihar, resembles *M. laetus* very closely, but there are the following differences. The fang is devoid of a tooth in the middle which is present in *M. laetus*. As regards the colour, this variety is distinctly pale yellow. The falces are pale yellow, with a blackish patch on the dorsal surface. The cephalic part is black dorsally, but laterally it is light brown like the thoracic part. The abdomen is yellowish anteriorly but black in the posterior two-thirds.

Myrmarachne providens (Peckham).

Peckham, *Ant-like Spiders*, 1892.

One specimen was collected by me at Navankot (Lahore). This species is very similar to *M. laetus* but differs in the smaller size of the falces, which are more strongly rounded towards their exterior margin.

Myrmarachne himalayensis, sp. nov.

(Plate xxxii, figs. 5*a-c*.)

Two ♂ specimens of this species were collected by Mr. Gravely at Ghumti in the Darjiling district, at a height of about 4000 ft. Unfortunately the abdomen is separated from the cephalothorax in both specimens; otherwise, the specimens are quite whole and all structures can be made out easily.

Measurements.

Total length 7 mm.
Cephalothorax: length 3·2 mm.; width 2 mm.; cephalic part 1·7 mm.
Falces 1·6 mm.
Legs 1432.

The cephalothorax is moderately high, the cephalic part being a little higher than the thoracic. The constriction between the cephalic and thoracic part is not so deep as in *M. tristis* or *M. laetus*, and it is only just indicated. The cephalic part is a little convex dorsally, almost flat, but rounded on the sides. The thoracic part begins a little lower than the cephalic and slopes gradually to its posterior margin which is fairly broad. The quadrangle of eyes is one-fourth wider than long and occupies about one-half

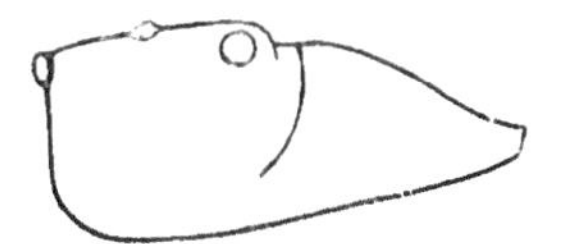

TEXT-FIG. 2.—Cephalothorax of *Myrmarachne himalayensis*, sp. nov., from the side.

of the cephalothorax. The anterior eyes are in a recurved row and are bent somewhat downwards. The middle eyes are situated about midway between the first and 3rd rows.

The falces (pl. xxxii, fig. 5*a*) are comparatively short and stout and are divergent. The unguis bears on its "marge inférieure" 5 minute teeth, but on its superior margin there are 7 larger teeth. The fang is bent almost at right angles just a little above its base, where it is also constricted. The lip is longer than wide and there is a constriction about its middle (pl. xxxii, fig. 5*c*). The relative position and shape of the lip, maxillary process of the palp and coxa are also shown in fig. 5*c*. In the palpus, both the tibia and tarsus are flattened and constitute the palpal organ. A ventral view of the right palpal organ is shown in pl. xxxii, fig. 5*b*.

The coxae of the 1st leg are separated by less than the width of the lip and are nearly approaching. The 1st femur is specially thick and the tibia of the 1st leg bears 2 rows of 6 long and strong spines on its underside. The 2nd tibia bears three shorter and thinner spines.

The sternum is long and narrow and is pointed both anteriorly and posteriorly. The pedicle is moderately long. The abdomen is long and oval with a constriction in the anterior third.

Colour.—The cephalothorax is dark brown, but black round the eyes. A number of white hairs arise about the anterior eyes and also from the clypeus. The falces are dark brown and rugose dorsally; ventrally the colour is lighter. The lip is darker in colour than the maxillary processes of the palps which are medium brown. The last two legs are darker than the 1st two, which are yellowish in colour. The 1st femur is dark brown. The sternum is medium brown.

The posterior two-thirds of the abdomen dorsally and laterally are shining and smooth and are of a testaceous colour; the anterior portion is of a dull greenish-brown tinge. In the mid-ventral line there is a broad yellowish band, while ventro-laterally there is a series of furrows and ridges running longitudinally.

Myrmarachne ramunni, sp. nov. (♂)

(Plate xxxii, figs. 4*a-c*.)

Some 13 ♂ specimens of this species were collected by Prof. Ramunni Menon at Madras and sent to the Indian Museum in two lots. They are referred to a new species on account of the peculiarities in the falces and the abdomen.

Measurements.

Total length 6 mm.
Cephalothorax: length 3 mm.; width 1·7 mm.
Falces 3 mm. long; 1 mm. wide.
Legs 4132.

The cephalic part is high and rounded on the sides. There is a constriction behind the dorsal eyes which cuts much more deeply into the sides of the cephalothorax than into the upper surface. The thoracic part is just a little lower than the cephalic; its highest part is in the anterior third, from which it slopes down in all directions, the slope being steeper on the sides than posteriorly. The posterior margin of the thorax is considerably narrower than the middle portion, where it is broadest. The quadrangle of eyes is more than a third wider than long and wider behind than in front. The first row of eyes is bent a little downward with the eyes close together; the 2nd row of eyes is about midway between the 1st and 3rd rows.

The most characteristic feature which distinguishes this species at once from others is the shape of the falces (*cf.* pl. xxxii, figs. 4*a*, 4*b*). They are long, stout structures with their proximal halves compressed from side to side, and elliptical in transverse section; while the distal halves are convexly flat dorsally and ridged ventrally and triangular in transverse section, the dorsal surface forming the base of the triangle. At the junction of the two halves, there is, so to speak, a regular twist through a right angle, the outer edge of the distal half being continued into the mid-dorsal ridge of the elliptical posterior half of the falx. Looked at from the side the falx is sinuous and possesses a short basal piece as in *M. lactus*. Ventrally there is a row of 9 small teeth on the outer edge and a row of 17 larger teeth on the inner edge of the falx. The fang is as long as the falx and has a curve at the base and a bend at the apex. The right palpus from below is shown in pl. xxxii, fig. 4*c*. The tibia of the 1st leg bears two rows of five spines each on its underside and the femur has one spine dorsally. The lip is longer than broad and the sternum is truncate anteriorly.

The abdomen also is characteristic. Out of 13 specimens almost all have got their abdomens flexed; in some it is only bent, while in others it is distinctly vertical, the posterior two-thirds bending on the anterior third. It is long and oval, but is not constricted. Dorsally it is convex and hard with chitin, while ventrally it is soft and flat.

Colour.—The cephalothorax is medium brown, the cephalic part with an olivaceous tinge dorsally. Both the cephalic and the thoracic parts are covered with short white hairs which also line the constriction behind the dorsal eyes specially towards the sides. The falces are dark brown in colour. The abdomen is brown and is covered with glistening yellowish-white hairs. There are white hairs on the sides at the anterior third. The posterior legs are darker in colour than the anterior. The metatarsus and tarsus in all the legs are darker than the other joints.

This species is closely allied to *M. manducator* (Westwood. *Mag. de Zool.* Anneé 1841, pl. i) from which it differs in the following points: the twist in the falx is characteristic of this species; the number of teeth on the "marge inférieure" is 17 and not 9

(5 anterior and 4 posterior) as shown by Westwood for *M. manducator*; the double curve of the fang is absent here and, lastly, the maxillary process of the palp has sharp bendings and is not rounded as in the other species.

Myrmarachne uniseriatus, sp. nov. (♂)

(Plate xxxii, figs. *6a-b*.)

This small spider belongs to a new species and was collected by Prof. Ramunni Menon at Madras.

Measurements.

Total length 4·2 mm.
Cephalothorax: length 2 mm.; width 1·1 mm.
Falces 0·8 mm.
Legs 4123.

The cephalothorax is moderately high; the cephalic part is limited behind by a shallow transverse depression and not by a sharp constriction as in most other species. Laterally there is a crescentic groove to separate the cephalic from the thoracic part. The anterior thoracic part is at about the same level as the cephalic, behind which the thorax slants posteriorly. As in most species the thoracic part narrows behind. The quadrangle of eyes is more than one-third wider than long and occupies two-fifths of the cephalothorax. The anterior eyes are close together in a recurved row, the middle being twice as large as the lateral. The 2nd row is nearer the first than the third row. The dorsal eyes are of the same size as the lateral.

The characteristic feature which distinguishes it readily from other species is that it has only one row of 10 teeth on the ventral side of the falces. These teeth are situated quite towards the inner margin and therefore belong to the "marge supérieure"; the teeth on the inferior margin are thus absent. The teeth present are larger towards the apex and smaller towards the base of the fang. It will be seen that in most of the species, as for example *M. laetus*, *M. himalayensis* and *M. ramunni*, the teeth on the inferior margin, or outer row, are smaller, both in number and size, than those of the superior margin. In the present species we have reached an extreme of this condition of the reduction of teeth on the "marge inférieure." Besides, the fang has an extra tooth on its underside somewhere about the middle of its length (pl. xxxii, fig. 6*a*). The lip is longer than wide and there are 2 rows of 4 spines each on the under side of the 1st tibia, and 2 rows of 2 spines on the 2nd tibia. The abdomen is long and oval and there is only an indication of a constriction at the anterior third—nothing like what we find in other species.

Colour.—The cephalothorax is light brown in colour except round the eyes, where it is black. There are white hairs both on the cephalothorax and the clypeus. The falces are brown, but the fangs are of a deeper colour. The legs are yellowish. Dorsally, the abdomen is covered with two chitinous pieces which bear some resemblance to the pieces of a carapace. The anterior piece occupies a little more than one-fourth of the abdomen and the posterior, which is larger, covers the rest of it. It is shining and olivaceous dorsally but white ventrally. It is sparsely covered with white hairs dorsally but thickly on its ventral side.

Myrmarachne manducator, Westwood. (♂)

(Plate xxxii, fig. 7.)

Salticus manducator, Westwood, *Mag. de Zool.*, 1841, pl. i.
Salticus luridus, Simon, *Bull. Soc. Zool. France*, 1885, p. 453.
Ascalus manducator, Thorell, *Spiders of Burma*, 1895, p. 323.

There is one specimen of this species in the collection sent by Mr. Mackenzie from Siripur, Saran (Chapra) in Bihar. It has already been recorded from Singapore and Tharawaddy (Burma). Westwood gives its locality as " India septentrionali."

The only contribution I have made is a diagram of the ♂ palpus (pl. xxxii, fig. 7) which is not found in the literature cited.

Myrmarachne paivae, sp. nov. (♂)

(Plate xxxii, fig. 8.)

This new species is described from a specimen collected by Mr. Paiva at Katihar in the Purnea district (Bihar). It is one of the largest ant-like spiders in the Indian Museum collection.

Measurements.

Total length 8·1 mm.
Cephalothorax : length 4 mm.; width 2 mm.; cephalic part 1·7 mm.; thoracic part 2·3 mm.
Legs 4132.

The cephalothorax is moderately high; the cephalic part is only a little higher than the thoracic. There is a constriction separating the cephalic from the thoracic part, which cuts much more deeply into the sides than dorsally. The thoracic part is distinctly longer than the cephalic and has a hump which slopes abruptly behind. The cephalic part is rounded dorsally and laterally and, being rather short, gives a rounded appearance as a whole. The quadrangle of eyes is one-fourth wider than long, wider behind than in front and occupies less than one-third of the cephalothorax. The anterior eyes are in a recurved row and the middle row is nearer the first than the third. The dorsal eyes are just a little larger than the lateral and are placed on the side of the cephalothorax.

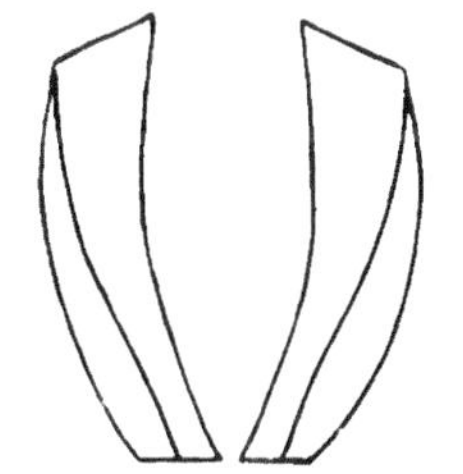

TEXT-FIG. 3.—Falces of *Myrmarachne paivae*, sp. nov., from above.

The falces are characteristic. They are long and horizontal; they are not flattened on the upper surface but the inner face of each falx slants downwards and inwards from the upper edge so that the two only meet along the line of their lower edges, not along the whole surface of the inner faces, as is usually the case. It agrees in this feature with *Salticus imbellis* (Peckham, *Ant-like Spiders*, 1892) In the present species, however, the inner edge of the falces is also curved like the outer and the outer edge at the distal extremity passes into a ridge situated on the upper face of the falces behind (*cf. M. ramunni* above). There are 11 large teeth in the inner row and 9 smaller teeth in the outer row of teeth of the falx. The lip is longer than wide and the sternum is clongated and pointed anteriorly as well as posteriorly. The tibia of the 1st leg has 6 pairs of spines and the tibia of the 2nd leg has 3 pairs of spines on their under surfaces.

Colour.—The spider is of a dark olivaceous colour dorsally, the cephalic part is darker, almost black, while the thoracic has a brownish tinge, the abdomen being paler towards its anterior third. There are white hairs about the anterior eyes and the clypeus; on the latter, they arise from the sides and are bent in towards the middle line. The falces are reddish-brown and are also covered with white hairs. The legs are of the same colour as the cephalothorax, except the first which has a much lighter colour. The coxa and trochanter of the first and the trochanter of the fourth are pale white. Ventrally, the abdomen is of a yellowish colour with longitudinal blackish lines. The 1st femur is black-lined anteriorly and posteriorly; the metatarsus and tarsus of the 2nd and 3rd legs are yellowish-white.

This species is closely allied to *Salticus imbellis* from which it differs in size, shape of the cephalothorax, disposition and size of the eyes and the colouration.

Myrmarachne satarensis, sp. nov. (♂)

(Plate xxxii, fig. 9.)

The description of this new species is based on a specimen collected by Mr. Gravely at Helvak, Koyna Valley in the Satara district (Bombay) at a height of about 2000 ft.

Measurements.

Total length 9 mm.
Cephalothorax: 3·5 mm. long; 1·6 mm. wide.
Pedicle 2·1 mm. long; 0·35 mm. wide.
Legs 4132.

The cephalothorax is moderately high; the thoracic part is dome-shaped and is as high as the cephalic, not lower, and is one-fifth longer than the cephalic. The cephalic part is separated from the thoracic by a constriction which cuts deeply into the sides. It is convex dorsally and is also rounded at the sides. The quadrangle of eyes is two-fifths wider than long, wider behind than in front and occupies about two-sevenths of the cephalothorax. The first row of eyes is a little bent downward and is recurved. The 2nd row is nearer the first than the 3rd row. The dorsal eyes are of about the same size as the anterior lateral. The pedicle is very long indeed, more than 2 mm. in length, the longest I have seen so far in these spiders; it is biarticulate. The falces are short and stout and a little oblique. The sternum is long and narrow and the lip is longer than wide. The abdomen is long and oval and has a constriction in the anterior third. The structure of the epigynum is shown in pl. xxxii, fig. 9.

Colour.—The cephalic part is of a deep blue colour; in strong light it gives a metallic, burnished lustre. The thoracic part, the pedicle and the falces are medium brown. The palps are also of a shining blue colour. The abdomen is darkish, olivaceous or dull black behind the constriction; anteriorly it is greyish-white. There are white hairs on the clypeus and also in the constriction between the cephalic and the thoracic parts. There are two white oblique bands, one on each side of the abdomen, running behind and from the abdominal constriction; they meet dorsally on the constriction. The last two legs are dark brown, but the 1st two are pale white in colour. The patella and tibia of the 1st leg and the trochanter, femur, patella and tibia of the 2nd leg are black-lined anteriorly. The tibia of the 1st leg bears 2 rows of 4 spines and that of the 2nd leg bears 2 rows of 3 spines on its under surface.

This species is allied to *M. praelonga* (=*Synemosyna praelonga*), Thorell (*Ann. Mus. Genova*, XXX, p. 64, 1890) from which it is easily distinguished by the depression and convexity of the cephalic part, the great length of the pedicle and also by the colour.

LIST OF LITERATURE.

1. Simon, E. .. *Histoire Naturelle des Araignées*, vol. II, Paris, 1897, pp. 496—505 and 866—867.
2. Peckham, G. W. and E. G. .. Ant-like spiders of the family Attidae, *Occasional Papers of the Nat. Hist. Soc. Wisconsin*, vol. II, No. 1, 1892.
3. Peckham, G. W. and E. G. .. Spiders of the *Homalattus* Group of the Family Attidae, *Occasional Papers of the Nat. Hist. Soc. Wisconsin*, vol. II, No. 3, 1895, pp. 171—172.
4. Simon, E. .. *Bull. Soc. Zool. de France*, X, 1885, pp. 440—441.
5. Thorell, T. .. *Abhandl. Senckenb. Naturf. Gesellsch.*, 1906.

6. Thorell, T .. *Spiders of Burma*, 1895, pp. 320—329.
7. Cambridge, O. P. .. *Ann Mag. Nat. Hist.* (4) III, 1869, pp. 68—69.
8. Westwood, J. O. .. *Magazin de Zoologie*, 1841, pl. i.
9. Strand, E. .. *Zool. Anz.*, XXXI, 1907, pp. 568—569.

EXPLANATION OF PLATE XXXII.

FIG. 1*a*.—*Harmochirus lloydii*, first leg as seen from the ventral side; 1*b*, falces from below: 1*c*, lip and the maxillary processes of the palps.

,, 2.—Epigynum of *Myrmarachne incertus*.

,, 3.—Epigynum of *Myrmarachne tristis*.

,, 4*a*.—*Myrmarachne ramunni*, falces of the ♂ from above; 4*b*, falx and fang as seen from the side; 4*c*, male palpus.

,, 5*a*.—*Myrmarachne himalayensis*, falces of the ♂ from above; 5*b*, male palpus; 5*c*, relative positions of the lip, maxillary process of the palp and the base of the first leg.

,, 6*a*.—*Myrmarachne uniseriatus*, falx from below; 6*b*, male palpus.

,, 7.—Male palpus of *Myrmarachne manducator*.

,, 8.—Male palpus of *Myrmarachne paivae*.

,, 9.—Epigynum of *Myrmarachne satarensis*.

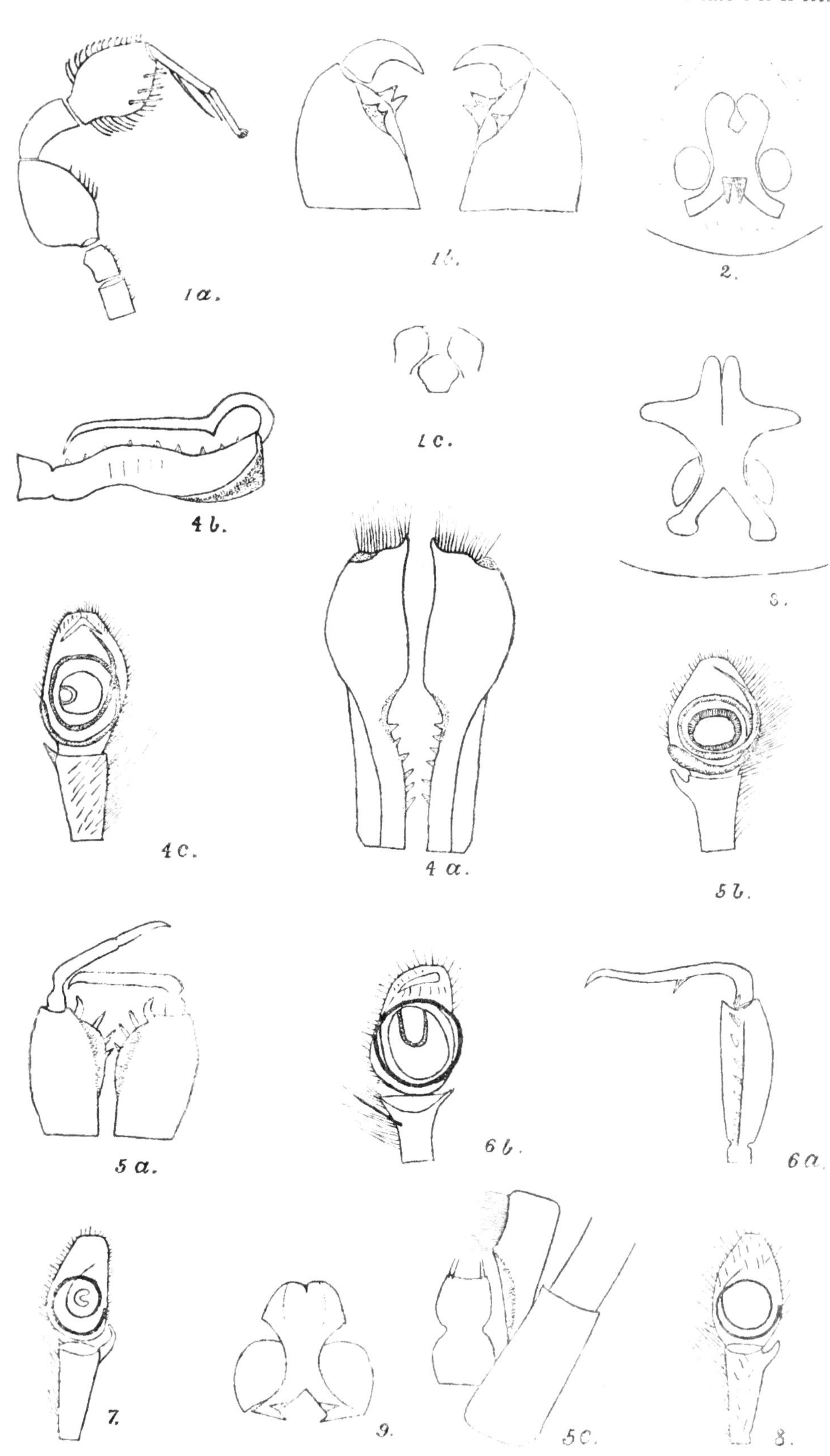

K. Narayan, del.

ANT-LIKE SPIDERS.

XXV. A CATALOGUE OF THE LUCANIDAE IN THE COLLECTION OF THE INDIAN MUSEUM.

By F. H. GRAVELY, *M.Sc., Asst. Superintendent, Indian Museum.*

(Plate XXIX.)

The size and variability of the mandibles of male Lucanidae have naturally attracted special attention from the early writers on this family; and the difficulty of correlating the sexes with certainty seems to have led to an undue neglect of the female. Leutner has, it is true, treated females as carefully as males in his monograph of the Odontolabinae; and a number of recently discovered species have been described from specimens of both sexes. But reasons for the traditional association of females with males seem in many species to have been too vague for record; and the most distinctive characteristics of the females of a number of well-known species seem still to remain undescribed.

In the following short account of our collection I have therefore paid special attention to females.[1] The determination of their subfamilies, and sometimes even genera, has been based on tradition, and I am doubtful whether any of my specific determinations are in disagreement with the associations commonly recognized in European museums. But so far as our material permits, reasons for the association have been found and recorded, and attention has been drawn to structural characters by which the females of various species may be recognized.

In one instance the consideration of female characters has led me to suggest a change in generic definitions. The genera concerned are *Hemisodorcus*, *Eurytrachelus* and *Dorcus* of Van Roon's catalogue.[2]

The female of *Hemisodorcus fulvonotatus* was found to differ from that of *H. nepalensis* in having a pair of tubercles on the head instead of a single one; and the female of *Dorcus suturalis*

[1] Except in the subfamily Odontolabinae, where Leutner has rendered this unnecessary, and in the subfamily Cladognathinae where our material, which is almost entirely Indian, is inadequate for this except in the genus *Cladognathus*. Outside this genus the association of opposite sexes of such species of Cladognathinae as we possess has been based on colour and locality. I have no doubt of the correctness of the determinations made, but have had no opportunity of considering the differentiation of females of species differing in structure but not in colour.

[2] *Coleopterorum Catalogus*, Pt. 8, Lucanidae, Berlin, 1910.

was found to differ from the other species of *Dorcus* in our collection in the opposite direction. Further investigation then showed that the prosternal process of both sexes of the latter species was elevated as in *Hemisodorcus*, not flattened as in *Dorcus*; and that the anterior plates of the head of the male of the former species resembled those of *Eurytrachelus*, not *Hemisodorcus*.

Thomson[1] defines these genera—including as a distinct genus *Platyprosopus*, which Van Roon unites in his catalogue with *Eurytrachelus*—by means of the structure of the mandibles and anterior plates of the head of the male, and the structure of the prosternum. *Hemisodorcus* is said to differ from the other three genera in having elongate mandibles, and this character appears to have been regarded by subsequent authors as being in itself diagnostic of the genus. But it is not correlated with the other so-called generic characters, and a consideration of the female points to the conclusion that it is of less importance than them.

The material in our collection does not enable me to determine whether these other characters are always sharply distinctive, or sometimes grade into one another, but it proves clearly that the extent of their development shows a considerable degree of variation in different species. Thus the posterior end of the prosternal process, though it is always much higher than the mesosternum and more or less abruptly truncate in *Hemisodorcus* and *Eurytrachelus*, and is depressed in the most typical species of *Dorcus* and *Platyprosopus*, is sometimes, in the genus *Dorcus* at least, distinctly convex immediately in front of a narrow depressed posterior margin.

In the table for the determination of these genera given below, I have found it convenient to attach primary importance to the structure of the prosternum, because this applies to both sexes.[2] But when dealing with males only it is often easier to consider first the character of the anterior plates of the head.

These plates, the clypeus and labrum, appear to be more or less fused in most, if not all, Lucanidae; and the plate thus formed—which may be termed the clypeolabrum—is often itself indistinguishably fused with the frons. Among the species before me the outlines of these plates are best seen in *Lucanus cantori*. In the female of this species the clypeus is less coarsely punctured than the frons, and is bounded behind by a tolerably distinct suture: it is very narrow and is keeled in front, overhanging by almost its whole width the larger, still more sparsely punctured, and much more hairy labrum.[3]

In the male the ridge between the clypeus and the labrum is much larger, but the boundary between the clypeus and the

[1] *Ann. Soc. Ent. France* (4) II, 1862, pp. 421-2.

[2] The genetic relationships of these genera can only be determined by the examination of a representative collection of their species from all parts of the world. I have examined scarcely any but Indian forms.

[3] In the females of most species the labrum is much smaller and the clypeus somewhat larger; consequently the labrum is much obscured.

frons appears only as an exceedingly obscure transverse convexity about one-third of the way from the clypeolabral ridge to the ridge between the anterior angles of the head. That this obscure convexity does really mark the junction of the clypeus and the frons is confirmed by the fact that in *Hexarthrius forsteri*—a species of which the female is unknown to me—it is replaced by a very distinct suture.[1] In *Hexarthrius forsteri* the clypeolabral ridge is replaced by a pair of angular processes, structures which reappear in many species of Lucanidae, and may be supposed when present always to represent this ridge.

The presence of these processes, or of the ridge from which they are derived, distinguishes the genera *Eurytrachelus* and *Platyprosopus* from *Hemisodorcus* and *Dorcus*. But when the whole ridge is present it is often very low, and there seems reason to think that it has sometimes been overlooked, with the result that species of *Eurytrachelus* have been placed in the genus *Dorcus*, and that the differences between the prosterna of these two genera have come to be regarded as of no importance.

The morphological anterior margin of the labrum is, however, densely fringed with hair, and the clypeolabral ridge is hairless. When, therefore, the apparent anterior margin of the clypeolabrum as seen from above is hairy, and no ridge or processes are seen, the specimen will belong to the genus *Hemisodorcus* or *Dorcus*. When, however, it is hairless, a closer examination will show that this margin is really the clypeolabral ridge and that the true anterior margin of the plate is hidden beneath it.

The genera *Hemisodorcus*, *Dorcus*, *Eurytrachelus* and *Platyprosopus* may then be distinguished thus :--

1. Posterior end of prosternal process abruptly rounded or truncate in both sexes, its horizontal surface raised well above surface of mesosternum ... ... 2.
   Posterior end of prosternal process lower in both sexes, either uniformly depressed, or convex in front of a narrow depressed border defined on the inner side by a marginal groove ... ... 3.

2. Clypeolabral ridge absent in male ; female with upper tooth of mandibles very strong and with a median cephalic tubercle ... ... *Hemisodorcus.*
   Clypeolabral ridge present in male (often as a pair of lateral teeth) ; female with upper tooth of mandibles very weak and with a pair of cephalic tubercles ... *Eurytrachelus.*

3. Antennae normal in both sexes ; clypeolabral ridge absent in male ; female with a pair of cephalic tubercles ... ... *Dorcus.*
   Seventh joint of antenna (the last before the three bearing pilose lamellae) with a slender polished anterior process as long as the lamella of the succeeding joint and tipped with a cluster of hairs (? in both sexes) ; clypeolabral ridge present in male ; female ? ... *Platyprosopus.*

[1] This suture is also present in the other species of *Hexarthrius* in our collection ; but in them the clypeus is disproportionally large, the labrum being reduced to a narrow strip along its outer margin.

The above diagnoses are necessarily provisional, being based on Indian species only; but I have found it convenient to adopt them in the following catalogue. Apart from this I have followed the classification, and with one or two exceptions the synonymy, adopted in Van Roon's catalogue (*loc. cit.*, above, p. 407, footnote), where references to literature will be found. As, however, many references are incorrectly given there, the following corrections of the inaccuracies I have noticed will facilitate use.

Lucanus cantori. Hope, *Trans. Ent. Soc. London* IV, 1845-7, p. 73; Hope, *Ann. Mag. Nat. Hist.* XII, 1843, p. 363.

Odontolabis burmeisteri. Hope, *Trans. Ent. Soc. London* III, 1841-3, p. 279, pl. xiii, fig. 3.

Metopodontus maclellandi. Hope, *Ann. Mag. Nat. Hist.* XII, 1843, p. 364; Hope, *Trans. Ent. Soc. London* IV, 1845-7, p. 74.

Metopodontus occipitalis = asteriscus. Add—Westwood, *Cab. Or. Ent.*, pl. x, fig. 4.

Metopodontus wentzel-heckmannae. Locality—N. Nyassaland, not Annam.

Prosopocoelus buddha. Hope, *Trans. Linn. Soc. London* XIX, 1843, p. 107; Parry, *Trans. Ent. Soc. London* (3) II, 1864, pl. xii, fig. 3, ♂.

Prosopocoelus bulbosus. Hope, *Trans. Linn. Soc. London* XVIII, 1841, p. 589, pl. xl, fig. 2.

Aegus parallelus. Locality—Khasi Hills, not Prince of Wales Island. The latter locality is recorded for *A. capitatus* (♂) = *A. sinister* (♀), see Westwood, p. 56 of Parry's Catalogue in *Trans. Ent. Soc. London* (3) II, 1864-6.

Nigidius elongatus. Boileau, *Naturaliste* XXIV, 1902, p. 205.

Nigidius vagatus. I have failed to trace this species at all.

Boileau's "Note sur Lucanides conservés dans les collections de l'Université d'Oxford et du British Museum" (*Trans. Ent. Soc. London*, 1913, pp. 213-272, pl. ix) is an important paper published since the catalogue.

When on leave in Europe in 1913, I took the opportunity of checking my identifications by comparison of a selection of our specimens with those in the British Museum and the Deutsches Entomologisches Museum. My thanks are due to Mr. Arrow and Dr. Horn for the facilities granted me. I have also to thank Mr. Arrow for information with regard to a number of specimens which I had not time to examine fully in London myself, and H. E. Lord Carmichael, Mr. E. E. Green, Mr. R. S. Lister, Mr. E. A D'Abreu, the Colombo Museum, the Bombay Natural History Society, and the Imperial Agricultural and Forest Research Institutes for the loan of specimens, a number of which have been added to the Indian Museum collection.

Localities not represented by either sex in the Indian Museum collection are marked with an asterisk (*). The types of *Metopodontus foveatus* subsp. *birmanicus* are now in the British Museum. Those of all other new forms described are in the Indian Museum.

Genus **PSEUDOLUCANUS**, Hope.

Pseudolucanus atratus, Hope.

E. Himalayas: Darjeeling (♂).

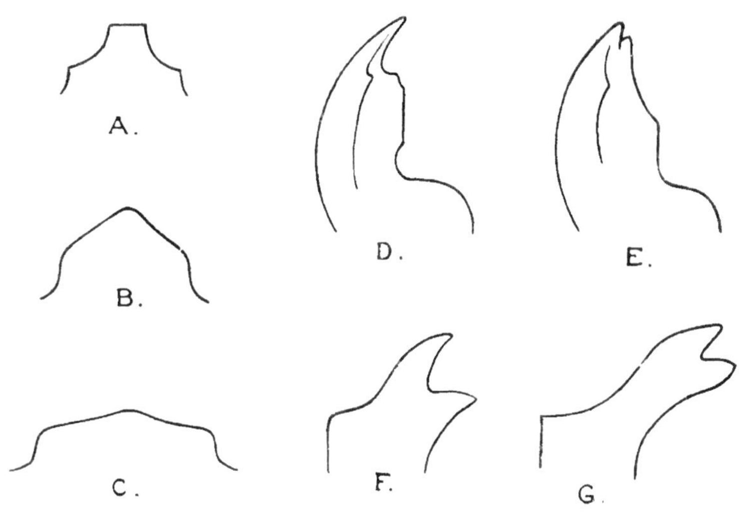

TEXT-FIGURE 1.

A. Clypeus of *Lucanus mearsi*, ♀.
B. ,, ,, ,, *lunifer*, ♀.
C. ,, ,, ,, *cantori*, ♀.
D. Left mandible of *Lucanus mearsi*, ♀.
E. ,, ,, ,, ,, *smithi*, ♀.
F. Distal end of right anterior tibia of *Lucanus mearsi*, ♀.
G. ,, ,, ,, ,, ,, ,, ,, ,, *westermanni*, ♀.

Genus **LUCANUS**, Scop.

Lucanus laminifer, Waterhouse.

Assam: Khasi Hills, 1000-3000 ft. (♂).

Lucanus cantori, Hope.

(Text-figure 1C.)

E. Himalayas: Darjeeling District—Darjeeling (♀); Kurseong, 5000 ft. (♂).
Khasi Hills: Shillong (♂).

The longitudinal yellow bands on the femora make the association of males and females of this species easy. In females the clypeus is transversely linear (text-fig. 1C), and the dorsal tooth of the mandibles is obsolete.

Lucanus lunifer, Hope.

(Text-figure 1B.)

W. Himalayas: Simla (♂): Dehra Dun* (♂); Mussoorie* (♂ ♀); Naini Tal (♂ ♀).

Assam: Khasi Hills (♂).

In the specimen from the Khasi Hills the upper fork of the mandibles is distally enlarged and truncate; in the rest it is normal as in the type specimen (see Boileau, *Trans. Ent. Soc. London*. 1913, p. 218).

The female closely resembles that of the preceding species, but the upper surface of the head is more convex; the clypeus (text-fig. 1B), whose anterior margin is strongly angular instead of almost straight, extends much further forward; and there is a small but distinct tooth on the dorsal surface on the mandibles. The association of this form of female with the male of the present species rests on its uniformly black legs, which separate it from the preceding species, and on its large size and apparent greater abundance in the Western than in the Eastern Himalayas which separate it from the three following species.

Lucanus mearsi, Hope.

(Text-figures 1A, D and F.)

W. Himalayas: Mussoorie (♂ ♀).

E. Himalayas: Darjeeling District—Darjeeling, 7000 ft. (♂ ♀); Kurseong, 5000 ft. (♂ ♀).

To one specimen is attached the label "Bores into dead sap wood of Kharani (*Symplocas* sp.); found at elevation 5000 to 6000 ft. in Sikkim—*G. Rogers*."

Females of this and of the two following species differ from those of the two preceding species in their smaller average size, and in the shape of the clypeus (text-fig. 1A) which, though produced in front as in *L. lunifer*, is always truncate or broadly rounded, instead of strongly angular, in the middle line. Females of the present species may be distinguished from those of the two following by the shape of their anterior tibiae, whose two distal marginal teeth are not specially elongated and are not fused at the base (text-fig. 1F). They are usually of a deep olivaceous colour rather than jet black, and fresh specimens are more or less completely covered with fine golden pile—characters which are shared by the male sex.

Lucanus smithi, Parry.

(Text-figure 1E.)

Van Roon gives *smithi*, Parry, as a synonym of *villosus*, Hope, in his catalogue. Boileau, however, states (*Trans. Ent. Soc. London*

1913, p. 219) that a male of *L. villosus* in the British Museum, possibly the type, closely resembles *L. lunifer* in structure: the measurements given in Gray's brief diagnosis are too large for any specimens of *L. smithi* known to me; the mandibles of *L. villosus* are described by Gray as unidentate, whereas those of *L. smithi* are at least tridentate except in very small specimens: Parry was acquainted with *L. villosus* when he described *L. smithi*.

L. smithi is represented in our collection from the following localities :—

E. Himalayas : Darjeeling District—Darjeeling, 7000 ft., (♂ ♀); Mungphu (♂); Kurseong, 5000 ft. (♂ ♀); Siliguri, in the Terai, a few miles south of the base of the hills (♂ ♀).

The thickening of, and multiplication of teeth on, the mandibles of the male a little beyond the middle is reflected in the great breadth of the mandibles of the female (text-fig. 1E) at about this point, beyond which the inner margin is straight or slightly wavy and blade-like instead of strongly excavate as in *L. mearsi* and *L. westermanni*. The surface is covered with pile as in *L. mearsi*, and the form of the clypeus also resembles that of this species. The anterior tibiae of *L. smithi* resemble those of *L. westermanni* rather than those of *L. mearsi*.

Lucanus westermanni, Hope.

(Text-figure 1G.)

E. Himalayas : Darjeeling District—Darjeeling, 7000 ft.* (♂ ♀); Kurseong, 5000 ft. (♂ ♀).

This species is less densely pilose in both sexes than are fresh specimens of either of the two preceding species. In very small males the mandibles are not forked distally, and the submedian tooth is minute. Small males of the two preceding species show a tendency in the same direction, and it is not impossible that in extreme cases the mandibles of *L. smithi* at least may be indistinguishable from those of the present species—in such cases the length of the pile on the reflexed margins of the elytra would afford a useful guide to identification. Fully hardened specimens of both sexes are jet black in colour, others are reddish. None are in any degree olivaceous.

The mandibles of the female resemble those of *L. mearsi*. The frons is more convex in the middle line in front than in that species. The clypeus differs from that of *L. mearsi* and *L. smithi* in being less abruptly truncate, often broadly rounded, in front. The two distal teeth of the anterior tibiae (text-fig. 1G) are united at the base; their length is about equal to the greatest breadth of the tibia exclusive of its teeth.

A female from Dehra Dun, in the Forest Research Institute collection, resembles *L. westermanni* in general appearance; but its clypeus is like that of *L. mearsi*, and its anterior tibiae are intermediate between those of these two species.

Genus HEXARTHRIUS, Hope.

Hexarthrius davisoni, Waterhouse.

Madras Presidency: Cuddapah* (♂); Palni Hills (♂ ♀).

Hexarthrius mniszechi, Thomson.

Assam: Sylhet (♂).

Hexarthrius forsteri, Hope.

Assam: Khasi Hills (♂).

Hexarthrius parryi, Hope.

Assam: Khasi Hills—Shillong, 3000-5000 ft.* (♂).
Sibsagar (♂).

The anterior parts of the elytra, though darker than the posterior, are not black in our specimen. They are not sharply marked off from the latter, as in Hope's figure, either in our specimen or in that from Shillong belonging to the Pusa collection.

Genus NEOLUCANUS, Thomson.

Neolucanus castanopterus, Hope.

W. Himalayas: Almora—Ramnee.
E. Himalayas: Nepal—Katmandu (♂).
Sikkim—Shamdang, 3000 ft. (♂).
Darjeeling District—Darjeeling, 6000 ft. (♂); Mungphu (♂); Kurseong, 6000 ft. (♂); Singla, 1500 ft.* (♂).
Bengal: Duars—Buxa, near Bhutan frontier (♂).
Assam: Khasi Hills (♂ ♀)—Shillong (♂).
Sibsagar (♂).

Neolucanus marginatus, Waterhouse.

Lower Burma: Amherst District of Tenasserim—Misty Hollow to Sukli, 2100-2500 ft., Dawna Hills (♀).

This specimen closely resembles *N. parryi*, but the anterior femora are scarcely denticulate though they do not look worn. The mentum, too, has only a rudimentary crescent-shaped crest.

Neolucanus lama, Olivier.

E. Himalayas: Darjeeling District—Lebong* (♀); Mungphu (♀); Kurseong, 5000 ft. (♂ ♀).
Assam: Khasi Hills (♂)

Genus **ODONTOLABIS**, Hope.

Odontolabis siva, Hope.

E. Himalayas: Darjeeling District—Darjeeling, 7000 ft*. (♂ ♀); Mungphu (♂ ♀); Pashok; (♂ ♀): Bhutan (♀).
Bengal: Duars—Buxa (♀).
Assam: Khasi Hills (♂ ♀)—Shillong (♂).
Sibsagar (♂).
Sylhet (♂).

Odontolabis cuvera, Hope.

S. India (♂ ♀); Slopes of Nilgiris (♂); Wynad (♂).
E. Himalayas: Darjeeling District—Darjeeling (♂ ♀); Lebong, 5000 ft.* (♂); Pashok* (♀): Mungphu (♂ ♀).
Assam: Khasi Hills (♂ ♀)—Shillong (♂): Cherrapunji (♂).
Naga Hills (♂).
Lower Burma: ? Rangoon*[1] (♂ ♀); near Sukli, Dawna Hills (Amherst District of Tenasserim), *ca.* 2000 ft. (♀); Tavoy (♀).

This species is evidently rare in South India. There is, however, a large male in the Madras Museum collection of South Indian insects; and in our collection are two intermediate males, one small male, and one female from South India.

The female can easily be distinguished from that of *O. delesserti*, which is much commoner in South India, by its broadly flattened mentum with distinct lateral keels close to the margin.

Odontolabis delesserti, Guerin.

S. India (♂ ♀); Travancore—High Range* (♀): Ponmudi* (♀); Malabar—Wynad (♀); Palni Hills* (♂ ♀); Cuddapah* (♀).

The mentum of the female is broadly concave in the middle and broadly convex laterally.

Odontolabis burmeisteri, Hope.

S. India: Coorg—Mercara* (♀); Travancore (♀).

The specimen from Travancore agrees in every respect with Leutner's description of this species, except that there is only a narrow band of yellow on the reflexed borders of the elytra. The other specimen belongs to the Pusa collection. It is very much

[1] This record is based on specimens in the collection of the Bombay Natural History Society. The male (a large one) is the only specimen of its sex that I have seen from Burma, and establishes the identity of the form found there with that found in Assam and the Himalayas. The specimens are unlikely, however, to have come originally from Rangoon itself, for the species appears to be confined to hilly country, and is not known to descend below 1500 or 2000 ft.

darker in colour and has no black on the reflexed borders of the elytra.

Odontolabis latipennis, Hope.

Malay Peninsula : Johore (♂).

Odontolabis aeratus, Hope.

Sumatra : Sinkep Island (♂ ♀).

Odontolabis carinatus, Linnaeus.

Ceylon : Central Province—Lindula* (♂); Maskeliya (♂ ♀).
Sabaragamuwa—Bulutota in Ratnapura District (♀).

Genus HETEROCHTHES, Westwood.

Heterochthes andamanensis, Westwood.

South Andamans (♂ ♀).

Genus CLADOGNATHUS, Burmiester.

Cladognathus arrowi, n. nom.

=*C. confucius*, *auct.*, *nec* Hope.

E. Himalayas : Darjeeling District—Darjeeling, 7000 ft. (♀); Singla, 1500 ft. (♂); Pashok (♂).
Assam : Sibsagar (♂).

The type specimen of *C. confucius*, Hope, is in the British Museum, and has proved to be a small specimen of the following species. A new name is therefore required for the present species in which the mandibles even of the largest males are straight and bear no very strong teeth. I have much pleasure in naming so fine an insect after Mr. G. J. Arrow, whose ever-ready help in the naming of Lamellicornia has greatly facilitated my work on the group.

In the female the head is very finely punctured; the anterior angles of the pronotum are scarcely truncate; and the terminal process of the anterior tibia is slender, being formed by the union of two spines only.

Cladognathus giraffa, Fabricius.

W. Himalayas : Dehra Dun (♂ ♀).
E. Himalayas : Darjeeling District—Singla, 1500 ft. (♀); Pashok (♂ ♀).
Bengal : Kaptai, Chittagong Hill tracts (♂ ♀).
Assam : Khasi Hills (♂).
Sibsagar (♂ ♀).
Andamans : Port Blair (♂).

The large tooth and double curve of the mandibles, characteristic of large males of this species, are both lost in intermediate and small forms; but the forked apex is indicated even in a specimen (in the Dehra Dun collection) in which scarcely a trace of any tooth except the basal remains. The wide separation of the tubercles on either side of the middle of the anterior margin of the head, and the sharp posterior margin of the lower surface of the anterior femora are characters which distinguish small as well as large males of this species from those of the last. Attention may also be called to three other differences, differences which, though slight in themselves, are worthy of note on account of their intensification in the female. They are: the faintly rougher average texture of the anterior parts of the present species; the slightly broader (oblique) truncation of the anterior angles of the pronotum; and the somewhat less slender terminal process of the anterior tibia. The female of the present species differs from that of the preceding species in having the head very coarsely punctured, the anterior angles of the pronotum distinctly (transversely) truncate, and the terminal process of the anterior tibia much stouter and composed of 3-4 teeth.

Genus **METOPODONTUS**, Hope.

Metopodontus foveatus, Hope.

(Text-figure 2.)

The type of this species, which has been re-examined by Boileau, is from Sylhet in Assam; our specimens of the typical form are from Assam and the adjacent Naga Hills. Boileau (*Bull. Soc. Ent. France*, 1911, pp. 63-5, 1 text-fig.) has described as a distinct species *M. poultoni* "nombreux specimens des deux sexes, de diverses provenances, mais principalement recus du Boutan (Sakiou, Maria Basti [1])." Our specimens of this form are all from the Eastern Himalayas. A third form is represented by a series of specimens from (?) Rangoon,[2] belonging to the Bombay Natural History Society and to the Agricultural Research Institute, Pusa. The large male of this series (text-fig. 2F) has one large tooth only on each mandible as in *M. foveatus* (text-fig. 2D), not two as in *M. poultoni* (text-fig. 2A); but this tooth is basal as in males of *M. foveatus* of moderate size (text-fig. 2E), not median as in large males of that form. These three forms, and possibly *M. cinnamomeus* from the Sunda Islands, should probably be regarded as local races of a single species. It seems doubtful whether any definite distinctions between them exist except in large males.

1. M. FOVEATUS subsp. POULTONI, Boileau.

W. Himalayas: Almora—Kimoli (♀).

E. Himalayas: Darjeeling Dist.—Darjeeling (♂). Kurseong (♂ ♀); Pashok (♂ ♀).

---

[1] Maria Basti (or Kaggia Monastery) is situated in the part of the Darjeeling District sometimes known as "British Bhutan." Sakiou is doubtless a misprint for Sakion(g), a few miles further west in the same district.

[2] Probably brought with timber from some hilly district.

The only male from Kurseong is a small one and, like all females, is identified on zoogeographical grounds.

2. METOPODONTUS FOVEATUS, Hope, *s. str.*

Assam (♀): Khasi Hills—Cherrapunji (♂).
Naga Hills (♂).

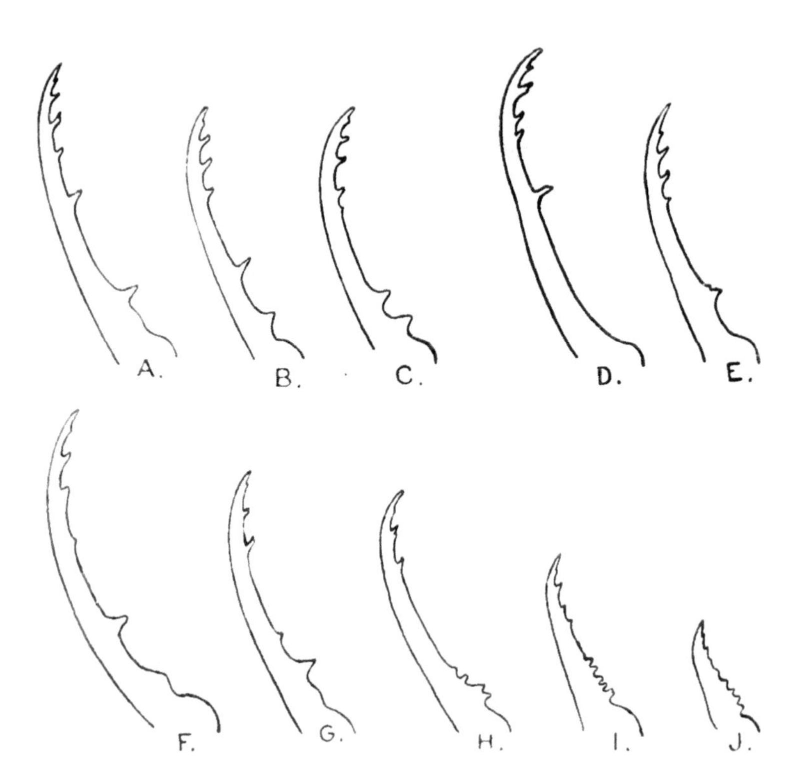

TEXT-FIGURE 2.

A–C. Left mandibles of macrodont males of *Metopodontus foveatus* subsp. *poultoni*, × 2.
D–E. Left mandibles of macrodont males of *Metopodontus foveatus*, s. str., × 2.
F–J. Left mandibles of males of *Metopodontus foveatus* subsp. *birmanicus*, × 2.

3. M. FOVEATUS subsp. BIRMANICUS, n. subsp.

The distinctive characters of this form have already been noticed (see previous page). All the specimens I have seen are somewhat dark in colour, therein resembling the Assamese race rather than the Himalayan.

Upper Burma: Chin Hills—Haka* (♂).
Lower Burma: ? Rangoon[1] (♂ ♀).

[1] See previous page, footnote 2.

Metopodontus maclellandi, Hope.

(Text-figure 3A.)

E. Himalayas: Darjeeling Dist.—Darjeeling (♂); Pashok* (♂).
Assam: Sibsagar (♂).

Metapodontus impressus, Waterhouse.

(Text-figure 3B.)

E. Himalayas: Darjeeling District (♀).
Abor Country—Kobo, 400 ft. (♂).

Our only specimens of this species are a macrodont male and a female. The former is very slightly smaller than our macrodont male of the preceding species. The armature of the

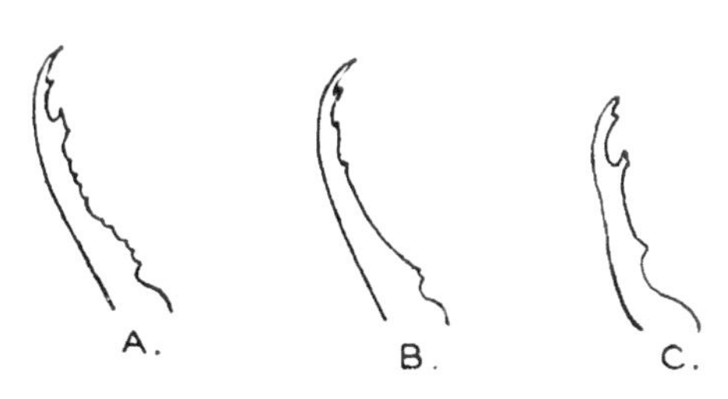

TEXT-FIGURE 3.

A. Left mandible of macrodont male of *Metopodontus maclellandi*. × 2.
B. ,, ,, ,, ,, ,, ,, ,, *impressus*, × 2.
C. ,, ,, ,, ,, ,, ,, ,, *biplagiatus*, × 2.

mandibles of these representatives of the two species is of one type, but is much weaker in *M. impressus* than in *M. maclellandi.* The obscure black median markings of the former species are absent in the latter, and on this character the identification of small males may probably be based. The hind tibiae are hairy on the inner side in both species, but not so strongly as in Westwood's figure of *M. jenkinsi*, nor are they enlarged as in *M. calcaratus* which Boileau believes to be identical with *M. jenkinsi*.

Metopodontus suturalis, Olivier.

E. Himalayas: Darjeeling District (♂)—Kalimpong, *ca.* 2500 ft. (♀).
Harmutti, base of Dafla Hills (♂).
Bengal: Duars—Buxa near Bhutan frontier (♂).
Assam: Dunsiri Valley (♂).
Andamans (♂).

Metopodontus occipitalis, Hope.

Lower Burma: Rangoon (♂ ♀).

The median black mark on the pronotum of the male occupies about one-third of the distance between the anterior and posterior margins. In the female it is larger and touches both these margins.

Metopodontus biplagiatus, Westwood.

(Text-figure 3C.)

1. METOPODONTUS BIPLAGIATUS, Westwood, *s. str.*

E. Himalayas: Darjeeling District—Pashok* (♂ ♀).
Assam: Cherrapunji (♂); Sibsagar (♂).
Lower Burma: Tavoy (♂).
Andamans: Port Blair (♂ ♀).

In the Sibsagar specimen the reddish areas of the pronotum are almost as dark as the black ones. One of the males from the Andamans is more highly macrodont than any that appear hitherto to have been described. The form of the mandible of this specimen is shown in text-fig. 3C.

2. M. BIPLAGIATUS subsp. NIGRIPES, Boileau.

Siam: Hills between Thaungyin and Me Ping, *ca.* 1000 ft. (♂).

The markings of this specimen agree exactly with those described by Boileau, but the slight structural characters mentioned are not clearly shown. The pale markings are distinctly yellower than in our specimens of the typical form.

3. M. BIPLAGIATUS subsp. INDICUS, n. subsp.

S. India: Mysore (♂).

We have only one specimen, a small male. The head, thorax, legs and lower surface of abdomen are uniformly black with a faint reddish tinge. The distal tooth on the mandibles is remarkably strong.

Genus PROSOPOCOELUS, Hope.

Prosopocoelus approximatus, Parry.

Siam: Raheng (♂) "came into bungalow during the day time."

Prosopocoelus buddha, Hope.

E. Himalayas: Darjeeling District—Singla, 1500 ft. (♂ ♀); Pashok (♂ ♀).

Prosopocoelus oweni, Hope.

Assam: Sibsagar (♂).

Prosopocoelus wimberleyi, Parry.

Andamans (♂ ♀).

In our smallest male (14·3 mm. long) the colour is very near that of *Metopodontus biplagiatus*, though the black markings are less clearly defined. Transitional specimens connect this with the large form. The female resembles *Metopodontus biplagiatus* in colour still more closely.

Prosopocoelus parryi, Boileau.

E. Himalayas: Darjeeling District—Ghumti (♀); Nagri Spur*.

Genus CYCLOMMATUS, Parry.

Cyclommatus tarandus, Thunberg.

Malay Peninsula: Johore (♂).

Genus PRISMOGNATHUS, Motschulsky.

Prismognathus subnitens, Parry.

E. Himalayas: Darjeeling District—Kurseong (♂ ♀).

The small male (16 mm. long, excluding mandibles) resembles the form described by Parry, but the head is squarer than it appears in his figure. The clypeus is small and strongly bilobed. In the large male (21 mm. long) the head is flatter between the jaws, clypeus very broad and less strongly bilobed. The mandibles are armed with two stout teeth, one just below the tip and one about half way between this and the base: between these teeth, and between the median tooth and the base, it is armed with about 6-7 smaller teeth.

The female (13·5 mm. long) is slightly darker in colour than the male. The mandibles have an upper, terminal, and lower tooth. The clypeus is undivided; behind it the anterior border of the head is steep and concave much as in the small male, but there is no angular projection of the canthus. On each side of the head, between and in front of the eyes, there is a rounded ridge followed by a pronounced depression bordering a broad anteromedian convexity—structures which have their counterpart in the male. The whole upper surface is evenly but not very closely punctured, the head more strongly than the pronotum, and the pronotum than the elytra.

Genus HEMISODORCUS, Thomson.

Hemisodorcus nepalensis, Hope.

W. Himalayas: Mussoorie (♂ ♀).
Tehri Garhwal—Balcha (♀).
Almora—Binsa* (♂ ♀).

E. Himalayas: Darjeeling District—Darjeeling, 7000 ft. (♂ ♀); Kurseong, 5000-6000 ft. (♂ ♀); Mungphu (♂).

The elytra of the female are scarcely if at all less glossy at the sides than above: the sides (especially the anterior angles) of the pronotum are finely roughened and are as a rule coarsely punctured in addition. The head is finely roughened in front of the tubercle, coarsely roughened on either side of a smooth median band behind it.

Hemisodorcus suturalis, Westwood.

W. Himalayas: ? Kashmir Valley, *ca.* 5000-6000 ft. (♀).
Dehra Dun District—Jaunsar* (♂ ♀).
Tehri Garhwal—Balcha (♂).

This species has hitherto been placed in the genus *Dorcus*, from which it is distinguished by the structure of the prosternum and the characters of the female. The female differs from that of the preceding species in the more uniform sculpturing of the upper surface of the head, in the less prominent canthus, and in the elytra which, like those of the male, are polished only in their anterior inner angles. In the Kashmir specimen the distinction between the dull and polished parts of the elytra is much less

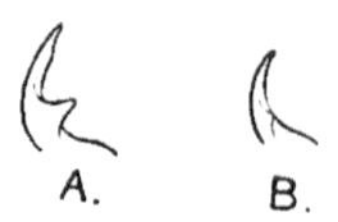

TEXT-FIGURE 4.
Left mandibles of large and small males of *Dorcus yaksha*, nat. size.

marked than in the others. This specimen may, therefore, belong to a distinct species or local race; or it may be in poor condition.

Genus DORCUS, MacLeay.

Dorcus yaksha, n. sp.

(Pl. xxix, fig. 1: text-figure 4.)

E. Himalayas: Darjeeling District—Kurseong, *ca.* 5000 ft. (♂ ♀).
Dafla Hills—Dikrang Valley (♂).

This species resembles *D. vicinus*, Saunders and *D. ratiocinativus*, Westwood, but differs from both in having the large tooth on the mandibles much smaller in proportion to the size of the insect, and situated more basally. It also resembles *D. antaeus*, Hope, but is much smaller, its tooth being proportionally larger than in that species. Females closely resemble those of *D. antaeus*, but are very much smaller.

From Kurseong we have two males, 33 and 29 mm. long respectively, and two females 28 and 22 mm. long respectively. They were all presented to us by Mr. N. B. Jahans, who collected them when he was at school there. We also have a male collected by Col. Godwin-Austen in the Dikrang Valley ; it is intermediate in size between the two males from Kurseong.

In the largest males there is a rudimentary tooth on the gently tapered distal part of the mandibles as in large specimens of *D. antaeus*, and the proximal tooth is large and conical. In the smallest male there is no trace of the distal tooth, and the proximal tooth is smaller and less acute. The mandibles of the female resemble those of *D. antaeus*. The upper surface of the head is glossy but very finely roughened in the male, and coarsely roughened in the female.

The outer margins of the prominent anterior angles of the prothorax are highly **S**-shaped in the large male, the concavity being situated behind the convexity In the small male and both the females it is entire In the small but not in the large male the lateral and posterior parts of the marginal groove of the pronotum are broad and coarsely punctured. In the female this puncturing is still more extensive.

The elytra are glossy in both sexes. They are smooth above and coarsely and closely punctured at the sides ; but the punctures are almost obsolete in the large male. In the smaller, and to a less extent in the larger of our two females, the smooth dorsal area is traversed by incomplete longitudinal rows of punctures, arranged after the manner ot the striations with which the elytra of *D. hopei* are marked. The posterior margin of the prosternal process is bordered by a groove, in front of which there is a distinct convexity in both sexes. The anterior tibiae are armed with about six teeth, the middle and posterior each with one small tooth.

Dorcus antaeus, Hope.

E. Himalayas : Darjeeling District, 4000 ft. (♂) ; Darjeeling, 7000 ft. (♂ ♀) ; Kurseong, 5000-6000 ft. (♂ ♀).
Upper Burma : Southern Shan States—Keng Dung* (♂).

The elytra of females and small males are obscurely punctured at the sides only. The Burmese specimen perhaps represents a distinct local race. It is about 54 mm. long (mandibles excluded), and may conceivably be the large form of *D. laevidorsis*, Fairmaire. It is the property of the Bombay Natural History Society.

Dorcus hopei, Saunders.

W. Himalayas : Dehra Dun District—Jaunsar* (♀).
E. Himalayas : Darjeeling District—Darjeeling, 7000 ft. (♂ ♀) ; Kurseong, 5000-6000 ft. (♂ ♀).
Assam : Khasi Hills (♂).

The elytra of females and small males are coarsely striato-punctate.

Genus EURYTRACHELUS, Thomson.

Eurytrachelus fulvonotatus, Parry.

E. Himalayas: Darjeeling District—Darjeeling, 7000 ft. (♂ ♀); Kurseong 5000-6000 ft. (♂ ♀).

This species is distinguished by the structure of its clypeus from the genus *Hemisodorcus* with which it has hitherto been associated.

The extent of the fulvous markings is very variable in both sexes; sometimes only the posterior spots on the pronotum and the posterior streaks on the elytra remain; usually their anterior counterparts are also present and these may fuse with them; the pronotum may be bordered by fulvous markings on all four sides. The female resembles the male in colour.

Eurytrachelus reichei, Hope.

(Pl. xxix, fig. 2.)

E. Himalayas: Darjeeling District—Darjeeling, 6000-7000 ft. (♂ ♀); Kurseong 5000-6000 ft. (♂ ♀); Siliguri, in the Terai a few miles south of the base of the hills (♂).

E. praecllens, Möllenkamp, also from the Himalayas, must be very near if not identical with this species. The thickness of the mandibles of relatively large males, and the width of the pair of teeth on the inner side of each, are very variable.

Females of this species closely resemble those of *E. tityus*. They are, however, distinguished by the sculpture of the posterior ends of the elytra. In females and small males of both species the elytra bear a series of deeply impressed longitudinal punctured grooves with smooth ridges between them, of which ridges the first and third and often the sixth from the suture are the broadest. In the female of *E. reichei* the sixth ridge is of approximately uniform width throughout, and tends to be enlarged at the posterior end where it bends inwards to meet the end of the third ridge (see pl. xxix, fig. 2). In small males of *E. reichei*, it is also of uniform width throughout but the posterior ends of all the ridges are obsolete.

Eurytrachelus submolaris, Hope.

(Pl. xxix, fig. 4.)

W. Himalayas: Murree (♂ ♀): Naini Tal* (♂).

This species is represented in the Dehra Dun collection by a short series of males without any locality record. The largest specimen answers closely to Boileau's account of the type (*Trans. Ent. Soc. London*, 1913, pp. 251-2, pl. ix, fig. 10). There is a similar specimen from Naini Tal in the Pusa collection.

The female is very like that of *E. reichei*, but the striation of the elytra is weaker (see pl. xxix, fig. 4, and Boileau, *Bull. Soc.*

Ent. France, 1904, p. 27—*Dorcus brachycerus* = *Eurytrachelus submolaris*, Boileau, *Trans. Ent. Soc. London*, 1913, p. 251).

Eurytrachelus tityus, Hope.

(Pl. xxix, fig. 3.)

W. Himalayas. Naini Tal (♂).

E. Himalayas: Darjeeling District—Darjeeling (♂ ♀); Kurseong, 5000-6000 ft. (♂ ♀); Soorn, 4000-5000 ft. (♀); Rungneet Tea Estate, 4500-5000 ft.* (♀); Siliguri, in the Terai a few miles south of the base of the hills (♂ ♀).

Assam: Cachar (♀).

In this species the third and sixth ridges of the elytra of females and small males taper away behind, and their union is only faintly indicated (see pl. xxix, fig. 3).

Eurytrachelus travancorica, n. sp.

(Pl. xxix, fig. 5.)

South India: Travancore—High Range, 6000 ft. (♂).

A single male of this species has been presented to us by the Agricultural Research Institute. It is very small (12·3 mm. long), but the dorsal tooth on the punctured and glossy mandibles is so long and slender (pl. xxix, fig. 5) that I think the specimen must be a large one of its kind. Below and slightly proximal to the dorsal tooth is an obsolete ventral tooth as in *Dorcus rugosus*, Boileau, from which the species may be distinguished by its black colour and imperfectly divided eye. The punctured and finely roughened clypeolabrum is keeled above the margin, but the keel is low and is not more pronounced laterally than medially; though perfectly distinct, it is not at all conspicuous. The prosternum, too, is that of a *Eurytrachelus*, not a *Dorcus* [1].

The anterior angles of the head are rounded and slightly prominent; the oblique anterior surface of the head is lightly concave and finely roughened between them; the remainder of the upper surface is glossy, and the whole is strongly punctured. The mentum is very coarsely punctured; it is roughly trapezoidal with strongly rounded anterior angles and very faintly concave anterior margin.

The pronotum is glossy and more coarsely punctured than the head; it is vaguely sulcate in the middle line. The anterior margin is convex in the middle. The anterior angles are acute and very strongly produced forwards by the side of the head. The sides are divergent and lightly convex. The posterior angles are replaced by a lightly concave margin. The posterior margin is faintly convex.

---

[1] Since the above was written a specimen of Boileau's species has been presented by Mr. H. E. Andrewes. It proves to belong to the genus *Eurytrachelus*, not *Dorcus*. *E. rugosus* and *E. travancorica* are practically identical in structure apart from the slightly shorter canthus of the latter.

The elytra are strongly rugose throughout, with indications of the striae found in small males of *E. reichei* and *E. titvus*. The prosternum is coarsely but not very closely punctured in front of the coxae; between them it is more closely and finely punctured, clothed with golden yellow hair, and medially concave; it is closely punctured and almost rectangularly truncate behind. The mesosternum and metasternum are coarsely and closely punctured and clothed with rather long hair. The abdominal sterna are less closely punctured and their hair is very short.

Genus **PLATYPROSOPUS**, Hope.

Platyprosopus titanus, Boisduval.

1. PLATYPROSOPUS TITANUS, Boisd., *s. str.*

Malay Peninsula: Penang (♂).

2. P. TITANUS subsp. WESTERMANNI, Hope.

E. Himalayas: Darjeeling District—Pashok (♂).
Dafla Expedition (♂).
Assam: Khasi Hills—Shillong (♂); Sibsagar (♂).

Genus **GNAPHOLORYX**, Burmeister.

Gnapholoryx velutinus, Thomson.

E. Himalayas: Darjeeling District—Darjeeling, 7000 ft. (♀).
Abor Country—Kobo, 400 ft. (♂).

Genus **AEGUS**, MacLeay.

Aegus adelphus, Thomson.

Malay Peninsula: Johore (♂).

Our specimen agrees in all respects with Deyrolle's figure of a specimen from Borneo (*Ann. Soc. Ent. Belg.* IX, pl. ii, fig. 8).

Aegus capitatus, Westwood.

Malay Peninsula: Johore (♂).

Aegus labilis, Westwood.

E. Himalayas: Dafla Hills—Dikrang Valley (♂); Dafla Expedition (♀).
Abor Country—Upper Rotung, under leaf-stem of plantain (♀).
Upper Burma: Southern Shan States—Reng Dung* (♂ ♀).
Andamans (♂).

The basal tooth is distinctly smaller than the dorsal in the large male (Dikrang Valley). In a smaller form (Andamans) it is smaller and median; in a smaller one still (Dikrang Valley) it is obsolete and very near the base; and in the smallest of all it has disappeared. The elytra are coarsely punctured marginally, especially in small males.

The clypeus of the female is very broad with concave anterior margin. The sloping anterior part of the head, which is lightly concave, meets the horizontal posterior part at a distinct angle. The pronotum and elytra are coarsely punctured at the sides, more finely above. In the female from Reng Dung, which belongs to the Bombay Natural History Society, the anterior part of the head is less markedly concave than in the Himalayan specimens. The male by which it is accompanied is of the small form in which the dorsal tooth on the mandibles is not developed.

Aegus impressicollis, Parry.

Sumatran Islands: Sinkep (♂).

Aegus chelifer, MacLeay.

Malay Peninsula: Johore (♂).

Aegus roepstorffi, Waterhouse.

Andamans: Port Blair (♂ ♀).
Nicobars (♀).
? Lower Burma: Rangoon (♀).

The clypeus of the female is narrower than in the preceding species. The separation of the anterior and posterior parts of the upper surface of the head is less abrupt. The punctures of the pronotum and elytra are much coarser and tend, even on the innermost ridge of the latter, to fuse together so as to produce a general rugosity of the surface.

Aegus kandiensis, Hope.

Ceylon: Central Province—Kandy (♀); Peradeniya (♂ ♀); Pundaluoya (♂); Talawakelle* (♂).
Sabaragamuwa—Kegalle (♀); Yatiyantota (♀).
Uva—Haldummulla (♂).

The female of this species closely resembles that of the last. It is, however, even more closely punctured, the difference—which is never very great—being clearest on the pronotum.

Genus **NIGIDIUS**, MacLeay.

Nigidius dawnae, n. sp.

(Pl. xxix, fig. 7.)

Lower Burma: Amherst District of Tenasserim—near Misty Hollow and Sukli, towards the top of the western and eastern slopes respectively of the Dawna Hills, 2000-2500 ft. (♂ ♀).

Siam: Meetaw Forest, west of Raheng.

The Burmese specimens of these beetles were found in the interior of two pieces of hard dry wood by the road-side. In one piece—that found on the eastern side of the hills—they were accompanied by larvae.

♂. Short and stout like *N. obesus.* Black, 13·0-21·5 mm. long. The long curved dorsal tooth of the mandibles is about as long as the part of the mandible in front of it. Its posterior margin bears a rounded laminar tooth which is sharper in small than in large specimens. The apex of both mandibles is tridentate, but the lowest tooth is rudimentary on the right side The labrum is glossy; its anterior margin is concave and its anterior angles are laterally produced and acute; the clypeus forms a transversely linear convexity above its whole basal width, and is bounded behind by a groove from which a row of erect hairs project in unworn specimens.

The anterior margin of the frons is convex in the middle. The upper surface of the head is flattened as a whole, and slightly undulating; its anterior part is always glossy, but its posterior part may be dull; its punctures are much finer in front than they are behind, where each is broad and flat with a raised ring in the centre. The canthus is abruptly constricted on the outer side about the middle of its length; the posterior part is broader and more convex than the anterior. It is coarsely and very closely punctured. The mentum is closely and very coarsely punctured; it is slightly wider in front than behind and distinctly wider than long; the anterior angles are rounded and the anterior margin is concave.

The pronotum is more than twice as broad as long; it is strongly convex across the middle in front. Its surface is glossy; it is coarsely punctured at the sides, along the anterior and posterior margins and in the median groove which does not nearly reach the anterior margin. The anterior margin is lightly concave on either side of a broad median convexity, the other margins are convex as a whole, but the anterior part of the narrow reflexed border is widened. In large specimens this widening is very abrupt (see pl. xxix, fig. 7).

The grooves of the elytra each contain one row of large shallow punctures, except the last, which contains about three such rows; and there is a row of fine punctures on each side of each of the somewhat narrow ridges between them.

All the sterna are somewhat coarsely punctured; the prosternum is lightly keeled between the coxae and descends gradually to the level of the mesosternum behind them. The metasternum has a median groove.

The basal piece of the genital tube is simple, and does not overlap the median lobe; it is not chitinized in the middle line above. The lateral lobes are very slightly concave on their inner sides The median lobe is cylindrical and is deeply bifid. The internal sac is laminar, elongate and parallel-sided. It is supported by a pair of strongly chitinized laminae of which one lies on either side of the ductus ejaculatorius in the tissues of the sac.

The tips of these supporting laminae are weak and unite with the chitinous support of the ventral margin and lateral walls of the funnel-shaped aperture of the ductus ejaculatorius. The left lamina, in addition, gives off a broad tongue-like branch just below the tip. This tongue extends slightly beyond the end of the body of the sac. The tongue is unarmed, but a band of fine backwardly directed teeth extends along the ventral distal margin of the body of the sac, and thence obliquely backwards on both sides to the dorsal surface, where the left hand portion of the band ends about opposite the unpaired tongue-like process, the right hand portion being about five times as long.

There is no flagellum.

The internal sac is permanently everted. When the genital tube is retracted this sac does not lie against the median lobe, but in a delicate sheath attached to the outer surface of the internal abdominal segments.

♀. Differs from males only in having somewhat smaller mandibles in proportion to its size, and in the structure of the genitalia.

This species appears to come very near *N. obesus*, Parry, but it is larger and the anterior angles of its pronotum are neither simple nor of the shape shown in Westwood's figure (*Trans. Ent. Soc. London*, 1874, pl. iii, fig. 5).

The anterior margin of the frons, too, is concave on either side of the median convexity, not evenly convex as shown in that figure. There are also slight differences in the shape of the canthus.

Nigidius himalayae, n. sp.

(Pl. xxix, fig. 6.)

E. Himalayas : Darjeeling District—Pashok (♂).

A male of this species was obtained by H. E. Lord Carmichael's collectors in the Darjeeling District. It is 20·0 mm. long, and was the only specimen I had seen when the following description was drawn up. More recently Mr. Lister has sent me specimens from Pashok which vary from 13·7 to 17·7 mm. in length.

In general appearance this species resembles the last, but it is distinctly slenderer. The lowest terminal tooth of the right mandible is absent. The clypeus is longer than and scarcely as wide as in the last species, and is bilobed. The middle part of the frons is less prominent than in that species. The canthus is less deeply cleft, and the posterior part is less prominent. There is a strongly marked depression in the middle line towards the back of the head, as well as a pair of depressions behind the anterior angles.

The pronotum is less than twice as broad as long. It is bordered in front by a very broad groove. This groove is marked with large shallow punctures, and is crossed in the middle line by a fine keel, which is terminated behind by a transverse keel of similar dimensions to itself. The broadly reflexed anterior parts of the lateral margins of the pronotum are well developed. The metasternum is not punctured in the middle. The abdominal sterna are also less uniformly punctured than in *N. dawnae*.

The basal piece of the genital tube is furnished with an elongate triangular mid-ventral lamella between the lateral lobes. Each of these lobes is strongly concave on the inner side, and is furnished with a large inwardly directed ventral lamina. Together these structures form an imperfect sheath in which the median lobe and the base of the internal sac are hidden. The exposed portion of the latter is ribbon-like with a rounded extremity; the terminal portion, though supported by the chitin accompanying the ductus ejaculatorius, is composed apart from this of a curious cellular material which when dry resembles dried vegetable tissue. The armature is confined to the region immediately preceding this terminal portion. Although the internal sac is permanently evaginated, as in the preceding species, it lies with the remainder of the genital tube inside the internal abdominal segments when at rest.

In other respects this species resembles *H. dawnae*.

Nigidius distinctus, Parry.

Bengal: Duars—Maindabari, Buxa Division (♂ ♀).
Upper Burma: N. Shan States—Hsipaw* (♀).

Mr. Beeson has sent me eight males and four females of what I take to be *Nigidius distinctus* from Maindabari. In both sexes there is some variation in the proportion of length to breadth, and in the puncturing. The specimens agree as well with Fairmaire's description of *N. oxyotus* from Tonkin (*Ann. Soc. Ent. Fr.* (6) VIII, 1888, pp. 339-340) and Boileau's description of *N. birmanicus* from Rangoon (*Trans. Ent. Soc. London*, 1911, pp. 446-449) as they do with Parry's description and figure of *N. distinctus* (*Trans. Ent. Soc. London*, 1873, pp. 341-2, pl. v, fig. 7); and I am unable to distinguish them from the specimen from Hsipaw in the Dehra Dun collection. The Hsipaw specimen is unfortunately a female, and I have been unable to examine the male genitalia of any Burmese or Malaysian specimens. Possibly they might afford distinctive characters as in the two species described above.

Nigidius impressicollis, Boileau.

Assam: Khasi Hills—Maflong, 5900 ft. (♂ ♀).

Adults and larvae of this species were found by Mr. S. W. Kemp in damp and thoroughly soft and rotten wood. The sexes are scarcely distinguishable externally.

Genus **FIGULUS**, MacLeay.

Figulus interruptus, Waterhouse.

Ceylon: Peradeniya.

Figulus scaritiformis, Parry.

Malay Peninsula : Johore.

Genus **CARDANUS,** Westwood.

Cardanus sulcatus, Westwood.

Malay Peninsula : Johore.

EXPLANATION OF PLATE XXIX.

FIG. 1.—*Dorcus yaksha*, n. sp. ♀ × 2.
,, 2.—*Eurytrachelus reichei*, Hope, ♀ × 2.
,, 3.— ,, *tityus*, Hope, ♀ × 2.
,, 4.— ,, *submolaris*, Hope, ♀ × 2
,, 5.— ,, *travancorica*, n. sp. ♂ (type) × 2.
,, 6.—*Nigidius himalayae*, n. sp. ♂ (type) × 2.
,, 7.— ,, *dawnae*, n. sp. ♂ (type) × 2.

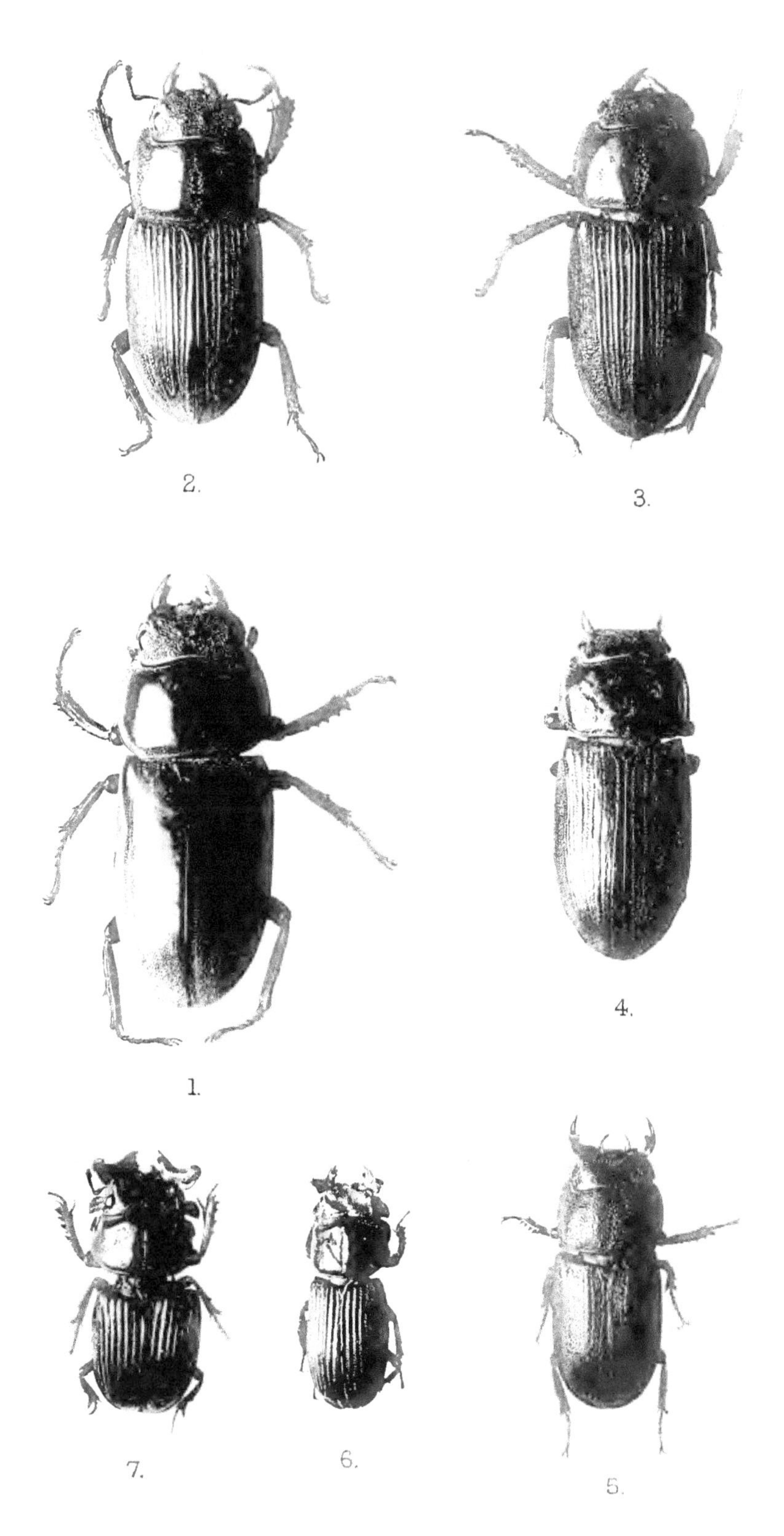

Bemrose, Collo, Derby

INDIAN LUCANIDAE.

XXVI. A REVISION OF THE ORIENTAL SUBFAMILIES OF TARANTULIDAE (ORDER PEDIPALPI).

By F. H. Gravely, *M.Sc., Asst. Superintendent, Indian Museum.*

(Plate XXXI.)

The species of Tarantulidae are rendered exceptionally difficult to separate and define by the insignificance of many of their most distinctive characters, and by the conspicuousness of others whose striking modifications indicate the age of a specimen rather than the species to which it belongs. It is only by the study of long series of specimens that the latter characters can be eliminated and the former recognized with certainty. Kraepelin's "*Revision der Tarantuliden*" (Abh. Ver. Hamburg, xiii [3] 1895, 53 pp., 1 pl.) has straightened out the synonymy of the family, and has gone a long way towards putting the classification into shape. But when this, and the volume of "*Das Tierreich*" by which it was followed, were written, the material available for study appears to have been somewhat scanty. A number of described species which are undoubtedly distinct had therefore provisionally to be united; and a number of species still remain unnamed.

I have now for several years been making special efforts to obtain adequate series of specimens from different parts of the Indian Empire, and whenever possible from beyond. In the present paper I propose to consider the Oriental species in the light of material recently obtained; and it seems best to complete the paper by references to all known members of the two subfamilies dealt with, although those found outside the Oriental Region are not well represented in the material before me.

I am indebted for help in getting material to Dr. Henderson, Mr. E. E. Green, Mr. Kinnear, Mr. T. Bainbrigge Fletcher and especially to Mr. B. H. Buxton who has presented to the Indian Museum a number of new species which he recently collected in the Malay Peninsula.

With the exception of *Stygophrynus moultoni*, of which the type is in the British Museum, the types of all new species described below are in the Indian Museum.

Subfamilies and General Structure.

The Oriental Tarantulidae fall into two very distinct subfamilies, which may be recognized thus :—

Pulvilli present; hand able to bend till it forms a right angle with the tibia, the terminal spines of which are directed sideways; prosomatic sterna small, more or less tuberculiform .. ..*Charontinae*, p. 435.
Pulvilli absent; hand unable to bend at less than an obtuse angle to tibia, the terminal spines of which are directed forwards in adults above base of hand; prosomatic sterna broadly expanded, lightly concave or flat ... ... *Phrynichinae*, p. 447.

The American subfamily Tarantulinae differs from the subfamily Charontinae chiefly in the absence of a pulvillus.

The structure of the arm and hand, though differing in detail in different species, is remarkably uniform in plan in the young of all the species of Tarantulidae whose development is known to me. Considerable changes, however, take place during the growth of individuals belonging to the larger species. This is especially the case with species of the subfamily Phrynichinae, the hand of which is so modified in the adult that each is capable of grasping prey without the aid of the other (see Gravely, 1915, pl. xxiv, fig. 28 of this volume). In this respect the Phrynichinae may be regarded as more highly specialized than the Charontinae, and as the structure of the arm and hand presents greater difficulties than does that of other organs, the Charontinae may conveniently be considered before the Phrynichinae.

In some respects, however, the former are probably more highly specialized than the latter. The jointing of the hind tibiae, for instance, which is often less marked, when it occurs, in young specimens than in old ones, is carried further in the Charontinae than in the Phrynichinae. And it is difficult to think that pulvilli can ever have been present in creatures with the habits of the Phrynichinae, when these are not found in them at the present day. For both Phrynichinae and Charontinae habitually live clinging to the underside of stones or logs of wood; and the latter, which have pulvilli, can cling in this position to polished glass, whereas not even the young of the former, which lack them, can do this.

The fundamental structure of the arms and hands of the Tarantulidae may now be described as it is to be seen, more or less distinctly, in the young probably of all species, and in the adults of many Charontinae. The modifications to which it is subject during the growth of the more highly specialized forms are all in the direction of the specialization of particular spines and the loss of others.

The anterior face of the trochanter is bounded above by a dorsal row of spines, and below by a ventral cluster; while between these is a middle group or longitudinal row.

The anterior face of the femur is flattened, and is bounded by a dorsal and a ventral row of spines. The tibia is similarly flattened in front and armed above and below, the spines of the distal half of the dorsal row always being much the longest.

The hand is armed with two spines above and one below. Occasionally additional spines are also present.

The finger may be armed at the base with 0, 1 or 2 dorsal spines, which remain throughout life. It is always unarmed ventrally.

Subfamily *CHARONTINAE*.

The structure of the second visible abdominal sternum and of the margin of the carapace opposite the lateral eyes, the relative lengths of the two dorsal spines on the hand and of those on the end of the arm, and the jointing of the finger and of the tibia of the fourth leg, appear to be the principal characters that have been used in the definition of genera.

The structure of the posterior margin of the second visible abdominal sternum seems to be very variable, and I am unable to attach any importance to it.

The segmentation of the hind tibiae is often less marked in the young of species in which it occurs, than in adults; it is sometimes variable within the limits of a single well-marked species, and it reaches its highest development in more than one genus, among them the specialized cavernicolous genus *Stygophrynus*. There can, I think, be little doubt that the extent of this segmentation is a mark of the degree of specialization in the species in which it occurs. Probably increased segmentation facilitates in some way the activities of the animal exhibiting it, and may appear independently in different branches of the subfamily. It is also found in the genus *Damon* of the subfamily Phrynichinae.

In species in which the hind tibiae are normally not more than 3-jointed, the tarsi (excluding the metatarsi) appear to be invariably 4-jointed. In most species in which the hind tibiae are 4-jointed the tarsi are 5-jointed. *Sarax javensis* is the only species known to me which appears to have both tibiae and tarsi of the hind legs 4-jointed, and as I have only one specimen before me the tibiae may be abnormal. The structure of the tarsi appears to be constant within the limits of each species, whereas in *Phrynichosarax cochinensis* and *singapurae*, and perhaps therefore in other species also, the structure of the tibiae is variable. Although, therefore, the structure of the hind tibiae is usually much easier to distinguish than is that of the tarsi, it seems best to use the latter rather than the former for the separation of genera.

The structure of the margin of the carapace appears to be of more fundamental importance from a taxonomic point of view than is the structure of the legs. By its means the subfamily may split into two distinct groups. One of these, which may be termed the *Sarax* group, includes only small species whose distribution extends from India through Malaysia as far as the Solomon Islands.

The other, which may be termed the *Charon* group, includes the small species found on the outskirts of and beyond this area from the Seychelles to the Galapagos Islands, together with the large and highly specialized species belonging to the genera *Stygophrynus* and *Charon*.

The *Charon* group is probably older than the *Sarax* group, its range being wider, the hind tibiae being 4-jointed in all except two species (in which they are 3-jointed), and the arms being comparatively long and slender, at least in well developed males, in most if not all species. The carapace of the former group, too, resembles that of all other subfamilies of Tarantulidae, and also, apparently, that of the newly hatched larvae of the only species—*Phrynichosarax cochinensis*—of the *Sarax* group whose larvae I have seen.

The hind tibiae of the *Sarax* group may be 2-jointed (occasionally even entire), and the proportion of the species in which they are 4-jointed appears to be smaller than in the *Charon* group. The arms are almost invariably short and stout even in males.

The relative lengths of the two dorsal spines on the hand, and of those on the end of the arm, may perhaps to some extent be correlated with cavernicolous habits in both groups. So far as I know, however, nothing is known of the habits of the genus *Charon*, one of the two genera of its group in which the spines tend to resemble those of the single exclusively cavernicolous genus *Catageus* of the *Sarax* group. A few species of the latter group belonging to the non-cavernicolous genera have, moreover, been found in caves.

The finger is jointed in all genera except *Charon*.

The genera of Charontinae may now be defined as follows:—

1.	Margin of carapace indentated beside lateral eyes ... ... (*Sarax* group)	2.
	Lateral eyes situated further from margin of carapace, which is entire ... (*Charon* group)	4.
2.	Longest spine on tibia of the arm the middle one of five well developed dorsal spines in adults, and of three in young; proximal dorsal spine of hand longer than distal	*Catageus*, p. 437.
	Penultimate well developed dorsal spine of tibia of arm the longest in all stages; distal dorsal spine of hand longer than proximal ...	3.
3.	Tarsi (exclusive of metatarsi) 4-jointed; hind tibiae 2- to 4-jointed (sometimes entire on one side) but normally 3-jointed (? always) ...	*Phrynichosarax*, p. 437.
	Tarsi (exclusive of metatarsi) 5-jointed; hind tibiae 4-jointed	*Sarax*, p. 441.
4.	Penultimate dorsal spine of tibia of arm the longest, the one next behind it longer than the one next behind that; distal dorsal spine of hand longer than proximal, not accompanied by additional spines	5.
	Penultimate dorsal spine of tibia of arm not longer than the one next behind it, often about equal to the one next behind that, sometimes even shorter; long spine on dorsal side of hand usually succeeded by several shorter ones,[1] a short spine often fused to it proximally at base	6.

[1] Always, so far as is known, except in *Stygophrynus moultoni*, for which a new genus ought perhaps to be established.

5. Tarsi (exclusive of metatarsi) 4-jointed; hind tibiae 3-jointed ... ... *Charinides*, p. 442.
   Tarsi (exclusive of metatarsi) 5-jointed: hind tibiae 4-jointed ... ... *Charinus*, p. 442.
6. Finger jointed, three dorsal spines of tibia of arm much longer than any others ... *Stygophrynus*, p. 443.
   Finger unjointed; two dorsal spines of tibia of arm much longer than any others ... *Charon*, p. 446.

Genus **CATAGEUS**, Thorell.[1]

Type *Catageus pusillus*, Thorell. No other species of the genus is known, *C. rimosus*, Simon, belonging in reality to the following genus.

Catageus pusillus, Thorell.[2]

(Plate xxxi, fig. 1.)

Catageus pusillus is only known from the Khayon ("Farm") and Dhammathat caves near Moulmein.

The Indian Museum collection includes specimens from both groups of caves They were found under stones, and their habits have already been described.[3]

The *carapace* of our largest specimen is 4·4 mm. across and 3·2 mm. long in the middle line. The *arms* (see fig.) are short and stout in all specimens. The finger is armed dorsally with two minute and slender spinules (see fig.). The *antenniform legs* are very variable in length, their femora being from about two to about three times as long as the carapace is broad. The femora of the first pair of *walking legs* are about 1·3–1·5 times as long as the carapace is broad. The metatarsi of the same pair of legs are about 1·2 or 1·3 times as long as the tarsi, and the first tarsal joints are about 1·4 or 1·5 times as long as the remaining tarsal joints.

Genus **PHRYNICHOSARAX**, n. gen.

Margin of carapace indentated beside lateral eyes; penultimate dorsal spine of tibia of arm longer than all others; distal dorsal spine of hand longer than proximal; hind tibiae normally composed of less than four pieces, tarsi of less than five.

Type *Phrynichosarax cochinensis*, n. sp.

Five species are known to me. They may be distinguished thus:—

1. Dorsal margin of finger armed with one spine ... 2.
   Dorsal margin of finger armed with two spines ... 3.
2. Spine of finger long; hind tibiae 2- to 3-jointed *P. cochinensis*, p. 438.
   Spine of finger minute; hind tibiae (? always) 4-jointed ... ... *P. javensis*, p. 439.
3. Spines on finger large and conspicuous, the distal one about twice as long as the proximal ... *P. buxtoni*, p. 439.
   Spines on finger small and of more nearly equal size ... 4.
4. Spines on finger small but quite distinct ... *P. singapurae*, p. 440.
   Spines on finger minute and inconspicuous ... *P. rimosus*, p. 440.

---

[1] *Ann. Mus. Civ. Genova*, XXVII, p. 530.
[2] *Ibid.*, pp. 531-8.
[3] *Journ. Asiat. Soc. Bengal* (n. s.), IX, 1914, p. 419.

Phrynichosarax cochinensis, n. sp.

(Plate xxxi, fig. 2.)

This species is common under stones in the evergreen jungles of the lower slopes of the Western Ghats in Cochin, and it is on account of the large number of specimens available for study that I have selected it as the type of the genus. The specimens were found near Kavalai, on the Cochin State Forest Tramway, at altitudes up to about 2500 ft. above sea level; at about 0-300 ft. above sea-level near the same tramway between miles 10 and 14; and at the base of the hills near Trichur. Specimens from the last-named locality differ from the others in that the legs (both kinds) tend to be much longer and slenderer, while the separation of the first and second joints of the hind tibiae is usually obscure or absent. In one specimen, indeed, the right hind tibia is entire.

As type of the species I have selected a female with young still adhering to her back in the preserved state. This specimen is from jungle beside the lower part of the State Forest Tramway.

The *carapace* is $1\frac{1}{2}$ times as wide as it is long in the middle line, or may be a little wider; its maximum width is slightly over 4 mm. It resembles that of *P. buxtoni* (below, p. 439) in general structure, but is finely and evenly granular throughout, and usually *looks* much broader in proportion to its length. The depression in the median groove behind the eyes is less defined, although the groove is well developed. The second radial grooves of the two sides are united across the middle-line, together forming an almost straight line in contact with the anterior part of the fovea.

The *arms* are always short and stout. The proximal dorsal spine on the hand is scarcely as long in proportion to the distal as in *P. buxtoni*. There is only one spine on the finger (see pl. xxxi, fig 2); it is situated close to the base of the dorsal margin, and is about as long as the ventral spine of the hand, which latter spine is situated close to the lower distal angle.

The *legs* are variable in length. The femora of the antenniform legs may be from scarcely $1\frac{1}{2}$ to fully $2\frac{1}{2}$ times as long as the carapace is wide. The femora of the first pair of walking legs may be from a little less than, to nearly $1\frac{1}{2}$ times as long as the carapace is wide. The metatarsi are longer than the tarsi, and the first tarsal joint of each leg is longer than are the rest together—very slightly so in short-legged specimens and much more so in long-legged. The hind tibiae may be more or less distinctly 2- or 3-jointed. In one specimen that of the right side is entire, that of the left side being 2-jointed. The extent of the jointing of the hind tibiae and the slenderness of the legs appears to be correlated with locality as noted above. All the localities from which the species is yet known are situated in one comparatively small area, over the whole of which comparatively uniform conditions probably prevail. Specimens from any one of these localities appear to exhibit a much smaller range of variation than the species as a whole, their extremes scarcely, indeed, overlapping. The fact that this much variation does, however, occur, and that specimens from other localities in the same neighbourhood may ultimately be proved to show similar ranges of variation which overlap extensively, seems to render it improbable that the Trichur form ought to be recognized as a definite race worthy of a subspecific name.

Phrynichosarax javensis, n. sp.

(Plate xxxi, fig. 3.)

Only one specimen is known to me. It is from Buitenzorg. It differs from *P. cochinensis* only in the minuteness of the spine on the finger (see pl. xxxi, fig. 3) and in the 4-jointed hind tibiae. The 4-jointed tarsi suggest that a larger series would be not unlikely to show that the hind tibiae were normally 3-jointed as in other members of the genus.

The carapace is 3·2 mm. broad by 2·2 mm. long in the middle line. The femora of the antenniform legs are 4·8 mm. long, those of the first walking legs 2·9 mm.

Phrynichosarax buxtoni, n. sp.

(Plate xxxi, fig. 4.)

Two specimens (one immature) were collected by Mr. B. H. Buxton in Kubang Tiga cave, Perlis, Malay Peninsula.

The *carapace* is heart-shaped. In the mature specimen (♀) it is 4·1 mm. broad by 3·3 mm. long in the middle line. Behind the lateral eyes it is bordered by a broad horizontal ledge. The fovea is deeply impressed, continuous with a pair of large lateral grooves directed slightly backwards, and with a short median groove behind it. In an anterior median groove, about two-thirds of the way from the fovea to the eye, is a hollow somewhat smaller than the fovea, with which, and with two pairs of lateral depressions together enclosing a rectangle, it forms an almost regular hexagon. The anterior sides of this hexagon are, however, a little longer than the posterior, and these than the lateral. A radial groove extends outwards and a little forwards from each member of the two pairs of lateral depressions, and between the posterior of these grooves and the lateral grooves connected with the fovea is a pair of short grooves extending from the margin about half way to the fovea. A single line of tubercles runs from the fovea outwards and backwards towards the margin between the last-mentioned grooves and those immediately behind them. The rest of the surface is ornamented with less definite bands and patches of tubercles.

The *arms* are short and stout. The proximal dorsal spine of the hand is little more than half as long as the distal; there is a somewhat shorter spine on the ventral margin. Even the ventral of the spines of the hand is, however, longer than either of the two spines with which the finger is armed. Both the spines on the finger are dorsal, and the proximal is less than half the size of the distal, being about equal in length to the distance from its base to the base of the finger or of the distal spine (pl. xxxi, fig. 4).

The femora of the antenniform *legs* are 8·6 mm. long in the adult specimen, those of the first walking legs being 5·5 mm. The anterior metatarsi are 2·6 mm. long, the anterior tarsi 2·0 mm. The tarsi are 4-jointed, the first joint distinctly longer in all legs than the other three together. The posterior tibiae are 3-jointed in both specimens.

Phrynichosarax singapurae (Gravely).[1]

(Plate xxxi, fig. 5.)

In view of what has been pointed out above with reference to *P. cochinensis*, it is very doubtful whether the proportions of the legs have any great taxonomic importance; and it was on these that my preliminary separation of the present form as a subspecies of *Sarax sarawakensis* was based. A more detailed examination has shown, however, that the armature of the hand and finger of the Singapore form differs markedly from that of the Sarawak form, and that the tarsi have one joint less.

Only one out of our series of eleven specimens from Singapore shows any trace of a fourth joint in the hind tibiae, though this joint is well developed in two specimens recently collected by Mr. B. H. Buxton in Lankawi (? main island) off the west coast of the Malay Peninsula. One of the Lankawi specimens has slenderer arms than any other specimen belonging to the *Sarax* group known to me.

This species is closely related to the next, from which it only differs in the larger size of the spines on the hand (compare figs. 5 and 6, pl. xxxi).

Phrynichosarax rimosus (Simon).[2]

(Plate xxxi, fig. 6.)

The Superintendent of the Cambridge University Zoological Museum has been good enough to send me the type specimen of this species for examination. It is an ovigerous female, and was found by a member of the "Skeat" expedition to the Malay Peninsula at Kuala Aring in Kelantan. The species is represented in our collection by two specimens (one probably, the other certainly, immature) collected by Mr. B. H. Buxton in Lankawi (? small island not far from main island) off the west coast of the Malay Peninsula.

The *carapace* resembles that of *P. buxtoni* rather than that of *P. cochinensis*, but the depression in the anterior part of the

---

[1] *Sarax sarawakensis* subsp. *singaporae*, Gravely, *Rec. Ind. Mus.*, VI, pp. 36-38.

[2] *Proc. Zool. Soc. London*, 1901, p. 77.

median groove is not distinct. The hand also resembles that of *P. buxtoni*. The finger is armed above with two spines situated as in *P. buxtoni*, but quite minute, each being about half as long as the shorter of the two found in that species (see pl. xxxi, fig. 6). In this character *P. rimosus* resembles species of the following genus.

Genus SARAX, Simon.[1]

Type *S. brachydactylus*, Simon.

In "*Das Tierreich*" Kraepelin recognized two species in this genus, *S. brachydactylus*, Simon, and *S. sarawakensis* (Thorell). A number of species have undoubtedly, however, been grouped together by various authors under the latter name, including some belonging to the genus *Phrynichosarax*.

S. brachydactylus is not known to me. The remaining species may be distinguished thus:—

Proximal spine of hand slightly more than half as long as distal ... ... *S. willeyi*, p. 441.
Proximal spine of hand scarcely half as long as distal ... ... *S. sarawakensis*, p. 441

Sarax brachydactylus, Simon.[2]

Simon records this species from Luzon in the Philippines, where it was found in the caves of Antipolo (Province Morong), San-Mateo (Province Manila) and Colapnitam (Province Camarines-Sur).

Sarax willeyi, n. sp.[2]

(Plate xxxi, fig. 7.)

Two specimens preserved in the Indian Museum were collected by Dr. Willey in New Britain. The only character by which they appear to be distinguished from *S. sarawakensis* has been noted in the above key (see also pl. xxxi, figs. 7 and 8). In both *S. willeyi* and *S. sarawakensis* the spines on the finger are extremely small. In this respect these species closely resemble *Phrynichosarax rimosus*, which *S. willeyi* also resembles in all other characters except the structure of the legs by which the genera *Sarax* and *Phrynichosarax* are separated.

A specimen from Narcondam Island in our collection, and one from Table Island (Andamans) in the British Museum collection, must belong to this species or to one not yet described ; but the spines on the finger are imperfect in both.

Sarax sarawakensis (Thorell).[3]

(Plate xxxi, fig. 8.)

This species was described by Thorell from Sarawak. Mr. Moulton has sent me from the Sarawak Museum two specimens

---

1 *Ann. Soc. Ent. France*, LXI, 1892, p. 43.
2 *Ibid.*, pp. 43-44.
3 *Charon sarawakensis*, Thorell, *Ann. Mus. Civ. Genova*, XXVI, 1888, pp. 354-358.

found on Klingkang summit, between Sarawak and Dutch Borneo. They differ from all other species of *Sarax* and *Phrynichosarax* known to me in the markedly greater difference in size between the two spines on the dorsal margin of the hand (see pl. xxxi, fig. 8). The spines on the finger are minute as in the preceding species.

The larger specimen is somewhat larger than the type, the carapace being 4·5 mm. in width.

Genus CHARINIDES, Gravely.[1]

Type *Charinides bengalensis*, Gravely.

The genus *Charinides* bears to *Charinus* the same relation as does the genus *Phrynichosarax* to *Sarax*. Both *Charinides* and *Charinus* resemble the preceding genera in general structure, and in the size to which specimens grow. They differ only in the structure of the ocular part of the carapace and in this they resemble the following genera, from the much larger adults especially of which they differ markedly in the structure of the arm and hand.

Only one species of *Charinides* is known.

Charinides bengalensis, Gravely.[1]

This species is only known from Calcutta and its immediate neighbourhood, where it is quite common under bricks in shady places where desiccation is not too severe.

The proximal spine on the dorsal margin both of the *hand* and of the *finger* is about half as long as the distal. These spines are long and slender on the finger as well as on the hand (see pl. xxiv, fig. 29 of this volume). They closely resemble those of *Phrynichosarax buxtoni* (pl. xxxi, fig. 4).

Genus CHARINUS, Simon.[2]

Type *C. australianus* (Koch).

The genus *Charinus* is represented in the Indian Museum collection by two specimens of *C. seychellarum*, Kraepelin.

Kraepelin distinguishes three species in "*Das Tierreich*":—*C. australianus* (Koch)[3] from Viti and Samoa, *C. neocaledonicus*, Simon,[4] from New Caledonia, and *C. seychellarum*, Kraepelin,[5] from the Seychelles. *C. insularis*, Banks,[6] has since been described from the Galapagos Islands.

This genus and the preceding include all the most primitive species of the group to which they belong, and it is noteworthy that they are only found north, east and west of the country inhabited by the following genera, genera of which the adults are much larger and have more highly specialized arms and hands.

---

[1] *Rec. Ind. Mus.*, VI, pp. 35-36, fig. 2B.
[2] *Ann. Soc. Ent. France*, LXI, 1892, pp. 43 and 48.
[3] *Phrynus australianus*, Koch., *Ver. Ges. Wien*, XVII, p. 231.
[4] *Abh. Ver. Hamburg*, XIII, p. 47.
[5] *Mitt. Mus. Hamburg*, XV, p. 41.
[6] *Proc. Washington Ac.*, IV, p. 67, pl. ii, fig. 8.

Genus **STYGOPHRYNUS**, Kraepelin.[1]

Type *S. cavernicola* (Thorell).

In this genus, as in all of the foregoing of which I have sufficient knowledge to speak with certainty, particular spines on the second appendages have proved to provide admirable characters for specific diagnoses, while others are absolutely worthless for this purpose. The granulation of the surface of these appendages, and of the carapace, is also important in this connection.

The following species may be recognized:—

1.	Armature of hand consisting of two long dorsal spines and one ventral one only	*S. moultoni*, p. 443.
	Hand armed above and below with one long spine succeeded by a series of short spines which increase in length distally	2.
2[2].	Adults pale in colour[2], rather small and very lightly built, with long slender arms; occular lobes of carapace finely and evenly granular, without tubercles	*S. cavernicola*, p. 444.
	Adults somewhat or much darker in colour, larger and more heavily built with much stouter arms, occular lobes of carapace more coarsely and less evenly granular, usually marked with a number of scattered tubercles	3.
3.	Distal of three long spines on dorsal margin of tibia of arm with a spine of nearly half its own length on either side of it	*S. longispina*, p. 445.
	Spines on either side of distal of three long spines on dorsal margin of tibia of arm quite short in adults, the proximal one short in young specimens also	4.
4.	Adults somewhat pale[2] in colour, not very strongly granular	*S. berkeleyi*, p. 445.
	Adults very dark and strongly granular ...	*S. cerberus*, p. 446.

Stygophrynus moultoni, n. sp.

(Plate xxxi, fig. 9.)

Mr. Moulton has sent me a single much broken specimen of this species. It was found on Klingkang summit, between Sarawak and Dutch Borneo. It is somewhat small, but appears to be mature or very nearly so. It is a male and is very distinct from all other species of the genus.

The *carapace* is 7·4 mm. broad by 5·7 mm. long in the middle line. It is somewhat pale in colour, finely granular and without tubercles, like that of *S. cavernicola*.

The *arms* (pl. xxxi, fig. 9) are somewhat slender, but are much shorter than in adult males of that species, the femur being no longer than the carapace is broad. The armature of the femur resembles that of *S. cavernicola*, but the spines are necessarily closer together. The tibia is also armed much as in that species, but the subsidiary spines among the longer spines of the ventral

[1] *Abh. Ver. Hamburg*, XIII, p. 44.

[2] The young of all species are pale in colour and have relatively short arms.

margin are obsolete, while on the dorsal margin the first of the three long spines is situated in about the middle of the length of the joint, as in females of *S. cavernicola*. This spine is preceded by an additional spine about half way to the base of the joint and nearly as long as the spine following the distal of the three long spines, which terminal spine is fully half as long as the three long spines. The granulation of the convex posterior surface of both femur and tibia is obsolete. The backs of the hand and finger are smooth. The hand is armed above by two spines of about equal length and not much shorter than the long spines of the upper margin of the tibia; it is armed below by one somewhat shorter spine opposite the distal of the two upper ones. The finger is armed above with three minute tooth-like spinules.

The *legs* are coloured in a similar manner to the rest of the body. They appear to have been long and slender, with the antenniform legs exceptionally long as in the other species of the genus, all of which are known to be cavernicolous. The hind femur scarcely exceeds the basal piece of the hind tibia in length by more than the length of the patella, which suggests that the remaining pieces, which are broken, may have been two instead of three in number.

Stygophrynus cavernicola (Thorell).[1]

The habits of this species have been described from specimens found, like the type specimen, in the Khayon or "Farm" caves near Moulmein.[2] The Indian Museum collection includes a number of specimens from the larger of these caves, and two from a small cave at Dhammathat. The species has been recorded from Saigon by Kraepelin.[3] Kraepelin had, however, insufficient material for the determination of specific characters[4], and geographical considerations render it very improbable that this determination is correct.

The *carapace* of specimens which are probably adult—no ovigerous females of this species ever appear to have been found—is about 9·5 mm. broad by 7 mm. long in the middle line. It is of a pale yellowish-brown colour, and is finely granular as in the preceding species.

The *arms* are always slender; in the female the femur is a little longer than the carapace is wide, in the male it is nearly twice as long. The femur and tibia are finely granular, with two smooth longitudinal bands on the convex posterior surface. There

[1] *Charon cavernicola*, Thorell, *Ann. Mus. Civ. Genova*, XXVII, 1889, pp. 538-542.

[2] *Journ. Asiat. Soc. Bengal* [n.s.], IX, 1914, pp. 418-9.

[3] *Bull. Mus. Hist. Nat. Paris*, 1901, p. 265.

[4] This statement is based on an examination of specimens from Mentawei and Java, which Prof. Kraepelin showed me in Hamburg. They, too, are distinct and Prof. Kraepelin very kindly promised to send them to me for description whenever I should be ready to deal with them, a promise whose fulfilment the war has unfortunately made impossible.

is sometimes a small spine between the three long distal spines on the dorsal margin of the tibia and the base of the joint, especially in the female. The three long spines are succeeded by a spine of about half their own length, but the spines between them are never well developed and are often absent. The hand is armed above and below with one very long spine, succeeded by a series of much shorter ones, of which the distal are longer than the proximal, the dorsal spines being somewhat longer than the ventral. The long dorsal spine bears at its base a strong backwardly directed spinule, and this is often succeeded in adults by a short row of very much smaller spinules on the margin of the long spine. The finger is unarmed.

The *legs* are pale in colour like the rest of the body.

Stygophrynus longispina, n. sp.

(Plate xxxi, fig. 10.)

Two male and two immature specimens were collected by Mr. Buxton in a cave on Langkawi Island off the west coast of the Malay Peninsula.

The *carapace* of the adults is about 12 mm. broad by 9 mm. long in the middle line. It is of a very dark brown colour, and is somewhat more coarsely and sparsely granular than is that of the preceding species, with a few strong tubercles among the granules.

The *arms* are very short and stout, their femora being little if at all longer than the carapace is wide. Their femora and tibiae are more coarsely granular than in *S. cavernicola* and the smooth bands on the convex posterior surface are invaded by scattered rows of granules. The three long spines on the dorsal margin of the tibia are followed, as in *S. cavernicola*, by a spine of about half their own length (perhaps a little shorter in the present species), and a similar but even longer spine occurs between the last two of them, serving to distinguish this from all other species known to me. The hand (pl. xxxi, fig. 10) is armed as in *S. cavernicola*, but is somewhat more coarsely and less extensively granular behind. The finger is unarmed as in that species.

The *legs*, especially the antenniform legs, are long and slender as in other species of the genus. They are dark in tint, harmonizing with the rest of the body though actually somewhat paler than the carapace and much darker than the abdomen.

Stygophrynus berkeleyi, n. sp.

(Plate xxxi, fig. 11.)

One male and several immature specimens were collected by Mr. Buxton in caves at Lenggong, Perak, Malay Peninsula. The species is named after Mr. H. Berkeley, the District Officer of Upper Perak, who greatly facilitated Mr. Buxton's work in the district.

The *carapace* of the adult male is 15 mm. broad by 10·5 mm. long in the middle line. It is paler in colour than is that of

S. longispina, but lacks the yellow tint of that of *S. cavernicola*. The immature specimens with it suggest that this is the normal colouration of the species. The granulation of the carapace is very coarse, and the tubercles are more numerous and more conspicuous than in *S. longispina*.

The *arms* are longer than in *S. longispina*, the femora being about 20 mm. in length (four-thirds as long as the carapace is broad), but are very stout. The granulation of the femora, tibiae and hands resembles that found in *S. longispina* (compare figs. 10 and 11, pl. xxxi). The only well-developed spines on the upper margin of the tibia are the three long ones near the distal end which are characteristic of the genus; all others are quite small, the contrast being more marked in the adult than in the immature specimens. The hand and finger resemble those of *S. longispina*.

The *legs* resemble those of other members of the genus, but the walking legs especially are of a much paler and more yellowish colour than in *S. longispina*, this colour difference between the two species being somewhat more marked in the legs than in the carapace.

Stygophrynus cerberus, Simon.[1]

(Plate xxxi, fig. 12.)

The habits of this species from the Jalor caves (Gua Glap or "Dark Cave", and Biserat) have been described elsewhere.[2] Cotypes have been presented to the Indian Museum by the Cambridge Museum.

This species closely resembles *S. berkeleyi*, but has all the integuments harder, much darker in colour, and more strongly granular (compare pl. xxxi, figs. 11 and 12).

Stygophrynus spp. indet.

In addition to the species from Saigon, Mentawei and Java already referred to (p. 444), mention may be made of "an animal allied to Phipson's Tarantula" found by Flower in the depths of the Batu Caves at Selangor,[3] which may well have belonged to this genus.

Genus CHARON, Karsch.

This genus is represented in the Indian Museum collection by one immature specimen of *C. grayi*, the only species recognized by Kraepelin in "*Das Tierreich*." *C. annulipes*, Lauterer,[4] does not appear to be referred to in that work, but it cannot be recognized either from the description or from the figure. It is compared with *C. australianus*, Koch, a species now placed in the genus *Charinus*. Its position must remain uncertain till the type is re-examined.

[1] *Proc. Zool. Soc. London*, 1901, pp. 76-7.
[2] *Journ. Asiat. Soc. Bengal* (n.s.), IX, 1914, p. 419.
[3] *Rep. Austr. Ass.* VI, 1895, pp. 413-4, pl. lii.
[4] *J. Straits R. Asiat. Soc.*, No. 36, 1901, p. 40.

Subfamily *PHRYNICHINAE.*

Kraepelin divides this subfamily into genera as follows :—

1. { Tibia of fourth leg 1-jointed ; hand of adult with basal of two dorsal spines rudimentary or absent.[1] *Phrynichus*, p. 447.
2. { Tibia of fourth leg 2-jointed ; both dorsal spines of hand strongly developed in adult ... *Damon*, p. 455.

Genus PHRYNICHUS, Karsch.[2]

Type *P. reniformis* (Linn.).

The generic identity of Linnaeus's *Phalangium reniforme*, which has an important bearing on the nomenclature of the subfamily, has been much in dispute. Kraepelin summarised the available evidence at the commencement of his " Revision der Tarantuliden "[3], and has given his final opinion as regards the correct nomenclature in " *Das Tierreich.*"[4] His conclusions have been confirmed by Lönnberg, who examined the type still preserved in the Zoological Museum at Upsala.[5]

The generic identity of *Phalangium reniforme* having been settled, its specific identity was for Kraepelin a simple matter, since, from the material at his disposal, he was unable to recognize more than two species in the genus. The rich material in the Indian Museum collection shows, however, that several of the names regarded by Kraepelin as synonymous with *Phrynichus reniformis* will have to be revived ; and that even these will not cover all the species to which the name *P. reniformis* may conceivably belong. The description of *P. reniformis* is generic rather than specific, and the identity of the species must, I am afraid, remain a matter of doubt until the type is redescribed. Lönnberg says, " To judge from the descriptions and from the table given by Pocock, the Linnean specimen most closely agrees with '*Ph. deflersi*,' Simon." But the value of the characters used by Pocock in diagnosing this species is perhaps open to question; and it is more likely that the Linnean specimen belongs to one of the two well-known forms called below *P. ceylonicus* and *P. nigrimanus* respectively, than to a species only known otherwise from a single specimen from Obock.

The description of *P. lunatus* (Pallas) is also generic rather than specific; and the figures with which it is accompanied are too rough to be of any help. The identity of this species also must therefore remain in doubt.

P. ceylonicus (Koch) is clearly a large species found in Ceylon. Only one such species is known to me, and I have accordingly applied the name to it.

P. scaber (Gervais) comes from the Seychelles (? and Mauritius). It is probably distinct from the Indian and Ceylonese species, but

---

[1] Except in *P. deflersi* (Simon).
[2] *Arch. Naturg.* XLV (1), 1879, p. 190.
[3] *Abh. Ver. Hamburg* XIII, 1895, pp. 1-53, 1 pl.
[4] See also *Zool. Anz.* XXVIII, 1904, pp. 201-203.
[5] *Ann. Mag. Nat. Hist.* (7) I, 1898, pp. 88-89.

the description is again generic, not specific, and as I have no specimens before me from these islands I cannot add to it.

P. nigrimanus (Koch) is from India. It is probably the species common in the Eastern Ghats, as has already been suggested by Hansen.[1] This is the only species known to me of which (spirit) specimens ever seem to resemble Koch's figure in colour.

P. deflersi (Simon) may be distinguished by the presence, even in large specimens such as the type of the species, of two well developed spines on the dorsal margin of the hand, as in the genus *Damon.* Both these spines are, however, present in the young of certain other species.

P. jayakari, Pocock, differs from all other known species in the presence of a pair of stout spines on the margin of the carapace in front of the lateral eyes.

P. phipsoni, Pocock, is a distinct species, apparently confined to the northern parts of the Western Ghats.

P. pusillus, Pocock, is a common Ceylonese form, allied to but distinct from *P. ceylonicus* (Koch) of which it may conveniently be regarded as a variety. It is much smaller than this or any other species of the genus known to me.

Phrynichus scullyi, Purcell, from S. Africa[2] is probably described from immature specimens, but as I have not seen any I cannot speak with certainty.

P. bacillifer (Gerst.) remains, of course, distinct.

The determinable species of the genus may be recognized thus:—

1.	Margin of carapace without strong spines ...	2.
	Margin of carapace with a pair of strong forwardly directed tooth-like spines in front of lateral eyes	*P. jayakeri*, p. 455.
2.	One spine only present on upper margin of hand of adult	3.
	Vertical basal spine as well as oblique spine distal to it persistent on upper margin of hand in adult	*P. deflersi*, p. 455.
3.	Anterior surface of femur of arm with 3-5 sharp spines, or simply granular; lower margin always with some sharp spines	4.
	Anterior surface of femur of arm with 2 or 3 blunt rounded bacilliform processes in the basal third; lower margin spineless	*P. bacillifer*, p. 455.
4.	A longitudinal row of granules present on lower surface of hand (pl. xxxi, fig. 14) ...	*P. ceylonicus*, p. 449.
	Lower surface of hand smooth (pl. xxxi, fig. 13).	5.
5.	Tibia of arm of adult with two long terminal dorsal spines preceded only by a minute tubercle[3]; basal dorsal spine of hand absent in adult, small or absent in young	*P. nigrimanus*, p. 453.
	Tibia of arm of adult with the two long terminal dorsal spines preceded by a short but well developed spine; basal dorsal spine of hand probably always well developed in young, represented by a tubercle in adult	6.

[1] *Ent. Med.* IV, 1894, p. 150. [2] *Ann. S. Afr. Mus.* II, 1900-1902, p. 206.

[3] This tubercle replaces a spine which is present in the young of this as of other species.

6.	Terminal ventral spine of tibia of arm of adult small, more or less conical, not decumbent (much as in *P. ceylonicus*, pl. xxxi, fig. 14) ...	*P. granulosus*, p. 454.
	Terminal ventral spine of tibia of arm of adult long, parallel-sided, decumbent (pl. xxxi, fig. 13)	*P. phipsoni*, p. 454.

Phrynichus reniformis (Linnaeus).[1]

The identity of this species can only be settled by a further examination of the type which is preserved in the Zoological Museum at Upsala (see above, p. 447).

Phrynichus lunatus (Pallas).[2]

Also an indeterminable species (see above, p. 447).

Phrynichus ceylonicus (Koch).[3]

(Plate xxxi, fig. 14.)

Three varieties of this species may be recognized as follows:—

A. ♂ & ♀: width of carapace of adult 15-18 mm.; $\frac{\text{length of femur of arm}}{\text{width of carpace}} = 2\text{-}2{\cdot}2$	*ceylonicus* (Koch), *s. str.*
B. ♀: width of carapace of adult 13-14·5 mm.; $\frac{\text{length of femur of arm}}{\text{width of carapace}} = 1{\cdot}5\text{-}1{\cdot}8$	var. *gracilibrachiatus*, Gravely.[4]
C. ♂: width of carapace of adult 10·5-13 mm.; $\frac{\text{length of femur of arm}}{\text{width of carapace}} = 1{\cdot}8\text{-}2{\cdot}3$	var. *gracilibrachiatus*, Gravely.[4]
D. ♂ & ♀: width of carapace of adult 8-10·5 mm.; $\frac{\text{length of femur of arm}}{\text{width of carapace}} = 1{\cdot}1\text{-}1{\cdot}5$	var. *pusillus*, Pocock.[5]

1. PHRYNICHUS CEYLONICUS (Koch), *s. str.*

This form is remarkable for its ability to live in comparatively dry surroundings; it seems to live mainly in jungles where the soil is specially porous or the climate not very moist, and in houses in moister regions. Specimens from the following localities in Ceylon are preserved in collections belonging to the Indian Museum, to the Colombo Museum, or to Mr. E. E. Green:—

North East Province: Horowapotama, *ca.* 200 ft.; Moha-Illuppalama, *ca.* 300 ft.
Western Province: Wennappuwa, 10 mls. from Negumbo.
Central Province: Nalanda, *ca.* 900-1000 ft.; Galagedara, *ca.* 800-2000 ft.; Haragama, *ca.* 1200 ft.; Kandy, *ca.* 1500-2000 ft.; Peradeniya, *ca.* 1500 ft.
Southern Province: Ambalangoda, 0-100 ft.; Kottowa, 0-100 ft.

[1] *Systema Naturae*, 10th ed., p. 619.
[2] *Spicilegia Zoologica*, fasc. IX, pp. 33-37, pl. iii, figs. 5-6.
[3] *Die Arachniden*, X, p. 336, fig. 776.
[4] *Spolia Zeylanica*, VII, p. 140.
[5] *Ann. Mag. Nat. Hist.* (6), XIV, p. 296.

There must, I think, be some mistake about a specimen in the Indian Museum collection that is supposed to have been collected by Major Beddome in South India.

This is the largest form of *P. ceylonicus* known, and full-grown specimens may easily be distinguished from the varieties *gracilibrachiatus* and *pusillus* by their size. Younger specimens may be distinguished by the loss, at a time when the size of the specimen is greater than that at which these changes take place in the permanently smaller forms, first of the bright and chequered juvenile colouration, and later of the first of the three spines on the upper surface of the distal end of the tibia of the arm. But in the smallest specimens of all there appears to be no certain means of distinguishing the different forms.

The fully grown female of var. *gracilibrachiatus* is the only other form at all likely to be confused with this typical form. It approaches the typical form much more closely in size than do either the male of the same variety or either sex of var. *pusillus*; and, except when their maturity is made evident by the presence of embryos under the abdomen, the identity of these forms is very difficult to establish unless by comparison with a good series of typical specimens in various stages of growth.

The presence of a pair of well-developed semilunar lobes on the posterior margin of the third abdominal sternum of *P. ceylonicus*, *s. str.*, is useful in checking the identity of immature specimens, as in the varietal forms these are always proportionally smaller than is usual in the typical one, and they are often apparently absent altogether. But as, in a long series, every stage can be found from their absence in the varieties to their full development in the typical form, their condition does not in itself fully indicate to which of the three forms a specimen belongs.

The following measurements of the mature or approximately mature specimens in the Indian Museum collection will serve to indicate the proportions borne by the arms to the width of the carapace in adults of this form:—

Sex.	♂	♀ (with embryos).	♀	♂	♂	♂
Width of carapace in mm.	18	17·5	17	16	16	15
Length of femur of arm in mm.	40·5	35·5	34	33	33	31

2. P. CEYLONICUS var. GRACILIBRACHIATUS, Gravely.

The habits of this form resemble, so far as is known, those of the next variety.

The Indian Museum collection contains specimens from the following places in Ceylon:—

Central Province : Nalanda, *ca.* 900-1000 ft.; Galagedara, *ca.* 800-2000 ft.; Kandy, *ca.* 1500-2000 ft.; Peradeniya, *ca.* 1800 ft.

The sexes of this variety differ from one another in a more striking manner than do those either of the typical form or of the other variety of the species, and but for certain indications of an identical geographical distribution for the two and the fact that I have seen no female which superficially resembles the male of this variety, and no male which resembles what I believe to be its female, it would hardly, perhaps, have occurred to me to regard them as a single form. Thus the adult male is small,[1] often closely resembling var. *pusillus* in the size of its body, though always distinguished therefrom by its relatively longer appendages, the arms especially being very noticeably longer and slenderer, bearing about the same proportion to the width of the carapace as they do in adults of *P. ceylonicus*, *s. str.*; whereas the female is large, being intermediate in size between *P. ceylonicus*, *s. str.* and var. *pusillus*, and has proportionally shorter arms. Specimens in which maturity is not clearly indicated by the presence of embryos under the abdomen may therefore be very easily mistaken for immature specimens of *P. ceylonicus*, *s. str.*, since the proportion borne by the arms to the width of the carapace increases with growth.

So far as I know it is impossible to distinguish immature specimens of either sex of var. *gracilibrachiatus* from those of var. *pusillus*; and from this it may be concluded that the arms of the male of the former become greatly lengthened at about the time when maturity is reached (as do those of the male of *Charinides bengalensis*) and that previously they are no longer than in the latter variety.

In practice there is never any difficulty in distinguishing the adult male of var. *gracilibrachiatus* from the form most like it—the male of var. *pusillus*. But to distinguish adult females of var. *gracilibrachiatus* from immature females of *P. ceylonicus*, *s. str*, of the same size is much more difficult except, as has already been pointed out, when the former bear embryos. The chief differences between the two are :—(1) the retention in (? all) specimens of the latter of a distinctly spiniform rudiment of the first of the three dorsal spines at the distal extremity of the tibia of the arm, a spine which has probably already disappeared in all specimens of the former; and (2) the size of the semilunar lobes on the posterior margin of the third abdominal segment, which are always present

---

[1] This difference in size and proportions shown by the two sexes is present in var. *pusillus* also, and probably in *ceylonicus*, s. str., as well; but in these two forms it is less striking, and only apparent in a series of measurements; whereas in var. *gracilibrachiatus* it is very noticeable at once—more so in fact than the measurements would lead one to suppose. The name *gracilibrachiatus* is an unfortunate one now that *pusillus*, Poc., has to be regarded as a variety and not a species; for it is from this form only that var. *gracilibrachiatus* is distinguished by the slenderness of its arms, and not from *P. ceylonicus*, s. str. It was as a variety of *P. pusillus*, Poc., that *gracilibrachiatus* was originally described.

and usually well developed in *P. ceylonicus*, *s. str.* but are either small or absent in var. *gracilibrachiatus.*

The following measurements (in mm.) were used in calculating the proportions given for this variety in the table on p. 449 :—

Sex.	♀ (with embryos).	♀	♀ (with embryos).	♀	♂ (type)	♂	♂	♂	♂	♂	♂	♂
Width of carapace.	14·5	13	13	13	13	13	12	11·5	11·5	11	10·5	10·5
Length of femur of arm.	25	23	22	20·5	29·5	28·5	25	24	24	22·5	20·5	19

3. P. CEYLONICUS var. PUSILLUS, Pocock.

This variety is unable to live long in the absence of moisture. It is only known from the Central Province of Ceylon, and was first described from specimens caught at Punduloya by Mr. E. E. Green, who tells me he got them at an altitude of about 4200 ft. above sea level. It is represented in the Indian Museum collection by specimens from Nalanda, *ca.* 900-1000 ft. ; Galagedara, *ca.* 800-2000 ft.; Kandy, *ca.* 1500-2000 ft. ; Peradeniya, 1600-2200 ft.

The best means of distinguishing this variety from the last and from the young of *P. ceylonicus*, *s. str.*, have already been dealt with (pp. 450-451). One other character remains, however, to be noted here : it is quite common to find the first of the three dorsal spines at the distal end of the tibia of the arm represented in mature specimens of this variety by a distinct spiniform process. The process is, however, always very much smaller than in young specimens of equal size of *P. ceylonicus*, *s. str.* Thus the greatest length for this spine seen in an ovigerous female of var. *pusillus* was about 0·5 mm. ; but in a specimen of *P. ceylonicus*, *s. str.*, of approximately the same carapace-width (former 8 mm. the latter 7·5 mm.) this spine is as much as 2 mm. long, and in a somewhat older specimen (carapace-width 9·5 mm.) it is 1·5 mm. long.

The following are measurements (in mm.) taken from the series of this variety in the Indian Museum collection :—

Sex.	♀ (with embryos).	♀	♀	♀	♀	♀ (with embryos).	♂	♀ (with embryos).	♀ (with embryos).	♂	♂	♀	♂	♀ (with embryos).	♀	♀ (with embryos).	♂	♂	♀ (with embryos).	♀
Width of carapace.	10·5	10	10	10	9·5	9·5	9	9	9	8·5	8·5	8·5	8·5	8·5	8·5	8·5	8	8	8	8
Length of femur of arm.	14·5	13·5	12·5	12	11	10·5	13·5	11·5	10	11·5	11	11	10·5	10	9·5	9	9·5	9	9	8·5

Phrynichus nigrimanus (Koch).[1]

The Indian and Madras Museums possess between them specimens from all but one[2] of the following localities, all of them on the eastern side of the Indian Peninsula :—

Orissa : Hills, 0·1000 ft., near Barkul, Chilka Lake ; Balugaon Chilka Lake.
Ganjam : Russelconda.
Vizagapatam District.
Nellore : Rambuga cave, Udyagiri droog.
Karnul : Bairani, Chelama Ry. Station, Nallamalais, *ca.* 2000 ft.
N. Arcot : Vellore.
Chengalpat : Pallavaram, 12 miles from Madras.
Salem : Shevaroy Hills.

Barkul is the only place where I have myself collected specimens of this species. They are quite common in the hills and in the jungle at the foot of them, but I failed to get any very large specimens or ovigerous females—though I went for this purpose in the rains, when *P. ceylonicus* breeds. None of the specimens found had lost the third spine on the dorsal surface of the distal end of the tibia of the arm ; but one of the largest of them, in which it was quite small (about 0·5 mm. long), did so on casting its skin after a few weeks' captivity, when the spine was reduced to a tubercle. The width of the carapace of the cast skin of this specimen is 12·0 mm., that of the specimen itself being 14·0. Probably mature specimens are at least 14 mm. across the carapace and live, as is more or less the case with other species, in the securest retreats.

In the hills further south the species attains a much greater size than at Barkul. This does not, however, appear to be the case near the coast since the width of the carapace of the Pallavaram specimen, in which the third spine on the dorsal surface of the distal end of the tibia of the arm is absent, is barely 13 mm. The third spine on the dorsal surface of the distal end of the tibia of the arm is over 2 mm long in the specimen from Rambuga cave, the width of whose carapace is 11 mm. ; and it is nearly 2½ mm. long in two specimens from the Shevaroys whose carapaces are nearly 11 and a little over 12 mms. broad respectively. The largest specimen I have seen is that from Bairani, whose carapace is 20 mm. broad. It appears to be a mature female. The length of the femur of the arm is 38·5 mm., and the third dorsal spine at the distal end of the tibia of the same appendage is tuberculiform. This specimen belongs to the Madras Museum. It is possible that this form and the one common at Barkul may ultimately have to be recognized as distinct varieties or subspecies.

[1] *Die Arachniden*, XV, p. 69, fig. 1464.
[2] The only specimen I have seen from Vellore belongs to Rev. J. E. Tracey, to whom my thanks are due for sending it. It is doubtless identical with the form described by Hansen (*Ent. Med.* IV, 1894) as common at Vellore.

Phrynichus granulosus, n. sp.

This species is represented in the collections of the Indian, Madras and Trivandrun Museums by specimens from the following localities:—

Cochin: State Forest Tramway 10th-14th mls., 0-300 ft.; Kavalai, 1300-3000 ft.

Travancore: Ponmudi, 2000-3000 ft.

The specimen which Pocock records in the "Fauna" from Trivandrum under the name *P. phipsoni* doubtless also belongs in reality to the present species.

This species, whose distinctive characters are given above (pp. 448-449), is intermediate in character between *P. nigrimanus* and *P. phipsoni*, resembling the former and *P. ceylonicus* in the shape of the terminal ventral spine of the tibia of the arm, and the latter in the other spines of both arm and hand. The integuments are more coarsely granular than in any other species with which I am acquainted. In this character the species presumably resembles *P. scaber* (Gervais) from the Seychelles. The male type—the largest specimen known to me—has a carapace 18 mm. broad, the femur of the arm being 31 mm. long. The female type has a carapace 15·5 mm. broad, the femur of the arm being 24·5 mm. long. Both these specimens are from jungle near the rubber estate between the 10th and 14th miles of the Cochin State Forest Tramway.

Phrynichus phipsoni, Pocock.[1]

(Plate xxxi, fig. 13.)

This species has been recorded by Pocock from Bombay and Trivandrum, and from various other localities by subsequent authors, who have apparently confused with it the earlier stages of other species, *i.e.* the stages which retain the third dorsal spine of the distal end of the tibia of the arm. I have little doubt that the Trivandrum specimen referred to by Pocock belongs in reality to the preceding species, and that *P. phipsoni* is confined to the more northerly parts of the Western Ghats.

Phrynichus scaber (Gervais).[2]

Gervais records this species from the Seychelles, and the same or an allied form from Mauritius. Its distinctive characters have yet to be described.

Phrynichus scullyi, Purcell.[3]

This species is recorded only from Cape Colony (Pakhuisberg in Clanwilliam Division, and Namaqualand). The specimens from which it was described were probably young, judging from their size and colour.

---

[1] *Ann. Mag. Nat. Hist.* (6), XIV, 1894, p. 295, pl. viii, fig. 4.
[2] *Histoire Naturelle des Insectes, Apteres,* III, p. 3.
[3] *Ann. S. Afr. Mus.*, II, 1902, p. 206.

Phrynichus bacillifer (Gerstaecker).[1]

This species, according to Kraepelin, occurs from Madagascar and Zanzibar to Mozambique, Tanganyika and Lake Rudolph.

Phrynichus deflersi (Simon).[2]

Described from a single specimen from Obock in French Somaliland.

Phrynichus jayakeri, Pocock.[3]

Described from two specimens from Muscat in Arabia.

Phrynichus spp.

The above record of the distribution of various species of Phrynichus by no means exhausts the localities given by previous authors. Most of the additional localities refer to the composite "species" to which Kraepelin applied the name *P. reniformis;* others refer to species which have clearly been wrongly named. These localities, and those of certain immature specimens in the Indian Museum collection, show the distribution of the genus to be wider than appears above, and may therefore be recorded here :—

Africa: Natal; Mozambique; Kondoa (? French Congo); Massaua (Somaliland); several localities in Central Africa (Albert Lake, Kossenje; Kirk Falls south-west from Albert Lake; plains below Semliki; Awakubi).
Madagascar (east coast).
Asia: Arabia—Aden.
Assam—Sibsagar.
Siam—Chantaboon.
Cochin China—Saigon.
Malay Peninsula—Penang.

Genus **DAMON**, Koch.[4]

Type *D. variegatus* (Perty).

I have nothing to add to Kraepelin's account of this genus. It is mainly African, but Kraepelin records *D. variegatus* from Arabia as well.

[1] *C. v. d. Decken's Reisen in Ostafrica*, III (2), p. 472.
[2] *Bull. Soc. Zool. France*, XII, 1887, p. 454.
[3] *Ann. Mag. Nat. Hist.* (6), XIV, p. 294.
[4] *Übersicht des Arachnidensystems*, V, p. 81.

EXPLANATION OF PLATE XXXI.

FIG. 1. Four distal joints of arm of *Catageus pusillus* from above (diagrammatic, showing armature of dorsal margin).

" 2. Basal joint of finger of *Phrynichosarax cochinensis*.

" 3. " " " " *javensis*.

" 4. " " " " *buxtoni*.

" 5. " " " " *singapurae*.

" 6. " " " " *rimosus*.

" 7. Four distal joints of arm of *Sarax willeyi* from above (diagrammatic, showing armature of dorsal margin).

" 8. Hand of *Sarax sarawakensis*.

" 9. Four distal joints of arm of *Stygophrynus moultoni* from above (diagrammatic, showing armature of dorsal margin).

" 10. Back of hand of *Stygophrynus longispina*.

" 11. " " " *berkeleyi*.

" 12. " " " *cerberus*.

" 13. Hand and distal part of tibia of arm of *Phrynichus phipsoni* from below.

" 14. Hand and distal part of tibia of arm of *Phrynichus ceylonicus*, *s. str.*, from below.

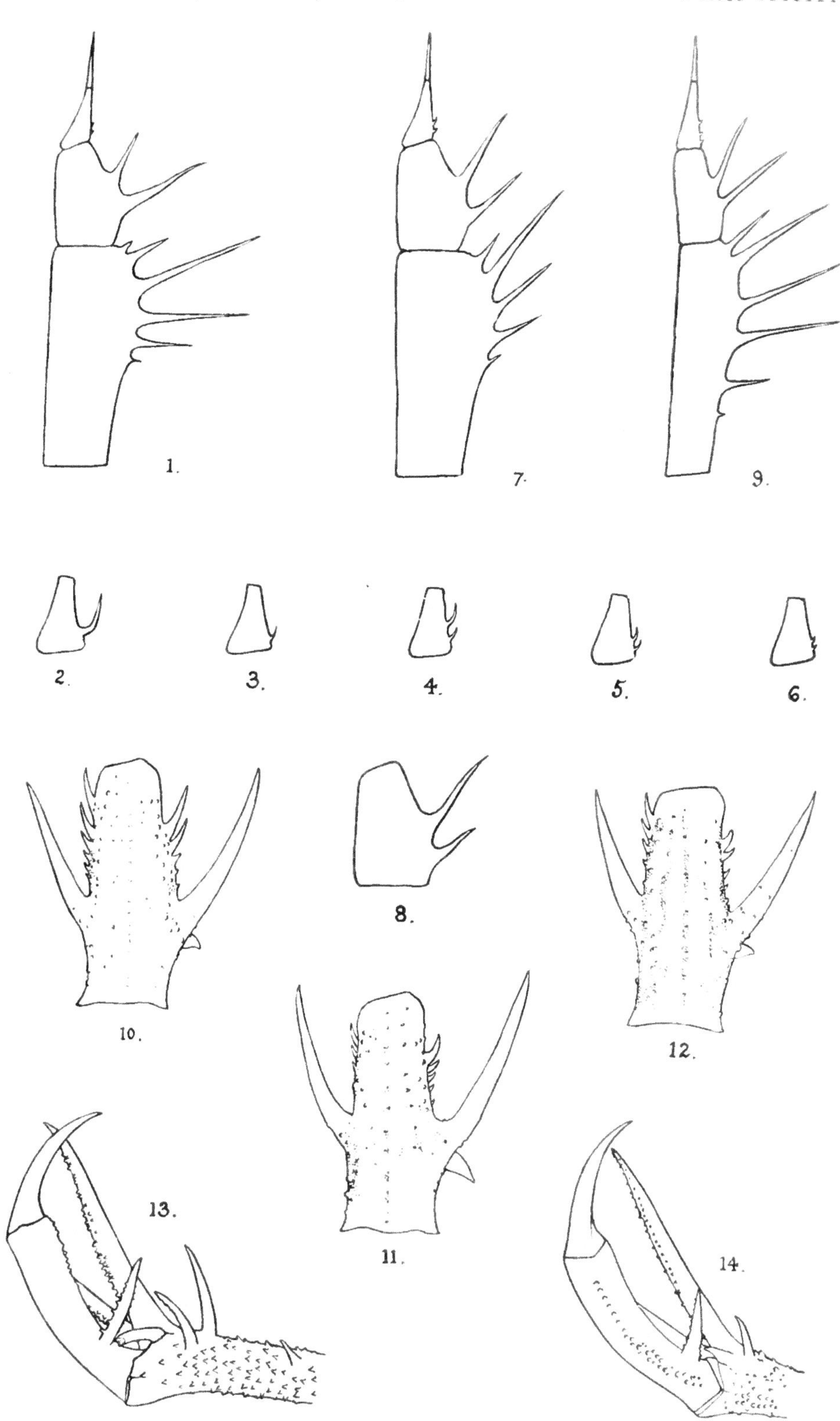

F. H. G. & D. N. Bagchi, del.

Oriental Phrynichidae.

XXVII. SOME SPONGES PARASITIC ON CLIONIDAE WITH FURTHER NOTES ON THAT FAMILY.

By N. ANNANDALE, *D.Sc.*, *F.A.S.B.*, *Superintendent*, *Indian Museum.*

(Plate XXXIV.)

In a recent paper on the Clionidae of Indian seas (*Rec. Ind. Mus.* XI, pp. 1-24) I referred incidentally to other sponges parasitic in their burrows. I now propose to give an account of these sponges adding some additional notes on the Indian Clionidae.

The systematic position of the different species may be considered first, in taxonomic order, and then their biological relationships.

Part I.—SYSTEMATIC.

The following is a list of the species to be considered; all belong to the order Tetraxonida.

Grade TETRAXONELLIDA.

Family PACHASTRELLIDAE.

Stoeba plicata var. *simplex* (Carter).

Family STELLETTIDAE.

Stelletta vestigium, Dendy.

Grade MONAXONELLIDA.

Family EPIPOLASIDAE.

Coppatias penetrans (Carter).
Coppatias investigatrix, sp. nov.

Family CLIONIDAE.

Cliona carpenteri, Hancock.
Cliona mucronata, Sollas.
Cliona quadrata, Hancock.
Cliona kempi, sp. nov.
Thoosa hancocci, Topsent.

Family DESMACIODONIDAE.
Subfamily *ECTYONINAE*.

Rhabderemia prolifera, sp. nov.

Family AXINELLIDAE.

Amorphinopsis excavans, Carter.
A. e. var. *digitifera*, nov

Family CHONDROSIIDAE.

Chondrilla nucula, Schmidt.
Chondrilla mixta, Schulze.
Chondrilla distincta, Schulze.

Grade **TETRAXONELLIDA.**

Family PACHASTRELLIDAE.

Stoeba plicata (Schmidt).

1868. *Corticium plicatum*, Schmidt, *Die Spong. d. Kuste v. Algier*, p. 2, pl. iii, fig. 11.
1880. *Samus simplex*, Carter, *Ann. Mag. Nat. Hist.* (5) VI, p. 60, pl. v, fig. 26.
1888. *Stoeba simplex*, Sollas, '*Challenger*' *Rep. Zool., Tetractinellida* (vol. XXV), p. 102.
1888. *Calcabrina plicata*, *id.*, *ibid.*, p. 281.
1889 (1887). *Samus simplex*, Carter in Anderson's *Faun. Mergui* I, p. 75.
1894. *Dercitus plicata*, v. Lendenfeld, *Denk. Ak. Wien.* LXI, p. 105, pl. ii, fig. 10, pl. iii, fig. 43.
1895. *Dercitus plicatus*, Topsent, *Arch. Zool. expérim.* (3) III, p. 531, pl. xxii, figs. 6-10.
1903. *Dercitus simplex*, Thiele, *Abh. Senckenb. Natur. Gesellsch.* XXV, p. 20, pl. ii, fig. 1.
1903. *Dercitus simplex & D. plicatus*, v. Lendenfeld, *Das Tierreich, Tetraxonia*, pp. 81, 82.
1905. *Stoeba simplex & S. plicata*, Dendy in Herdman's *Ceylon Pearl Fisheries*, III, pp. 71, 230.

Carter mentions spicules of his *Samus simplex* as being among those he extracted from a specimen of dead coral in the late Dr. Anderson's Mergui collection. From the same specimen I have been able to extract numerous pieces of this sponge in sufficiently good condition to study its general structure and spiculation; the latter is evidently more varied than either Carter himself or Sollas realized and is apt to be not fully understood because certain spicules are practically confined to certain parts of the sponge. I am of the opinion that Topsent's suggestion (1895, p. 536) as to the specific identity of the species with Schmidt's *Corticium plicatum* is fully justified by the specimens I have examined.

The sponge, as it exists in dead coral, forms small oval or globular masses which entirely fill corresponding cavities. From these are given out slender, cylindrical or flattened branches, some of which join them to other similar masses, while others terminate in flattened and often ramifying lamellae. The latter make their way among interstices of the calcareous material. The larger masses contain a dense crowd of well-formed triaenes arranged with their sharply pointed shafts pointing outwards, but in the connecting branches the macroscleres are more scanty and more delicate in form, while they are practically absent in the distal parts of the lamellae. In the proximal parts thereof they have precisely the form of the small slender spicules figured by Topsent (1895, pl. xxii, *o'*, *d'*), whereas in the larger masses they agree equally well with the figures *o* and *d* on the same plate. The proportions of all these types of spicules also agree with Topsent's description. Spicules of the "calthrops" type are extremely scarce in my specimens. Indeed, I was for some time of the opinion that they were altogether absent. After a prolonged search through spicule-preparations, however, I at last succeeded in finding one. The microscleres are a little larger than in Topsent's European specimens, measuring about 0·0162 mm. in length, and their spines are much shorter and more slender than is indicated in Schmidt's original figure. They are extremely numerous in the ectosome all over the sponge, but almost absent from the choanosome. The large cells containing brown granules to which Topsent and other authors refer are still conspicuous, after about 28 years in spirit.

S. plicata is common in dead coral in Indian seas, but in all the specimens I have examined seems to be associated with some species of *Cliona*. In places where the coral is of a crumbling consistency the external surface of the sponge is often covered with small calcareous granules of irregular form, while the larger masses of sponge often contain in their interior larger granules of a similar nature. These granules are larger than those produced by the activities of *Cliona*. The more slender processes of the *Stoeba* are as a rule in contact with the *Cliona* and often contain *Cliona*-spicules in their ectocyst and choanosome.

In consideration of its method of life and growth this Indian form of *Stoeba plicata* is perhaps worthy of a varietal name and should be known as *S. plicata* (Schmidt) var. *simplex* (Carter), for Topsent (1895) in his elaborate account of the species, as it occurs in the Mediterranean, makes no mention of the peculiarities noted in the preceding paragraph.

Family STELLETTIDAE.

Stelletta vestigium, Dendy.

1905. Dendy in Herdman's *Ceylon Pearl Fisheries*, III, p. 78, pl. ii, fig. 7.

My specimens of this species are from the same fragments of dead coral as those in which the specimens of *Stoeba plicata* var. *simplex* described above were found. They permeate the coral in a fine network of slender strands and in part, at any rate, occupy the excavations of *Cliona viridis* (Schmidt), spicules of which adhere to their ectosome. The original specimen is described as "irregular in shape, massive, encrusting, and containing many foreign bodies." Possibly it commenced its growth in the same manner as the example from Mergui, which agrees with it closely in spiculation and, so far as it is possible to say, in general structure.

The species is only known from Ceylon and Tenasserim.

Grade **MONAXONELLIDA**.

Family EPIPOLASIDAE.

Coppatias penetrans (Carter).

1880. *Tisiphonia penetrans*, Carter *Ann. Mag. Nat. Hist.* (5) VI, p. 141, pl. vii, figs. 44*a–d*.

1905. *Coppatias* (*Tisiphonia*) *penetrans*, Dendy in Herdman's *Ceylon Pearl Fisheries*, III, p. 231.

A minute sponge of which the spicules agree well with Carter's description and figures occurs in abundance in dead reef-coral from Port Mouat in the Andamans, occupying the burrows of various boring organisms and in particular those of *Cliona ensifera* and *C. lobata*. The form of the sponge is precisely that of the cavity it occupies. It is of solid structure, the natural cavities being small except when occupied, as is often the case, with fragments of calcareous matter. Specimens treated with acid are apt to appear cavernous owing to these fragments being dissolved. The ectosome, which is in contact with the wall of the burrows occupied, is thin but somewhat impenetrable by liquids and it is difficult to clear specimens in oil of cloves. The whole structure of the organism is on so minute a scale that it could only be elucidated properly by means of sections of specially preserved material, which I do not possess.

Coppatias investigatrix, sp. nov.

(Plate xxxiv, figs. 1, 2.)

This sponge is closely related to *C. penetrans*, with which it agrees in habits, but the macroscleres are as a rule spined near the tips and the microscleres exhibit much greater diversity of form. Unlike *C. penetrans* it is a deep-sea species.

Sponge.—In its early stages the sponge consists of minute masses of an irregularly oval form. These penetrate into the burrows of Clionids in shells, then increase in size and assume the shape of the spaces they occupy; before doing so completely, they give out relatively slender, blunt, finger-like processes. The internal structure appears, so far as can be seen, to resemble that of *C. penetrans*.

Spicules.—Both macroscleres and microscleres are very variable. The majority of the latter are slender, spindle-shaped amphioxi about 15 to 30 times as long as broad, smooth for the greater part of their length, but bearing scattered, sharp, erect spines near the two extremities, the actual tips being smooth. Smaller absolutely smooth amphioxi also occur.

The microscleres are of three kinds, *viz.* (*a*) oxyasters with spined tips, (*b*) spherasters with spined tips, and (*c*) smooth spherasters. Intermediate forms occur, however, in all cases.

The spiny oxyasters have as a rule six cladi, but may have only four, or occasionally more than six. The tips are sharply and gradually pointed and bear sharp erect spines scattered rather densely. There is no distinct central nodule and the bases of the cladi are smooth.

The spiny spherasters are merely more compact forms of the same type, with a larger number of shorter and stouter cladi fused together at the base. They are, as a rule, smaller than the oxyasters, but every intermediate form of spicule can be found.

The smooth spherasters have still shorter and more numerous cladi than the spiny ones and a relatively larger central sphere.

The degree to which the spines are developed on the spherasters is, however, as variable as the proportions of their several parts

Measurements of Spicules.

Length of spiny macroscleres	..	0·098—0·205 mm.
Length of smooth macroscleres	..	(average) 0·115 ,,
Diameter of oxyasters	..	0·0126—0·0252 ,,
Diameter of spiny spherasters	.	(average) 0·0126 ,,
Diameter of smooth spherasters	..	,, 0·0115 ,,

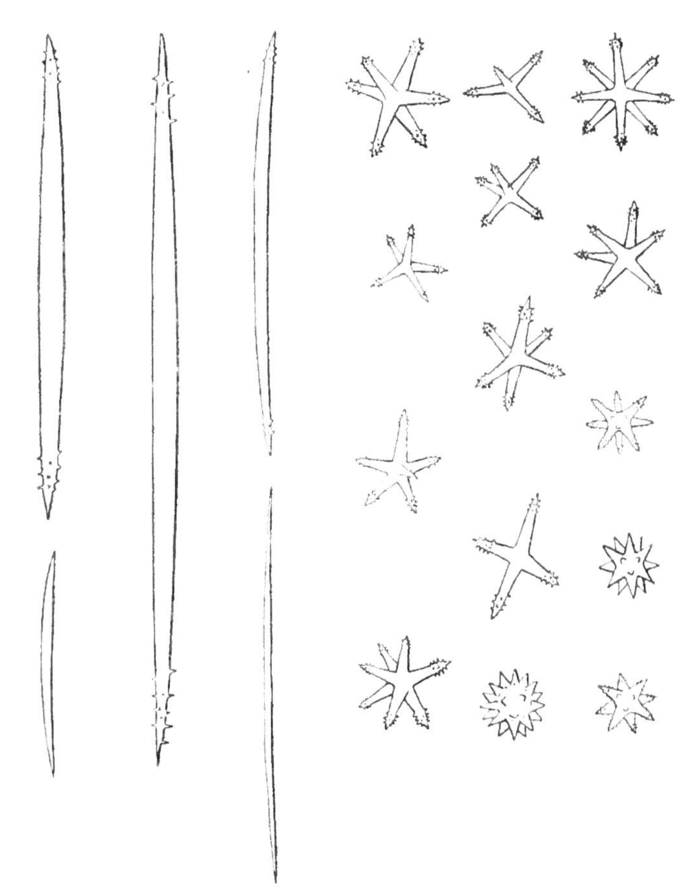

FIG. 1.—Spicules of *Coppatias investigatrix*, sp. nov.

Type.—No. 64 5/7 ZEV, *Ind. Mus.*

Locality.—Off Ceylon in 703 fathoms: with *Thoosa investigatoris* in a dead Gastropod shell (in alcohol).

At points at which the *Coppatias* comes in contact with the *Thoosa*, the latter secretes a thick horny covering through which the tips of its own macroscleres penetrate (pl. xxxiv, fig. 2).

Family CLIONIDAE.

The following notes on the Clionidae are based on a small collection recently made by Mr. S. W. Kemp at Port Blair in the Andamans. All the specimens are from shallow water and, except the first, from dead reef-coral.

Cliona carpenteri, Hancock.

Shells of edible oysters (*Ostrea viriginiana*, Gmel.) from the head of Port Blair harbour are riddled with the galleries of this sponge, precisely as shells of the same species of oyster are riddled with those of *C. vastifica* in lagoons on the east coast of continental India.

Cliona mucronata, Sollas.

Well preserved specimens of this peculiar sponge occur in fragments of dead reef-coral with those of the two following species. They agree closely with Sollas's original figures in respect of the structure of the characteristic diaphragms.

Cliona quadrata, Hancock.

1849. *Cliona quadrata*, Hancock, *Ann. Mag. Nat. Hist.* (2) III, p. 344. pl. xv, fig. 6.
1881. *Cliona warreni*, Carter, *ibid.* (5) VII, p. 370, pl. xviii, fig. 6.
1900. *Cliona quadrata*, Topsent, *Arch. Zool. expérim.* (3) VIII, p. 54.

Topsent is undoubtedly right in regarding Carter's *C. warreni*, which came from the Gulf of Manaar, as synonymous with Hancock's species of unknown *provénance*. Well-preserved specimens are present in Mr. Kemp's collection.

Cliona kempi, sp. nov.

This species is closely allied to *C. lobata*, Hancock and *C. michelini*, Topsent, but is distinguished from both by the complete absence of microscleres.

The galleries are almost cylindrical but swell out slightly at intervals. They branch sparingly, giving off slender lateral branches that bifurcate acutely. The whole growth is slender and sparse. Diaphragms containing spicules that lie transversely occur at irregular intervals. The galleries lie completely in one plane, parallel to and only a short distance below the surface of the coral.

The papillae are numerous but of very small size. They are each guarded by a dense mass of upright spicules which, at any rate in the centre of the papilla, have a somewhat spiral arrangement.

There are numerous large cells in the parenchyma that contain granules of a comparatively pale brown colour.

The *spicules* are small, moderately slender and all of one kind. They are by no means numerous except in the papillae; in the galleries, except in the diaphragms, they lie parallel to the surface. They are somewhat variable in form, but are all tylostyles with well-developed heads. These are usually subglobular but may be trilobed or irregular; occasionally they contain a single relatively large dilatation of the axial canal There is occasionally a projecting annulus a short distance below the head. The shaft is as a rule slightly curved; its curvature may be of a general nature or confined to the uppermost third. Immediately below the head the shaft is slightly constricted; lower down it swells slightly but never becomes quite as broad as the head; the broadest part is usually situated in the upper half and the lower half tapers very gradually to a fine point.

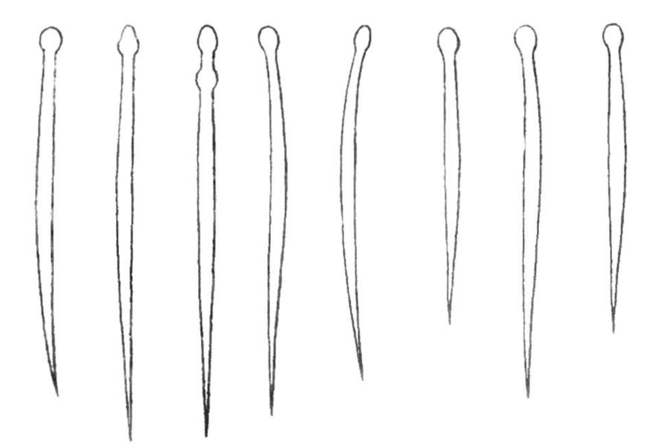

FIG. 2.—Spicules of *Cliona kempi*, sp. nov.

Measurements of Spicules.

Length of spicule ..	..	0·127—0·205 mm.
Greatest breadth of shaft	..	0·0041—0·0082 ,,
Diameter of head ..	..	0·0082—0·0125 ,,

Type.—No. 6956/7 ZEV, *Ind. Mus.* (on slide in Canada balsam).

Locality.—Port Blair, Andaman Is., Bay of Bengal: in dead reef-coral with *Cliona lobata* and *C. mucronata*.

Thoosa hancocci, Topsent.

1915. *Thoosa hancocci*, Annandale, *Rec. Ind. Mus.* XI, p. 21.

The species is evidently common in dead coral in the Andamans. Specimens in Mr. Kemp's collection all possess nodular amphiasters, but these spicules, which are confined to the papillae, are present only in very small numbers. In some papillae they are altogether absent, and there are never more than about half a dozen in any one papilla. These specimens, therefore, which are well preserved in spirit and had evidently reached their full or about their full development, on the whole bear out what I have said in the paper cited on the possible disappearance of the nodular amphiasters in certain phases of the species.

Family DESMACIODONIDAE.

Subfamily *ECTYONINAE.*

Rhabderemia prolifera, sp. nov.

(Plate xxxiv, fig. 3.)

The *sponge* forms an excessively thin film, much less than 1 mm. thick, and coats the burrows of *Cliona* in dead coral. Its surface bears numerous small rounded buds, each of which con-

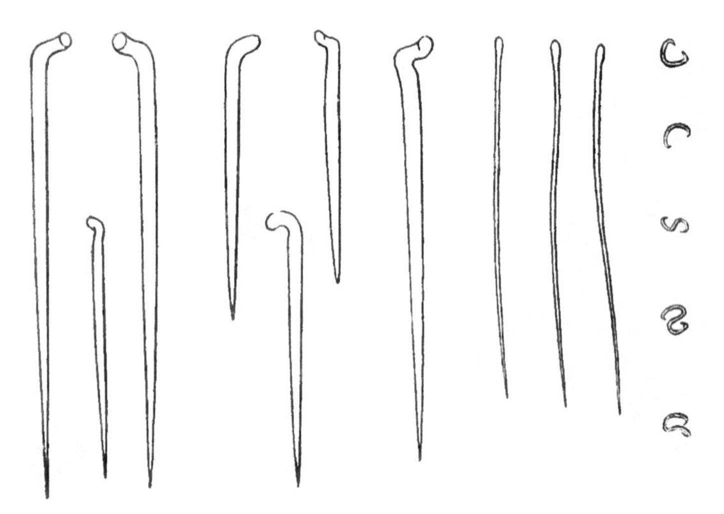

FIG. 3.—Spicules of *Rhabderemia prolifera*, sp. nov.

tains in its centre a particle of calcareous matter. In dried specimens the surface is hispid, but this character may be artificial. The apertures are very small and cannot be detected with certainty in dried specimens. There is a very thin, colourless basal membrane. Owing to the manner of growth, specimens extracted from the coral by the use of acid often appear to be turned completely inside out, or else to contain large irregular cavities in their interior; both appearances are easily explained if the small size of the chambers occupied by the sponge is remembered, and also its own filmy form. The masses that seem to be inside out are merely hollow membranes that have lined the walls of small chambers of corresponding form and size and the surface exposed when the coral is dissolved away is the basal surface of the sponge that was in contact with the wall, while the existence of relatively large spaces of irregular shape in masses in which the true external surface is outermost is due to the fact that they have grown round projecting fragments of coral at the angles of the Clionid's galleries and that these fragments have disappeared owing to the action of the acid.

Spicules.—There are three kinds of spicules, *viz.* (*a*) comparatively stout, smooth styli of the type called rhabdostyles by Topsent[1] in his definition of the genus, (*b*) much more slender, almost hair-like tylostyli, which are shorter than the longest rhabdostyles and (*c*) small, much contorted sigmata.

The rhabdostyles are perfectly smooth and have their heads almost truncate, not at all swollen and as a rule spirally contorted in two whorls. They are actually rather slender and vary greatly in length; indeed, two series may perhaps be distinguished as regards size, but intermediate forms occur. Those of the larger series are on an average about 0·176 mm. in length; those of the smaller series not more than 0·099 mm. The shaft tapers gradually to a fine point.

The dermal tylostyles are curved or sinuous, perfectly smooth, very slender and almost hair-like in appearance; they are longer than the shorter rhabdostyles. Their heads are of an elongate oval form and often not at all clearly differentiated.

The sigmata are fairly uniform in size, small and slender, variable in shape but never having a complete twist or knot in the centre and never enlarged at the extremities.

Skeleton.—The skeleton is very degenerate and the number of spicules present is comparatively small. The rhabdostyles stand separate and semi-erect, with their contorted heads resting on the basal membrane and their shafts pointing obliquely upwards. The tylostyles lie horizontal in ill-defined bundles, which are often comparatively broad and sometimes form as a whole well-marked curves, but are never reticulate. The sigmata are scattered sparsely without definite arrangement. The slender tylostyles are more numerous than either of the other two kinds of spicules.

Measurements of Spicules.

Length of rhabdostyles	0·0902—0·209 mm.
Diameter of shaft of rhabdostyles ..	0·0057—0·0082 ,,
Length of slender tylostyles ..	0·147 mm.
Length of sigmata	0·0123 ,,

Type.—No. 6420/7 ZEV, *Ind. Mus.* (mounted in Canada balsam on a slide).

Locality.—Port Mouat, Andaman Is., Bay of Bengal ('*Investigator*').

The type-specimen occupies the galleries of *Cliona viridis* in a piece of dead Madreporarian coral. The external surface of the

[1] *Rés. Camp. Sci. Monaco,* fasc. II (Spongaires de l'Atlantique Nord), p. 115 (1892).

coral is much eroded owing to the attacks of various burrowing organisms and part of the galleries excavated by the *Cliona* have completely broken down, leaving a fairly large open cavity. The growth of the *Rhabderemia* appears to have commenced in this cavity and then to have proceeded inwards along the excavations of the other sponge, parts of which it had completely surrounded and was apparently in the act of engulfing.

The buds to which reference has been made are merely portions of the sponge that have grown over projecting fragments of coral in the angles of the galleries and have then become constricted at the base.

The specimen, though dry, is in good condition, having originally been preserved in spirit.

The species is very closely related to *R. pusilla* (Carpenter)[1], of which it should perhaps be regarded as a variety. It is distinguished, however, by its larger sigmata, which are of a slightly different type, its longer slender styli (or tylostyli), and its stouter and more variable rhabdostyli. Topsent describes *R. pusilla* as an excessively thin "éponge jaune pâle revêtante." The only Indian sponge hitherto referred to the genus *Rhabderemia* is Dendy's *R. indica*[2] from Ceylon. It has short roughened styli and sigmata that are often twisted into a complete knot in the centre; the skeleton is reticulate.

Family AXINELLIDAE.

Genus **Amorphinopsis**, Carter.

1887. *Amorphinopsis*, Carter, *Journ. Linn. Soc. London* (*Zool.*) XXI, p. 77.
1896. *Spongosorites*, Topsent, *Mém. Zool. Soc. France* IX, p. 117.
1900. *Spongosorites*, *id.*, *Arch. Zool. expérim.* VIII, p. 265.
1905. *Spongosorites*, Dendy in Herdman's *Ceylon Pearl Fisheries*, III, p. 182.

In examining a fragment of the piece of dead sponge-riddled coral described by Carter in 1887 I came across a small sponge that afforded me much difficulty, until I had compared my preparations with others made from the material sorted out and named by that author. On making a comparison I could not remain in doubt that this sponge was the same as the one named by him *Amorphinopsis excavans*; indeed, it was probably a schizotype of that species. Carter's descriptions are as a rule remarkably clear and accurate, but this was not the case in the present instance, in which his figures are actually misleading. He gave no separate description of *Amorphinopsis*, the generic characters of which he left to be inferred mainly from his specific diagnosis.

The sponge agrees with Topsent's description of *Spongosorites*, except in the fact that its spicules are not "biangulate." In Car-

---

[1] *Microciona pusilla*, Carter, *Ann. Mag. Nat. Hist.* (4) XVIII, p. 239, pl. xvi, fig. 51 (1876) and Topsent, *Mém. Zool. Soc. France* 1889 (II), p. 41, fig. 7.

[2] In Herdman's *Ceylon Pearl Fish.* III, p. 180, pl. xii, fig. 10 (1905).

ter's figure the amphioxi are shown as having a regular curve, but this is by no means always the case and though they are not swollen in the middle they are often distinctly geniculate at or near that point. With Dendy's redefinition of *Spongosorites* the species agrees precisely. All this is made abundantly clear when *A. excavans* is compared with the form here described as *A. excavans* var. *digitifera*.

The genus *Amorphinopsis* may now be redefined as follows:—

> Axinellidae of encrusting, reticulate or massive shape, sometimes bearing upright branches or conuli; the skeleton composed of stout spicule-fibres containing little horny material and forming a coarse and irregular reticulation. The fibres consist of large, smooth styli or amphioxi, or of a mixture of smooth styli and amphioxi, lying parallel to one another. Smaller spicules of the same types surround the fibres and as a rule form a horizontal layer in the ectosome. Some or all of the spicules are geniculate in the middle; sometimes they are also inflated at this point.

Amorphinopsis excavans, Carter.

1887. *Amorphinopsis excavans*, Carter, *op. cit.*, p. 77, pl. v, figs. 12-15.

The sponge in Carter's specimen consists of a thin external crust and a network of fine cylindrical basal branches that ramify in the excavations of Clionidae in dead coral. The external crust is remarkable for the curious little prominences or bosses with which it is ornamented and for the strands of spicules that radiate from them. The regularity of their arrangement is somewhat exaggerated in Carter's figures. The prominences seem to me to be no more than incipient, or possibly abortive, conuli or branches. Each probably contains an osculum obliterated by contraction. There is a dense external covering of smaller spicules lying horizontally and matted together on the surface of the sponge. The internal or basal branches rarely contain more than a single stout strand of spicules, but they ramify and anastomose in accordance with the ramifications and anastomosings of the cavities they occupy. A horizontal reticulation of fibres occurs near the surface.

The spicule-fibres, whether on the surface or inside the coral, consist mainly of the larger spicules, which are for the most part true amphioxi. Occasionally a slender sub-stylote spicule is to be found amongst them, while a comparatively large number of true styli are also present. The last are as stout as the stoutest amphioxi, but usually shorter. All the amphioxi are more or less curved and most are crescentic in form; a few are, however, distinctly geniculate at or near the middle and forms (which must be regarded as mere abnormalities) may be found in which there is a regular angle near one end.

The smaller spicules, which surround the fibres in an irregular manner as well as forming a layer on the surface of the sponge, comprise both amphioxi and styli. The former resemble the

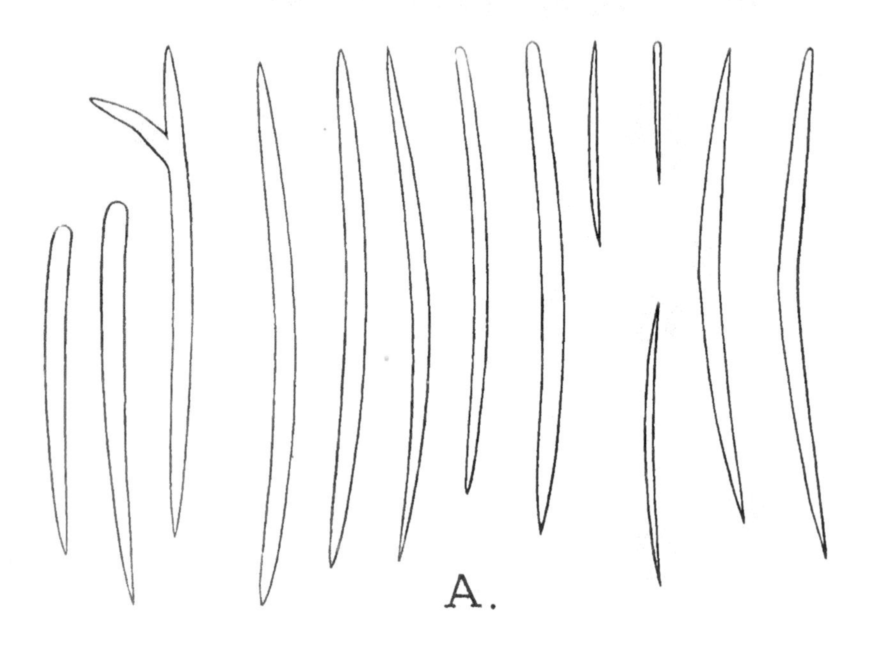

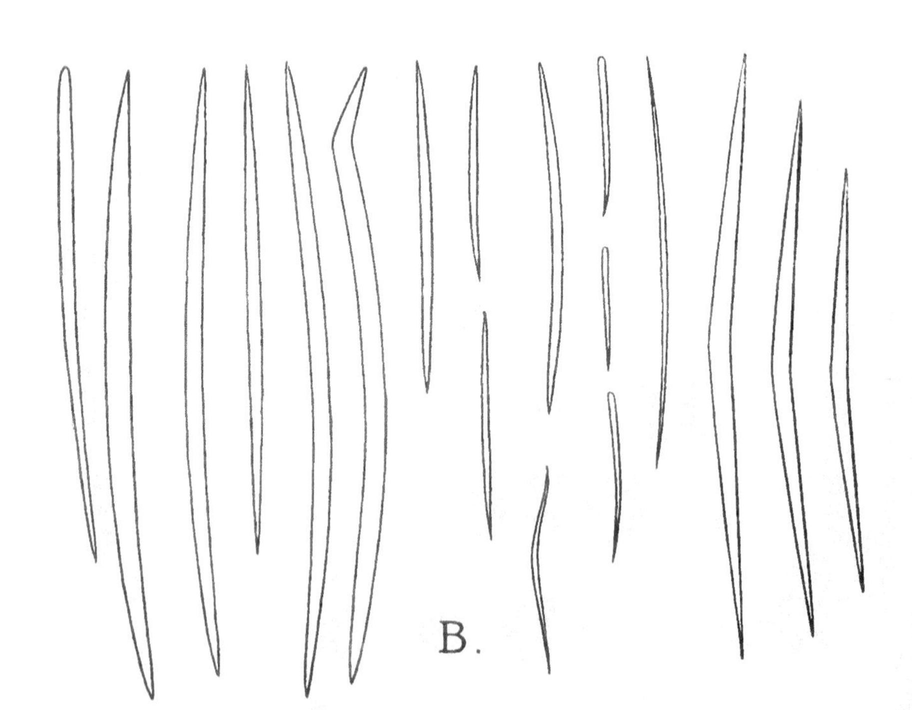

FIG. 4.—Spicules of *Amorphinopsis excavans*, Carter.

A. From the type of the species.
B. From the type of var. *digitifera*, nov.

larger amphioxi in shape and proportions, but the latter are usually straight or nearly so.

Schizotype.—No. 6597/7 ZEV, *Ind. Mus.* (dried specimen).

Spicules of *Cliona* often occur in the parenchyma and films of *Chondrilla* sometimes envelop the basal branches.

var. **digitifera**, nov.

1913. *Spongosorites* sp., Sewell, *Journ. As. Soc. Bengal* (n.s.) IX, p. 346.

I have had by me for some years a sponge that I identified provisionally for Capt. Sewell as a new species of *Spongosorites*. A comparison with Carter's *A. excavans* shows an absolute identity of skeletal structure, though the external form is very different and slight differences in spiculation can be detected. I propose therefore to regard this sponge as a variety of *A. excavans*, of which it may be no more than a growth-phase.

The sponge consists of a number of short, pointed, somewhat compressed upright branches of rather irregular outline, united by means of a crust in which are embedded numerous small stones (non-calcareous) and dead shells of Lamellibranchs and Balanidae. The longest branches are about 30 mm. long and about 14 mm. broad at the broadest point; their thickness is about 7 mm. The shortest axis is directed towards the centre of the mass. The whole specimen is about 100 mm. long by 40 mm. broad, but has probably formed part of a larger mass. In spirit the colour is dirty white. The sponge is rather hard but can be torn easily.

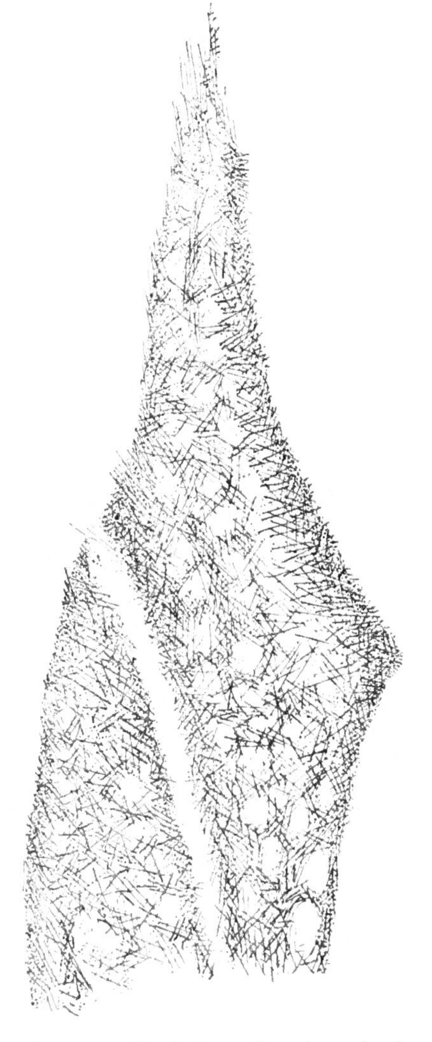

FIG. 5.—Vertical section through the skeleton of the distal part of a branch of *Amorphinopsis excavans* var. *digitifera* (enlarged).

The external surface is in places obscurely and minutely reticulate, elsewhere distinctly hispid. No external orifices can be detected and it is probable that both oscula and pores are highly contracted.

The internal structure of the sponge is somewhat cavernous and several large canals run vertically up each branch, one situated in the middle being as a rule of greater calibre than the others. Probably the oscula are situated near the tips of the branches and the pores on the hispid parts of the surface.

The skeleton forms a dense, irregular network. In the branches its fibres curve upwards and outwards towards the external surface; as a rule they are directed mainly towards the inner side of the branch. They frequently fuse together to form strands of great thickness, but seem to contain little or no horny matter. There is a horizontal reticulation of fibres below the external layer of small spicules. The larger spicules are closely packed together in the fibres and lie quite parallel to one another. The external layer of small spicules is horizontal over the greater part of the surface but in the hispid parts the spicules are vertical and little upright bunches can sometimes be detected that project through the dermal membrane. The bunches are arranged with considerable regularity at fairly equal distances. Sometimes they coincide in position with the terminations of skeletal strands, but this is not always so.

The spiculation differs from that of the typical form in the complete absence of large stout styli and in the fact that the large amphioxi are on an average considerably shorter.

Type.—No. 5010/7 ZEV, *Ind. Mus.* (in spirit).

Locality.—Rock-pool at Fisher Bay, Tavoy I., off the coast of Tenasserim.

This sponge approaches *Dactyella*, Thiele[1] in structure and fully bears out Dendy's[2] suggestion as to a possible relationship between the two genera. Indeed, I doubt whether they are distinct.

Family CHONDROSIIDAE.

Chondrilla nucula, Schmidt.

1862. Schmidt, *Spong. Adriat. Meeres*, p. 39, pl. iii, figs. 22, 22*a*.
1877. Schulze, *Zeitschr. Wiss. Zool.* XIX, p. 108, pl. ix, figs. 11-18.
1881. Carter, *Ann. Mag. Nat. Hist.* (5) VII, p. 384.
1889 (1887). ? *Id.* (*Cliona stellifera* ?) in part, Anderson's *Fauna of Mergui* I, p. 62.
1891. Keller, *Zeitschr. Wiss. Zool.* LII, p. 327.
1892. Topsent, *Rés. Camp. Sci. Monaco*, fasc. II, p. 54.

A re-examination of part of Carter's original material leaves no doubt that the provisional species he described in 1889 under the name of *Cliona stellifera* ? was founded on the association of spicules of a *Cliona* with those of a *Chondrilla*. The *Cliona* was in all probability *C. viridis*, while the *Chondrilla* was either *Ch. nucula*, *Ch. mixta* or *Ch. distincta*, if it was not composed of all

[1] Süd. ü. pacif. Spongien, *Bibl. Zool.* XXIV (i), p. 55 (1898).
[2] In Herdman's *Ceylon Pearl Fisheries*, III, p. 182 (1905).

three species. *Cliona viridis* is particularly abundant in the masses of dead coral from which he extracted the spicules on which he based his description, or rather indication, and it is frequently covered by a thin film, of one or other of the Chondrillae. The only other species of *Cliona* [1] present is *C. ensifera*, which Carter distinguished from " *stellifera*."

The specimens of *Ch. nucula* I have examined from this material consist of extremely thin films much less than 1 mm. thick and spread out over the surface of *Cliona viridis*, *C. ensifera* and *Stoeba simplex* in their excavations in dead coral. The spicules correspond well with the figures cited above and agree in dimensions (diameter 0·01 to 0·2 mm.) with those of a specimen from the Red Sea examined by Keller. They are densely crowded in the ectosome and frequently touch one another in that part of the sponge. The colour, after some 28 years in spirit, is pale brown. The film is usually uniform, but sometimes reticulate. The fragments extracted have been very imperfect.

Ch. nucula is cosmopolitan in distribution. It has been recorded from the Mediterranean, the Azores, the Red Sea and the Gulf of Manaar. The specimens referred to above are from King I. in the Mergui Archipelago, which lies off the coast of Tenasserim, the southern extension of Burma.

Chondrilla mixta, Schulze.

1877. Schulze, *op. cit.*, p. 113.
1891. Keller, *op. cit.*, p 327.

In the same fragments of dead coral, and in precisely similar conditions, I find imperfect examples of another *Chondrilla* which agrees well enough with Schulze's description of *Ch. mixta* so far as the shape and arrangement of the spicules are concerned. The film it forms in these circumstances is still more delicate than that formed by *Ch. nucula* and is quite colourless. The spicules include both oxyasters and spherasters, the largest of both of which are not more than 0·012 mm. in diameter.

Distribution.—Red Sea (*Schulze*); Mergui Archipelago, Burma.

Chondrilla distincta, Schulze.

(Plate xxxiv, figs. 4, 4*a*.)

1877. Schulze, *op. cit.*, p. 133, pl. ix, fig. 19.
1903. Thiele, *Abh. Senckenb. Natur. Gesellsch.* XXV, p. 67, pl. iii, fig. 20.

Still in the same fragments of coral from Burma a third species of *Chondrilla* occurs, in the same circumstances. It is undoubtedly *Chondrilla distincta*, Schulze, with which its spicules agree in every respect.

---

[1] Not having Carter's full material in my hands, I have been unable to find the spicules he associated in his provisional species *Cliona sceptrellifera* (. In any case this species, if it exists as such, is clearly not a *Cliona*.

Owing to the more robust form of this sponge it has been possible to extract larger and more complete pieces, which exhibit its manner of growth in the burrows of *Cliona*. The specimens were found in the centre of a piece of coral about 4 cm. thick. No part of the sponge was visible on the surface of the coral. It consisted of irregular cylindrical, ramifying and even reticulate masses, the component branches of which were about 2 mm. thick. The colour was deep purple-brown, except at the extremities, where it was much fainter, if not altogether absent. The surface was for the most part smooth, but crater-like pits surrounded by a particularly dense zone of spicules occurred sparingly. Large oval cells containing brown pigment granules could be detected in the choanosome. At many points the greater part of the ectosome was entirely concealed by spicules, mostly spherasters. Oxyasters occurred sparingly in the choanosome.

The most interesting feature of the sponge, however, consisted in little tentacle-like club-shaped branches (pl. xxxiv, fig 4*a*) the free extremities of which were densely covered with spherasters, while the cylindrical portions were bare of spicules or almost so. In some cases the tips of these branches were in contact with the surface of other sponges or of tubes constructed among them by Polychaete worms. Wherever this occurred the tip was splayed out and, if the sponge touched was a *Cliona*, the latter was protected by a dense layer of its own macroscleres and by a chitinous sheath (pl. xxxiv, fig. 4). Some cases were seen in which the expanded tip of a branch of the *Chondrilla* was actually spreading out in a thin, colourless film over the surface of another sponge or of a worm-tube. We have here proof of actual aggression on the part of the *Chondrilla*, and evidence of the methods by which *Cliona* defends itself against such aggression. This subject is discussed later (p. 476). In every case, on the other hand, in which *Stoeba plicata* is the sponge attacked by this or other species of *Chondrilla* its ectosome, with the microscleres abundant in that part of the sponge, had disappeared where the attacking sponge had covered it.

Part II.—BIOLOGICAL.

The large proportion of the sponges referred to in this paper were found in two small pieces of dead Madreporarian coral, neither weighing more than a few ounces. One piece came from the Andamans, the other from the Mergui Archipelago. The former is a portion of a somewhat larger specimen examined by Carter many years ago and described by him in his account of the sponges collected by the late Dr. John Anderson. He found in it examples of no less than 8 species of sponges and yet it is clear that his examination was not exhaustive, for (in addition to the majority of the species he noticed) the fragment now in the Indian Museum contains at least four others. There seems to be a stage in the decay of the more solid Madreporarian corals at which their skeletons become peculiarly attractive to a large number of small sponges, some of which are true excavators, while others are primarily thin encrusting forms able to exist on a solid even surface but preferring an irregular one, and capable of penetrating into its interstices. Sponges of both kinds play an important part in the final disintegration of both corals and calcareous algae.[1]

I have recently pointed out elsewhere[2] that sponges which excavate their burrows in molluscan shells are often liable to be killed by the growth of encrusting forms. The association of such species as *Cliona vastifica* and *Laxosuberites aquaedulcioris*, though it may be physically intimate, is evidently quite fortuitous; the *Laxosuberites* merely happens to grow on the surface of the oyster-shells in which the *Cliona* has burrowed, and its presence, though ultimately fatal, is not correlated with the presence of the other sponge; it grows on many shells that the Clionid has not attacked and is in no way prejudiced by so doing.

Off the coast of Orissa and the north of the Madras Presidency oyster-shells are often attacked by another species of *Cliona*, recently described as *Cliona acustella*,[3] which ultimately eats away the entire surface, leaving it deeply and densely pitted. Apparently the excavator retires deeper into the shell when this occurs. The roughened surface it has produced is, however, attractive to at least two kinds of very thin encrusting sponges, both of which belong to the genus *Eurypon*. They are not content with the surface, however, but pursue the *Cliona* into its retreats, coating the walls of its galleries and apparently driving it before them. In other Lamellibranch shells (of *Ostrea*, *Malleus* and *Tridacna*) from Indian seas I have found the remains of sponges of similar habits that belong to allied but probably undescribed genera and have little doubt that the species originally described by Hancock as *Cliona purpurea*[4] is a form of the kind There is no evidence that any of these Desmaciodonid sponges actually attack the Clionid with which it is associated, and I have never found spicules of the latter family embedded in the substance of one of the former; they merely overwhelm them or suffocate them and usurp their place. Unfortunately the remains of sponges of this kind now in my hands are insufficiently preserved to justify technical descriptions.

The Tetraxonellid sponge *Stelletta vestigium* (*antea*, p. 459) goes a little further. It is a more massive species than those alluded to in the preceding paragraph and makes its way into the burrows of Clionidae, not by merely growing along their walls, but by thrusting practically solid processes into them. When these processes come in contact with the rightful owner of the burrow

[1] Carter has described a collection of boring organisms from calcareous algae from the Gulf of Manaar. See *Ann. Mag. Nat. Hist.* (5) VI, p. 150 (1880).

[2] *Mem. Ind. Mus.* V, p. 35 (1915).

[3] *Rec. Ind. Mus.* XI, p. 14 (1915).

[4] See Topsent, *Arch. Zool. expérim.* (4) VII, p. xvi (1907).

its spicules adhere to and are even incorporated in what we may call for this purpose the "skin" of the aggressor.

Amorphinopsis excavans has similar habits, but takes the borrowed spicules into its own internal parts.

Stoeba plicata var. *simplex* differs from these species in that it possesses independent powers of excavation and only uses the burrows of Clionidae as the basis of its own operations. It adapts and enlarges these burrows and at the same time not merely attaches the spicules of its host to its own surface, but takes them into its own inner parts and possibly even utilizes them in strengthening its own attenuated and delicate terminal processes.

Coppatias investigatrix—and possibly also *C. penetrans*—attack in a similar manner, but its parasitic character is more marked, in that, having once penetrated into the burrows of a Clionid, it is content with them and so far as its external form is concerned becomes a mere cast of them. Moreover, it enters the burrows at a comparatively early stage of development and appears to have only a short-lived and very inconspicuous encrusting phase.

All these sponges may be classed, in greater or less degree, as parasites, in that they appropriate the fruit of the labours of other species and even possibly make use in some cases of the spicules of the sponges they attack. There is no evidence, however, that they feed on the bodies of their victims. In the case of the three species of *Chondrilla* and of *Rhabderemia prolifera* it is possible that the attacking species does so, for the Clionid is actually overwhelmed and engulfed, not merely thrust before the invader. The method of attack is not the same in the case of the Chondrillae as in that of the *Rhabderemia*. The former give rise to peculiar capitate tentacle-like processes when they approach the Clionid or any other body with which they may come in contact. The heads of these processes, which are armed with spicules, spread out over any surface that they happen to touch. If they do so on the surface of another sponge they surround it and absorb it completely.

The *Rhabderemia*, on the other hand, which forms a much thinner film as a whole, spreads bodily round portions of the Clionid, which it ultimately absorbs in a similar manner.

It is noteworthy that the great majority of all these parasitic sponges are known to have free encrusting phases or varieties, which are able to exist independently of the labours of other species. *Coppatias penetrans* and *C. investigatrix*, and possibly *Rhabderemia prolifera*—if the latter is to be regarded as specifically distinct from *R. pusilla*—are apparently exceptions. They seem to have become specially adapted for a parasitic life, but it is very desirable that further investigations should be made into their minute structure.

Most of these sponges are probably able to enlarge the burrows that they occupy, though there is no evidence that *C. investigatrix* and *C. penetrans* do so, by the mere expansion of their growth. If the material into which they have penetrated is at all soft or crumbling this causes it to split or even to fall in pieces, and the final result of the parasitism of most of the invading sponges must be to produce a state of affairs in which it is necessary for them, unless they are to perish altogether, to assume again an independent form of existence. Sooner or later they destroy the walls of their retreat and so are once more exposed.

The species of *Stoeba* and *Coppatias* do not depend solely on expansion as a means of penetration, for they are able to break off fragments of calcareous matter. These are more or less rounded in form and are stored up in the interior of the sponge. How the fragments are broken off we do not know, but it is evident that the sharp points of the spicules play an important part in the operation. Even in the case of the Clionidae the precise method by which the burrows are excavated is not yet by any means clear. It has been shown[1] that the action of acid is absent, and it seems most probable from the disposition of the spicules in the growing points of the sponge that little pieces of shell or coral are broken off, not merely by impact of the spicules, but also by a rotary action. The points of a number of macroscleres are probably directed in a circle covering a small area of the surface on which they are to work. The heads of these spicules may be then rotated by what would be called in an animal more highly organized than a sponge, muscular action. The fragments observed in the interior of *Coppatias* and *Stoeba* are as a rule larger and of less regular shape than those produced by the activities of *Cliona* or *Thoosa*, and it seems probable that the operation by which they are produced is of a less specialized nature than in the case of the Clionidae. Moreover, the manner in which the spicules are arranged appears to be much more haphazard, and we can only suppose that their action is less concerted.

The fragments of calcareous matter removed by *Rhabderemia prolifera* are certainly separated by an entirely different process. The species of *Coppatias* and *Stoeba* that invade Clionid burrows grow forwards as bodies that are practically solid, whereas the *Rhabderemia* merely coats the walls of the excavations it invades as an extremely thin film. This film grows round projecting fragments of coral and separates them from the walls by constricting itself round their bases. There is no evidence that the contained particles of calcareous matter are of any utility to the other species, but to this sponge they are probably directly useful. The film that surrounds each fragment contracts away from the main body of the sponge and forms a bud that separates itself from its parent and doubtless aids in the distribution of the species by so doing. The fact that it has a solid core of relatively heavy material must aid it considerably by causing it to fall away more readily.

[1] For a full discussion see Topsent, *Arch. Zool. expérim*, (2) V ?, pp. 59-71 (1887).

The fact that a considerable number of small encrusting sponges are in the habit of invading and occupying the excavations of Clionidae to the detriment of the latter is quite clear from the foregoing notes, and I have abundant evidence that the parasitic species described form only a very small proportion of the sponges of similar habits that exist in Indian seas, more particularly on the decaying parts of coral reefs. The question naturally arises, How do the Clionidae protect themselves? No direct observations on this point have been made in the field but in the case of *Thoosa investigatoris* and *Coppatias investigatrix* the fact that the invading sponge was evidently in a comparatively early developmental phase enabled some interesting deductions to be made. Fig. 1 on pl. xxxiv shows a young sponge of *C. investigatrix* which has just penetrated into the outer part of a burrow of *T. investigatoris.* The shell has been dissolved away and one sees in the lower part of the figure the base of an exhalent papilla from below, the middle of the figure is occupied by the *Coppatias*, while in the upper part a confused mass of spicules belonging to the Clionid is shown. The invading sponge appears to have made its way through an inhalent papilla that has degenerated into a mere confused mass; it is shown in the upper part of the figure. The *Coppatias*, however, has not merely penetrated the papilla, for it contains small cavities that apparently represent fragments of calcareous matter detached by itself. Fragments of precisely the same shape and size were observed *in situ* in preparations in which the action of the acid used in extracting them from the shell had not gone so far.

There are several points of interest to be noted in this preparation. Firstly, the Clionid has secreted a horny membrane [1] (*h.c.*) wherever it is in contact with the invading sponge. Secondly, the exhalent papilla (*e.p.*) at the base of which the invading sponge has entered the shell is distorted and has its armature of macroscleres greatly extended and increased. Thirdly, the inhalent papilla through which the *Coppatias* has apparently made its way is as already stated completely disorganized. Fourthly, the invader is very minute and forms a compact mass that does not spread out over the surface of the shell.

Fig. 2 represents a later stage in the attack in the same case. The *Coppatias* has penetrated well into the burrows of the Clionid and has to some extent adapted itself to their form. The Clionid has shrunk considerably in its excavation and has secreted round itself a thick horny coat, not merely where it is in actual contact with the *Coppatias*, but also at those points at which it was liable to

---

[1] It is noteworthy that there are none of the characteristic nodular amphiasters present in the parenchyma of the Clionid. As I pointed out in my original description of this species (*Rec. Ind. Mus.* XI, p. 20), these spicules often occur in great abundance in association with a horny membrane covering projecting parts of the sponge, in circumstances that suggest that they are utilized in excavating fresh papillae. It is now evident that the secretion of horny substance is not necessarily correlated with their development.

be attacked by a flank movement. A number of its macroscleres project through the horny covering into the body of the invader.

When *Cliona ensifera* or *C. viridis* is attacked by a *Chondrilla* a similar horny coating is produced and a mass of macroscleres is formed lying parallel to the tranverse axis of the part with which the attacking sponge is in contact. This also occurs when *C. viridis* is attacked by *Rhabderemia prolifera*, but the horny coating is very thin.

It therefore appears that the mode of defence adopted by the Clionid is not always precisely the same, even in cases in which it can be adduced with practical certainty from observations made on preserved material. There are other methods of defence that can only be surmised from general considerations. One of these is possibly the production of diaphragms in the galleries of the Clionidae. In *C. mucronata* these structures are remarkably well developed and are protected by highly specialized spicules. It is perhaps more than a coincidence that I have not found any examples of this species that were overwhelmed or even attacked by other sponges.

I have pointed out elsewhere[1] that the gemmules of the Clionidae are possibly useful in permitting regeneration after the parent sponge has been suffocated by the growth of encrusting forms over its papillae. The production of gemmules in *C. annulifera* and *Thoosa investigatoris* at a depth of over 700 fathoms is particularly interesting, because at depths of such magnitude it is probable that conditions remain identical, so far as temperature, currents, etc., are concerned, throughout the year. It is only in a very few species of Clionidae that resting bodies of the kind have been discovered and I am convinced that they are not as a rule produced in Indian species other than the two just mentioned and the shallow-water form *C. vastifica*. In both the deep-sea species the gemmules are of a highly specialized character. In *C. annulifera* they are provided with spicules of a type that does not occur in the vegetative part of the sponge. These spicules are microscleres of an unusually large size; they cover one surface of the somewhat lens-shaped gemmule in a dense horizontal layer, forming a regular shield, but are entirely absent from the other surface. The surface that they protect is the one in contact with the parent sponge, that is to say the one with which an invading sponge would come in contact if it made its way along the galleries already excavated. The naked surface is in contact with the walls of the excavations, which protect it in the natural position.

The gemmule of *T. investigatoris* is very different from that of *C. annulifera*. It has neither a horny covering nor spicules of any kind, but is hidden away in a special chamber excavated in some unknown manner for its reception, and is only connected with the parent sponge by an extremely fine strand of living matter enclosed in a narrow canal.

---

[1] *Mem. Ind. Mus.* V, p. 35 (1915).

Both these Clionids are known only from specimens taken in a single haul of the 'Investigator's' net, and it is impossible therefore to say much about their enemies. We know, however, that *T. investigatoris* is attacked by *C. investigatrix*, and I have not been able to find any example of the latter that is drawn out into a sufficiently fine filament to make its way into a gemmular chamber of the Clionid.

The information conveyed in the foregoing biological notes may be summarized as follows:—

1. The Clionidae are liable to be attacked in their burrows by a large number of small sponges belonging to several different families.

2. The majority of these invading species are known to exist also as ordinary encrusting forms but in a few instances (*e.g.* that of *Coppatias investigatrix*) the sponge has possibly become a pure parasite.

3. In most cases the invader merely occupies the burrow of the Clionid, which it thrusts before it, but in some instances it is possible that it actually engulfs and digests the proper occupant.

4. Different species of Clionidae protect themselves against invasion in slightly different manners, but all secrete a horny coat where the invader comes in contact with them.

5. The production of transverse diaphragms in the galleries of the Clionidae is possibly a means of protection against invading sponges, especially in the case of *C. mucronata*, in which these diaphragms are of an unusually elaborate nature.

6. The production and elaboration of gemmules in the Clionidae is perhaps another means of defence against similar enemies, particularly in the case of the deep-sea species *C. annulifera* and *T. investigatoris*.

7. The cases of invasion investigated represent only a small proportion of those in which similar phenomena occur.

EXPLANATION OF PLATE XXXIV.

Figs. 1, 2.—*Thoosa investigatoris* attacked by *Coppatias investigatrix*.

1. A young *Coppatias* that has just made its way into the burrows of the *Thoosa* in a Gastropod shell, seen from below (× 16).

2. Portion of an older sponge of the same species in contact with the *Thoosa*, seen from the side (× 65).

A, A′ = the *Thoosa*: B = the *Coppatias*: S = Gastropod shell in section: *c* = cavity from which calcareous matter has been removed by acid: *e p.* = exhalent papilla of the *Thoosa*: *h.c.* = horny coat secreted by the *Thoosa*.

In fig. 1 the young invading sponge has apparently made its way through an inhalent papilla of the *Thoosa*, which is represented by a confused mass of spicules (A′). The adjacent exhalent papilla (A) is distorted and greatly enlarged.

Fig. 3.—*Cliona viridis* attacked by *Rhabderemia prolifera*.

A = *C. viridis*; B = *Rh. prolifera*: *c* = cavity from which calcareous matter has been removed by acid: *c′* = passage between two calcareous masses coated with the sponge.

Figs. 4, 4a.—*Chondrilla distincta* attacking *Cliona ensifera*.

4. A mass of the *Chondrilla* sending out tentacle-like branches to envelop the *Cliona* in dead coral (× 75).

4*a*. A single tentacle-like branch more highly magnified (× 255).

A = *C. ensifera*: B, B′ = *Ch. distincta*: C = cavity from which calcareous matter has been removed by acid: *t* = tentacle-like branch.

At B′ a tentacle-like branch has grown out from behind over the surface of the *Cliona*, which it is enveloping.

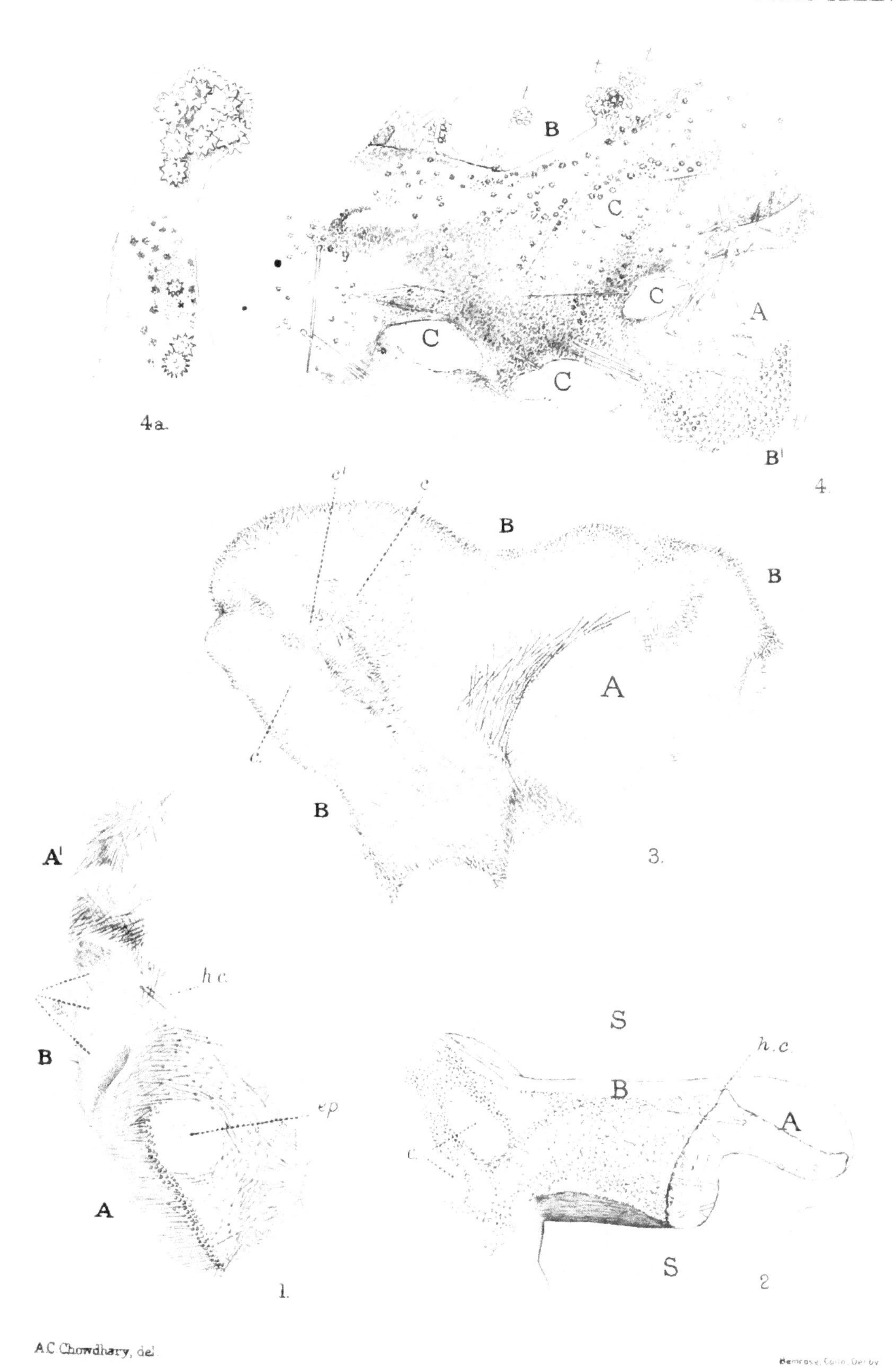

A.C. Chowdhary, del.

Bemrose, Collo, Derby.

SPONGES PARASITIC ON CLIONIDAE.

XXVIII. REPORT ON A COLLECTION OF MOLLUSCA FROM THE OUTSKIRTS OF CALCUTTA.

By H. B. Preston, *F.Z.S.*

Class GASTROPODA.

Order *PULMONATA.*

Suborder *GEHYDROPHILA.*

Family Auriculidae.

Scarabus plicata, Fér.

Prodrome, p. 101.

Chingrighatta, outskirts of Calcutta.

Order *PROSOBRANCHIATA.*

Suborder *PECTINIBRANCHIATA.*

Family Nassidae.

Nassa denegabilis, Preston.

Rec. Ind. Mus., X, 1914, pp. 297-298.

Canal near Chingrighatta, outskirts of Calcutta (a single young specimen).

Nassa orissaënsis, Preston var. **ennurensis,** Preston. (MS.)

Canal near Chingrighatta, outskirts of Calcutta (a single specimen).

Nassa fossae, sp. n.

Shell allied to *N. orissaënsis,* Preston, but differing from that species in its larger size as compared with the type and both larger size and much broader form as compared with the above variety; the subperipheral band is of a whitish colour and the spiral lirae are considerably coarser and very much more numerous, the aperture is much broader and the columella margin is distinctly curved; it is 6 whorled.

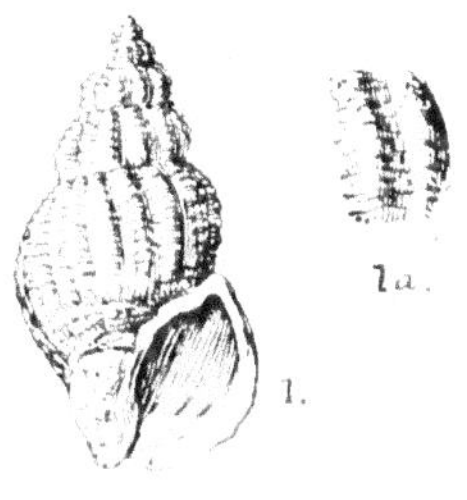

1a.

1.

Fig. 1.—*Nassa fossae,* sp. n., × 3.
„ 1a.— *do.,* sculpture, × 6.

Alt. 9·5, diam. maj. 5·5, diam. min. 4·5 (nearly) mm.

Aperture: alt. 4, diam. 2·75 mm.

Hab.—Canal near Chingrighatta, outskirts of Calcutta.

Family TIARIDAE.

Tiara (Striatella) tuberculata (Müller).

Hist. Verm. 1774 (as *Nerita*).

Canal near Chingrighatta, outskirts of Calcutta (a small form).

Tiara (Tarebia) lineata (Gray).

Wood., *Index Test. Supp.*, 1828, fig. 68 (as *Helix*).

Canal near Chingrighatta, outskirts of Calcutta.

Family RISSOIDAE.

Iravadia princeps, sp. n.

Shell imperforate, elongately fusiform, in dead condition whitish; whorls 7, the first smooth, submanmillary, the remainder sculptured with fine, acute, regular and slightly distant, spiral lirae, the interstices being occupied by very fine, transverse riblets; suture impressed; columella margin porcellanous, narrowly outwardly expanded and reflexed, continuous with the labrum which is varicosely thickened, bevelled behind and rather markedly angled at each termination of the spiral lirations, aperture a little oblique, rather broadly ovate.

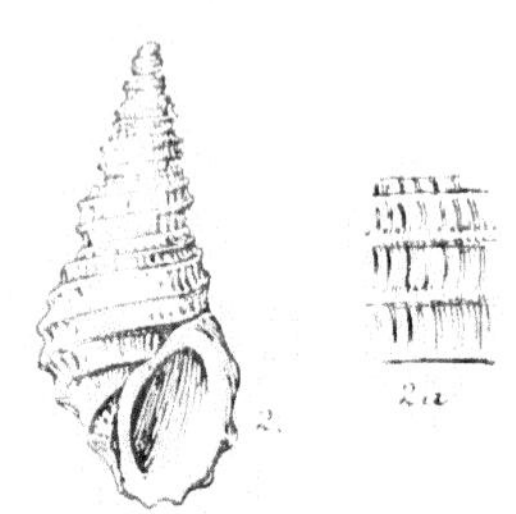

FIG. 2.—*Iravadia princeps*, sp. n., × 6.
,, 2a.— *do.*, sculpture, × 15.

Alt. 6·5, diam. maj. 3, diam. min. 2·25 mm.

Aperture: alt. 2, diam. 1·5 mm.

Hab.—Canal near Chingrighatta, outskirts of Calcutta.

Family ASSIMINEIDAE.

Assiminea francesiae, Wood.

Ind. Test. Supp., 1828; *A. fasciata*, Benson, *Zool. J.*, 1835, p. 463.

Canal near Chingrighatta, outskirts of Calcutta.

Family NERITIDAE.

Septaria crepidularia, Lamarck.

Anim. s. vert., VI, 2, 1822.

Canal near Chingrighatta, outskirts of Calcutta.

Septaria depressa (Reeve).

Con. Icon., *Neritina*, 1855, sp. 86, pl. xviii, figs. 86*a*, *b*.

Canal near Chingrighatta, outskirts of Calcutta.

The above two species seem to the author to be very doubtfully separable.

Class LAMELLIBRANCHIATA.

Order *TETRABRANCHIA*.

Suborder *MYTILACEA*.

Family MYTILIDAE.

Brachydontes emarginata (Reeve).

Con. Icon., Modiola, 1858, sp. 60, pl. x, fig. 73.

Canal near Chingrighatta, outskirts of Calcutta.

Suborder *CONCHACEA*.

Family VENERIDAE.

Sinodia jukes-browniana, sp. n.

Shell irregularly trigonal, inflated, yellowish-white, both valves closely concentrically ridged throughout; umbones small, curved

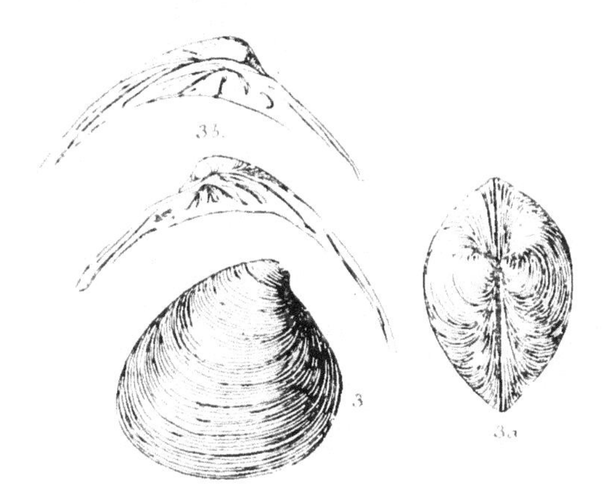

FIGS. 3, 3*a*.—*Sinodia jukes-browniana*, sp. n. (nat. size).
,, 3*b*.— ,, ,, ,, hinge, × $\frac{1}{2}$.

inwards, not prominent; lunule large, cordiform; dorsal margin sharply arched; ventral margin anteriorly slopingly rounded, posteriorly gently rounded; anterior side sloping above, rather sharply rounded in the median part; posterior side slightly produced below, steeply sloping and very gently rounded above; teeth in both valves normal; interior of shell pinkish, shading to pure white towards the ventral, anterior and posterior margins.

Long. 27·5, lat. 25·5 mm.

Hab.—Canal near Chingrighatta, outskirts of Calcutta.

Family CYRENIDAE.

Cyrena bengalensis, Lamk.

Anim. s. vert., Cyrena, 10.

Salt Lakes near Chingrighatta, Calcutta.

Order *DIBRANCHIA*.

Suborder *TELLINACEA*.

Family TELLINIDAE.

Macoma gubernaculum, Hanley.

Proc. Zool. Soc., 1844, p. 142.

Canal near Chingrighatta, outskirts of Calcutta.

Suborder *ANATINACEA*.

Family CUSPIDARIIDAE.

Cuspidaria annandalei, Preston.

Rec. Ind. Mus., XI, 1915, p. 308.

Canal near Chingrighatta, outskirts of Calcutta.

Family ANATINIDAE.

Anatina induta, sp. n.

Shell small, oblong, cuneiform, gaping anteriorly, thin, whitish, covered by a very thin, transparent, very pale brownish periostracum, smooth, but for somewhat distant, concentric growth lines; umbones small, somewhat flattened, dorsal margin gently arched : ventral margin almost straight, a little contracted in the median part ; anterior side rounded ; posterior side produced, sharply rounded, wedgelike

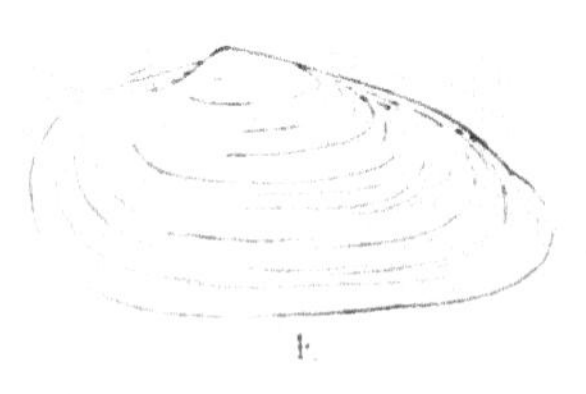

FIG. 4.—*Anatina induta*, sp. n., × 6.

Long. 4, lat. 8·75 mm.

Hab.—Canal near Chingrighatta, outskirts of Calcutta.

XXIX. NOTES ON THE HABITS OF INDIAN INSECTS, MYRIAPODS AND ARACHNIDS.

By F. H. GRAVELY, *M.Sc., Assistant Superintendent, Indian Museum.*

(Plates XXII—XXV).

The preparation of a course of popular lectures during the summer of 1914 necessitated the completion, so far as opportunity permitted, of a number of more or less casual observations that I have chanced to make from time to time on the habits of insects and spiders of Calcutta, and the production of figures to illustrate them. The present, therefore, seems a favourable opportunity of putting on record both these and certain observations made in other parts of India, in Burma and in Ceylon during the last five or six years, incomplete though they are in some cases.

Although a number of notes on the habits of Indian insects have been published from time to time, they are still regrettably few, considering the richness and interest of the fauna with which they deal; and they are so scattered that the discovery of their existence, by anyone in a position to make use of them, is a matter of great uncertainty.

In order to bring all these notes together search would have to be made through a number of European journals; but the results of such a search would probably be very small in comparison with the amount of time it would occupy. Indeed, the time would probably be better employed in making fresh observations.

Since, however, observations on living Indian animals must almost necessarily be made in India, many of them will naturally be recorded in Indian journals, which are comparatively few. And I have tried, in the following pages, to combine with the record of my own observations such references to those of others as I have been able to find in journals, chiefly Indian, up to the end of 1914. The necessity for this became more and more apparent as the work of compilation progressed; for I found that several of my own observations were simply confirmatory of those of others; and that in several instances observations having a very definite bearing upon one another were recorded by different authors, sometimes in different parts of the same journal, without any reference to one another

It has been extremely difficult to find convenient limits to the subject in hand; for notes on habits pass by almost imperceptible gradations into notes on mimicry, development, crop-protection, sanitation, etc. I have not attempted to go through the rapidly increasing literature on Indian "economic" entomology; because, although it undoubtedly contains much that is of scientific interest, I doubt whether the records obtained would be worth the time involved—especially as a large proportion of these have already been brought together in Lefroy's "*Indian Insect Life*", Patton and Cragg's "*Text-book of Medical Entomology*", Fletcher's "*South Indian Insects*", and Stebbing's "*Indian Forest Insects*", text-books all of comparatively recent date.

Nor have I attempted to go through all the literature on Indian Butterflies, a very large proportion of which appears in the Journal of the Bombay Natural History Society. Much of the earlier work done on this group was brought together in Marshall and de Nicéville's well known "*Butterflies of India, Burma and Ceylon.*" It may therefore be mentioned here that the Indian Museum possesses the latter author's file copy of this work, extensively interleaved with published and unpublished notes and figures, and continued in manuscript to deal with Pierinae and Papilioninae. The remaining parts were sent to Col. Bingham for use in connection with the unfinished butterfly volumes of the "Fauna of British India" series, and unfortunately appear to have been lost at the time of his death.

Such observations on butterflies and their larvae and pupae as have come under my notice have been carefully sifted, and only those that seem likely to be of general interest have been referred to below. But in other groups recorded observations are so comparatively few that even the most trivial often seems worth noting; and I have thought it best to include as wide a range of them as possible. I am indebted to Dr. N. Annandale, Mr. T. Bainbrigge Fletcher, Mr. C. Beeson and Mr. E. E. Green for a number of references. I am also indebted to these and other observers for several original notes, each of which is separately acknowledged.

INSECTA.

THYSANURA.

Cunningham ("*Plagues and Pleasures*"[1], p. 190) notes that "fish-insects" prefer "size" to paper, but eat the latter also. Lefroy (*J.B.N.H.S.*[2] XIX, pp. 1006-7), who used *Acrotelsa collaris*, Fabr., as food for the larvae of *Croce filipennis*, Westw., reared the former from the egg, feeding it entirely on paper. The eggs, which were white, soft, and of an oval shape, were laid loosely among the paper.

---

[1] "*Plagues and Pleasures of Life in Bengal*", by Lt.-Col. D. D. Cunningham, C.I.E., F.R.S. (London, 1907).

[2] *Journal of the Bombay Natural History Society.*

DERMAPTERA.[1]

Burr notices the habits when attracted to light of *Labidura lividipes* (*J.A.S.B.*[2] [n.s.], II, p. 391), and the feeding habits, etc., of *Diplatys gladiator* (*J.A.S.B.* [n.s.], VII, p. 772). The attraction of a giant stinging neetle for various kinds of earwigs is also noted in the latter place.[3]

The habits and development of *Diplatys longisetosa* and *D. nigriceps* are described by Green (*Trans. Ent. Soc. London*, 1898, pp. 381-390, pl. xviii and xix).

Willey records the maternal instincts of a Ceylonese earwig (*Spolia Zeylanica*, VI, p. 53).

ORTHOPTERA.

Blattidae.

Annandale notes that *Pseudoglomeris flavicornis* lives under the bark of trees (*Mem. A.S.B.*, I, p. 207).

C. Drieberg notes that cockroaches are common in beehives in Ceylon and appear to attack the combs (*Spolia Zeylanica*, IV, p. 33).

Annandale (*J.A.S.B.* [n.s], II, pp. 105-106) and Shelford (*Rec. Ind. Mus.*, III, p. 125-7) refer to the amphibious habits of cockroaches of the geuus *Epilampra*. These cockroaches are common among stones at the edge of streams in many parts of India.

Green (*Spolia Zeylanica*, VI, p. 135) and Annandale (*Rec. Ind. Mus.*, V, pp. 201-2) describe cockroaches (*Periplaneta australasiae* and *americana* respectively) preying upon winged termites.

Leucophaea surinamensis is ovo-viviparous. When the egg-capsule is protruded it splits along one side, and the young (about 30 in number) at once escape, leaving what looks like a mass of exuviae behind with the capsule. This observation was made at Peradeniya.

Phasmidae.

The development and habits of *Phyllium scythe* and *Pulchriphyllium crurifolium* are described by Murray (*Edinburgh New Phil. Journ.* [n.s.], III, pp. 96-111, pl. vi-viii), Morton (*Bull. Soc. Vaud. Sci. Nat.*, XXXIX, pp. 401-7, pl. iii), St. Quentin (*Entomologist*, XL, pp. 73-75 and 147, pl. iv) and Leigh (*Proc. Zool. Soc. London*, 1909, pp. 103-113, pl. xxviii; *Rep. and Trans. Manchester Ent. Soc.* 1912, pp. 22-29). Green records an authenticated case

---

[1] See also Annandale "Notes on Orthoptera in the Siamese Malay States", *Ent. Rec.*, XII, 1900, pp. 75-77 and 95-97.

[2] *Journal of the Asiatic Society of Bengal.*

[3] The statement in the same place that *Labidura riparia* and *bengalensis* occur under stones between tide-marks by the Chilka Lake must not be taken to imply that this insect is amphibious, for this lake has since been found to have its level so greatly affected by winds and by flooding that regular tides can scarcely be said to exist. High-water marks at all times are more likely to be due to the most recent flood than to a tide (see *Mem. Ind. Mus.* V, pp. 10-11, pl. i).

of parthenogenesis in the latter species (*Spolia Zeylanica*, VII, p. 54).[1]

T. V. Ramakrishna Aiyer describes the life-history of stick-insects hatched from eggs laid in a group on a wooden rafter instead of singly and loose (*J.B.N.H.S.*, XXII, pp. 641-3, 1 pl.).

I have never seen any record of the fact that, in some Phasmids at least, copulation and oviposition go on simultaneously. This is certainly so in the case of a large stick-insect[2] common near Kurseong in the rains. The union continues for several days on end, perhaps longer; and eggs are protruded from an apperture ventral to that occupied by the penis of the male.

Mantidae.

Anderson (*Proc. A.S.B.*, 1877, pp. 193-5) and Willey (*Spolia Zeylanica*, II, pp. 198-9, 2 pl.) describe the floral simulation of *Gongylus gongylodes*. Willey (*Spolia Zeylanica*, III, pp. 226-7) describes the stridulation of this species, and an account of its development has been published by Williams (*Trans. Ent. Soc. London*, 1904, pp. 125-137). Annandale (*Proc. Zool. Soc. London*, 1900, pp. 839-854, 2 text-figs.) gives an account of the habits of the flower-mimicing *Hymenopus bicornis* and of other Malayan species.

Browne (*J.B.N.H.S.*, XII, pp. 578-9) records the killing of a sunbird, *Arachnechthra minima*. by a large mantis, "probably *Hierodula bipapilla*." An immature specimen of a large green mantis was recently sent to the Indian Museum by Mr. Matilal Ganguli, who had found it surrounded by six or seven sparrows that were attempting to kill it. When they tried to peck at it, it ran very fast towards the assailants, making darts at them which caused them to withdraw. The struggle was still in progress when the specimen was captured.

The food of mantises, with an account of the gradual eating of the male of an American species by the female during and without interfering with copulation is described by Mosse (*J.B.N.H.S.*, XX, pp. 878-9) and Coleman (*J.B.N.H.S.*, XX, pp. 1167-8). I have seen newly hatched young of a big green mantis feeding on minute Chloropid flies (*Pachylophus adjacens*, Brun., MS.) on a bush of *Zizyphus jujuba* on the Calcutta maidan—a bush which always attracts these flies during the rains, when they sit about on its leaves in large numbers.

Acridiidae.

Alcock (*J.A.S.B.*, LXV [II], pp. 539-540; reprinted *J.B.N.H.S.*, XI, pp. 149-150) records the behaviour of a bear towards *Aularches*

[1] Concerning parthenogenesis in Phasmidae see also Hanitsch, *J. Straits R. Asiatic Soc.*, July 1904, pp. 35-38 (*Eurycnema herculanea*). Fryer records polymorphism in a Ceylon stick-insect (*Journal of Genetics*, III, pp. 107-111, pl. iii).

[2] Belonging apparently to the subfamily Lonchodinae. The female is a very heavily built stick-insect, the male more moderately stout.

miliaris (Linn.) as an instance of the natural repellent effect of "warning colours." This species when irritated, besides exuding a pungent-smelling frothy fluid, makes a curious hissing sound. Precisely how it does so I have been unable to determine. Legs and wings commonly vibrate synchronously with the production of this sound when the insect is held by the body; but when any or all of these appendages are prevented from moving the sound may still be produced, though the insect is usually less readily disposed to produce it under these conditions. There is no perceptible vibration of the body wall such as occurs when a fly or wasp buzzes. The breeding and other habits of this locust are described by Green (*Cir. R. Bot. Gardens, Ceylon*, III, pp. 227-235).

The "terrifying attitude" assumed by a grasshopper (*Acridium violascens*) when attacked by a myna (*Acridotheres tristis*) is described by Manders (*Spolia Zeylanica*, VII, pp. 204-5).

Kershaw gives a note on the habits and development of a Chinese "*Mastax* or *Eumastax*" (*J.B.N.H.S.*, XXII, pp. 416-7, pl. B, part).

Mr. Fletcher informs me that when he was in Coorg last year he found an Acridiid eating a large spider, a curious reversal of the normal course of events.

Cotes and others between 1890 and 1907 contributed a series of notes to the Journal of the Bombay Natural History Society, many of them of considerable length, on the habits, and especially on the migrations, of Indian locusts.

Locustidae.

Green has described the stridulation of the common green locustid of Peradeniya (*Spolia Zeylanica*, VII, p. 56). A very similar but slightly stouter insect occurs in Calcutta. It has a different note, which has been described by Cunningham ("*Plagues and Pleasures*", p. 171). This note is, however, not unlike the last syllable of the Peradeniya insect, though somewhat harsher and less prolonged. When the insect is in full song in the open a distinct click is audible alternating with the somewhat rapid succession of these notes. *Mecopoda elongata* has a somewhat similar note which it repeats indefinitely in a similar manner, but this note is louder and still more raucous. All three of these insects are nocturnal.

Concerning the habitual attitude assumed by *Sathrophyllia rugosa* ("*Acanthodis ululina*") see Willey (*Spolia Zeylanica*, II, p. 199, 1 fig.) and Annandale (*Mem. A.S.B.*, I, p. 209).

Green (*Spolia Zeylanica*, VI, pp. 134-5) has described the habits of a leaf-rolling species of *Gryllacris*, presumably a close ally of, if not identical with, a species—*Gryllacris aequalis*—found in the Calcutta Botanical Gardens by Wood-Mason (see Griffini, *Atti. Soc. Ital. Sci. Nat.* LII pp. 237-239, where references to other nest-making Orthoptera will be found[1]). Gryllacrids are also sometimes to be found in holes in trees, under loose bark, and under the eaves of buildings.

Annandale and Gravely have described the habits of the Stenopelmatinae found in Burmese and Malay caves (*J.A.S.B.* [n.s.], IX, p. 413). In the Cochin Ghats Stenopelmatids are common under logs of wood.

Alarming colour and attitude in *Capnoptera*, spp., and a possible use of the spines on the thorax of *Eumegalodon blanchardi*, are described by Annandale (*Proc. Zool. Soc., London*, 1900, pp. 854-5 and 866).

Gryllidae.

The habits of a noisy burrowing cricket—doubtless *Brachytrypes portentosus* ("*achatinus*"[2])—are described at length by Cunningham ("*Plagues and Pleasures*," pp. 161-170). I have never seen "molehill-like heaps of loose earth cast out of the mouths of almost finished diggings" of these crickets. Sometimes there is a small and untidy collection of loose earth, but I have usually found the burrows somewhat difficult to locate in spite of the vigour with which the insects proclaim their whereabouts.[3]

A cricket closely resembling *Brachytrypes achatinus*, but much smaller, often flies to light in Calcutta. Like many still smaller species it has a way of partly unfolding its wings and then rapidly vibrating them. Why it should do this I have been unable to determine. The action, which is performed equally by both sexes, looks like stridulation, but only the faintest rustling sound is produced, and the male stridulates loudly in the ordinary way.

Mr. Fletcher tells me that *Liogryllus bimaculatus* is neither exclusively vegetarian nor exclusively carnivorous, feeding on both vegetable matter and dead insects when both are offered.

Nothing yet appears to have been recorded of the Calcutta house-cricket. It is a fair-sized, mottled, grey-brown insect, flightless in both sexes. The female is entirely wingless, but the male has well-developed elytra provided with a stridulating organ of the usual Gryllid type, with the aid of which he sings even more persistently, though fortunately more quietly, than *Brachytrypes portentosus*, going on from evening far into the night. This familiar song is, however, not the only one that he is capable of pro-

---

[1] See also Rutherford, *Spolia Zeylanica*, X, p. 77.

[2] For synonymy see Kirby's "*Synonymic Catalogue*" (British Museum).

[3] Mr. Bainbrigge Fletcher tells me that most of the burrowing is done before the insect becomes mature and begins to sing. Concerning the singing he says "The male first looks out of its burrow, then runs out rapidly and retreats again as quickly, having apparently brought up a little earth; sometimes it repeats this two or three times. Satisfied that the coast is clear, the cricket runs boldly out onto the little platform of earth outside its burrow, turns round facing its hole and with its head almost in the entrance, raises itself on its legs which are well spread out, slightly opens out its tegmina and commences to shrill. A slight quivering of the tegmina is all that can be seen, the motion apparently being too rapid for the eye to follow."

ducing: and when courting a female he changes it for a low whirring sound accompanied at regular short intervals by an abrupt squeak.

The first occasion on which I heard this was early in the rains of last year. I had three or four adult males in a glass jar. They stridulated as usual till I chanced to catch a couple of females which I put with them, when a change in their behaviour was at once apparent. First one and then another would approach one of the females and commence his courting notes, vibrating his elytra to produce the continuous whirring sound to all appearance just as when producing his ordinary song, but giving them periodic jerks which synchronized with the sudden squeaks. And this in spite of the fact that the females were all in their penultimate stage, and so failed to respond to any advances.

Some time later I heard these peculiar notes under different circumstances. On entering my office on a holiday, when the room was quite quiet, I heard what I at first took to be the squeaking of an electric fan. But it came from a direction where there were no fans, and on following it up I became aware of a low whirring sound accompanying it which suggested that I might be on the track of a pair of crickets, courting under natural conditions, although it was still early in the afternoon. The noise was located in a narrow covered space open at both ends, and on inserting a stick at one end a pair of common house-crickets soon appeared at the other. Unfortunately one of them escaped, so I was unable to make further observations upon them.

On another occasion, when attracted by the normal note of a male, I found him to be accompanied by a female to whose posterior end a small white body—presumably a spermatophore—was attached. So it may be customary for the male to entertain his mate for a time with his normal song after the pairing is over. Shortly afterwards I saw the female put her head between her legs, seize the spermatophore in her jaws and devour it. She was in a jar with several males, and I chanced to notice during the next morning that another spermatophore had been attached. This disappeared soon after, but I do not know how.[1]

EMBIOPTERA.

The first Indian Embiid whose habits appear to have attracted any attention was *Oligotoma michaeli*, of which specimens were transported from India to England in orchid roots, among which they lived in silken tunnels, and to which they proved destructive (see Michael in *Gardener's Chronicle*, Dec. 30, 1876 and M'Lachlan *J. Linn. Soc. London, Zool.* XIII, pp. 373-384, pl. xxi). The first observations made in India appear to be those of Wood-Mason on

[1] Changes in the notes of American locusts, and their association with courtship, are noted by Allard (*Ent. News*, XXV, 1914, pp. 463-466). They have, I believe, been noted in other Orthoptera saltatoria also, but I do not know where the observations have appeared.

Oligotoma saundersi published in 1883 (*Proc. Zool. Soc. London*, pp. 628-634, pl. lvi). Lefroy published a short note on this species in 1910 (*J.B.N.H.S.*, XIX, pp. 1009-1010). In 1911 Imms published an account of the habits and life-history of *Embia major* (*Trans. Linn. Soc. London, Zool.* XI, pp. 167-195, text-figs. 1-6, pl. xxxvi-xxxviii.)

ISOPTERA.[1]

A brief note on the tapping noises made in unison by termites was published by Fedden (*Proc. A.S.B.* 1866, p. 19). A paper by Bugnion (*Bull. Soc. Ent. Suisse*, XII, Berne 1912, pp. 125-139, pl. ix) deals with the same subject. These noises were frequently heard last year on some trellis-work in the Indian Museum compound, though I failed to notice any rythmic unison in their production. The trellis was covered with the mud shelters of termites, and when approached or tapped myriads of faint clickings were clearly audible. The sound at first suggested the cracking of the mud; but it was to be heard in the morning before the sun fell on the trellis (which faced west) as well as in the evening. If, moreover, the mud were broken away while the clicking was in progess, termites were always found beneath: whereas if the disturbance causing the clicking were kept up for a few minutes the clicking would cease, and then no termites would be found. This clicking is of course quite different from the clicking of *Capritermes*, which appears to be produced by the combined action of the remarkable jaws of the soldier, and sounds like the sudden cracking of a piece of thin glass. The force expended by *Capritermes* in producing it often flicks the producer up into the air.

A note on the repairing of a damaged termite nest was published by Millett in 1902 (*J.B.N.H.S.*, XIV, pp. 581-2), and one on strange mortality of termites among tea bushes by Green in 1905 (*J.B.N.H.S.*, XVI, pp. 503-4). Doflein, in his paper on termite truffles published in the same year (*Ver. Deutschen Zool. Ges.*, XV, pp. 140-149, 2 text-figs.; translated, *Spolia Zeylanica* III, pp. 203-9), notices the food of termites. In 1906 Petch noticed the habits of some Ceylon termites in his paper on the fungi of certain termite nests (*Ann. R. Bot. Gard., Peradeniya*, III, pp. 185-270, pl. v-xxi). Green in 1907 recorded the occurrence of two queen termites in one royal cell (*Spolia Zeylanica*, IV, p. 191). Escherich's "*Termitenleben aus Ceylon*" (Jena, 1911) deals extensively with habits. In 1913 Assmuth described the habits of many species of termites in his paper on "Wood-Destroying Termites of the Bombay Presidency (*J.B. N.H.S.*, XXII, pp. 372-384, pl. i-v), and Petch described those of *Eutermes monoceros*, the black termite of Ceylon (*Ann. R. Bot. Gard., Peradeniya*, V, pp. 395-420, pl. vi-xiv). During the same year Green's "Catalogue of Isoptera recorded from Ceylon" (*Spolia Zeylanica*, IX,

---

[1] See also Escherich, "Die Termiten" (Leipzig 1909); and Wasmann's "Neue Beiträge zur Kenntnis der Termitophilen und Myrmecophilen", *Zeitschr. wiss. zool.* CI., 1912, pp. 70-115, pl. v-vii.

pp. 7-15), and John's "Notes on some Termites from Ceylon" (*Spolia Zeylanica*, IX, pp. 102-116) were published, and in both the habits of a number of species are referred to. The most recent paper on the habits of Ceylon Termites appears to be by Bugnion[1] (*Bull. Mus. Hist. Nat. Paris*, 1914, no. 4, pp. 3-37, pl. i-viii).

Termites usually "swarm" in the rains; but in some species at least winged adults are ready to emerge even in the cold weather, and need only the stimulus of rain to bring them out. The cold weather of 1914-5 was remarkable in Calcutta for several periods of exceptionally damp and chilly weather. On each occasion numbers of termites were seen flying above the Maidan. On one occasion (16-i-15) I found a dense swarm emerging from a nest and collected specimens, which have been identified by Mr. Fletcher as a species of *Odontotermes*, probably new.

PSOCOPTERA.

Green describes the habits of Scaly-Winged Psocids (*Spolia Zeylanica*, IV, pp. 123-125) and of Psocids which combine to spin extensive webs on trees (*Spolia Zeylanica*, VIII, 1912, p. 71, 1 pl., 2 text-figs.).

The habits of Psocids, and the occurrence of fatal epidemics among gregarious species, are referred to by Cunningham ("*Plagues and Pleasures*", pp. 151-5).

ODONATA.

Observations on the food of dragonflies have been recorded by Young (*J.B.N.H.S.*, XV, p. 530), Lefroy (*J.B.N.H.S.*, XX, pp. 236-8), Fulton (*J.B.N.H.S.*, XX, p. 876), and Green (who publishes information supplied him by Mr. John Pole, *Spolia Zeylanica*, VIII, p. 299). The oviposition of dragonflies is described by Cunningham ("*Plagues and Pleasures*", pp. 133-5). The vitality of dragonfly larvae out of water form the subject of a note by Green (*Spolia Zeylanica*, V, pp. 104-105).

NEUROPTERA (*s. str.*).

Annandale notices the habits of an Indian *Sisyra* larva (*J.A.S.B.* [n.s.], II, pp. 194-5, pl. i, fig. 3).[2]

---

[1] Other papers by this author are scattered in various journals. The following list of those dealing to some extent with the habits of Oriental Termites is compiled from reprints sent to Mr. T. Bainbrigge Fletcher:—"Le Termite noir de Ceylan", *Ann. Soc. Ent. Fr.* 1909, pp. 271-281, pl. viii-x; another paper with the same title, *Bull. Soc. Vaud. Sci. Nat.*, XLVII (173), pp. 417-437, figs. 1-5; "Observations relatives à l'Industrie des Termites", *Ann. Soc. Ent. Fr.*, 1910, pp. 129-144; "*Eutermes lacustris*, nov. sp. de Ceylan", *Rev. Suisse Zool.*, XX, 1912, pp. 487-505, 1 text-fig., pl. vii-viii; "*Le Termes Horni*, Wasm. de Ceylan, *Rev. Suisse Zool.*, XXI, 1913, pp. 299-330, 1 text-fig., pl. xi-xiii, "Les Termites de Ceylan", *Le Globe, Organe Soc. Geogr. Genève*, LII, 1913, pp. 2-36, pl. i-viii.

[2] Not fig. 2 as stated in the text of the paper. This probably represents the larva of a Trichopteron, not a beetle.

The habits of Myrmeleonid and Ascalaphid larvae from tree-trunks are described by Gravely and Maulik (*Rec. Ind. Mus.*, VI, pp. 101-3, pl. v). Perhaps the "ant-lion" which Ryves found dead in a spider's web in a mango tree (*J.B.N.H.S.*, X, pp. 152-3) belonged to a species with similar habits.

The life-history of *Helicomitus dicax* is described by Ghosh (*J.B.N.H.S.*, XXII, pp. 643-8, 1 pl.). The larva of this Ascalaphid lives on the ground and covers itself with dust.

"The Indian Nemopterid and its food" is the title of a note by Lefroy on the larva of *Croce filipennis* (*J.B.N.H.S.*, XIX, pp. 1005-7, 1 text-fig.). Further studies on *Croce* have since been published by Ghosh (*J.B.N.H.S.*, XX, pp. 530-532, 1 pl.) and Imms (*Trans. Linn. Soc. London, Zool.* XI, pp. 151-160, pl. xxxii).

TRICHOPTERA.

A viviparous caddis-fly is described by Wood-Mason under the provisional name *Notanatolica vivipara* (*Ann. Mag. Nat. Hist.* [6], VI, pp. 139-141, text-figs. *a-b*).

HYMENOPTERA.

Miscellaneous.

The habits of various Indian Hymenoptera are very briefly referred to by Wroughton (*J.B.N.H.S.*, IV, pp 26-37). The habits of a number of Indian Aculeata are described by Dutt (*Mem. Dept. Agric. India, Entom. Ser.* IV, pp. 183-267, pl. xi-xiv, 22 text figs.).[1]

Chalcidae.

Cunningham devotes the third chapter of, and an appendix to, his "*Plagues and Pleasures of Life in Bengal*" to fig-insects. The particular insects whose habits are described are those which are associated with *Ficus roxburghii* in Calcutta, and his observations are clearly the result of his work on the fertilization—if such it may be called—of this fig by these insects (*Ann. R. Bot. Gardens, Calcutta*, I, Appendix 2, 1889, 37 pp., 5 pl.).

The habits of *Syntomosphyrum indicum* are described by Silvestri in *Div. Ent., Hawaii Board Agric. and For.*, No. 3, pp. 125-127.

Ichneumonidae.

Ramsay describes the oviposition of a species of *Rhyssa*—probably a species found in the Himalayas (*Entomologist*, XLVII, pp. 20-22, 3 text-figs.).

Braconidae.

A note on a species of *Apanteles* parasitic in the caterpillar of a Death's Head Moth has been published by Green in *Spolia Zeylanica*, V, p. 19, 1 pl.

---

[1] A note on the capture of a leaf-mining caterpillar by a wasp is contributed by Ridley to *J. Straits R.A.S.*, July 1905, pp. 227-8.

Chrysidae.

Bingham refers to the habits of *Chrysis fuscipennis* in *J.B.N.H.S.*, XII, p. 586, Cretin to those of *Stilbum splendidum*, *J.B.N.H.S.*, XIV, pp. 823-4.

Mutillidae.

Wroughton refers to the habits of an Indian Mutillid (*J.B.N.H.S.*, VI, p. 118).

Pompiliidae.

The food of several members of this family is recorded by Bingham, who also describes the capture of a Galeodid by *Salius sycophanta* (*J.B.N.H.S.*, XIII, pp. 178-180).

Sphegidae.

Bingham mentions the food of several species of Sphegidae, and a remarkable concentration of the nests of a variety of *Sphex umbrosus* (*J.B.N.H.S.*, XIII, pp. 177-8).

Notes on the habits and food of *Sphex lobatus* are contributed by Lefroy ((*J.B.N.H.S.*, XV, pp. 531-2), and by Beadnell (*J.B.N.H.S* , XVII, p. 546).

Wickwar describes the habits of *Sceliphron violaceum* (*Spolia Zeylanica*, VI, p. 179), Cory those of *S. intrudens* (*J.B.N.H.S.*, XXII, p. 648), and Field those of *S. coromandelicum* (*J.B.N.H.S.*, XXIII, pp. 378-9).

Eumenidae.

Concerning *Eumenes conica* see Bingham (*J.B.N.H.S.*, XII, pp. 585-6), and Ramakrishna Aiyer (*J.B.N.H.S.*, XX, pp. 243-4). The former author deals with the construction of the nest and with pugnacity displayed towards a parasitic *Chrysis*, and the latter with breeding habits and development.

For notes on *Eumenes dimidiatipennis* see Cretin, *J.B.N.H.S.*, XIV, pp. 820-824.

Odynerus punctum is recorded as cleaning out and using empty cells of *Eumenes dimidiatipennis* (Cretin, *J.B.N.H.S.*, XIV, p. 824).

Vespidae.

Battles between wasps and bees are recorded by Hewett (*J.B.N.H.S.*, IV, p. 312) and by Drieberg (*Spolia Zeylanica*, IV, p. 33). In the former case the wasps were *Vespa magnifica* and the bees "the large jungle bees" (? *Apis dorsata*). A battle between two kinds of wasp, apparently *Vespa cincta* and *Polistes hebraeus*, is recorded by Cunningham ("*Plagues and Pleasures*", p. 31). The habits of the former wasp are dealt with on pp. 29-33 of the same book, and of the latter on pp. 23-28. Mr. Fletcher has given me the following additional note on this subject: "Last

July, when travelling by train, a specimen of *Vespa cincta* flew into the carriage carrying a *Polistes hebraeus* which it had captured. *V. cincta* and various other large *Vespa* spp. are determined captors of honey-bees as these enter or leave the hive."

The capture of a small Pyralid moth by *Vespa cincta* is recorded by Green (*Spolia Zeylanica*, II, p. 197).

Apidae.

In addition to the notes just referred to recording battles between wasps and bees, the following references to bees may be given.

Douglas contributes information about the hive-bees indigenous to India and the introduction of the Italian bee (*J.A.S.B.*, LV [II], pp. 83-96).

Storey records the poisonous action of the nectar of *Lapindus emarginatus* on bees (*J.B.N.H.S.*, V, p. 423).

Eardley-Wilmot refers to an instance of a man who, having disturbed a bees' nest, was attacked by its inhabitants, and later in the day was singled out from his companions for attack by bees from other nests which he chanced to approach (*J.B.N.H.S.*, XI, pp. 741-2).

Bingham describes the habits of *Megachile disjuncta* and its parasite *Paravaspis abdominalis* (*J.B.N.H.S.*, XII, p. 587).

Several parasites from the nests of *Xylocopa tenuiscapa* have been recorded by Green (*Ent. Mo. Mag.* [2], XIII, pp. 232-3). In an article in *Spolia Zeylanica* (I, pp. 117-9) on the mites which inhabit the remarkable abdominal pouch of this species, references to two other papers dealing with these mites are given. These are Perkins, *Ent. Mo. Mag.*, [2], X, pp. 37-9; and Oudemans, *Zool. Anz.*, XXVII, pp. 137-9. The latter contains further references.

A note on the effects of the sting of *Xylocopa tenuiscapa* is contributed by Green (*Spolia Zeylanica*, VI, p. 134).

Notes on the habits of *Apis dorsata* are contributed by Willey (*Spolia Zeylanica*, VI, p. 181, 1 pl.).

The characteristic odour of leaf-cutting bees is described by Green (*Spolia Zeylanica*, VII, p. 55).

Castets contributes an article entitled "Les Abeilles du sud de l'Inde" to the *Revue des Questions Scientifiques* (Brussels, Oct. 1893). He deals with the habits of the three Indian species of *Apis* and of *Mellipona iridipennis*. An abstract of this article will be found in the *Tropical Agriculturalist* (XXX, 1908, pp. 48-54).

The peculiar way in which a bee "painted in alternate bands of shining black and the brightest, purest cobalt"—doubtless an *Anthophora*—collects pollen, and its way of resting for the night, are described by Cunningham ("*Plagues and Pleasures*", pp. 37-8).

The burrows of *Anthophora* (or *Podalirius*) *pulcherrima* are described by Annandale (*Rec. Ind. Mus.*, III, p. 294, 1 text-fig.), who notes that they open in a direction which prevents rain from entering them to any great extent.

Formicidae.

The habits of a number of different species are referred to by Rothney (*Trans. Ent. Soc. London,* 1889, pp. 347-374 ; reprinted *J.B.N.H S.*, V, pp. 38-64), Wroughton, (*J.B.N.H.S.*, VII, pp. 13-60 and 175-202, pl. A-D), and Cunningham ("*Plagues and Pleasures*", pp. 40-54).

The care of Lycaenid larvae by ants is described by de Niceville (*J.B.N.H.S.*, III, pp. 164-8, pl. 26-7).

Bingham contributes a note on the habits of *Diacamma* (*J.B.N.H.S.*, XII, pp. 756-7).

Green describes the web-spinning of *Oecophylla smaragdina* (*Proc. Ent. Soc. London*, 1896, p. ix and *J.B.N.H.S.*, XIII, p. 181). Some earlier papers on this subject, and the fact that *Oecophylla smaragdina* does not spin a cocoon in which to pupate, are noticed by Green (*Spolia Zeylanica*, I, pp. 73-4), and the matter forms the subject of notes by Fletcher (*Spolia Zeylanica*, V, p. 64), Ridley (*J. Straits R. Asiatic Soc.* No. 22, Dec. 1890, pp. 345-7) and Shelford (*J. Straits R. Asiatic Soc.* June 1906, pp. 284-5).

A living chain of *Oecophylla smaragdina* spanning a gap of 3 inches is described by Green (*Spolia Zeylanica*, VII, pp. 53-4). The capture of a living butterfly (*Catopsilia crocale*) by this species is recorded by Henry (*Spolia Zeylanica*, IX, pp. 142-3). A lengthy note on the habits of the same species in the Malay Peninsula will be found in *Fasciculi Malayenses*, Zool. III, pp. 27-30.

A remarkable illustration of the very large quantities of grain carried away and stored by ants is given by Fraser (*J.B.N.H.S.*, XX, p. 877).

The carrying away of a partially disabled caterpillar by a party of ants is described by Sladen (*J.B.N.H.S.*, XXII, p. 649).

COLEOPTERA.[1]

Passalidae.

I have already once gathered together as much information as I could obtain about the habits of Indian Passalidae (see *Mem Ind. Mus.*, III, pp. 339-340). Since then Mr. T. Bainbrigge Fletcher has taken *Episphenus neelgherriensis* at light in Coorg, and has obtained eggs of *Macrolinus rotundifrons* from under a log at Peradeniya where they were found "in a circular chamber partly filled with gnawed wood." In view of the suggestion made in the "*Fauna of British India*" (Lamellicornia, I, p. 20) that the Passalidae are a viviparous family the latter observation is of great interest. It may not be out of place to note here that when, during my visit to Berlin in 1913, I called the attention of Dr. Ohaus to the suggestion, he immediately refuted it by the production of eggs of American species preserved in his fine private collection.

---

[1] Concerning stridulation in this Order, with which several of the following notes are concerned, see Gahan, *Trans. Ent. Soc. London*, 1900, pp. 433-452, pl. vii ; and Arrow, *Trans. Ent. Soc., London*, 1904, pp. 709-750, pl. xxxvi.

My suggestion (*Mem. Ind. Mus.*, III, p. 215) that *Pleurarius brachyphyllus* is probably not a gregarious species has proved to be incorrect. This species is abundant in the evergreen jungles of the lower western slopes of the Western Ghats in Cochin. Occasionally isolated pairs were found in a log, but usually numbers were found together. It is scarcely possible that insects of this species are able to fly; for although the wings are well developed the elytra are fused. How this fusion takes place I was unable to determine, as only one pupa was found, and no stages intermediate between this and the almost fully blackened adult. The elytra are not fused in the pupa.

The conclusion that *Episphenus indicus* is to some extent gregarious, and that *E. neelgherriensis* is not, was confirmed by my observations in Cochin. All of the three last mentioned species burrow more deeply into logs than does *Leptaulax bicolor* which, together with its larvae and pupae, was only found close under the bark. *Pleurarius brachyphyllus*, especially, makes galleries well below the surface, a fact which probably accounts for its comparative rarity in the collections I had previously seen. It often burrows in somewhat hard wood and is very difficult to dig out; but I found it even commoner in Cochin than *Episphenus indicus*, a species which was distinctly commoner than *E. neelgherriensis*.

The larvae of *Pleurarius brachyphyllus* and *Episphenus indicus* —I got very few of *Episphenus neelgherriensis* and *Leptaulax bicolor* —were commonly found widely separated from adults. In some cases no adults at all could be found, and it is curious, in view of Ohaus's observations on American species, that although all the larvae which I attempted to keep thrived for a time, whether associated with adults or not, only those without adults survived the journey to Calcutta; and that of these one or two lived for between one and two months. I regret now that I did not make an effort to keep single families by themselves. This was, however, rendered almost impossible, firstly by the difficulty of recognizing a single family as it occurred scattered along one or more of the groups of burrows made by the various members of the colony, and secondly by an insufficient supply of separate tins.

Stridulation in adults of both *Episphenus* and *Pleurarius* is brought about by movements of the abdomen, and is faintly audible at a yard or two's distance from the ear. In larvae it is much fainter. I never saw any indication of its being used as a means of communication, and this agrees with Mr. Kemp's experience of species found in the Abor Country. Adults, at least, appear to stridulate whenever they are disturbed, presumably in order to drive off the enemy.

The stridulating organs resemble those of *Popilius* (*Passalus*, auct.) *cornutus* and *Pentalobus barbatus*[1] described by Babb (*Ent.*

[1] The abdominal part resembles that of *Proculus goryi* also; but the wings are not reduced as in that species. I cannot understand Schulze's statement that in *P. goryi* the abdominal part is situated on the fifth segment, for his figure (in which the first segment is omitted) clearly shows it on the sixth, where it is

News, XII, p. 271 [1]) and Schulze (*Zool. Anz.*, XL, pp. 209-216, figs. 5-7). The organs to which Ohaus attributed stridulatory functions (*Stettin Ent. Zeit.* 1900, pp. 167-169) are also well developed. In a footnote to the first page of Schulze's paper Ohaus says, "Was die von mir l. c. beschriebene Bildung für eine Bedeutung hat, ist bis jetzt noch nicht festgestellt. Sie findet sich bei den meisten, vielleicht allen, holzbewohnenden Lamellicorniern und hat vielleicht den Zweck, das Eindringen von Wasser, vielleicht auch von Schmarotzern, in die Räume zwischen den Tergiten und Flügeln zu verhindern. Speziell die Passaliden sind an den Rändern der Tergite häufig mit Milben besetzt."

In order to test the possible stridulating powers of the two sets of organs I removed the wings of a *Pleurarius*. Although the abdomen subsequently moved as if trying to stridulate no sound was produced. A good deal of fluid escaped, however, from the places where the wings had been inserted, which might have affected the vibrations; so I then cut off the ends of the wings of another specimen of the same species. Its abdomen moved vigorously but only a very faint sound was produced, a sound which I attribute to a small portion of the stridulating surface of the wing having escaped removal. I then took a specimen of *Episphenus indicus*, in which the elytra are not fused and can consequently be opened, and found that so long as the folded wings were pressed down on to the abdomen by a needle the insect could stridulate as well as before, even though the elytra were held right away from the sides of the abdomen.

I have never heard any Passalid emit notes of more than one kind, and all have been fainter than those produced artificially by rubbing the end of the wing of a softened specimen of *Proculus goryi* on the plate beneath it.

Lucanidae.

Nigidius dawnae lives in hard dry pieces of wood on the higher slopes of the Dawna Hills. Both adults and larvae were found in one such piece (see *Rec. Ind. Mus.*, XI, pp. 427-429). Mr. Kemp informs me that *N. impressicollis* lives, in both the larval and adult condition, in thoroughly damp and rotten wood. Mr. Beeson informs me that *N. distinctus* [2] lives in dead wood of Malatta (*Macaranga pustulata*) in the Duars.

Dynastinae. [3]

The stridulating ability of *Xylotrupes gideon* has been recorded by Cunningham ("*Plagues and Pleasures*", pp. 126-7, pl. ii,

situated in the specimen of *P. goryi* that I have examined, and also in *Pentalobus barbatus*, *Pleurarius brachyphyllus*, etc.

[1] See also Sharp, *Ent. Mo. Mag.* (2) XV (XL), 1904, pp. 273-4.

[2] Concerning the identity of this species see *Rec. Ind. Mus*, XI, p. 430.

[3] Attention may be called here to the occurrence, in a paper on Paussidae,

fig. 5; quoted in *Fauna of British India*, Lamellicornia, I, p. 265).[1]

Concerning the action of the stridulating organs of *Oryctes rhinoceros* nothing yet seems to have been published. I have had great difficulty in obtaining any evidence as to the use of the so-called stridulating organ found in the larva (pl. xxii, fig. 1). When a specimen is tightly held by the head, however, it may be seen to move the mandibles and maxillae in a manner likely to bring the organ into action, and a faint rasping sound may sometimes be heard if the specimen be brought close to the ear. No definite vibrations have been felt, and the movements of the mandibles and maxillae are those which would probably be used, in order to free itself, by any insect similarly placed. Pressure on the body does not seem to induce any such movements, but they are sometimes indulged in by larvae which find themselves on their backs on a hard surface in the open. The movements are often greater in extent than their use for stridulatory purposes requires; the mandibular part of the organ is, indeed, sometimes fully exposed at intervals, and could not then be scraped at all by the maxillary portion. The rasping seems, nevertheless, to be produced only when these movements occur. It is therefore probable that it is produced by the organs in question, and it is noteworthy that the movement of the mandibles and maxillae is often very small—as it should be to keep the two parts of the organ in contact—and that this does not interfere with the sound produced.

The pupa, in which no stridulating organs appear to have been described, stridulates quite audibly when disturbed. The sounds are produced as the result of backward and forward movements of the abdomen movements which cause a pair of scrapers situated on the dorsal part of the anterior margins of segments 2-6 to rub over the faintly ridged surface of the hard chitinous walls of oval depressions on the posterior parts of segments 1-5 (pl. xxii, figs 2-3), producing vibrations through the whole pupa, as well as sound. The organs are very conspicuous in living specimens, but in preserved ones they are apt to be largely hidden between the terga. The organ between segments 6 and 7 is rudimentary.[2]

I have heard the adult stridulate, but not loudly. The sound appears to be produced by the rubbing of the well-known ridges on the posterior end of the abdomen against the posterior ends of the elytra (pl. xxii, fig. 4), but I have not yet been able to investigate this as fully as I would like.

---

where it is most unlikely to attract the attention of those interested (*J.A.S.B.*, XII, pp. 421-437), of a coloured figure of *Eupatorus hardwickei* from the summit of the Gogur Range, 9000 ft., in Kumaon.

[1] See also Rutherford, *Spolia Zeylanica*, X, p. 77.

[2] Similar structures are present in the pupae of several other beetles—e.g. *Adoretus* (Rutelinae) and *Hectarthrum* (Cucujidae)—but they do not appear to be stridulatory on any segments in them.

Rutelinae.

Leaves of *Lagerstroemia* bushes in the Indian Museum compound are frequently eaten extensively by a nocturnal insect, and by searching among them after dark a few Melalonthids and a large number of Rutelids have been obtained. All of the latter belong to the genus *Adoretus*, and Mr. Arrow has identified almost all of them as *A. versutus* Occasional specimens have been found feeding on *Bauhinia*, *Canna*, and a leguminous shrub (? *Cassia*); but they are found in much greater abundance on *Lagerstroemia* than on anything else.[1]

Mr. Arrow informs me that nothing is yet known of the manner of feeding in this genus, and I have been able to make the following observations.

At night, after emerging from the ground in which it has been buried all day and to which it returns before morning, the beetle flies to a leaf, and settles either on the upper or under side, usually the latter. It never settles on the edge. The claws of two or three tarsi, often all on the same side of the insect, grasp the edge; the others rest on the surface.

In beetles of this genus, the mouth is divided into two by a median process of the labrum (pl. xxii, fig. 5). When the insect wishes to take a bite, therefore, it turns its head slightly on one side; and although the mouth-parts of both sides work simultaneously, the bite is effected by those of one side only.

The strongly toothed extremity of the maxilla forms the principal biting organ. When a specimen begins to feed both mandibles and maxillae are opened widely. Then the maxillae are exserted between the mandibles and the median process of the labrum, the maxilla of whichever side of the head has been turned nearest the leaf scooping out a small quantity of the soft tissue of which the leaf is composed between the principal veins. This tissue does not appear to offer the slightest resistance to the maxilla, which seems to scoop it up as easily as if it were soft wax; and so far as I have been able to see the beetle makes no special effort to keep the leaf from being pushed away instead of cut into. I do not even think that the median process of the labrum is lowered against it, as I have been unable to see this organ during the process. Had it been lowered it must, I think, have come into view.

Three or four bites are required to make a hole right through the leaf, after which bigger bites can be made. The general method is the same, but the end of the maxilla is passed through the hole, and as far beyond the edge as it will go, so that it bites each time through the whole thickness of the leaf. Here again

---

[1] A few specimens of *A. duvauceli* have also been found on *Lagerstroemia* and of *A. lasiopygus* on *Hibiscus*. *A. versutus* has been found in great abundance on Cannas since the above was written, and its larvae and pupae have been found among their roots.

the tissues of the leaf appear to offer no resistance, but as the maxilla passes back into the cavity between the mandible and labral process close to the concave part of the serrate margin of the latter, there is probably some amount of scissor action between them.

The mandible closely follows the maxilla in all its movements, and forms a sheath above it. Precisely to what extent it functions as a biting organ is most difficult to see with certainty. Its smooth dorsal face works along the serrate transverse edge of the labrum from end to end, and its distal end slides past the serrate longitudinal edge of the labral process as it follows the maxilla into the cavity behind. Probably the mandible makes all the transverse cuts that are required, and it could no doubt make longitudinal cuts as well should the maxilla fail to work properly; but no transverse cutting seems to be left for it under ordinary circumstances, and the tracks of the maxillary teeth can be distinctly seen on each freshly bitten surface. One of the chief functions of the mandible appears to be to protect these teeth, when they are not in use, by closing in the cavity between the labrum and the labium, in which they lie when at rest.

Intervals of varying length between the bites are devoted to mastication. During this process the gnathites of the two sides work simultaneously as before, but the maxillae are not exserted —i.e. they remain in the cavity between the mandibles, their extremities being exposed between (and posterior to) the mandibles and the labral process each time they are opened. At the same time the labral process and labium are alternately separated a little and brought together again. Mastication presumably takes place chiefly between the large molar teeth, situated one at the base of each mandible (pl. xxii, fig. 6), the triangular thickened area on the inner side of the labium (pl. xxii, fig. 6), and the somewhat similar convexity on the inner side of the labrum. It is possible, however, that the terminal teeth of the maxilla take some part in it also, for those of opposite sides are not quite alike, and when pressed together after removal of the labrum, the teeth of one side may be seen to fit into the spaces between the teeth on the other, although the teeth can never be brought together thus during the process of biting.

This method of feeding differs in several respects from the method of feeding observed by Ohaus (*Stett. Ent. Zeit.*, 1909, pp. 12-13) in Rutelinae of the *Geniates* group, South American insects whose mouth is also divided longitudinally into two parts. *Geniates* and its allies always cling to both sides of the leaf at the same time instead of to one side only, commencing to feed at the edge instead of on the upper or under side; they also exude such large quantities of saliva that it escapes from the mouth and stains the bitten margin of the leaf—a thing which has never been observed in *Adoretus*.

The difference in the method of biting the leaf is associated with differences in the structure of the mouth parts. *Geniates* and its allies[1] are said to take the edge of a leaf in their mouths and cut a piece out by a scissor-like action between the mandibles and labrum and the maxillae and labium. The mandibles are likely, therefore, to have a very strongly developed cutting edge in front. This is the case in *Geniates impressicollis*, a species in which there is in addition a narrow posterior part, forming an imperfect sheath for the maxilla, at right angles to the cutting part. The cutting portion of the mandible of *Geniates* appears to be homologous with the greater part of the sheathing mandible of *Adoretus*. The maxillae of *Geniates impressicollis* are well developed, but are prismatic in form rather than scoop-like; they presumably aid in cutting only by dragging the leaf down over the sharp edge of the labium.

The mouthparts of *Leucothyreus trochantericus*, the only other species of the *Geniates* group that I have been able to examine, are more difficult to understand. The mandibles are so massive that it is difficult to see how the edge of a leaf is ever introduced into the mouth. Presumably, however, this must be the manner of feeding; for the species whose feeding habits were actually observed by Ohaus included some of the genus *Leucothyreus*. The maxillae are small and are not in any way sheathed by the mandibles, whose anterior edge appears to overlap the edge of the labrum when closed and so to be useless for cutting. Presumably the cutting is done by the leaf being dragged down across the edges of the labrum and labium by the main mass of the mandible, though even this is a little difficult to understand.

Coprinae.

The stridulating habits of *Heliocopris mouhotus* are described by Annandale (*Fasciculi Malayenses*, Zool. I (II), p. 283).

Specimens of a somewhat smaller species of *Heliocopris*—probably *H. bucephalus*, Fabricius—sent to me by Mr. Bainbrigge Fletcher, stridulated loudly, but with the hind, not the middle, coxae I failed to associate any form of stridulation with the middle coxae although these moved as freely as the others in life, and an exceedingly faint sound could be produced by moving them artificially after death. The front legs produced strong but inaudible vibrations, but whether in the coxal cavities or between the coxae and femora, I was unable to determine. I have been unable to reproduce these vibrations on dead insects.

Cicindelidae.

Notes on the habits of a number of species are recorded by Annandale and Horn in the "*Annotated List of the Asiatic Beetles in the Collection of the Indian Museum*," Part I (Calcutta, 1909). The habits of some tiger-beetles from Orissa form the subject of a

[1] *Leucothyreus* and *Bolax* appear to have been the actual genera observed (*loc. cit.*, pp. 18-21).

note by Gravely (*Rec. Ind. Mus.*, VII, pp. 207-9). In both places the habits of the littoral *Cicindela biramosa* are mentioned. An earlier communication with regard to this species was made by Fletcher (*Spolia Zeylanica*, V, pp. 62-3), who has recently published a note on tiger-beetles from Coorg (*J.B.N.H.S.*, XXIII, p. 379).

The breeding places of common Indian Cicindelidae have been discussed by Lefroy (*J.B.N.H.S.*, XIX, pp. 1008-9) and the life-history and habits of *Collyris emarginata* in the Sunda Islands by Koningsberger (*Med. 'Slands Plant.*, XLIV, p. 113, fig. 59) and Shelford (*J. Straits R. A. S.*, June 1906, pp. 283-4).

Carabidae.

Calosoma orientale is recorded as an enemy of locusts by Cotes (*J.B.N.H.S.*, VI, p. 416).

Paussidae.

Some Indian representatives of this family form the subject of a paper by Boyes, in which some account of their habits is given (*J.A.S.B.*, XII, pp. 421-437, 3 pl.).

Malacodermidae.

The flashing in unison of swarms of fire-flies is discussed by Cameron, Clark, Fry and others (*Proc. Ent. Soc. London*, 1865, pp. 94-5 and 101-2, the former reprinted in *J.A.S.B.*, XXXIV [II], pp. 190-2); Theobald (*J.A.S.B.*, XXXV [II], pp. 73-4; reprinted *Proc. Ent Soc. London*, 1866, pp. xxvii-xxviii); Fedden (*Proc. A.S.B.*, 1866, p. 19); Severn (*Nature*, XXIV, p. 165); Annandale (*Proc. Zool. Soc. London*, 1900, pp. 864-5); and Cunningham ("*Plagues and Pleasures*", pp. 129-130). I have only once seen an example of this phenomenon. I was walking after sunset near Dhammathat on the Gyaing River above Moulmein when I noticed that all the fire-flies of the neighbourhood seemed to have congregated round an isolated tree, and were flashing in unison with wonderful effect.

Aquatic fire-fly larvae are described by Annandale (*Proc. Zool. Soc. London*, 1900, pp. 862-4, and *J.A.S.B.* [n.s.], II, pp. 106-7).

Green notices the luminosity of *Harmatelia bilinea* and *Dioptoma adamsi* (*Trans. Ent. Soc. London*, 1912, pp. 717-719, and *Spolia Zeylanica*, VII, pp. 212-4, 1 pl.).

A glowworm with nine pairs of lights has been recorded from Ceylon by "M" (see *Proc. Ent. Soc. London*, 1865, p. 101).[1]

The large yellow-edged black larvae of *Lamprophorus tenebrosus* are luminous, but do not shine as brilliantly as do the mature females, which are uniformly yellowish in colour. The female may

---

[1] See also Rutherford, *Spolia Zeylanica*, X, pp. 72-74 (*Dioptoma adamsi*).

sometimes be found at dusk in vegetation by the road-side at Peradeniya, sitting curled up on the ground with the tail erected so as to expose her light to the best advantage. Males fly up with a loud buzzing sound, but without lights, and drop close to her. They then become faintly luminous and run round about her. When copulation takes place the female uncurls, and her lights die down till they give only a faint ventral glow. If the pair be separated the female lights up again at once. Males are often attracted to lights in houses, when they emit a steady bluish glow from the posterior part of the abdomen (see also Green, *Trans. Ent. Soc. London*, 1912, p. 719).

Cleridae.

The habits and life-history of a Clerid near *Thanasimus nigricollis*, which is predaceous on Scolytidae, is described by Stebbing (*J.A.S.B.*, LXXII [II], pp. 104-110).[1]

Anthicidae.

Ant-mimicry by a *Formicomus* is the subject of a note by Fletcher (*J.B.N.H.S.*, XXII, p. 415).

Meloidae.

Blistering powers are recorded in *Cantharis rouxi* by Coleman (*J.B.N.H.S.*, XX, pp. 1168-9).

Green notices that *Cissites debeyi* lays its eggs in masses inside the galleries made by the Carpenter Bees with which the species is associated (*Ent. Mo. Mag.* [2], XIII, pp. 232-3).[2]

Cerambicidae.

Saunders states that adults of *Batocera rubus* feed on the round buds, but not on the leaves, of the Pipal tree (*Trans. Ent. Soc.*, I, 1836, pp. 60-61). The development and habits of several Longicorns which bore in *Ficus elastica* are described by Dammerman (*Med. Afd. v. Plantenz.*, No. 7, Batavia, 1913, 43 pp., 3 pl.).

Larvae of *Stromatium barbatum* attack furniture in Calcutta. *Xystocera globosa* was present in large numbers in a tree which died recently on the Calcutta Maidan. All stages of *Logaeus subopacus* were found in a rotten log at Kavalai, *ca.* 2000 ft., in the Western Ghats in Cochin on Sept. 26, 1914. Similar larvae were abundant in rotten wood both there and at the base of the Ghats at about the same time of year, but later stages were only found in the one instance.

---

[1] Since named *T. himalayensis*, Stebbing. See *Indian Forest Insects*, London, 1914, p. 186.

[2] See also E. Bugnion "Le *Cissites testaceus*, Fabr. des Indes et de Ceylan. Métamorphoses—Appareil Génital", *Bull. Soc. Ent. Egypte*, 1909 (Cairo, 1910), pp. 182-200, pl. i-iii.

Scolytidae.

The supposed effect of moonlight on the attack of the "shot-borer" is discussed in the *Journal of the Bombay Natural History Society* by Troup (XVII, p. 526), Barton-Wright (XVII, pp. 1026-7) and Stebbing (XVIII, pp. 18-26).[1]

Strohmeyer (*Ent. Blät.*, 1914, pp. 103-107) suggests that the group of bristles and processes on the head of the female of *Spathidicerus thomsoni* serves for the transport of ambrosia fungus spores. Mr. Beeson tells me that he has found inside the frontal processes of the swarming female of *Diapus furtivus* bunches of small cell-like bodies of similar appearance to the clusters of ambrosia which occur in its galleries. They stain with cotton blue, but he has been unable to germinate them. The male of this species, he tells me, possesses a group of minute pores near the apex of each elytron, which secrete a white wax The wax is moulded into a cylindrical tube which projects about a third of an inch from the entrance-hole in the bark of the host-tree. The male brings up the pellets of excrement from the sapwood galleries, in which the larvae live, into the wax tube and, collecting a mass of material in a deep concavity at the posterior end of the abdomen, suddenly jerks the body outward and shoots the pellets for a distance of several feet from the trunk of the tree.

Mr. Beeson also tells me that the large concavities in the front of the head and the lateral processes on the antennal scape of the female of *Crossotarsus bonvouloiri*, and the processes on the mandibles of the female of *Diapus quinquespinatus*, are used for picking up the eggs and carrying them about in the galleries.

Curculionidae.

How a leaf-rolling weevil (*Apoderus* sp.) rolls up leaves and lays its eggs is recorded by Sage (*J.B.N.H.S.*, VI, pp. 263-4).

The habits and life-history of an aquatic weevil are described by Annandale and Paiva (*J.A.S.B.* [n.s.], II, pp. 197-200, figs. 1A-F).

Alcides collaris is noticed by Lefroy as a gall-producer (*J.B.N.H.S.*, XIX, p. 1007).

Notes on the habits and life-history of *Cyrtotrachelus longipes* are given by Witt (Indian Forester, XXXIX, pp. 265-272, pl. v).

Concerning the development and habits of *Aclees birmanus*, a borer in *Ficus elastica*, see Dammerman (*Med. Afd. v. Plantenz.*, No. 7, Batavia, 1913, pp. 29-30, 1 text-fig., pl. i, figs. 10*a*-*b*).

STREPSIPTERA.

Green records the occurrence, in the Jassid *Thompsoniella arcuata*, of parasites belonging to this Order (*Spolia Zeylanica*, VII, p. 55).

[1] Mr. Beeson informs me that the species referred to in this discussion is *Platypus biformis*, Chap.

LEPIDOPTERA.

Rhopalocera.

Cases of butterfly migration are noted in the *Journal of the Bombay Natural History Society* by Aitken (XI, pp. 336-7 and XIII, pp. 540-1), Prall (XI, p. 533), Dudgeon (XIV, pp. 147-8) Nurse (XIV, p. 179), and Andrewes (XIX, p. 271); and in *Spolia Zeylanica* by Wickwar (III, pp. 216-8), Green (III, pp. 219-220), Fletcher (IV, pp 178-9), Daniel (V, pp. 106-7) and Willey (V, pp. 186-8).[1]

Prall records the rate of flight of certain butterflies (*J.B.N.H.S.*, XI, pp. 533-4).

Henderson records the occurrence of *Melanitis ismene* at sea (*Spolia Zeylanica*, IX, pp. 45-6).

Ormiston contributes a note on the length of life of butterflies as winged insects (*Spolia Zeylanica*, IX, p. 143).

The enemies of butterflies are discussed in the *Journal of the Bombay Natural History Society* by Nurse (XV, pp. 349-350), Lefroy (XV, p. 531) and Aitken (XVI, pp. 156-7).

The capture of *Huphina remba* by a Lycosid spider is recorded by West (*Spolia Zeylanica*, V, p. 105).

Green refers to "the habits of the leaf-butterfly" (*J.B.N.H.S.*, XVI, p. 370), and Cave publishes "a note on *Kallima inachus*" (*Spolia Zeylanica*, V, p. 142).

The climatal changes of *Melanitis*, etc., are discussed by Manders (*J.B.N H.S.*, XVII, pp. 709-720); and Aitken (*J.B.N.H.S.*, XVIII, pp. 195-197).

Some effects of moisture on the behaviour of butterflies are described by Cunningham ("*Plagues and Pleasures*", pp. 103-8).

Green describes the oviposition and early larva of *Jamides bochus* (*Spolia Zeylanica*, II, pp. 204-5), and the gregarious habits of the larva of *Parata alexis* (*Spolia Zeylanica*, III, p. 157).

An account of the habits of the leaf-cutting caterpillar of *Suastus gremius* is given by Willey (*Spolia Zeylanica*, VI, pp. 124-130, 7 text-figs.), who further notes (p. 125) the ability of the adult of this species to emit a loud clicking sound.

A note on the development and larval habits of *Aphnaeus hypargyrus* is contributed by Fraser (*J.B.N.H.S.*, XX, pp. 528-530, 1 pl.).

Mimicry in unpalatable caterpillars (*Papilio polytes*) is the subject of notes in *J.B.N.H.S.*, IV, by Hart (pp. 229-230) and Aitken (p. 317).

Carnivorous habits and cannibalism in caterpillars of butterflies are recorded in *J.B.N.H.S.*, XVIII, by Fischer (pp. 510-1), and Lefroy (pp. 696-7).

---

[1] See also Shelford, *J. Straits R.A.S.*, June 1903, pp. 203-4 (*Cirrochroa bajadeta*, Moore).

Heterocera.

Lefroy records carnivorous habits and cannibalism in the larvae of moths (*J.B.N.H.S.*, XVIII, pp. 696-7). The coccidiphagous habits of *Eublemma* larvae, which are mentioned in this note, have also been recorded in *J.B.N.H.S.*, XIII, by Dudgeon, (pp. 379-380), and Green (p. 538).

The aerial dissemination of the larvae of a wingless moth is noted by Aitken (*J.B.N.H.S.*, V, p. 421).

Troup records a plague of the web-spinning caterpillars of *Naxa textilis* var. *hugeli* on the Silang tree, *Olea fragrans* (*J.B.N.H.S.*, XII, pp. 775-6).

Certain *Drosera*-eating larvae and their habits are described by Fletcher (*Spolia Zeylanica*, V, pp. 26-27, figs. 3-7 and pp. 95-97).

The larval habits of the Tineid moth *Melasina energa* form the subject of a note by Fryer (*Trans. Ent. Soc. London*, 1913, pp. 420-422, pl. xxi).

Green gives an account of the curious Scolopendriform caterpillar of *Homodes fulva* (*Spolia Zeylanica*, VII, pp. 166-7, figs. 8*a-b*), of a Geometrid caterpillar (*Comiboena biplagiata=Uliocnemis cassidara*) which disguises itself by attaching small pieces of leaves and withered flowers to paired fleshy processes of the body (*Spolia Zeylanica*, I, p. 74), and of the efficacy of the hair of a small Lithosiid caterpillar as a protection against ants (*Spolia Zeylanica*, VI, p. 135). Wise states that the hair which *Nepita conferta* works into its cocoon serves the same purpose, and calls attention to the male-attracting power possessed by females of this species (*J.B.N.H.S.*, II, pp. 54-5). Aitken, however, shows that the hair of the larvae of *Nepita conferta* does not protect them against toads (*J.B.N.H.S.*, XI, pp. 337-8).

The method by which certain Saturniidae cut their way out of their cocoons is described by Kettlewell (*J.B.N.H.S.*, XVII, pp. 541-2).

Meyrick (*Ent. Mo. Mag.* [2], XXV, p. 220) records Fletcher's discovery of a moth, to which he gives the name *Brachmia xerophaga*, symbiotic with *Stegodyphus* at Guindy near Madras. I have examined specimens of the spider with which it was found, and have identified them as *Stegodyphus sarasinorum*. More recently I have myself obtained the same species of moth from nests of the same species of spider near Balugaon in Orissa. All the moths I saw were on the outside of the nest, but the caterpillars were inside.

Fletcher mentions the occurrence of several specimens of *Ophideres fullonica* and *Cephonodes hylas* at sea (*Spolia Zeylanica*, III, p. 202). He also contributes a note on the significance of the stridulation of the Death's Head Moth (*Spolia Zeylanica*, IV, pp. 179-180).[1]

[1] See also Rutherford, *Spolia Zeylanica*, X, pp. 77-78. For collected observations on the stridulation of European Death's Head Moths see Tutt "British Lepidoptera," IV, pp. 406-8 and 447 (larva), 432 (pupa) and 444-453 (imago). According to a notice long exhibited in the insect gallery of the Indian Museum

Stridulating organs on the wings of certain Indian moths have been described by Hampson (*Proc. Zool. Soc. London*, 1892, pp. 188-193, 6 text-figs.).

With regard to the supposed stridulating organ found in males of the genus *Arcte* (pl. xxiii, fig. 7) Mr. Henry has sent me the following note on an observation he made a few years ago in the Matale District of Ceylon. "I was walking through jungle at dusk and noticed two dark moths with light patches on the under-wings, which I am sure were *Arcte caerulea*, flitting up and down and round each other, and producing a curious clicking noise. Unfortunately at that time I was not specially interested in moths, so I neglected to preserve the specimens or to make a note of the occurrence. I was merely struck by the curious fact of moths producing a sound. It may have been a pair of males fighting (and I incline to this opinion) or a male courting a female."

Alarming colour and attitude, and also mimicry in certain caterpillars, are described by Annandale (*Proc. Zool. Soc. London*, 1900, pp. 855-857).

DIPTERA.

Psychodidae.

Concerning *Phlebotomus minutus* see Howlett, *Ind. J. Med. Res.*, I, pp. 34-8, 1 fig.

Cecidomyidae.

Stebbing describes the life-history and habits of a Cecidomyid which produces false cones on *Pinus longifolia* (*Indian Forester*, XXXI, pp. 429-433, pl. xxxviii).

Chironomidae.

The habits in all stages of the Colombo Lake Fly—since described by Kieffer (*Rec. Ind. Mus.* VI, pp. 136-137) under the name *Chironomus ceylanicus* (see Green, *Spolia Zeylanica*, VII, p. 50)—are referred to by Green, *Ind. Mus. Notes*, V (3), pp. 191-193, and Chalmers, *ibid.*, pp. 195-197.

The larva of a Chironomid, since described by Kieffer as *Chironomus fasciatipennis*, is recorded by Annandale as feeding on —and in its very early stages sometimes feeding—*Hydra orientalis* (*J.A.S.B.* [n.s.], II, pp. 112-116; see also *Fauna of British India*, Freshwater Sponges, Hydroids and Polyzoa, pp. 155-6). Other Chironomid larvae (*Chironomus* and *Tanypus* spp.) are recorded by the same author as living in association with *Spongilla carteri* (*J.A.S.B.* [n.s.], II, pp. 190-4, figs. 2A-B). He also notices some Indian blood-sucking midges (*Rec. Ind. Mus.*, IX, pp. 246-7).

the Indian species of Death's Head Moth stridulate by rubbing the tip of the proboscis on the ridged lower surface of the same appendage. I have had no opportunity of investigating this on a living specimen, but found no difficulty in artificially producing sound in this way on a freshly killed specimen that I once received.

Culicidae.

Ridley records the breeding of mosquitoes in pitchers of *Nepenthes* (*J. Straits R.A S.*, No. 22, Dec. 1890, p. 430).

MacDougall notices the habits of *Corethrella* (=*Ramcia*) [1] *inepta* (*Spolia Zeylanica*, VIII, p. 71).

The habits of *Toxorhynchites immisericors* are described by Green (*Spolia Zeylanica*, II, pp. 159-164, 1 pl.; and *Rec. Ind. Mus.*, VII, pp. 309-310) and Paiva (*Rec. Ind. Mus.*, V, pp. 187-190).

Green has seen *Culex vishnui* sucking a syntomid moth (*Spolia Zeylanica*, IV, p. 180).

Paiva records the habits of *Aediomyia squammipenna* (*Rec. Ind. Mus.*, V, p. 202).

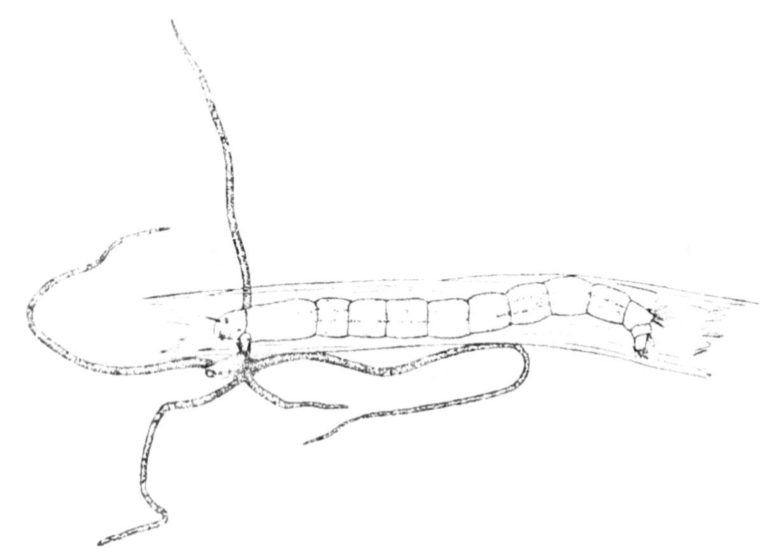

Chironomus larva attacking *Hydra*.

Tipulidae.

Conosia irrorata usually sits with the front legs and middle femora stretched forwards, the distal parts of the middle legs bent outwards at a right angle, the hind legs stretched backwards, and the body and wings pointed obliquely upwards. All the legs lie flat on the supporting surface. In this position the fly looks more like a scrap of rubbish caught in a cobweb, than like a fly.

Tabanidae.

Annandale gives an instance of adaptation in the habits of a Tabanid (*Rec. Ind. Mus.*, IX, pp. 245-6).

---

[1] This synonymy is based on information sent by Mr. F. W. Edwards to Dr. Annandale.

Asilidae.

Notes on the food of Asilidae are contributed by Bell (*J.B.N.H.S.*, XVII, p. 807) and Annandale (*Mem. A.S.B.*, I, p. 213) Notes on their oviposition are contributed by Kershaw (*J.B.N.H.S.*, XXI, pp. 610-3, pl. A-B) and Sen (*J.B.N.H.S.*, XXI, pp. 695-7, 1 fig.).

Phoridae.

Apiochaeta ferruginea, a fly capable of reproducing and developing in the alimentary canal of living human beings, is the subject of two papers by Brunetti (*Rec. Ind. Mus.*, VII, pp. 83-86 and 515-6).

Muscidae (*s. lat.*).

Limosina equitans, Collins (*Ent. Mo. Mag.*, 1910, pp. 275-279), was described from specimens found by Fletcher on a living Coprid beetle. See also Green (*Spolia Zeylanica*, IV, p. 183, and VII, p. 107).

Howlett describes the attraction of citronella oil for male specimens of two species of *Dacus* (*Trans. Ent. Soc. London*, 1912, pp. 412-418, pl. xxxix-xl).

A species of *Anthomyia*[1] is recorded by Cotes as parasitic on the eggs of locusts (*Proc. A.S.B.*, June 3, 1891, p. 94: and *J.B.N.H.S.* VI, p. 416).

Attacks of numbers of *Ochromyia jejuna* on a swarm of winged termites are noticed by Nangle (*J.B.N.H.S.*, XVI, p. 747),[2] Green (*Spolia Zeylanica*, III, p. 220 and IV, pp. 183-4) and Poulton (*T. Ent. Soc. London*, 1906, pp. 394-6). The observation that this fly has been seen taking away grains of sugar from large ants suggests that it may have been this insect which I several times saw taking the food of big ants in Cochin. On one occasion I saw a specimen flying about with a piece of food attached to its proboscis and a big ant attached to the other side of the piece of food.

Interesting observations on the feeding habits of certain blood-sucking Muscidae are recorded by Patton and Cragg (*Ind. Journ. Med. Res.*, I, pp. 11-25).

HEMIPTERA.

Pentatomidae.

The reaction of a Loris to *Aspongopus singhalanus* suggest that the taste of this bug, though at first startlingly pungent, is distinctly agreeable. The odour of the bug, though also pungent,

[1] *Anthomyia peshawarensis* (Bigot *nom. nud.*), Cotes, *Ind. Mus. Notes*, III, pp. 34-5—notes and figures but no description.

[2] Concerning the identity of the fly mentioned in this note see *Spolia Zeylanica*, IV, p. 184.

somewhat resembles essence of jargonelle (Green, *Spolia Zeylanica*, I, p. 73).

Canthecona furcellata eats Noctuid, Saturniid and Limacodid larvae (Antram, *J.B.N.H.S.*, XVII, pp. 1024-5).

Kershaw and Kirkaldy contribute biological notes on *Antestia anchorago* (*J.B.N.H.S.*, XIX, pp. 177-8, pl. B), on *Zicrona coerulea* (*t.c.*, pp. 333-6, 2 pl.), and on *Erthesina fullo* (*t.c.*, pp. 571-3. 2 pl., 1 text-fig.).

Concerning *Anasida orientalis*, *Plautia fimbriata*, *Nezara viridula* and *Aspongopus janus* see Mann (*J.B.N.H.S.*, XX, pp. 244-5 and 1166-7, 2 text-figs.).

Concerning *Coptosoma cribraria* see Ramakrishna Aiyar (*J.B.N.H.S.*, XXII, pp. 412-4, 1 pl.).

Coreidae.

Concerning the development and habits of ? *Dalader acuticosta* see Annandale (*Trans. Ent. Soc. London*, 1905, pp. 55-59, pl. viii).

Serinetha augur and *abdominalis* are said to be predaceous (*Indian Insect Life*, pp. 684-5). Green, however, points out (*Trop. Agric.*, Dec. 1909, pp. 482-3) that they suck fruit and seeds, and are preyed upon by a mimetic Pyrrhocorid *Antilochus nigripes*.

Mr. Beeson informs me that *Serinetha augur*, Fabr. is attracted to Kusum oil in October and December in Dehra Dun.

Lygaeidae.

Concerning *Lygaeus equestris* see Paiva (*Rec. Ind. Mus.*, I, p. 174).

Kershaw and Kirkaldy describe the development and habits of *Caenocoris marginatus* (*J.B.N.H.S.*, XVIII, p. 598, pl. figs. 1-7).

Galls formed on *Clerodendron phlomidis* by *Paracopium cingalense* are described by Fischer (*J.B.N.H.S.*, XX, pp. 1169-1170, 4 figs.).

Pyrrhocoridae.

Ipomoea seed is recorded by Paiva as a food of *Lohita grandis* (*Rec. Ind. Mus.*, I, p. 175, 1 fig.).

Kershaw and Kirkaldy describe the development and feeding habits of *Dindymus sanguineus* (*J.B.N.H.S.*, XVIII, pp. 596-7, 4 text-figs., 5 pl. figs.).

Henicocephalidae.

Green's observations on the habits of *Henicocephalus telescopicus* are recorded by Distant (*Fauna of British India*, Rhynchota, II, pp 194-5).

Henicocephalus basalis lives under bricks with small red ants, on which I believe it to feed.[1] Females, usually winged but

[1] I have never managed to see this species feeding, but on one occasion a wounded ant was introduced unnoticed into a killing tube with one of them, and I have little doubt that it was introduced on the tip of the proboscis, from which it must have fallen off later.

occasionally apterous, do not appear to venture out at all by day, but males are sometimes to be found running about in the evening or early morning near bricks frequented by females.[1]

In Cochin I found a specimen of *Henicocephalus* sp. sucking a termite. The colony from which this termite was taken has been identified for me by Mr. Bainbrigge Fletcher as belonging to the genus *Anoplotermes*.

Reduviidae.

Kinnear gives an instance of blood-sucking propensities in *Nabis capsiformis* (*J.B.N.H.S.*, XIX, pp. 534-5).

Concerning the occurrence of *Conorhinus rubrofasiatus* as a parasite of man see Green (*Spolia Zeylanica*, VII, p. 50).

Harpactor flavus ("*chersonesus*") when on the wing resembles a small bee, *Melipona vidua*, on which it has been seen to feed (*Fasciculi Malayenses*, Zool. II, p 263).

Millipedes are recorded as the food of *Physorhynchus linnaei* (*Spolia Zeylanica*, III, p. 159 and VII, pp. 55-6) by Mr. E. E. Green, who tells me that, of all the Ceylon millipedes, pill-millipedes appear to be the only ones which are able to withstand the attacks of this bug. I have seen large millepedes killed and eaten by *Physorhynchus* in Ceylon and in Cochin.

Physorhynchus linnaei stridulates by rubbing the tip of its proboscis between its front legs (Green, *Spolia Zeylanica*, VIII, p. 299). I have observed this mode of stridulation in *Conorhinus rubrofasciatus*, *Ectomocoris cordiger*, *Pirates arcuatus*, *Pirates affinis* and *Isyndus pilosipes*.[2] The stridulating organs of *Conorhinus rubrofasciatus* and *Ectomocoris cordiger* are shown on pl. xxiii (figs. 23-24). That of the latter insect, in which the posternum is greatly prolonged between the anterior coxae, is more finely striated and produces a louder sound than that of the former.

A specimen of *Isyndus pilosipes* was found in June, 1914, near Darjeeling, sitting on a leaf with its proboscis inserted into the carcass of a small Elaterid beetle. As I approached with a view to capturing it with its prey, it quickly took fright; but instead of flying away it struck a menacing attitude, and, standing as high as possible on its middle and hind legs, it raised the front legs into a more or less horizontal position, extending them obliquely forwards and outwards ; the antennae, which were similarly extended, were rapidly vibrated; and the proboscis, which had been withdrawn from the body of the Elaterid, was brought well into view by being bent downwards to its greatest possible extent.

---

[1] Most of my observations on this species were made in Mr. Green's garden at Peradeniya. After I left he noticed that males were much more abundant in the early morning than in the evening.

[2] See also A. Handlirsch "Zur Kenntniss der Stridulationsorgane bei den Rhynchoten. Ein Morphologisch-biologischer Beitrag" (*Ann. K. K. Naturhist. Hofmus. Wien*, XV, 1900, pp. 127-141, 15 text-figs., 1 pl.; and E. A. Butler "Stridulation in British Reduviidae" (*Ent. Mo. Mag.* (2) XXIII, 1912, p. 65).

I gathered the leaf, with the bug still standing in this attitude over its prey, and watched it for some minutes. Occasionally the front legs were lowered for a moment to grasp the edges of the leaf, the posterior end of the insect being on these occasions even further elevated than before, but they were never allowed to remain down long.

Finally I seized the bug by the thorax in order to transfer it to a killing bottle, when it at once set up a faint but distinct squeaking noise, something like that produced by longicorn beetles. The beats of this noise were found to correspond in time to the movements of the proboscis, whose tip was being rubbed vigorously up and down a well marked median longitudinal groove on the prosternum; and the noise was evidently produced by these movements.

Mr. C. A. Paiva tells me that a specimen of *Acanthaspis rama*, which he once found in a fissure of a large tree at Katihar in the Purnea District of Bihar, struck a menacing attitude when he tried to catch it. This species also possesses a stridulating organ between its front legs, and so do many other Indian Reduviids.

The habits of bugs belonging to the genus *Eugubinus* are very peculiar. The genotype (*E. araneus*, Distant [1]) is said to have been "found living in the nest of a spider (*Theridium* sp.)" at Uran near Bombay (Distant, *Fauna of British India*, Rhynchota, II, p. 207). I have found specimens at Ernakulam in Cochin (*E. intrudans*, Distant [1]), and in the Salt Lakes area near Calcutta (*E. reticolus*, Distant [1]). In both cases they were found in webs of *Cyrtophora ciccatrosa*, an Argiopid spider which spins a dome-shaped web. The web of this spider is really a horizontal orb-web pulled out of shape by a supporting framework of numerous irregular strands; it presents an appearance very unlike that of the orb-webs characteristic of other genera of Argiopidae, and superficially very like the irregular webs characteristic of the Theridiidae. Conspicuous web-spinning Theridiidae, though common round about Kandy and in the Cochin Ghats, seem to be comparatively rare in most parts of India, where *Cyrtophora ciccatrosa* is usually abundant; and it may be doubted whether one of the solitary webs of the Theridiidae would supply the bug with sufficient nourishment for development. I am inclined to think therefore, that the web from which the genotype was taken must also have belonged to *Cyrtophora ciccatrosa* and not to a Theridiid.

Eugubinus, like many other bugs of the sub-family Emesinae to which it belongs, is an excessively slender insect. It was originally described as being apterous and having two-jointed tarsi; but these are larval characters. So far as my observations go the adult is always winged and has three-jointed tarsi. It flies well, but does not appear to take flight very readily. When it settles

[1] *Entomologist*, Jan. 1915, pp. 8-9.

on a *Cyrtophora* web, instead of getting entangled it seems quite at home. When, however, it wishes to make its way into the inner parts of the framework, its long legs appear to be much in the way. If it cannot find room to get between the strands in the direction in which it wishes to go, it proceeds to cut some of them with its raptorial front legs; but these seem ill-adapted for the purpose, and progress is often very laborious and slow. Presumably, therefore, the unusual habits of the genus have been somewhat recently acquired.

Cyrtophora ciccatrosa is inclined to be gregarious, and although each spider makes for itself a separate dome, the frameworks of several webs are usually united. Males (which are minute) and young live in small domes in the common framework of the group and each female arranges her pear-shaped egg-cocoons in a string above the centre of her dome.

Eugubinus is often seen making its way towards the string of egg cocoons, and I suspect that their contents form its staple food. A specimen let loose in some webs in the Museum compound was seen more than once, soon after mid-day, with its proboscis inserted into one of the cocoons. This is not, however, the only food that it is able to take; for when I introduced some sweepings from among grass into a cage containing specimens that had had little or no food for several days, they began to investigate even grass seeds, and finally one of them made a meal off a moribund spider (? *Tetragnatha* sp.). Perhaps the ancestors of *Eugubinus* found insects caught in the outer parts of the frame-work of *Cyrtophora* webs an easy prey, and later found their way to the eggs in the interior.

The excessively slender body and legs of *Eugubinus*, and their variegated colour, make the bug somewhat difficult to distinguish among the strands of the webs of *Cyrtophora*, especially as only webs in shady situations seem to be frequented. But this alone seems insufficient to explain why the bug is allowed to destroy the spider's offspring. When specimens were let loose in webs in the Museum compound they shook the webs somewhat as they fell upon them. A spider immediately rushed out to one of the bugs, ran half way along its body, and seemed just about to strike when, instead of the bug writhing in its grasp as I expected, the spider fled back to its dome. I supposed that the bug must have emitted something highly distasteful to the spider; but next morning this very spider was seen making a meal off one of the bugs!

Green records the frequenting of the webs of *Archiopsocus* sp. by *Ploiariola polita*, and believes this Reduviid to be predatory on the Psocids in the webs (*Spolia Zeylanica*, VIII, p. 71).

Cimicidae.

The bat *Scotophilus kuhli* is recorded by Kunhikannan as a host of *Cimex rotundatus* (*J.B.N.H.S.*, XXI, p. 1342).

Cragg describes fertilization in *Cimex* (*Ind. J. Med. Res.*, II, pp. 698-705). The spermatozoa are introduced through an aperture on the fourth abdominal segment.

Cicadidae.

In the preface to his "Monograph of Oriental Cicadidae" Distant refers to the natural enemies of the group, and to the voices of the males. Later (p. 1) he gives references "to most of the published information respecting the structural details of the wonderful sound-producing organs" (p. vi). Observations on the production of sound by Indian species have been recorded by Middlemiss (*Nature*, XXXIII, pp. 582-3).

Annandale describes the habits of *Dundubia intemerata* and *Huechys sanguinea* (*Proc. Zool. Soc. London*, 1900, pp. 859-862).

The liquid discharge made by Cicadas is noticed by Biscoe (*J.B.N.H.S.*, X, pp. 535-6).

A captive specimen of *Lemuriana apicalis*, which was recently watched in the Indian Museum, emitted from time to time a jet of colourless liquid with considerable force from its hinder end, while feeding on the sap of a piece of the tree on which it had been caught. The note of this cicada is not unlike that of a cricket, and may frequently be heard in trees round about (and even in) Calcutta during the rains; but this is almost the only specimen I have seen and the only one I have managed to catch.

Huechys sanguinea is sometimes plentiful on *Zizyphus* bushes near Calcutta in the spring. Dr. Annandale tells me that when in the Malay Peninsula he noticed that this cicada frequented bushes rather than big trees.

Dracott describes the emergence of cicadas from their nymphal skins, and the nymphs from the ground (*J.B.N.H.S.*, XXIII, pp. 379-380). His observations were made at Gangtok in Sikkim at an elevation of about 6000 ft. above sea level, on a plot of ground from which large numbers of specimens have been seen to emerge year after year.

Fulgoridae.

Annandale shows that the peculiar prolongation of the head found in certain Fulgoridae is probably of use in jumping (*Proc. Zool. Soc. London*, 1900, pp. 866-868).

Concerning *Saturnis marginellus*, *Geisha distinctissima*, and *Neomelicharia furtiva* see Kershaw, *J.B.N.H.S.*, XXI, pp. 607-9, pl. A B.

Concerning *Phromnia marginella* see Imms, *Mem. Manchester Lit. Phil. Soc.*, LVIII (4), 12 pp.; 2 pl.

Membracidae.

Chatterjee describes the development and habits of *Oxyrhachys tarandus* (*Indian Forester*, XL, pp. 75-79, pl. iii-iv).

Several species of Membracidae are common on a number of different kinds of shrubs. They are usually sluggish insects and slip round the branch on which they are seated when disturbed. Only as a last resort do they jump or fly, although they can do both quite well. The posterior end of the female is armed with two pairs of fine lancets in a protecting sheath With the aid of these the eggs are laid in rows embedded horizontally in the bark of the twig, from which only one end of each protrudes (pl. xxiii, figs. 16-17). The larvae (pl. xxiii, figs. 20-22) are brown or black, with an eversible reddish appendage at the posterior end of the body (fig. 22). They are commonly more or less gregarious. Even adults (pl. xxiii, figs. 17-19) seem to scatter little if at all when not compelled to do so. Consequently very large numbers are usually found living together on an infected bush. They are generally attended by big black ants (pl. xxiii, fig. 17).

Cercopidae.

The habits of *Machaerota guttigera* have been described by Westwood from notes supplied to him by Mr. S. Green (*Trans. Ent. Soc. London*, 1886, pp. 329-333, pl. viii).

The habits of *M. planitiae*[1], which is common on *Zizyphus jujuba* in Calcutta (pl xxiii, fig. 13), are very similar. The larva (pl. xxiii, figs. 9-12) always lives head-downwards in its tube, which, though closed at the base, is not entirely shut off from the twig to which it is attached. I have never seen the larva come out to feed, as Westwood supposed that of *M. guttigera* must do: and it is so helpless when removed from its tube that I doubt if it could safely do this It must, I think, obtain all its nourishment from the supporting twig through the pore at the base of the tube, through which its stylets may sometimes be seen to protrude when the tube is separated from the twig.

As Green watched the commencement of tube-building by some newly-hatched larvae of *M. guttigera*, he felt that "it must be a close fit by the time they are ready to assume the perfect state." The difficulty is overcome by each larva producing two tubes—first a small one, and then a larger one. A separate small tube is always found at the base of each big one (pl. xxiii, fig. 8), I have seen the larva of another tubicolous form, protected only by a frothy fluid, at work commencing the latter at the base of the former.

The habits of another insect, *Hindoloides indicans*, Distant[2], which is common here on *Zizyphus jujuba*, are similar to those of *Machaerota*. I have, however several times watched the emergence of its adult at about sunset. In *Machaerota guttigera*, according to Green, emergence occurs shortly after sunrise, and I think this is probably also the case with *M. planitiae*.

---

[1] I am indebted to Mr. Distant for this identification.
[2] *Ann. Mag. Nat. Hist.* (8) xv, pp. 506-507.

The tubes of the *Hindoloides indicans* (pl. xxiii, fig. 14) are easy to distinguish from those of *Machaerota planitiae* by their more wrinkled appearance, and by their form, the free portion being shorter and less straight. This may perhaps account for the curious fact that although adult of *Machaerota* are often much more abundant than those of *Hindoloides*, the reverse is the case with their tubes; for most of these tubes are always found to be old and empty, and presumably the long straight distal portion of the tube of *Machaerota* soon gets broken off.

The larvae of the two genera are much alike; in their later stages they can, however, be distinguished by the size of the developing process of the scutellum (compare figs. 12 and 15, pl. xxiii).

Machaerota planitiae is recorded in "*Indian Insect Life*" (p. 733) from *Zizyphus jujuba* (ber), from *Aegle marmelos* (bael) and from cotton, as well as from "other plants in India." Early in February of this year I found its tubes common at Pusa on ber and on cotton. On the former it was accompanied by tubes of some species of another genus of which I failed to obtain adults. Only *Machaerota* appeared to occur on cotton. I doubt whether any of the tubes found on bael belonged to this genus. As the mixture of genera on *Zizyphus* has been so long unnoticed, it is not unlikely that the genera and species to be found on different plants will prove to be greater than has hitherto been supposed.

Jassidae.

Annandale notes that the phenomenon of "weeping trees" is sometimes due not to Cercopidae, but to Jassidae (*Rec. Ind. Mus.*, III, pp. 293-4). Other families of Homoptera appear also to take part in its production.

Lefroy (*J.B.N.H.S.*, XX, pp. 235-6) notes that the Mango Jassid *Idiocerus* appears only to breed when mango trees shoot freely. In the district where his observations were made this occurred in March only for five years in succession. During the sixth year, however, an exceptionally wet season caused the trees to shoot again in September. "Whether from this reason or not, the *Idiocerus* also bred and one distinct brood was produced at a season when we have never before observed it breed at all." A similar restriction of the breeding season may perhaps account for the freedom of Calcutta during the greater part of the year from the notorious "green-fly" (*Nephotettix bipunctatus* and *apicalis*) which appears every year towards the end of the rains in such myriads that it is often almost impossible to approach a glowing arc-lamp near the open maidan.

Concerning the eggs of *Tettigoniella spectra* see Lefroy, *J.B.N.H.S.*, XX, p. 236.

Aleurodidae.

Peal describes the function of the vasiform orifice of the Aleurodidae (*J.A.S.B.*, LXXII [II], pp. 6-7).

Coccidae.

Imms records the occurrence of *Dactylopius citri* in ants' nests (*Rec. Ind. Mus.*, VI, p. 111).

MYRIAPODA.

The habits of a number of Malay Myriapods are described by Flower (*J. Straits R. Asiatic Soc.* No. 36, 1901, pp. 1-25).

Concerning the food of Scolopendridae see Wells-Cole, Okeden and Cumming (*J.B.N.H.S.*, XII, p. 214; and XV, pp. 135, 1 pl., and 364-5 respectively).

Mr. G. Mackrell tells me that *Ethmostigmus pygomegas* is common on his tea garden in Sylhet just below the surface of the ground. Specimens are sometimes found in earth round the roots of grass growing at the bottom of a bush; but they more often crawl into light soil leaving no visible hole, or in between clods of earth. In captivity they appear to be nocturnal, and to shun light. Their food consists chiefly of worms and small insects, and they seem to be fond of Acridiids and Gryllids.

I have already described the stridulation, apparently to attract attention away from the creature that had cast it, of a detached leg of *Scutigera decipiens* (*J.A.S.B.* [n.s.], IX, pp. 415-6). More recently I noticed a detached leg of a Cochin species moving in the same manner, and on holding it to my ear was able to hear it squeaking. In this case, however, the squeak was much fainter, and was produced by legs which remained *in situ* as well as by others.

The occurrence of purplish-red millipedes in herds is noticed by Cunningham ("*Plagues and Pleasures*", p. 193). I have sometimes seen such herds on the muddy banks of the Havildars' Tanks on the Calcutta maidan, especially, if I remember rightly, in the spring.

Some of the larger species of Indian millipede exude an evil smelling coloured fluid when disturbed. The common big black species in the Cochin Ghats does this, but nevertheless falls an easy prey to its enemy the Reduviid bug *Physorhynchus*. A still larger black species, in which the middle of each segment is girdled with extra thick chitin, and the caudal horn is exceptionally long, emits a particularly virulent fluid which not only smells, but also stains and burns one's hands. This species was only found near the head of a small valley at Kavalai (Cochin Ghats) where, however, it was much commoner than the common form of the district. I regret that I did not also try *Physorhynchus* with it.

Arthrosphaera aurocincta[1], Pocock, a pill-millipede common in the Cochin Ghats—the commonest round Parambikulam at the end of the State Forest Tramway—surprised me by the vibrations which it usually set up when caught. On holding it to the ear a squeaking noise was heard. The noise would, however, have passed unnoticed but for the vibrations which called attention to it. I would have put this forward in support of Arrow's theory (*Fauna of British India*, Lamellicornia, I, p. 14) that the object of stridulation is often not noise, but vibrations that will bring discomfort to an enemy on contact, but that the pill-millipede seems to require no greater protection than its excessively hard carapace. Pillmillipedes, as already pointed out (p. 511), appear to be the only millipedes capable of withstanding the attacks of *Physorhynchus*. The fact, however, that stridulation always took place when the animal was seized and rolled itself into a ball points to its association with the instinct of defence. I have only noticed stridulation in the one species, although I specially looked out for it in other species found in Cochin. I never heard or felt it in an open specimen; consequently I found it impossible to locate the organ which produced the sound.

ARACHNIDA.

XIPHOSURA.

Notes on the habitat and breeding habits of *Limulus moluccanus* and *L. rotundicauda* are contributed by Annandale (*Rec. Ind. Mus.*, III, pp. 294-5). Sewell (*Rec. Ind. Mus.*, VII, pp. 87-8) records the capture of the former species in a surface townet in water of about 10 fms. depth. See also Flower, *J. Straits R. Asiatic Soc.*, No. 36, July 1901, p. 26.

SCORPIONIDEA.

Parturition in a scorpion is the subject of a note by Dreckman (*J.B.N.H.S.*, III, pp. 137-8, fig. facing p. 69). The species dealt with is incorrectly named, and evidently belongs to the genus *Heterometrus*, perhaps to the species *H. phipsoni*.

Pocock describes the habits in captivity of *Parabuthus capensis* and *Euscorpius carpathicus* (*J.B.N.H.S.*, VIII, pp. 287-294). Neither of these are, however, Indian species.

Newnham (*Nature*, LVI, p. 79; reprinted in *J.B.N.H.S.*, XI, pp. 313-4) records the carrying off of a large flower by *Parabuthus liosoma* one evening at Aden. The scorpion was holding the flower over its back in one of its claws. When camping at the foot of the Ghats in the Ratnagiri District I once saw a scorpion in the same way carry off a piece of white paper that had fallen from the table at which I was working in the open after dark.

---

[1] I am indebted to Dr. F. Silvestri for this determination.

The stinging power of scorpions forms the subject of notes by Green, Coomaraswamy and Drieberg (*Spolia Zeylanica*, III, pp. 197 and 215-6, and IV, p. 33).

Concerning the habits of *Archisometrus mucronatus* and other Malaysian species see Flower, *J. Straits R. Asiatic Soc.*, No. 36, July 1901, pp. 30-36.

PEDIPALPI.

Thelyphonidae.

The earliest mention of the habits of Indian Thelyphonidae appears to be by Stoliczka (*J.A.S.B.*, XLII [II], p. 127), who records his own observations and those of Mr. Peal. Peal's observation that "the *Thelyphoni* are generally found underneath the bark of decayed wood in groups, rarely singly" is somewhat surprising. All the specimens presented by him to the Indian Museum are, however, *Uroproctus assamensis*, a species whose habits have rarely come under my observation.

The next reference is by Wood-Mason (*Proc. A.S.B.*, 1882, pp. 59-60).[1] Observations of a similar nature to those made by Stoliczka and Wood-Mason are recorded by Oates (*J.A.S.B.*, LVIII (II), pp. 4-5).

Flower refers to the habits of the Siamese *Thelyphonus schimkewitschi* (*J. Straits R. Asiatic Soc.*, No. 36, July 1901, pp. 37-9).

A brief note by Green on the habits of *Thelyphonus sepiaris* will be found in *Spolia Zeylanica*, IV, p. 181, and one by myself on those of *Labochirus proboscideus* in *Spolia Zeylanica*, VII, pp. 44-46, fig. B.

A further contribution to the subject is made by Fischer, who describes the courtship dance of *Thelyphonus sepiaris* (*J.B.N.H.S.*, XX, pp. 888-9).

The habits of *Uroproctus assamensis* as observed by Kemp during the Abor Expedition are referred to in my note on the Pedipalpi collected on that Expedition (*Rec. Ind. Mus.*, VIII, p. 127).

I am now able to describe in greater detail the habits of several Indian species of Thelyphonidae. My earliest and most extensive observations were made on *Labochirus proboscideus*, and these will be described first.

Labochirus proboscideus is not uncommon under logs of wood and large stones in the jungles of the Kandy district of Ceylon: but it is only to be found when the ground has been thoroughly

[1] Stoliczka does not refer to the fluid of his Thelyphonids as *inodourous*, as stated by Wood-Mason, but as not having any *offensive* odour. In some species it is violently pungent and resembles acetic acid. In others it is more like essence of jargonelle and, although not very pleasant, is by no means pungent—individual opinions would probably differ as to whether it was offensive or not. In Wood-Mason's specimens the odour was "exactly like that of a highly concentrated essence of pears, but . . . . when deeply inhaled had all the characteristic smell and pungency of strong acetic acid." Compare pp. 509-510, above.

wetted by the rains, or occasionally near water in the dry weather. Thus before the rains I only obtained two or three specimens, and these were all found under stones on moist (but not swampy) ground within a few yards of the Mahawelli Gunga.

Specimens are always found on the ground, never on the under side of their shelter. When first uncovered they usually remain quite still for a time before attempting to hide. Sometimes a burrow is found under the shelter. In this case the *Labochirus* usually sits facing it, and disappears down it as soon as any attempt at capture is made. In other cases any burrow there is must be throughout its length in contact with the shelter.

In dry weather, when *Labochirus proboscideus* is difficult to obtain, it presumably burrows till it reaches soil that it finds comfortably moist and then remains there. If unable to find moisture it dies in a few days; and I found it impossible to keep this species in captivity for any length of time unless the floor of its cage was kept covered with moist soil, when no difficulty was experienced.

Both sexes construct burrows in which to live, digging the soil away with their second pair of appendages. As the excavation deepens they enter it head first, collect some soil between the second appendages, and then back out and deposit it at a little distance from the entrance. The tibial apophyses seem to enable them to carry more soil than would otherwise be possible. Of two very young (probably one year old) specimens kept in captivity one made a U-shaped burrow with one entrance under cover and the other exposed; but I have not been able to recognize any other instance of this being done.

Labochirus appears to be incapable of inflicting any injury on man. When irritated it usually extends its pedipalps to their fullest extent, and would no doubt use them in defence against a sufficiently small opponent; but it is a nervous creature and prefers retreat. It will not attack even a defenceless cockroach if it is very large, but will gladly kill and eat small ones.

The stink-glands are no doubt of service in self-defence. On two occasions I have seen the fluid ejected as a small cloud, but this is rare; one of the specimens noticed was a female, the other was almost certainly a male but escaped capture. According to Wood-Mason (*Proc. A.S.B.*, 1882, p. 60) the stink-glands are larger in female Thelyphonids than in males. The apertures of these glands are easily seen on each side of the medially situated caudal appendage (dorsal) and anus (ventral). If some object is placed near these appertures when the creature is irritated the drop of fluid ejected will be found upon it. It has "all the characteristic smell and pungency of strong acetic acid", but in this species I have never noticed any odour "like that of a highly concentrated essence of pears" (Wood-Mason, *l.c.*).

It is almost impossible to observe the feeding habits of these nocturnal animals in their natural haunts; and even in captivity they are very shy of any light that may be brought to bear upon them when they have emerged from their hiding-places in search of food. What their usual food is, and how often they get it under natural conditions I am unable to say. In captivity they appear to feed as often as suitable food is given, suitable food being winged termites, small locustids, blattids, etc., especially when these are disabled. But of larger insects and of very active ones they are easily frightened. A disabled locustid will be snatched at eagerly if held in front of a specimen : when however, it is presented alive and kicking *Labochirus* will extend the second pair of appendages as if to seize it, but in reality as a menace, and will then back away.

Food is seized between the second appendages and held between them and the head. In the male of this species little or no use is made of the chelae, though at times the long movable finger may be embedded in the prey and bent over so as almost to meet the tibial apophysis. The very long gnathobases of the second appendages appear to be of some use in supporting the food above the ground and keeping it in the neighbourhood of the mouth. In the female the second appendages are much shorter and stouter than in the male, and the form of the tibial apophysis renders it scarcely possible that the movable finger of the chela should be brought into apposition to it ; the gnathobase is also very much shorter. How far these structural differences affect the mode of feeding I am unable to say, as the only female kept in captivity, was, I believe, damaged when caught, and died after a few days without having taken any food.

Concerning the part played by the chelicerae in feeding I am also unable to say anything, as they were always obscured by the anterior end of the carapace and by the food itself; presumably they are used much as in *Phrynichus* (see below). Another function of the chelicerae was, however, repeatedly seen, namely the use of the brushes with which they are provided for cleaning the terminal joints of the legs. Hansen (*Arkiv för Zool.*, II [8], p. 8) says "The function of such hairs, 'blood hairs', is no doubt to intercept the blood of the prey when this has been cut to pieces." Doubtless they function to some extent in this way, but their use for cleansing purposes is manifested every time a specimen gets its feet a little soiled.

In *Labochirus*, and probably in all the Thelyphonidae, the antenniform legs are ordinarily held directed forwards and usually somewhat outwards in an arched posture. As the animal moves along they are lowered from time to time till the tip comes in contact with the ground, and then raised again, but the two are lowered alternately, not simultaneously.

I found it very difficult to determine whether these creatures drink water, as so many Arachnids do. I believe, however, that they do so, and on one occasion I saw a specimen apply its mouth to water placed in its cage on a leaf, although it refused to take any notice of this until a lamp that was near had been removed. I could not see whether any sucking movements were set up or not.

With regard to the breeding habits of this species my information is of the scantiest, but I believe, from the evidence of dissections, that the time for egg-laying was rapidly approaching when I left Ceylon in August. The young appear to attain a length of about 1 cm. (exclusive of the tail) during the summer after they are hatched, and to take two years more to come to maturity; but the evidence for this is not so extensive as in the case of *Charinides* and *Phrynichus* (see below, pp. 531-532).

It was not until August[1] of last year that I saw the courtship "dance" of a whip-scorpion. The specimens concerned were a pair of *Thelyphonus sepiaris* which I caught in Orissa and brought back with several others alive to Calcutta. Their positions, when I first noticed what was going on, were those described by Fischer (*J.B.N.H.S.*, XX, pp. 888-9). They are shown in pl. xxiv, fig. 25. The left antenniform leg of the female was crossed above the right, and about three joints of the tarsus of each of these legs were left exposed by the chelicerae of the male. The pair walked slowly round the cage in which they were confined, the male going backwards and the female following him. Once or twice they passed an unattached male, when the mated male left go his hold of the antenniform leg of the female on the side next the possible rival and seemed to prepare for defence. But none was needed.

Soon this type of "dance" ceased, the female raised her abdomen in the air, and the male commenced stroking her genital segment with his antenniform legs. These legs usually passed between the third and fourth legs of the female but sometimes behind the fourth; their tips were usually crossed, the right being above the left as a rule. The chelae of the male were held open and were kept slightly moving over the dorsal surface of the abdomen (pl. xxiv, fig. 26).

I expected this to lead up to the culminating action. But the female was a small one, and her genital segment was not fully developed. Probably she was immature. For this reason, perhaps, the first type of "dance" was soon resumed and continued till I went to bed. It was in progress at about 7 A.M. next day, but ceased soon afterwards. Next night it was repeated. After that the female died.

Thelyphonus sepiaris is much better able to withstand draught than is *Labochirus proboscideus*. It lives in much drier situations, and will live in a dry cage without water for several weeks at least, without apparent discomfort. It seems, too, to be of a somewhat less timid disposition. Green (*Spolia Zeylanica*, IV, p. 181) says that it emits an odour resembling strong acetic acid. The defensive odours of the Thelyphonids I have met with vary in character from this to something closely resembling essence of

---

[1] Mr. Fischer informs me that his observations (see above, p. 519) were made after dusk in June. Rain had fallen and brought out the Thelyphonids, which climbed about his tent. The dance took place on his writing table in the tent.

jargonelle, and I regret that I have not made notes of the scents of the various species.

Mr. H. N. Ridley has sent me several specimens of *Thelyphonus linganus.* He compares their defensive odour to that of chlorine gas, and says he knows of no other animal able to emit so widely diffused and powerful an odour for its size. This species eats crickets, woodlice, etc., and Mr. Ridley once found a specimen eating a cricket in the day-time, though usually the species is nocturnal. The modification of one or more of the sixth to eighth tarsal joints of the antenniform legs of the female of this and allied species of *Thelyphonus* is perhaps due to the development of special organs for use during courtship; for it is at about this point that these appendages are held in the chelicerae of the male during that process in the only species of *Thelyphonus* in which it has been described.

The odour emitted by *Uroproctus assamensis* resembles essence of jargonelle. This species lives in a damp region, and does not seem able to withstand drought in the way that *Thelyphonus sepiaris* can.

Mr. G. Mackrell tells me that *Hypoctonus oatesi* exudes a fluid smelling like acetic or formic acid, though perhaps a little more pungent. It inhabits country where stones are not to be found, living in the banks of roads and cuttings and in the vacated burrows of ants. In June specimens usually have to be dug out from a depth of about 18 inches, but in August they are often found at the entrance of their holes. On one occasion two females were found in a nest swarming with ants. Both had young clinging to their abdomens.

I have found several species of *Hypoctonus* under stones in Burma. I have not been able to study their habits in captivity, but there seems every reason to believe that they are very like those of *Labochirus proboscideus* and *Thelyphonus sepiaris.*

I do not think any of the Thelyphonids I have studied can be luminous as suggested by Sorensen (*Ent. Med.* 1894, pp. 175-177); and I can hardly believe that the sting described by Flower (*Journ. Straits R. Asiatic Soc.*, July 1907, pp. 38-39) was really due to the *Thelyphonus schimkewitchi* that he was handling when he received it. I have handled other species frequently without receiving any harm.

Schizomidae (Tartarides).

The only species which I have myself found in any abundance are *Schizomus* (*s. str.*) *crassicaudatus*, *S.* (*Trithyreus*) *peradeniyensis*, and *S.* (*T.*) *vittatus*, all from Ceylon, and it is only to these that the following account refers.[1] The Calcutta form—*S.* (*T.*) *lunatus* —I have found under bricks on somewhat moist stiff clay; and so

[1] A preliminary note on the habits of these species appeared in *Spolia Zeylanica*, VII, p. 46, fig. C. Nothing else beyond a brief note on a species from caves near Moulmein (*J.A.S.B.* [n s.], IX, p. 417) appears to have been written about the habits of Indian species.

far as I know its habits are much the same as those of the Ceylon species.

Schizomus crassicaudatus was only found under bricks, etc., on or close to open ground (usually grassy lawns) more or less shaded by trees, while the other two Ceylonese species named were found only among dead leaves, especially where these formed a layer of considerable depth and were matted together by fungal hyphae, in the midst of the shrubberies in the Peradeniya Botanical Gardens. *Schizomus crassicaudatus* was never found in company with the other two.

The ground was more or less moist, especially in the shrubberies, during almost the whole time I was at Peradeniya; but shortly before the break of the rains I found specimens (probably *Schizomus crassicaudatus*) on a very dry slope a few yards away from a little stream. A specimen of *S. peradeniyensis* subsequently lived in a corked tube in Calcutta without food or water for about three months. When a drop of water was placed near it by means of a fine pipette the soil was so dry that it was some time before the water began to be absorbed, but the *Schizomus*, after examining it carefully with the mobile tips of its antenniform legs, took no further notice, and made no attempt to drink.

I have only once seen a Tartarid take food. On this occasion a minute white centipede (*Scutigerella*) was seized by the second pair of appendages in much the same way as a fly is seized by the chelicerae of a spider. When secured it was carried off into a burrow to be eaten.

On one occasion I put two specimens (both female) of *S. peradeniyensis* in a tube partly filled with soil, and two of *S. vittatus* in another. Before very long the individuals of both pairs were found to be facing each other, their antenniform legs extended obliquely forwards, those of the one crossing those of the other. This futile hostility continued for several hours, after which the specimens were separated. Apparently neither dared either to attack the other from in front, or to leave her rear unguarded for a moment. When, however, a larger number of specimens are similarly confined, but on the slippery glass bottom of a tube without any soil, they frequently become panic-stricken and then attack each other in the same way as they attack their natural prey. But even a large number of specimens may safely be collected together in a tube when loose soil is provided for them to run about on.

When touched from in front *Schizomus* usually tries to escape by giving a sudden jump backwards. The stink-glands are no doubt used for defence; and when a number of specimens are caught and put together in a tube they may be observed to emit a distinct odour of acetic acid.

The chelicerae and second pair of appendages, besides being used for offensive and defensive purposes, are used for cleaning the feet; and the form of the second appendages allows them to be used simultaneously with the chelicerae for rubbing the feet, and not only for holding them in position.

A specimen of each of the three Ceylon species was kept alive by itself in a separate tube about one-third full of carefully packed soil. Each made two or three burrows before very long; but they rarely entered any of them even by day, and when disturbed they never seemed to know where to find them though the whole diameter of the tubes was little more than two centimetres. It is therefore rather difficult to believe that these burrows are used as permanent homes, and this is borne out by the following facts concerning the habits of *Schizomus crassicaudatus* in the open.

Whenever a specimen of this species was discovered by the removal of the brick under which it had been hiding, it would dart spasmodically about looking for somewhere to hide again, with no more idea than one of the captive specimens just mentioned as to where to find a suitable hole; and the hole into which it finally disappeared seemed to me to be as a rule a wormtrack or something of that kind. Further, it is apparently possible to go on collecting specimens from under one brick two or three times a week for an indefinite period, each time removing every specimen found; which seems to prove conclusively that they can have no fixed abode, but wander about from place to place among the roots of the grass not far from which they are always found.

In the case of the shrubbery forms which occurred in extensive layers of dead leaves it was impossible for such observations to be made; but if these species habitually lived in burrows it is difficult to understand why they were most abundant among the leaves and not in the soil below them; and why a specimen in captivity entirely without cover very rarely entered any of the two or three burrows that it made.

Of what use, except for reproductive purposes, the burrows can be it is difficult to see; but of the three specimens which made burrows in captivity only one—*Schizomus crassicaudatus*—produced eggs. This one constructed a little cavity against the side of the glass tube in which it was confined, at a depth of about 15 mm. below the surface of the soil (pl. xxiv, fig. 27). As far as I could see this nest had no opening, and the *Schizomus* never left it to my knowledge till the eggs disappeared. It was lined with soil cemented together in some way; when this lining was shaken free from the glass (to which it was similarly cemented) the damage was quickly repaired; but unfortunately I never saw this being done. The eggs were seven in number, subspherical (flattened at the poles), of a glistening white colour, and were neither tightly pressed together nor enclosed in a brood-pouch of any kind. They were arranged so that one of them was above and one below the centre of a ring composed of the five others: the general shape of the mass as a whole was approximately spherical, and it appeared to be attached to the abdomen only in the region of the genital aperture. The abdomen was carried at an angle to the rest of the body as shown in the figure. As a rule the creature rested on one of the sides of the nest with the thorax vertical and the abdomen horizontal, but it was impossible to see it sufficiently clearly in that position for it to be drawn. The nest and eggs were first noticed on Sept. 12, about three weeks after the creature's arrival in Calcutta, but they may have been produced a few days earlier; no changes were seen to take place in them, and eventually they disappeared and the mother left her nest. The mother had been captured and placed by herself in a tube at Peradeniya on August 14.

Tarantulidae.

The habits of the cavernicolous Charontinae, *Stygophrynus cavernicola*, *S. cerberus*, and *Catagius pusillus*, have been described by Annandale and myself (*J.A.S.B.* [n.s.], IX, pp. 417-420).

A small species of Charontinae, *Charinides bengalensis*, is common in Calcutta. I have been able to study its habits in greater detail than those of the cavernicolous species. They closely resemble those of the Ceylonese species of *Phrynichus* on which a preliminary note has already appeared (*Spolia Zeylanica*, VII, pp. 43-4, fig. A) and also those of *Phrynichus nigrimanus* [1], a species not uncommon in the Eastern Ghats. The habits of all of these may now be considered together, the few differences between them being noticed as occasion arises. There is no reason to suppose that the habits of cavernicolous species differ in any essential points from those of these species, apart from the fact that *Stygophrynus* does not habitually live under stones or logs of wood, but on the walls of caves.

All species that I have observed [2], except those of the genus *Stygophrynus*, live in crevices among bricks or stones, or under logs of wood, where there is room for them to move about freely. They are almost always found on the under side of the object beneath which they hide. *Charinides*, and doubtless other Charontinae also, having pulvilli on its feet, can walk up a vertical piece of polished glass, or even across its lower surface; but *Phrynichus*, which has no pulvilli, cannot do this. It is unlikely that any Tarantulids can burrow; and a specimen of *Phrynichus ceylonicus* that was brought to me after being dug out of a hole it had been seen to enter, can hardly have made the burrow for itself.

Phrynichus ceylonicus, *s. str.*, appears to be a regular inhabitant of bungalows; but its variety *pusillus* and *Charinides bengalensis* seem to visit them rarely, and it is very doubtful, on account of the inability of these species to live in the absence of moisture,

[1] This and other species of *Phrynichus* have been provisionally grouped by Kraepelin under the one name *P. reniformis*, Linn. See *Rec. Ind. Mus.*, XI, pp. 447-448.

[2] Mr. Ridley informs me that *Sarax singapurae* is found under bricks and among dead leaves; he thinks the latter form its usual home. When specimens are collected from under the bricks, others quickly take their places.

whether they ever make them a permanent abode.[1] That *P. ceylonicus, s. str.*, does this there can, I think, be no doubt. The first specimen I saw alive was found in Mr. Green's workshop after dark; it was sitting on the wall close to a large bookcase behind which it retreated as soon as I attempted to catch it. It must have been several days at least since it first came there, for a cast skin, which from its size appeared to have belonged to it, was found close by upon the wall, with cobwebs attached to it; and the animal itself had already become thoroughly hard and dry. On the following night the specimen (I have no doubt it was the same) was found again in the same place and captured; it lived healthily in a bare breeding cage for over two months, when I preserved it prior to leaving Ceylon. These striking creatures are also well known to residents in the island who frequently mention some particular room as one in which a specimen is often seen.

The difference between *Phrynichus ceylonicus* and its variety *pusillus* in their ability to stand dryness is very marked.[2] The former will live healthily for at least a fortnight, and usually longer, in a bare cage with a wooden base and frame, glass sides and a perforated zinc top, whereas the latter always dies in a few days if not supplied with constantly moist soil. *P. ceylonicus*, variety *pusillus*, appears to be confined to the moist jungles of the lower hills of Ceylon. *P. ceylonicus, s. str.*, on the other hand, seems to be most abundant in places where climate or a porous soil produce drier conditions. On the only occasion on which I was able to test the capacity of *P. ceylonicus*, variety *gracillibrachiatus*, one specimen of this form and one of variety *pusillus* were put into a bare cage in which two specimens of *P ceylonicus, s. str.*, were living, after giving them ample opportunity of satisfying their thirst. Variety *pusillus* was found dead next evening and variety *gracillibrachiatus* on the following evening.

It may be noted here that *P. ceylonicus*, variety *pusillus*, besides requiring damper surroundings than *P. ceylonicus, s. str.*, appears to be a much more thirsty animal; and I am inclined to think that small specimens of the latter are of a more thirsty disposition than old ones, though the evidence for this needs amplification.

Charinides bengalensis also requires a certain amount of moisture in its surroundings, and no doubt it is on this account that it always choses for its abode some pile of bricks in a sheltered place where the ground is moistened nightly in the dry cold weather of Calcutta by a heavy dew.

Tarantulids, like other Pedipalpi, are nocturnal feeders; by day they hide themselves away. When one is exposed by turning over its shelter, it crouches flat down, and when eventually it darts suddenly away it rarely tries to escape to other stones, how-

---

[1] I have no evidence on this point with regard to *P. nigrimanus* and other species.

[2] *P. nigrimanus* probably resembles *P. ceylonicus* in this respect, but the evidence is somewhat conflicting.

ever good the cover they may offer. I have several times chased a specimen of *P. ceylonicus*, variety *pusillus*, round and round a stone in this way for some minutes before being able to catch it, when the stone was resting on a mass of others among which the creature could have got away with the greatest ease, had it thought of doing so. *P. ceylonicus*, *s. str.*, and *Charinides bengalensis* appear to be equally prejudiced in this respect.

When caught, even the large *P. ceylonicus* is apparently quite incapable of inflicting any injury. All species of *Phrynichus* are, however, able to give a distinct (but painless) nip between the terminal finger of the second appendages and the distal spines of the tibia, and when caught are apt to claw viciously at one's hands as often as they get the chance.

No stink-glands are known in Tarantulids, and I have never noticed any particular smell associated with them.

What Tarantulids live upon when left to find their own food I cannot say; but in captivity cockroaches, crickets, and sometimes a green locustid will be taken by *Phrynichus*. *P. ceylonicus, s. str*, naturally manages larger specimens than *P. ceylonicus*, variety *pusillus*, can do. I believe that *Charinus bengalensis* will take very small cockroaches (? and woodlice) and have seen it eat swarming termites that have shed their wings. Tarantulids are extremely nervous beasts and winged termites are far too active for them. A specimen of *P. ceylonicus, s. str.*, became pitifully panic-stricken when one or two of these were placed in its cage, raising itself upon its legs with a start every time a termite touched it.

The following account, based on a particular instance, will serve to show how Tarantulids obtain their food.

A recently captured *Phrynichus* was sitting under a tile where it had been hiding all day, when it became aware of a wingless cockroach (*Dorylaea rhombifolia*) engaged in feeding upon a lump of bread in one corner of the cage. The *Phrynichus* left its retreat and cautiously approached to within a short distance of the cockroach when, after extending both arms, it made a sudden grab; but only the bread was secured, and this was not appreciated. For a few moments the *Phrynichus* waited in a defiant attitude, slightly raised upon its long legs, with its arms partially extended: then it subsided flat on the ground again. In the meantime the scared cockroach had retreated into another corner of the cage, where it was soon followed by the *Phrynichus* which made another grab at it. This time it was caught and brought within reach of the chelicerae, with the assistance of which it was finally demolished. Another cockroach was killed and partly eaten later on during the same night, after which the *Phrynichus* fasted for six days, when it ate another cockroach. This fast of several days after each meal appears to be the normal habit of all the species of *Phrynichus* I have studied; and the remarkable eagerness which newly caught specimens always show for food leads me to believe that in the wild state they are rarely able to secure as much food as they would like. This suggestion is further supported by the difficulty which they always seem to experience in the capture of active prey.

It is very difficult to observe the method of capture on account of the rapidity of this action; but repeated observations have convinced me that although both arms are shot forwards in any attempt to sieze the prey, the actual capture is usually between the terminal claw and the spines near the end of the second appendage of one side only. As will be seen on reference to fig. 28 (pl. xxiv) these spines are so arranged as to form a very effective hand, the terminal claw being apposable to the proximal of the two long dorsal spines at the distal end of the tibia, and the spine on the penultimate joint to the distal of these. As the claw and all three spines are rigid and sharply pointed it would be not unnatural to suppose that when grasping the prey they enter its body in such a way as to render its escape quite impossible. As a matter of fact, however, the strength necessary for this is apparently absent, and I have seen even a soft-bodied cricket unsuccessfully attacked time after time as its movements brought it within reach of a *Phrynichus*; and although once or twice it appeared to have been secured by one hand the other was never used to assist in holding it, with the result that it escaped before it could be brought within reach of the chelicerae.

Once within reach of these appendages, however, all chance of escape disappears. The prey, which remains alive for a time, is held between the two hands, often with the terminal finger embedded in its tissues, whilst parts of it are scooped into the region of the mouth by the terminal joint of the chelicerae, the sharp saw-like armature of their under surface perhaps being of use in severing pieces of a suitable size from the main mass. When such a piece had been secured by the chelicerae it is thoroughly masticated by vertical, combined with slight longitudinal, movements of these appendages, which rub it against each other and against the gnathobases of the second appendages. As the terminal joint is apparently kept closed except when required to scoop in a fresh piece of the edible material it is difficult to see any use to which the double row of teeth on the basal joint can be put. The long anteriorly projecting sternal spine no doubt assists in keeping the food from falling to the ground when it passes into the immediate neighbourhood of the mouth.

I have only once seen *Charinides* feed. Unlike *Phrynichus* it captured its prey between the two second appendages, not in one hand, the terminal claw and the spines of the hand and finger being unable to close against the spines at the end of the tibia (pl. xxiv, fig. 29). The terminal claw pressed into the body of the prey, probably penetrating the tissues, and other spines appeared to help to some extent. The capture was extremely sudden, and the details were only seen by repeatedly removing the captured prey until the *Charinides* became nervous, and acted more slowly.

It has already been mentioned that *Phrynichus ceylonicus*, variety *pusillus*, is a much more thirsty animal than *P. ceylonicus*, *s. str.* When thirsty, however, specimens of both species behave much alike if water is sprinkled into their cage. As long as small drops only are met with these are caught up between the spines of the hand, whose arrangement, when the hand is closed, is admirably adapted for this purpose. The drop is conveyed by the hand to the chelicerae which suck it off with movements like those of mastication. When, however, a small pool is found on a leaf or other receptacle the chelicerae are inserted directly into this and with the same movements proceed to suck it up. Once when I attempted to give water to a specimen of *P. ceylonicus*, variety *pusillus*, which did not want it, a drop that had been placed on the chelicerae was drawn off into one of its hands, and flung aside by a sudden movement of the arm. I have not seen *Charinides* drinking.

The brushing up of the other appendages by the chelicerae may often be seen in all species. I am inclined to think that this is sometimes done chiefly for the sake of the moisture upon them, small though this must be; at least the evident relish with which it is sometimes done after water has been sprinkled about, and before the creature has found any separate drops, emphatically suggests this. When the feet have to be brushed they are supported in position by the hands. The great care which is taken to keep the tips of all the appendages free from dirt is very striking. In the case of the second appendages this is probably due to the presence on the two terminal segments of an elaborate system of spines, clubbed hairs and pits, which may perhaps constitute an organ of taste or smell, functioning as a test of the suitability for food of anything that is captured. In the case of the feet it is probably necessary for the pulvillus in *Charinides*, and the pulvillus-like pad in *Phrynichus*, as well as the claws, to be kept perfectly clean if they are to be used effectively; and it is not unlikely that tactile organs may be concentrated in this region.

That the antenniform legs should be kept clean, not only at the tip (as are the other appendages), but also throughout a considerable part of their length, is clearly necessary on account of their great service to the animal as feelers. When *Phrynichus* moves sideways (at it usually does) these legs are extended outwards, their mobile extremities feeling cautiously about in all directions; when it moves forwards they are extended forwards somewhat as in the Thelyphonidae; and I have seen them, too, when the animal was at rest, extended straight outwards and then slowly rotated, the one forwards and the other backwards, so as to sweep as large an area as possible round the body. When undisturbed their position is rarely very different from that shown in *Spolia Zeylanica*, VII, p. 43, fig. A.

To what extent the sense of vision has been replaced in these creatures by senses localized in the anteniform legs I am unable to state with certainty. When a specimen is first found and exposed to light by turning over the stone on which it is resting, instead of at once trying to escape it crouches close down upon the stone as already noted, and remains motionless for a time, no doubt trusting to its inconspicuousness when in this position for safety. When eventually it darts round to the under side of the stone this may be due to its preference for an inverted position and not necessarily to its dislike of light. A touch from a foreign body upon the antenniform legs or other part of the animal instantly puts it to flight. That *Phrynichus* is sensitive to light, however, becomes inconveniently evident as soon as one attempts to study its habits; and I am inclined to believe that its sight is of use in seeking for prey, though I have not been able to apply any conclusive test of this.

All Indian and Ceylonese species probably breed at about the same time of year.[1] I first found egg-laden females of *P. ceylonicus*, variety *pusillus*, on July 20th at Peradeniya, but the eggs were all well advanced in development. This was also the case with similar specimens of variety *gracillibrachiatus* found at Nalanda a week later. The embryos are carried under the abdomen, where they are supported by a membrane secreted for the purpose. All egg-laden specimens kept in captivity died before the young were hatched, but it is probable that the maternal habits closely resemble those of *Charinides*, which are described below. The number of eggs carried by a female *Phrynichus* appears to be about fifteen for *P. ceylonicus* variety *pusillus*, about 40 for variety *gracillibrachiatus* and about 60 for *P. ceylonicus*, *s. str.*

Charinides bengalensis breeds in July and August, and sometimes earlier. A specimen in captivity produced eggs on June 26. An egg-laden female, caught on Aug. 29 and kept in captivity, hatched its young on Sept. 23. These were six in number. Before the evening of the day on which they appeared all of them had freed their appendages and climbed on to the dorsal surface and sides of the abdomen of the mother; they were entirely white, though their bodies became faintly darker next day, and their second appendages lacked the characteristic spines of the adult. Two or three eggs which failed to hatch remained for a time attached to the abdomen of the mother as before. On the night of Sept. 27-8 all six of the young ones cast their skins; they then assumed a distinctly greenish colour, and the characteristic spiny armature of the second appendages appeared. The young and their cast skins remained upon the mother during that day, but by the next morning the former had wandered away by themselves, and the latter had disappeared.

Immediately after the first moult the carapace is a little over 1 mm. in width. From a series of over twenty specimens

[1] *P. nigrimanus* appears to breed somewhat later than some others. I failed to find any ovigerous females at Barkul in Orissa when I went to look for them at the beginning of August, and saw no young ones likely to have been newly hatched.

collected one evening in September, it appears that during the first year its width increases to about 2 mm., during the second year to about 2·5 mm., and during the third to about 3 mm. At this period the arms of the male assume their distinctive length, and maturity is probably reached. Whether large specimens reach their maximum size at the same time, or whether this maximum (about 3·5 mm. thorax-breadth) is only reached subsequently, I am unable to say.

In *Phrynichus ceylonicus*, variety *pusillus*, and probably in other Phrynichinae also, the period of growth appears to be about the same; but the difference between the one-year-old specimens (thorax-width about 4 mm.), which are of a chequered green and yellow colour with conspicuously banded legs, and the much larger two-year-olds (thorax-width 7-8 mm. as a rule), of darker and more uniform colour, is very much more marked than between these and the adults, which as a rule are very little larger than them.

SOLIFUGAE.

C. E. C. Fischer contributes some notes on the habits of *Galeodes indicus* to the *Journal of the Bombay Natural History Society* (XX, pp. 886-7).

ARANEAE.[1]

Miscellaneous.

An instance of an apparently unprovoked assault by a spider on man is given by Cunningham ("*Plagues and Pleaures*", p. 209). The spider was dislodged with difficulty, but no after effects of the bite are recorded.

A spider's web suspended from a beam, and apparently kept taut only by the weight of a stone over four feet from the ground, is recorded by Sawrey-Cookson (*J. Sarawak Mus.*, I, p. 123).

Aviculariidae.[2]

Wood-Mason's account of the discovery of stridulating organs in a member of this family—*Chilobrachys* ("*Mygale*") *stridulans*—is recorded in the *Proceedings of the Asiatic Society of Bengal* for 1875 (p. 197). It has been reprinted in the *Annals and Magazine of Natural History* ([4] XVII, p. 96). A more detailed account is to be found in the *Transactions of the Entomological Society of London* (1877, pp. 281-2, pl. vii). In view of the fact that the stridulating organs of *C. fumosus* appear to be of a more primitive type than those of *C. stridulans* (*J.A.S.B.* [n.s.] X, pp. 416-417, pl. xxi, figs. 5-6) I noticed with interest when at Kalimpong that,

[1] See also Workman, "*Malaysian Spiders*" (Belfast, 1896).

[2] Theraphosidae, Pocock, *Fauna of British India*, Arachnida. I have followed Simon's classification and nomenclature (*Histoire Naturelle des Araignées*, I and II, Paris 1892 and 1897) on account of its completeness.

although the former species behaves like the latter when angry, it can only produce a faint rattling sound. The latter is said by Wood-Mason to hiss loudly.

Notes on the habits of large "bird-eating" spiders, doubtless of this family, are contributed by Macpherson and by Morris (*J.B.N.H.S.* I, pp. 28-29, and IV, pp. 69-70). In both cases the species is identified as "*Mygale*" *fasciata*, a name now confined to a Ceylonese species of *Poecilotheria*. The subjects of the former note lived in burrows, so probably belonged to the genus *Chilobrachys*. The genus of the subjects of the latter note may possibly have been *Poecilotheria*, but the species was an Indian one.

Pocock (*J.B.N.H.S.* XIII, pp. 121-2) discusses notes on the habits of "bird-eating" spiders, and records observations showing that the genus *Poecilotheria* is arboreal. Annandale (*Mem. A.S.B.* I, p. 216) states that *P. striata* is apparently not uncommon on *Acacia arabica* near Pamben. Flower refers to the habits of the Malaysian *Coremiocnemis cunicularius* (*J. Straits R. Asiatic Soc.* No. 36, July 1901, p. 42).

Walsh notices the habits of a trapdoor spider from Orissa, which he describes as *Adelonychia nigrostriata* (*J.A.S.B.* LIX [II], pp. 269-270). Specimens which he sent to O. P. Cambridge were described by the latter at about the same time under the name *Diplothele walshi*, by which the species is now generally known.

The nests of *Sason cinctipes* were pointed out to me at Peradeniya by Mr. Green. They are constructed on flat moss-covered rocks and walls, and are roofed in by two rounded flaps of equal size, hinged together to form a single 8-shaped structure, and attached to the lower part of the nest at either end of the hinge. These flaps are covered with fragments of moss, etc. and are flush with the moss growing round about them. Consequently they are very hard to see. The spider sits beneath them, and can get out by raising either of them.

All species with which I am acquainted belonging to the subfamily Ischnocoleae live under stones and logs of wood. They do not appear to make burrows.

Cyriopagus ("*Melopoeus*") *minax*[1] lives in silk-lined burrows which are not closed by a door of any kind.

Uloboridae.

Most of the Indian Uloboridae known to me live in groups, often in association with a web spinning spider of some other family. In Cochin I found a nest of the common gregarious spider, *Stegodyphus sarasinorum*, with the orb-webs of a small Uloborid spread around it. When disturbed the Uloborids retreated in among the *Stegodyphi*. I gathered the branch and destroyed most of the Uloborid webs in so doing; but the spiders

[1] Concerning the generic name of this spider see *Rec. Ind. Mus.* XI, p. 281.

came out during the night and made fresh ones, in the centres of which they remained so long as they were not disturbed.

Uloborid webs were several times seen grouped round the web of *Psechrus alticeps* in Cochin. One species seems to be definitely associated with *Cyrtophora ciccatrosa* in the Indian Museum compound, though I never saw it here till towards the end of last year. A somewhat larger species lives in groups of orb-webs in the Museum compound, but does not appear to be specially associated with any other spider. This species, and perhaps one other of similar habits, occurs also at Pusa, where Uloborids are found in association with *Nilus* sp., *Gasteracantha brevispina*, and *Cyrtophora ciccatrosa*. Another species found at Pusa spins solitary orb-webs over slight depressions on the sunny side of the trunks of smooth-barked trees.

One Uloborid found singly in Cochin spins two horizontal orb-webs, one above the other. The upper web is flat and finely spun, with a meshed hub on the under side of which the spider rests. The lower web is funnel-shaped with open hub, and is of coarser build.

Psechridae.

Psechrus ? singaporensis has been found in the Batu Caves, Selangor, out of reach of daylight (Flower, *J. Straits R. Asiatic Soc.* No. 36, July, 1901, p. 45).

Psechrus alticeps spins a large irregular web not unlike that of a Theridiid. One end of this web is always in contact with a tree-trunk, stone or bank of earth, to which the spider escapes with extraordinary rapidity when disturbed. The species is very common in the Cochin timber forests, but very difficult to catch.

A species of *Psechrus*, common at Pashok and Kalimpong in the Darjeeling District (*ca.* 4000 ft.), constructs a large and coarsely spun tunnel, the upper part of which spreads out to form a somewhat extensive roof-like snare—something like a *Pardosa* web (see below) inverted, but coarser in every way. The structure of the snare resembles that of the snare of *Stegodyphus* described below. I have not been able to examine a fresh snare in detail, but a microscopic examination of an old one showed that the silk was also very like that of *Stegodyphus* both in structure and in variety.

Eresidae.

The genus *Stegodyphus* is represented in India by several species; but only one, *S. sarasinorum*, seems to be abundant. The abundant and widely distributed form whose habits are described by Jambunathan (*Smithsonian Misc. Coll.* XLVII, pp. 365-372, pl. L.[1]) must presumably therefore have belonged to this species, on the habits of which Fischer has since contributed a few notes (*J.B.N.H.S.* XVIII, pp. 206-7).

[1] Banks has added a useful bibliography of the literature on social spiders to this paper.

Fischer's observations on the way in which these social spiders treat their victims is not in accordance either with Jambunathan's observations or my own. Instead of the prey being left to die as stated by Fischer, I have always found that anything caught in the web, whether by day or by night, is at once attacked. Under suitable conditions, however, very extensive snares may be spun, and it is possible that insects caught in the part of a large snare furthest from the nest may not attract attention so readily as do those caught near the nest.

The nest in which the spiders hide by day is a tough mass of cobweb mixed with dead leaves, insect integuments, etc. (pl. xxv, fig. 30). At dusk the spiders come out and float threads of silk from the upturned abdomen in the usual way, till one or more of them gets attached to some object at a little distance from the nest. These threads are then strengthened, and form the foundations of irregularly meshed snares (pl. xxv, figs. 30-31). The foundations of these snares are composed of one or more strands of relatively coarse silk. Between, and often along, these strands, strands of another kind are laid (pl. xxv, figs. 31-33). These are broader and have a softer and more woolly appearance; when carefully examined they are found to consist of a fine central thread overlaid with irregularly twisted threads and a sticky foam-like silk that I suppose to be the product of the cribellum. The hind legs may always be seen working against this organ when these strands are produced. The second type of strand is not only very sticky but also very elastic. Strands of this type unite the foundation lines in all directions, and when an insect gets caught among them they give before its struggles, and are not liable to break. Snares seem, however, to suffer from the weather; and many strands may be broken as the captured prey is dragged along towards the nest to be eaten.

Repairs and extensions are always carried out after sunset; but the spiders seem to be ready for food at any time. A fly thrown into the snare always brings them out at once. How the presence of prey in the snare is detected I have not been able to determine with certainty. If the snare is disturbed by one's finger the spiders hastily retreat into the nest. This suggests that sight is of use in this connection; but a dead fly was found to attract no attention. A specimen of the Rutelid beetle *Adoretus*, on the other hand, was at once pounced upon when placed in the snare although its movements were quite unlike those of the fly. Perhaps, therefore, touch calls the attention of the spiders to the presence of something in the snare, and sight determines their action towards it. The slaying of a large and strongly chitinized insect like *Adoretus* is a much more difficult feat for the spiders than is the slaying of a fly; but they persist until it is accomplished.

Associations of other animals with African species of *Stegodyphus* have been recorded by Marshall (*Zoologist*, [4] II, pp. 417-422) and Pocock (*Ent. Mo. Mag.*, [2] XIV [XXXIX], pp. 167-170).

Marshall describes a dormouse which lives in *Stegodyphus* webs and ultimately drives out the spiders; and both authors refer to a Microlepidopteron which lives with the spiders in their nests. Such a moth has recently been found associated with *Stegodyphus sarasinorum* in India (see above, p. 506). A Uloborid spider makes use of the webs of *Stegodyphus* (see above, pp. 533-534). Other associations, probably of a more casual nature, may also occur; and when pulling *Stegodyphus* nests to pieces in Orissa in order to obtain lepidopterous larvae I found in addition the following animals alive within them: one cribellate spider (? Dictynidae), one centipede (? Geophilidae), one large Lepismatid and two minute beetles (Anthicidae and Clavicornia).

Pholcidae.

Concerning *Artema atlanta* see Flower, *J. Straits R. Asiatic Soc.* No. 36, July 1901, p. 43.

At least two species of Pholcidae are common in Calcutta, one much larger than the other. The former is usually found hanging by its long legs from its untidy web. The latter, which probably belongs to the genus *Pholcus*, is often seen in a similar position, but seems to be more of a wanderer, and I have twice seen it using its extraordinarily delicate legs as a snare for insects. On the first occasion the captive was an earwig (*Nala lividipes*). The earwig seemed much the stronger of the two antagonists, but was encircled by the spider's legs, which were too slender to be seized, and the spider's body was raised out of reach of danger by them. From time to time the spider lowered its body and struck at the earwig; but it was always restrained from effecting its purpose by a flourish of the earwig's abdomen. When the earwig tried to escape the spider went with it, taking care not to let it go from between its legs. Finally, however, the earwig's patience proved greater than the spider's, and it got away. On the second occasion the captive was a small Tipulid, and I have little doubt that the spider would soon have been victorious, but that the contest took place on a wall, from which both combatants ultimately lost their hold.[1]

Argiopidae.

An instance of a small bird being caught in the web of a large Argiopid—doubtless *Nephila maculata*—is recorded by Sherwill (*J.A.S.B.* XIX, pp. 474-5). The "young spiders (about eight in number and entirely of a brick-red colour) feeding upon the carcass" were probably either males or parasitic Theridiids.

[1] More recently I have seen a hard-shelled jumping Chrysomelid beetle of considerable size (about 5 mm. long) similarly attacked by this spider. In this instance the spider had succeeded in attaching its silk to the victim and was busy spinning over it by the time it was seen. The spider stood high above its victim as usual, and appeared to be arranging the threads with its hind legs.

Further notes on the habits of *Nephila maculata* (? and other species) are given by Cunningham ("*Plagues and Pleasures*", pp. 203-5 and 210) and Fischer (*J.B.N.H.S.* XX, pp. 526-528 and 887-8, pairing habits). Brief notes on the habits of this species and of a few other Argiopids found in the Malay Peninsula are recorded by Flower (*J. Straits R. Asiatic Soc.*, No. 36, July 1901, pp. 43-4).

Argiope catenulata seems to be confined, near Calcutta,[1] to the Salt Lakes area, where it spins an orb-web surrounded by numerous irregular lines between bushes of the low holly-like mangrove, *Acanthus ilicifolius*. *Argiope pulchella*, which is common among larger bushes in the Salt Lakes area, as well as elsewhere, spins a simple orb-web.

Orsinome marmorea spins large and more or less horizontal webs between rocks above rapidly running streams at an altitude of about 1,500 ft. in the Cochin Ghats. Several webs are usually grouped together; often they are stretched above waterfalls. When the spiders are disturbed they fall into the water, which washes them away. When they reach a rock they cling to it, and remain an inch or two below the surface till danger is over. Males and females were sometimes found together in the middle of a web with their heads in contact. Presumably they were pairing, but I had not time to investigate this fully.

Herennia ornatissima spins its orb-web close to a tree-trunk. Dr. Sutherland tells me that at Kalimpong he has found the female in the middle of her web, and sometimes the male in the web with her. In Cochin I always found the female[2] in a small silk-lined concavity on the tree-trunk near the web; and when a male was present it was in a similar nest close beside that of the female. The specimens were brownish in colour and they were very difficult to distinguish from irregularities of the bark.

The Indian species of *Cyrtophora*, like those of other countries, spin more or less dome-shaped webs; and most of them are more or less gregarious. The web of *C. citricola* is spun horizontally in the midst of an irregular mass of supporting lines, and has the characteristically fine mesh and delicate texture of the domes constructed by its allies; but it is scarcely raised in the centre. Mr. W. H. Phelps informs me that the web is always made at night. First radial lines are constructed as if for an ordinary orb-web. Then the spiral is commenced, and fresh radii are run out from time to time between the others, this filling in of the radial spaces being done piece by piece, not by a succession of complete whorls. When the web is about half finished the centre is raised as far as it is intended to be by the attachment of lines from above, after which it is completed.

---

[1] It is also recorded from places such as Peradeniya, where conditions are quite unlike those of the Calcutta Salt Lakes and there is no *Acanthus ilicifolius*.

[2] I did not however, find many specimens.

On one occasion Mr. Phelps saw a specimen of *C. ciccatrosa* construct its cocoon. A soft loose sheet about an inch and a half long and a quarter of an inch wide was prepared first, on this the eggs were laid, after which it was rolled up and suspended above the web. The eggs hatch about three weeks after they are laid, and the spiders develop very rapidly.

The occurrence of a predaceous bug in the webs of *C. ciccatrosa* has already been noticed (above, pp. 512-513). The eggs of *C. feae* are parasitized by a Hymenopteron, both sexes of which are winged.[1]

Thomisidae.

Trench describes how "a lemon-coloured spider with a tri angular body and long yellow legs" sitting "on one of those virulent mauve zinnias" where "there was no effect whatever of any protective colouration" captured a bee-hawk moth (*J.B.N.H.S.* XX, p. 876). The spider was presumably a Thomisid.

Mrs. Drake has sent me from Serampore a number of specimens of *Amyciaca* sp., a mimic of the red ant *Oecophylla smaragdina*, together with a specimen of a bug, *Armachanus monoceros*, which mimics both, resembling the spider even more closely than it does the ant. Of the habits of a female *Amyciaca*, which she kept in captivity, she writes: "It is interesting to watch her method of securing a red ant. When one is put under her glass it at once goes towards the spider, who backs away from it or lets herself down by a cable if there is no room to draw back, after which she follows up and springs on it from behind Then comes the curious part. She does not retain her hold but leaps down and waits. Next, cautiously advancing to its head, she walks round it as if to make quite sure it is dead, and finally, after lightly touching it, she begins her meal, every now and then moving on with her prey held up just as the ants carry their finds. Her extreme carefulness looks as if she had instinctive knowledge of the power of the ant's jaws, for I suppose had she herself been bitten she would not have survived. I had a male killed by a red ant the other day. I have only seen this kind of spider near red-ant settlements, and of the ten seen at different times nine were eating red ants, and the tenth was letting itself down by a cable just over a red ants' road. I believe these drop lines help to entangle a stray foraging ant, and while it strives to free itself the waiting spider springs upon it."

Clubionidae.

The habits of the common Calcutta house-spiders of this family are described by Cunningham ("*Plagues and Pleaures*", pp. 206-

[1] Bugnion and Popoff have described from Ceylon *Baeus apterus*, a parasite with wingless females obtained from the eggs of a spider determined as *Argiope aetherea*, Walck. from which those of *A. catenulata*, Dol., were infected in captivity (*Rev. Suisse Zool.* XVIII, 1910, pp. 729-736, pl. v). The former spider has perhaps been incorrectly identified, for *A. aetherea* is a Papuan species which does not appear to be known from the Oriental Region (see Thorell, *Ann. Civ. Mus. Genova*, XVII, 1881, pp. 68-71).

208). These spiders are *Spariolenus tigris* and *Heteropoda venatoria*. They are probably about equally common, but the former is more often seen than the latter, as it seems to be less sensitive to light and the female often makes her home on the whitewashed wall of a staircase or bathroom, where she may be found day after day for weeks together. The male of *Spariolenus tigris* seems to be much rarer than the female, although the two sexes of *Heteropoda venatoria* are about equally common. Both species kill and eat cockroaches and crickets, which in some instances at least are not killed immediately they are bitten. Concerning *Heteropoda venatoria* see also Flower, *J. Straits R. Asiatic Soc.*, No. 36, July 1901, pp. 46, where *H. thoracica* is recorded as cavernicolous.

Lycosidae.

A species of *Pardosa*, common in hedge-bottoms in the plains of Cochin, spins a silken tube open at both ends, the upper end leading out on to a silken platform. Mature females may often be seen at the entrances of their tubes, each with a male (sometimes two) keeping guard on the platform outside. When disturbed the female disappears into the tube followed by the male. Egg-laden females are not attended by males.

Attidae.

Notes on ant-mimicing spiders are given by Rothney (*Trans. Ent. Soc. London*, 1889, p. 354; reprinted, *J.B.N.H.S.*, V, pp. 44-45) and Walsh (*J.A.S.B.* LX [II], pp. 1-4).

The mimicry of Mutillids by spiders is recorded by Green (*Spolia Zeylanica*, IV, pp. 181-2—spider *Caenoptychus pulchellus*, see *Spolia Zeylanica*, V, pp. 190-1; and *Spolia Zeylanica*, VIII, pp. 92-3, 1 pl.) and by myself (*Rec. Ind. Mus.* VII, p. 87). I take this opportunity of pointing out that my observations were made in Orissa, not in Calcutta as stated by Green.

Acari.

The habits of *Trombidium grandissimum* are described by Annandale (*Mem. A.S.B.* I, pp. 216-7).

EXPLANATION OF PLATE XXII.

Oryctes rhinoceros.

FIG. 1.—Mouthparts of larva from below, with left maxilla removed and turned over to show supposed stridulatory structures.

,, 2.—Pupa from above showing stridulating organs.

,, 3.—Left anterior stridulating organ of pupa more highly magnified.

,, 4.—Posterior end of abdomen of adult, with left elytron removed, showing the ridges with the aid of which the insect appears to stridulate.

Adoretus spp.

,, 5.—Mouthparts of *A. versutus*—the labrum slightly raised, the mandibles widely and maxillae more moderately opened.

,, 6.—Mouthparts of *A. lasiopygus*—the labrum removed, the mandibles and maxillae very widely opened to show the molar teeth of the former. This figure is more highly magnified than the last.

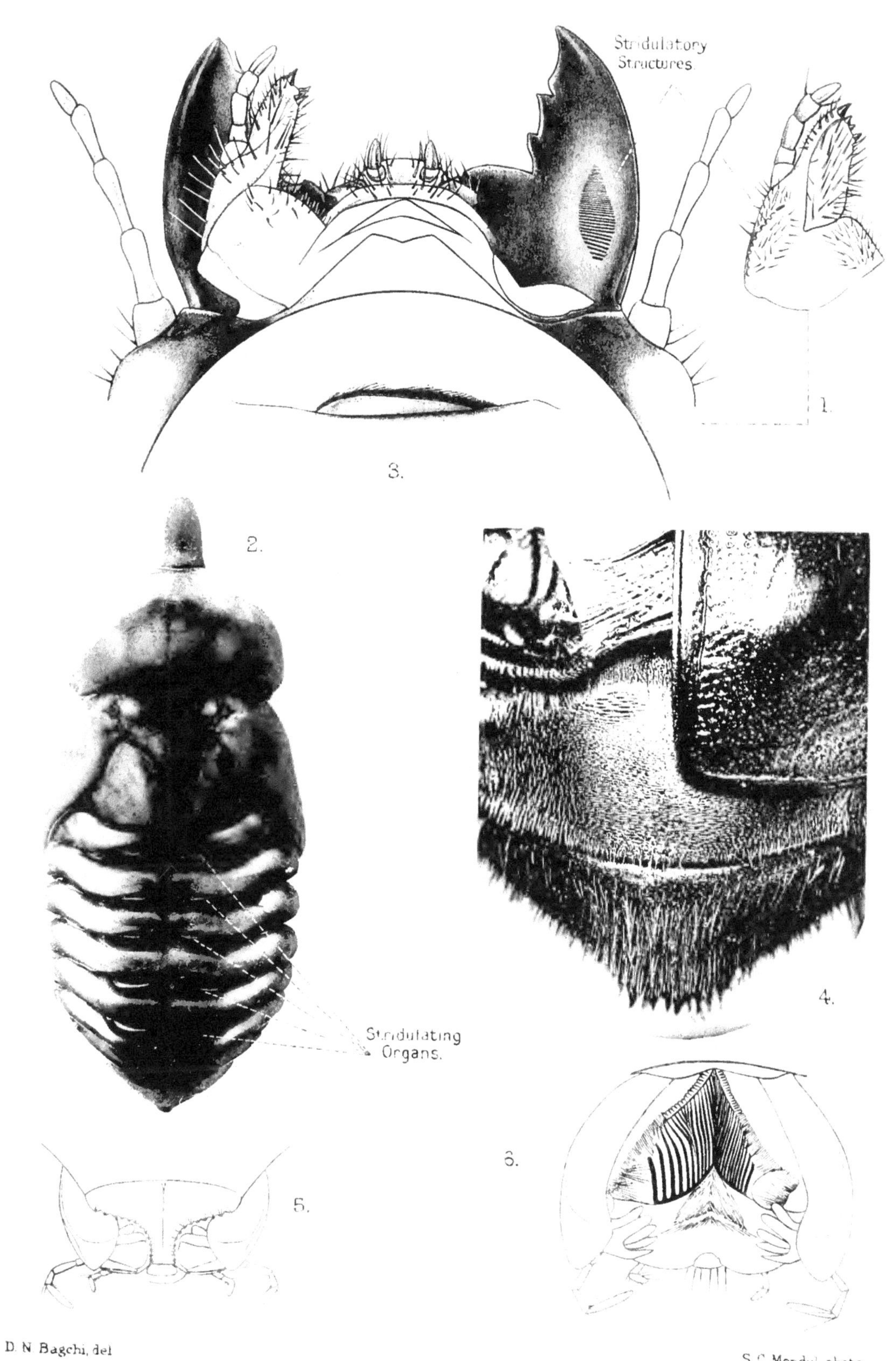

D. N. Bagchi, del.

S. C. Mondul, photo.

LAMELLICORN BEETLES.

EXPLANATION OF PLATE XXIII.

Arcte caerulea.

FIG. 7.—Posterior part of abdomen of male from above, after removal of the scales, to show the stridulating plate.

Machaerota planitiae.

,, 8.—Tubes on a twig of *Zizyphus jujuba.*

,, 9.—Larva.

,, 10.—A more advanced larval stage.

,, 11.—Last larval stage from above.

,, 12.—Last larval stage from the side.

,, 13.—Adult.

Hindoloides indicans.

,, 14.—Tube with larval exuvium and newly emerged adult.

,, 15.—Last larval stage from the side.

Otinotus oneratus.

,, 16.—Eggs in bark of *Zizyphus jujuba*, seen as a transparency.

,, 17.—Adult females with eggs, on stem of *Bauhinia varians*, accompanied by black ants (*Camponotus compressus*).

,, 18.—Adult from the side.

,, 19.—Adult from above.

,, 20.—Larva from above, with caudal filament partly exserted.

,, 21.—Larva from the side, with caudal filament retracted.

,, 22.—Larva from the side with caudal filament exserted.

Conorhinus rubrofasciatus.

,, 23.—Head and prothorax from below, showing stridulating organ.

Ectomocoris cordiger.

,, 24.—Head and prothorax from below, showing stridulating organ.

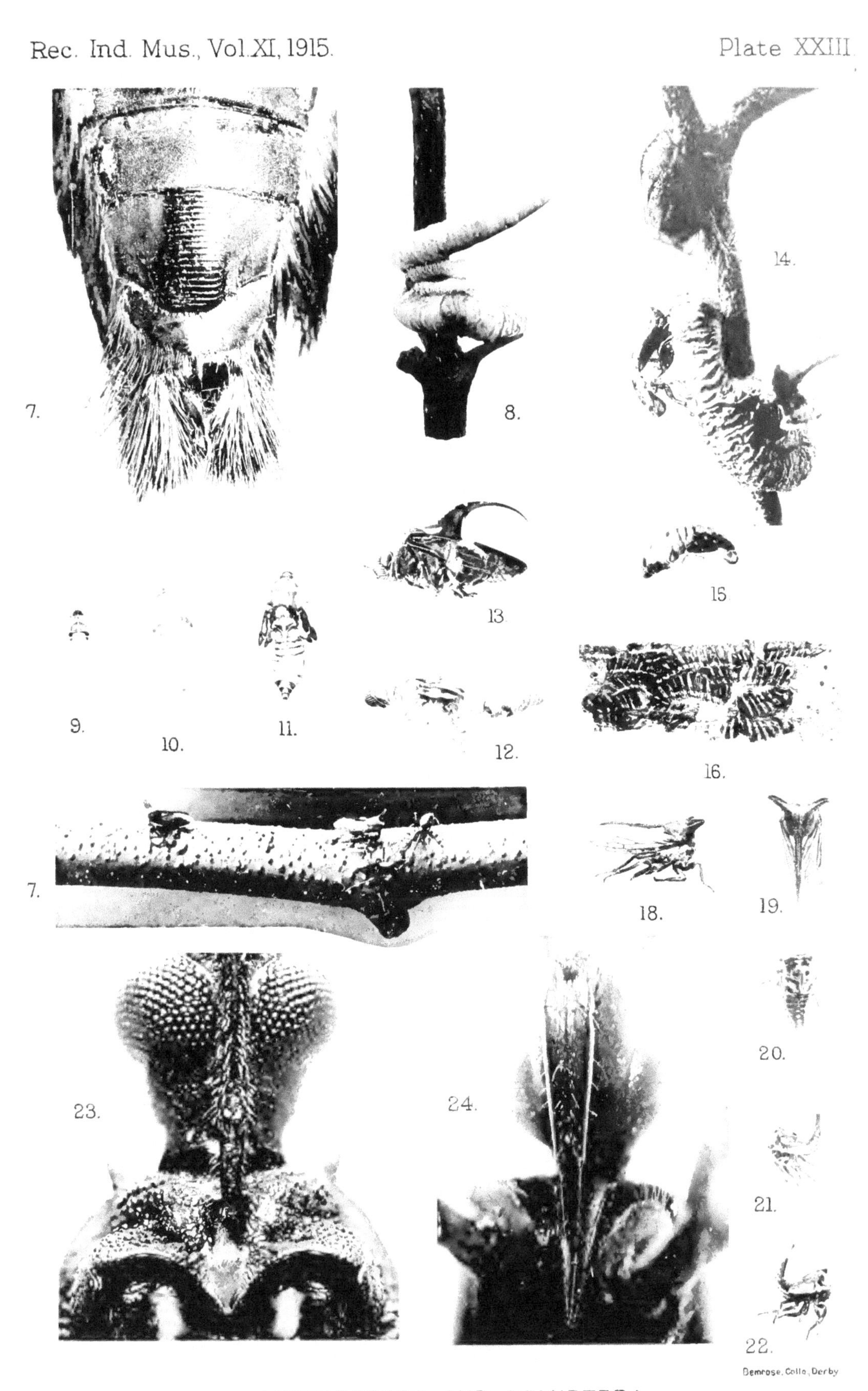

Bemrose, Collo, Derby

LEPIDOPTERA AND HEMIPTERA.

EXPLANATION OF PLATE XXIV.

Thelyphonus sepiaris.

FIG. 25.—Courtship "dance"; first position.
,, 26.—Courtship "dance"; second position.

Schizomus crassicaudatus.

,, 27.—Female with eggs in underground nest.

Phrynichus ceylonicus, s. str.

,, 28.—Closed hand from above.

Charinides bengalensis.

,, 29.—Closed hand from above.

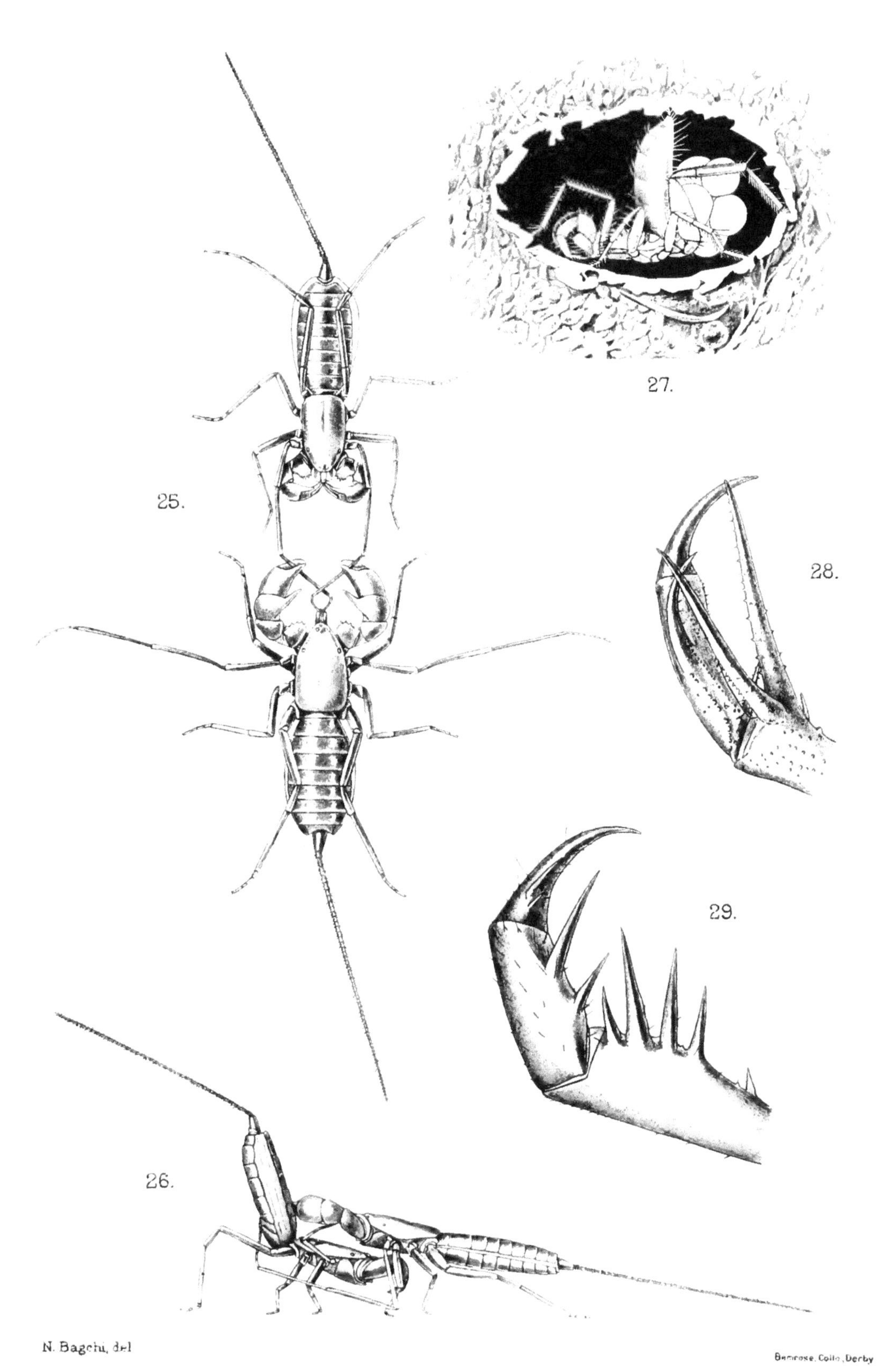

N. Bagchi, del. Bemrose, Collo, Derby

PEDIPALPI.

EXPLANATION OF PLATE XXV.

Stegodyphus sarasinorum.

FIG. 30.—General view of a nest, with small snares stretched between some of the outstanding leaves and the lower part of the stem of the plant on which it is constructed. (Much reduced).

,, 31.—Part of a snare, magnified to show its structure.

,, 32.—A double line of sticky silk, more highly magnified to show its component parts.

,, 33.—A double foundation line overlaid by sticky silk, still more highly magnified, showing the various thicknesses of silk used in making the snare.

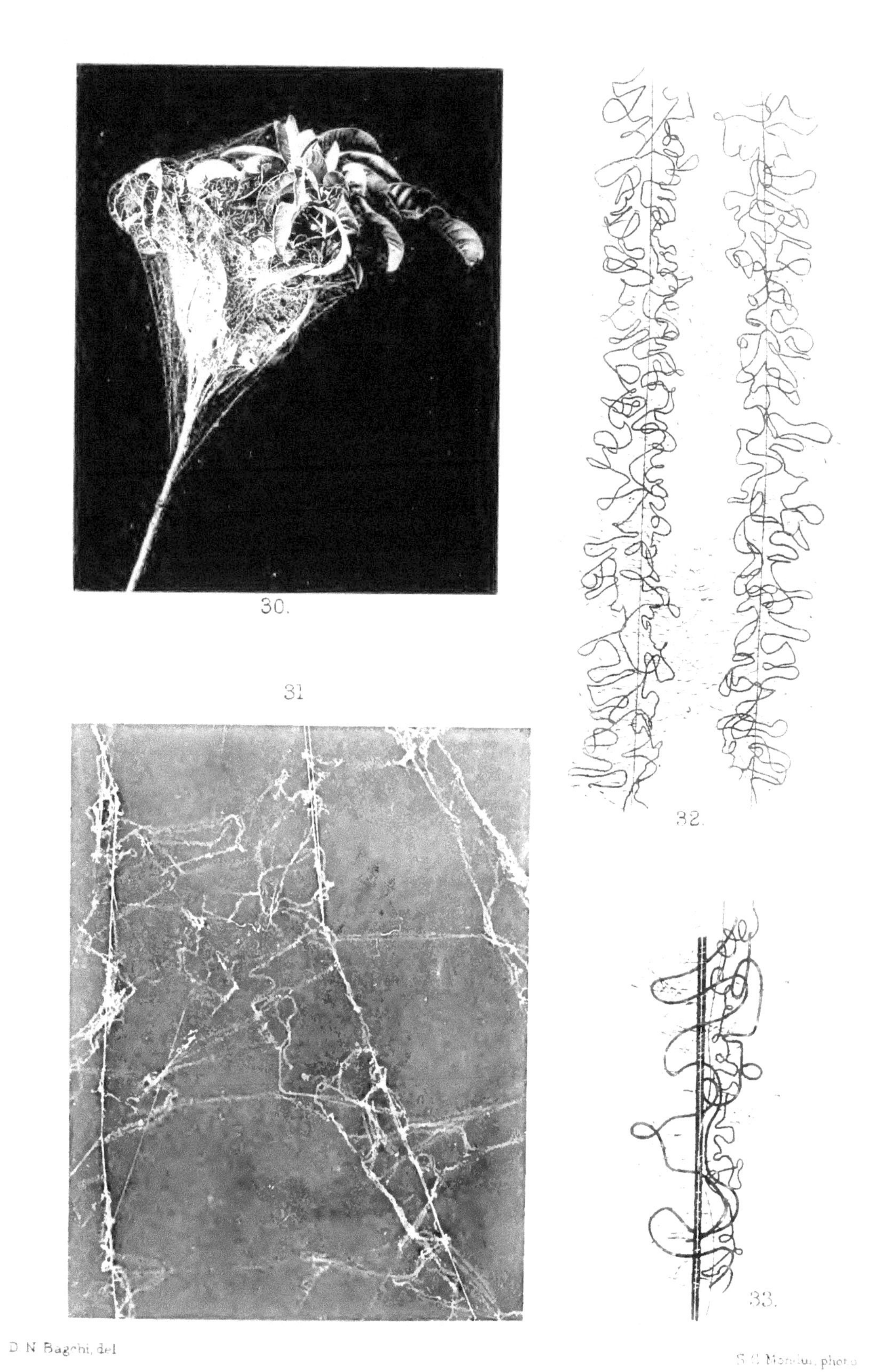

D. N. Bagchi, del. S. C. Mondul, photo.

STEGODYPHUS SARASINORUM.

XXX. THE HYDROIDS OF THE INDIAN MUSEUM.

II.—*ANNULELLA GEMMATA*, A NEW AND REMARKABLE BRACKISH-WATER HYDROID.

By JAMES RITCHIE, *M.A.*, *D.Sc.*, *The Royal Scottish Museum, Edinburgh.*

(Plates XXX, XXXa.)

CONTENTS.

The hydroid described in the following pages stands in several respects by itself. It combines in its structure and life-history peculiar features which are either new or have hitherto been found isolated in different species. Among these peculiarities are to be reckoned the occurrence in the vegetative hydroid phase of an alternation of free and fixed stages, the adoption of such singular methods of multiplication as the setting free of planula-like buds and of a mode of transverse fission associated with the basal-bulb, and the structural uniqueness of the tentacles and of the chitin-covered basal-bulb itself.

The examples of this minute but interesting species occurred in a valuable and extensive collection of Hydroids received from the Trustees of the Indian Museum for identification. It was first observed and collected by Dr. Nelson Annandale, during his investigation of the brackish water fauna of India; and to the fortunate fact that some living examples were kept for a short period in an aquarium is due Dr. Annandale's record of the free-swimming medusoid generation.

PROVENANCE AND HABITAT.

So far the species has been found in only one locality—Port Canning, Lower Bengal; and in that locality it seems to be confined to a shallow brackish pond. At any rate careful search of material from other places in the neighbourhood of Port Canning has failed to reveal any trace of its presence. In the brackish water it occurs growing upon a delicate branched weed, the surface of which also bears many clusters of Acinetaria. The specimens were collected in the months of December, 1909 by Dr. N. Annandale, and in March, 1910 by Mr. F. H. Gravely and Dr. B. L. Chaudhuri, and the latter are registered in the Natural History collections of the Indian Museum under the number ZEV 3702/7.

DESCRIPTION OF THE HYDROID.

HABIT.

The individuals are solitary, growing as a rule far apart from, and independently of each other. In very rare cases two individuals may appear to be attached at their bases; but this is due to imperfect separation of their basal masses, which are held together in a common growth of mucus. There is no coenosarcal connection between such individuals, nor has any semblance to colonial development been observed.

Consideration of the structures of this curious hydroid leads me to believe that the attached stage is merely a temporary phase in the life-history. This stage is, however, repeated again and again, each two periods of attachment being separated by an interval during which the hydroid is free. Whether during the free periods it floats in the water of the brackish ponds, or creeps upon the bottom, I do not venture to guess; but the analogy of *Hypolytus peregrinus* suggests that the Indian species may yet be captured in a tow-net, floating at the surface. In such case its minuteness would render difficult its detection in a miscellaneous plankton collection.

The following facts point to the alternation of free and fixed stages. A hydroid individual in its attached stage consists of a hydranth or polyp, with a long stalk-like extension of the body in older examples, and a unique basal development which I shall call the "basal bulb." The basal bulb, which alone is protected by perisarc, is the organ of fixation, actual adhesion being apparently due to a loose mass of debris-laden mucus which surrounds the bulb and spreads out upon the substratum. No part of the polyp, in its simplest condition, secretes perisarc or mucus. As will be found more fully described in a later section (p. 553) the bulb represents a method of vegetative reproduction, and is a temporary structure. Basal bulbs have been observed, both by Dr. Annandale and by myself, isolated and without any attached polyp. In such a case the polyp must either have disintegrated or have broken apart and become free. That the latter is the actual case is borne out by the condition of the isolated basal bulbs, which contain well-preserved coenosarc; and by the discovery of a polyp which has recently broken away from its base (pl. xxx, fig. 6).

Further, at the breaking-off period the released polyp possesses no means of attachment, although in course of time the lower end of the body secretes both perisarc and mucus, and gradually becomes modified into a new basal bulb. The details of these processes, so far as they have been traced, will be described in the paragraphs dealing with reproduction. The above more general observations, however, are sufficient to suggest that at certain phases the polyp is released from its old attachment, and that a period of freedom intervenes before a new organ of fixation has developed.

Structure of the Hydranth.

Form and Dimensions (see plate xxx, figs. 1-3).

An individual consists simply of an isolated polyp. There is no definite hydrocaulus, although the proximal end of the hydranth, especially in the more fully developed specimens, is extended into a stalk-like portion. Nor is there any stolon or hydrorhiza in the ordinary sense of the term, the functions of such being performed by the basal bulb.

In its living state, Dr. Annandale informs me, the hydroid is colourless.

The form of a normal adult resembles an Indian-club. The head of the club is ovate with a broad median zone on which the tentacles are placed. On both sides of the tentacle-zone the hydranth tapers gradually away: distally into a large conical hypostome on the truncated summit of which a shallow depression marks the position of the mouth; proximally into the long almost parallel-sided handle of the club. The total length of a well-grown individual varies from 0·63 mm. to 0·98 mm., the length of the "head" from 0·28 mm. to 0·52 mm., and the diameter of the tentacle-zone (the greatest diameter of the hydranth) from 0·16 mm. to 0·28 mm. In the youngest examples I have seen there was no proximal extension of the hydranth, and the "head" was sessile or rested upon the basal bulb (pl. xxx, fig. 1). In such cases the hydranths measured 0·15 mm. to 0·27 mm. in length, and their greatest diameter varied from 0·14 to 0·21 mm.

The following table showing the dimensions of various representative polyps gives at the same time an idea of the progressive development of an individual, the relative age or developmental stage being approximately indicated by the number of tentacles. All the polyps, with the exception of the first, were collected at the same place and time, and were killed under identical conditions. The measurements are in millimetres.

DIMENSIONS OF *ANNULELLA GEMMATA*.

Tentacles.		*Polyp.*		*Stalk-like prolongation.*		*Basal Bulb.*	
Number.	Maximum length.	Total length.	Maximum diameter.	Length.	Diameter.	Depth.	Horizontal diameter.
4	0·17	0·27	0·21	none		0·04	0·1
5	0·13	0·15	0·14	none			0·08 no perisarc
6	0·44	0·63	0·16	0·35	0·11	broken	
8	0·75	0·84	0·21	0·42	0·1	0·15	0·21
9	1·46	0·81	0·25	0·35	0·17	0·13	0·15
12	1·38	0·98	0·28	0·46	0·14	0·2	0·25

Tentacles.

The tentacles are confined to a somewhat prominent median zone on the hydranth. Over this they are irregularly scattered, at least three or four distinct levels being recognisable. Their number varies from 4 and 5 on the youngest individuals observed to 12 in the largest, but the average seems to centre about 6.

The appearance of the tentacles is characteristic and beautiful. They bear throughout their length, at fairly close and regular intervals, batteries of cnidoblasts aggregated in large projecting rings, or globular masses which resemble beads strung upon the axis of the tentacle (see pl. xxxa, fig. 7). These rings or globes have a diameter averaging three times that of the tentacle proper. Between the larger batteries there are occasionally smaller clumps of cnidoblasts in narrow rings or tiny circular groups. A globular battery terminates each tentacle, but since its size does not much exceed that of the cnidoblast rings the capitate condition is not always very evident, especially in contracted tentacles.

The detailed structure of the tentacles was examined in serial sections (see pl. xxxa, fig. 8). The typical cell-layers are represented by a solid endoderm, an exceedingly thin mesogloea, and an ectoderm of greatly varying thickness.

The *solid endoderm* is composed of many thin-walled cells, with sparse protoplasmic content which often simply lines the cell-wall and includes a small oval nucleus. The cells appear to be arranged, but somewhat irregularly, in four radial series of hexagonal cells, the bases of which rest upon the mesogloea, while the pyramids which form their apical regions interlock towards the centre of the tentacle. A longitudinal median section of a tentacle, therefore, generally exhibits a series of lateral walls of endoderm cells at right angles to the mesogloea, and in the centre a zigzag line representing the junctions of the pyramidal apices. Both in the character of its cells and in their arrangement the solid endoderm of this form differs very markedly from the solid endoderm of general occurrence in the tentacles of hydroids. Instead of thick-walled ("notochordal") cells arranged with great regularity in a single series lying along the long axis of the tentacle, as is the general rule, there are here delicate, thin-walled, multiserial cells.

The *mesogloea* of the tentacle-cells calls for no remark except that it is of extreme tenuity scarcely exceeding 1μ in thickness throughout the whole length of the tentacle.

The *ectoderm* of the tentacles falls into two distinct zones, the ring-like or globe-like swellings, which I shall designate *nodes*, and the spaces between them (see pl. xxxa, figs. 7 and 8). In the inter-nodes or inter-annular zones the ectoderm, even when the tentacle, in contraction, is at its stoutest, consists of a very thin layer of much flattened epithelial cells. In an extended tentacle this layer owing to its tenuity becomes scarcely visible. A rare cnidoblast, similar to the lesser variety in the nodes, forms the only inclusion in the internodal cells.

The nodes are composed of a zone of large oval cushion-shaped cells, closely appressed to each other laterally in a single row. Occasionally, however, incomplete zones or isolated individuals of these cushion-shaped cells occur in the inter-nodal areas. At the junction of nodes and inter-nodes the internodal ectoderm conforms to the outline of the nodal cells, being banked up against their curved walls with a gentle slope. The size of the nodal cells varies with the contraction of the tentacle, but the short diameter (parallel to the long axis of the tentacle) usually lies between 15μ and 27μ, while the height varies from 12μ to 22μ. Apart from inclusions the nodal cells contain little cytoplasm, the greater part of their interior being occupied by a large vacuolar space. Upon the base of the cell, however, there lies a thin layer of cytoplasm, and a median nucleus, 5μ by 3μ in diameter, containing a small nucleolus and surrounded by a sparse coat of cytoplasm whence delicate strands radiate outwards. The whole structure of the cell appears to be organised in relation to its function as a battery cell. I shall, therefore, discuss here the arrangement and structure of its cnidoblasts.

Cnidoblasts of the nodal cells.

In surface view of a tentacle-node a regular arrangement of cnidoblasts is apparent (see pl. xxxa, fig. 7). Round an individual of large size circles a group of smaller cnidoblasts. The latter are set singly and more or less regularly on an imaginary circumference,

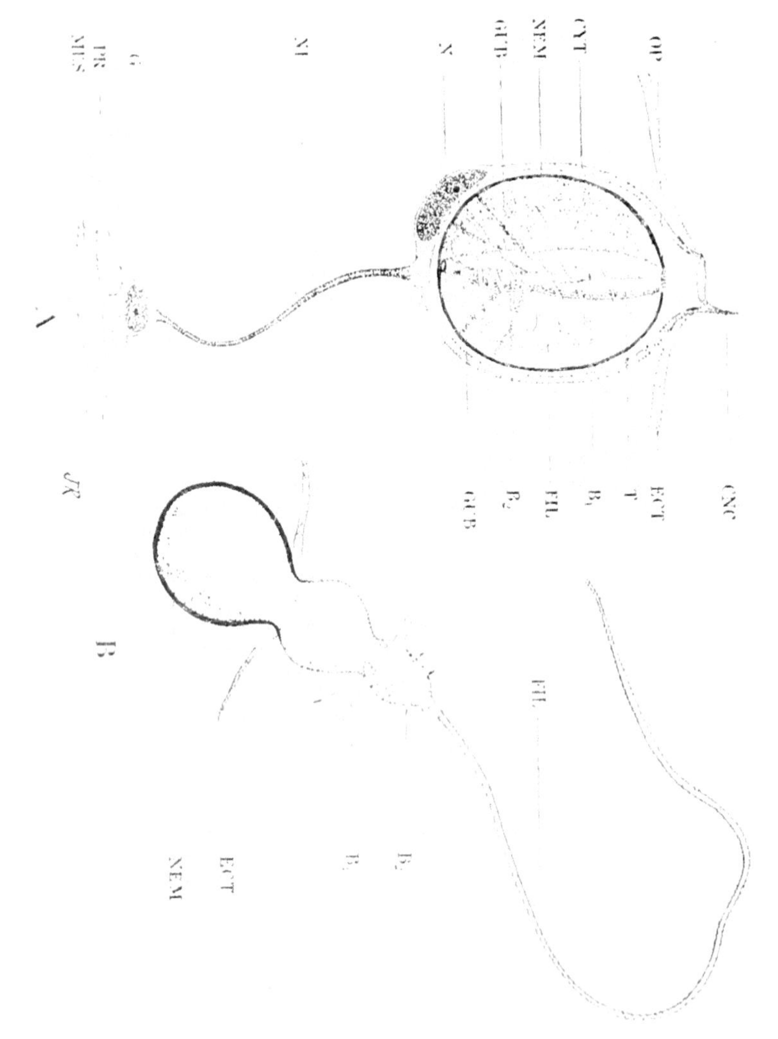

TEXT-FIG. 1.

Large type of cnidoblast (*macrocnide*) from nodal cell.

A. Undischarged cnidoblast and its connections *in situ* (× approximately 4000 diameters).

B. Discharged nematocyst (× approximately 3000 diameters).

B_1, B_2, major and minor barbs; CNC, cnidocil; CYT, cytoplasm; ECT, external wall of ectodermal cell; FIL, filament; G, ganglionic mass, GUB, gubernaculum,—protoplasmic strand supporting central tube; MES, mesogloea of tentacle; N, nucleus; NEM, nematocyst; NF, nerve fibril; OP, operculum; PR, layer of protoplasm lining bottom of cell; T, central tube.

with generally seven but sometimes as many as eleven examples. In a single node about 8 to 10 large cnidoblasts are present with their attendant satellites—a total of approximately 80 individuals. As a rule each cell contains only one complete group, but occasionally the cell-wall appears to become obliterated so that several groups come to lie within a cell's boundaries. In median longitudinal section of a tentacle (pl. xxxa, fig. 8) the nematocysts are seen to lie at the periphery of the cell, radially inclined outwards from the mid-point of its base. There are considerable differences in the structures of the two types of cnidoblasts.

The larger individuals (*macrocnides*) (text-fig. 1) consist of an almost spherical nematocyst, 7μ to 8μ long by 6μ in diameter, surrounded by a thin and uniform covering of cytoplasm which pushes up the cell-wall, and is produced into a short delicate cnidocil. At some point in the proximal portion of this cytoplasm lies an elongated nucleus the inner profile of which conforms to the outline of the nematocyst. Contrary to the experience of Schneider (1890, p. 332) as regards *Hydra fusca*, the nucleus contains a distinct nucleolus. At the proximal end of the cnidoblast a delicate thread, the nerve fibre, leaves the cytoplasm, and passing across the vacuolar space of the cell merges with the cytoplasm surrounding the nucleus that lies on the floor of the cell. The nucleus and its surrounding cytoplasm seem to constitute a ganglionic mass to which radiate the nerve filaments of many or all the cnidoblasts in a group. The ganglion mass is in its turn connected with the layer of cytoplasm which covers the floor of the cell. None of the macrocnides possessed a simple broad cytoplasmic peduncle such as forms an attaching structure in examples from several other species of Hydrozoa.

The interior of the nematocyst is filled with a highly refractive fluid, which renders accurate observation of the internal structures a matter of some difficulty. At the distal pole of the nematocyst is a circular area—the operculum—of consistency different from that of the nematocyst wall. From this area, whence the filament escapes on its discharge, a cylindrical tube of considerable diameter projects into the cyst, passing along its longitudinal axis almost to the proximal wall. In its upper half the tube contains a prominent opaque triangle pointing upwards and almost reaching the distal wall of the cyst, and this represents the single whorl of three major barbs which encircles the lower portion of the ejected filament. Similar smaller and less well-defined structures are sometimes apparent in the lower half of the tube. The lower section of the tube is kept in position by a series of exceedingly delicate gubernacula—protoplasmic strands which attach it to the wall of the nematocyst, and which are to be observed only under specially favourable conditions of staining and lighting. The proximal portion of the tube narrows rapidly and at its base is continuous with the filament which lies near to the wall of the cyst in an ascending spiral of some six loose coils.

The everted portion of a discharged nematocyst is some three to three and a half times the length of the nematocyst, and consists of a smooth-walled basal bulb, a second and smaller bulb furnished with about four whorls of barbs of which the proximal whorl contains three large individuals, while those of the distal whorls are more numerous and insignificant. From the second bulb proceeds the filament which throughout its length is armed by a close spiral of exceedingly minute barbules ascending in a contra-clockwise direction (see text-fig. 1B).

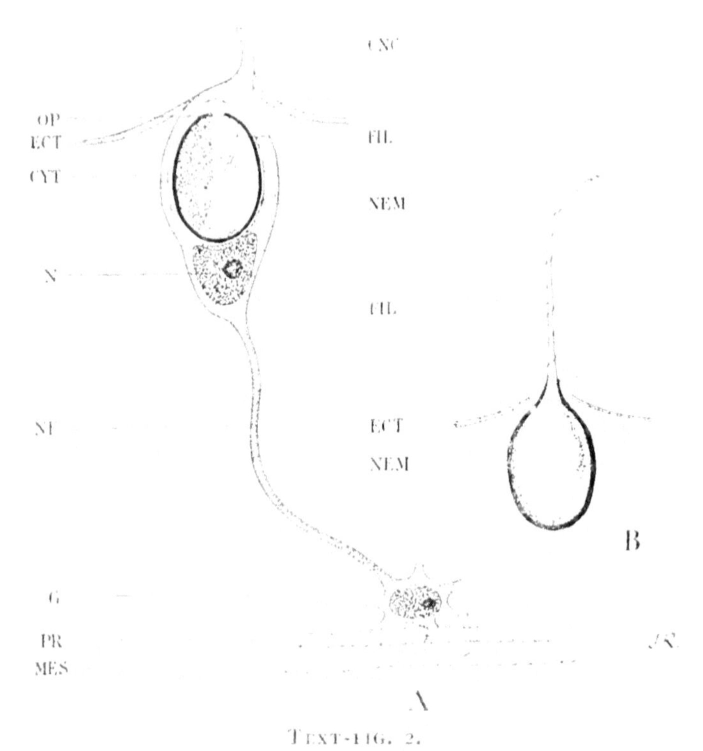

TEXT-FIG. 2.

Lesser type of cnidoblast (*microcnide*) from nodal cell.

A. Undischarged cnidoblast and its connections *in situ* (× approximately 4000 diameters).

B. Discharged nematocyst (similarly magnified).

Lettering as in text-fig. 1.

The second and smaller type of cnidoblast—*microcnide*—is of simpler structure. The nematocyst is similar in shape but is approximately half the linear dimensions of that of the large type, 4μ or 5μ by 3μ. The cnidocil is stouter and longer than in the macrocnides, and although the general arrangement of cytoplasm is the same the cytoplasmic coat is drawn out into an elongate oval shape to include a very large nucleus which lies against the proximal wall of the nematocyst. The nucleus varies in shape, but its inner surface is always closely moulded upon the nematocyst wall, and it frequently assumes the deep helmet-shape shown in section in text-fig. 2A. It contains a large nucleolus. The cytoplasm is connected by a long delicate nerve fibre with the ganglionic mass on the floor of the cell. Occasionally, however, the connection becomes a comparatively broad protoplasmic strand resembling the peduncular attachment of some cnidoblasts.

The internal structure of a microcnide differs much from that of a macrocnide The former contains only the filament which proceeds directly from the operculum at the distal pole of the cyst in a loose descending spiral of about three small coils. These appear to encircle a central pillar of delicate consistency which may, however, be simply one of those phenomena of refraction which render so difficult the exact observation of the contents of nematocysts. In a discharged microcnide (text-fig. 2B) three points strike one as characteristic : the shortness of the simple filament, the length of which is only twice that of the nematocyst; the directness with which the filament projects from the nematocyst, for it invariably lies in line with the long axis of the nematocyst and is straight, except for a very regular curve towards the tip; and, lastly, the openness of the spiral of minute barbules, which performs only about ten revolutions in its contra-clockwise ascent.

Hydranth Body.

The *ectoderm* of the hydranth-body consists of a layer of irregular epithelial cells, between which lie small interstitial cells. The bases of the epithelial cells are produced into longitudinal muscle fibres which rest upon the mesogloea. The ectoderm averages in thickness some 7·5μ, but especially in the lower prolongation of the hydranth is arranged in slight horizontal ridges. The cells contain large rounded nuclei, and in parts a large number of cnidoblasts, but only some of the latter possess cnidocils and are functional, the remainder being under process of formation. The cuticle, if present, is of extreme delicacy, and no perisarc is secreted. At certain stages of development, however,—when a new basal bulb is being formed (see p. 555)—a number of hydranth cells take part, along with the cells of the basal bulb, in the secretion of a thick coating of hyaline mucus. In portions of this secretion masses of diatoms and other debris become entangled, and it is interesting to note that in this condition the diatoms appear to have continued a symbiotic existence, for the greater number show evidence, in their well-preserved protoplasm, of recent active metabolism. A similar state of symbiosis has been noted by Schaudinn in the case of the diatoms and algae which surround the body of *Haleremita* (Schaudinn, 1894, p. 226).

In the neighbourhood of the tentacle-zone and around the margin of the mouth the ectoderm is thickened. In the latter region it contains a close array of functional cnidoblasts, similar to the macrocnides of the tentacles. In the tentacle-zone, however, the majority of the cnidoblasts—macro- and micro-cnides—lie some distance below the surface and are in process of formation.

The great numbers of developing nettling organs in this region clearly indicate it as a localised manufacturing area whence migration of fully developed nematocysts to the tentacles takes place, conditions apparently of general occurrence in the Hydromedusae (see Hadzi, 1911).

The *mesogloea* is colourless and of almost uniform thinness of 2μ.

Endoderm.—The endoderm cells present more variety in their shape and in their inclusions than the ectoderm. As in many other hydroids they fall into three indefinitely bounded regions, in all of which, however, a few longitudinal ridges of elongated cells project into the coelenteron.

The hypostome endoderm consists of a series of regular, elongated, narrow, palisade cells resting upon the mesogloea and lacking inclusions. Between the distal ends of these cells are inserted many clavate gland cells, with a large nucleus resting in the wedged-in narrowing portion of the club, and a content of finely granular cytoplasm.

In the region of the tentacle-zone the endoderm is considerably deeper. The palisade layer of the hypostome is replaced by several irregular rows of small highly-vacuolar cells. Upon these rest large clavate nutritive cells, containing oval nuclei and coarsely-granular secretory products, as well as foreign bodies the recognisable portions of which consist mainly of the frustules of diatoms. Throughout the endoderm in this region there are scattered dark oval cells containing excretory products.

Lastly in the prolonged basal portion of the hydranth the cell-varieties of the former regions disappear, and the endoderm consists of a network of regular, highly vacuolar cells in which the cytoplasm and nuclei are ranged along the cell-walls. Here the cells are almost devoid of inclusions, only a rare individual with excretory products being observable; and although a narrow central lumen penetrates the region it is clear that the cells lining it take little part in the secretory or digestive functions.

The structure of the basal bulb will be discussed in the section dealing with reproduction (p. 553).

METHODS OF REPRODUCTION.

I. SEXUAL REPRODUCTION.

There is no conclusive evidence of the occurrence of a sexual type of reproduction in the specimens which I have examined, although in one there is present, arising from the tentacle-zone, a very small globular bud (0·045 mm. in diameter) composed of ectodermal and endodermal elements, which might possibly have developed into a sporosac or medusoid gonophore. Its position within the tentacle-zone agrees with the position rather of the sexual bodies than of the simple buds of most other hydroid species.

But Dr. Nelson Annandale, who kept examples of the hydroid alive in an aquarium for some time wrote in a note accompanying some of the specimens: "The gonosomes, which develop into free medusae, are borne in a circle round the hydranth, below the tentacle"; and again, in reply to a request for examples of the medusa or for further information "I am sorry that I have not any specimens of the medusae of the little Hydroid from Port Canning. The only one I have seen escaped in my aquarium. It was so small that I could only just see it with a very powerful hand lens."

In view of the minuteness of the structures concerned it is possible, but unlikely, that a naturalist even of Dr. Annandale's acumen and experience, might have mistaken one of the planula-like buds to be afterwards described, for a medusa.

In any case the elucidation of the sexual phase must be left to new collections of material gathered possibly at a different season of the year.

II. ASEXUAL REPRODUCTION.

Lateral Budding.

A few hydranths possess lateral buds in various stages of development. The buds arise from the region below the tentacle-zone, and between it and the gentle narrowing which indicates the beginning of the stem-like basal prolongation. But they are not common on my specimens which were collected in the month of March, few hydranths possessing even a single bud, and two being the greatest number on any one hydranth.

The buds are of the simplest structure (see plate xxx, fig. 4). They arise as small hollow projections of ectoderm and endoderm, which increase in length much more rapidly than in breadth. So there is formed an elongated hollow sac with thin walls of single-layered ectoderm and endoderm. The base of the sac becomes much constricted at its point of junction with the hydranth, but the internal cavity retains connection with the coelenteron of the hydranth by a narrow passage. In due course the connecting neck of the bud becomes ruptured, and the bud, which is now vermiform and closely resembles the planula of many hydroids (except that it lacks cilia), breaks away and commences a free life.

A released and therefore mature bud contains no traces of sex cells, and it must be assumed that it gives rise directly to a new hydranth. The only free example which I have observed seems recently to have broken loose from the hydranth (plate xxx, fig. 5). It is almost cylindrical in shape, 0·30 mm. in length by 0·085 mm. in maximum and 0·065 mm. in minimum breadth, slightly narrower in its median region and widening gently to its rounded extremities. The resemblance in shape to the planula of *Cordylophora lacustris*, as figured by Allman (1871, pl. iii, fig. 5*a*) is very marked. The proximal extremity still retains an opening representing the lumen which connected the bud cavity with that of the hydranth, and as a slight ridge well furnished with nematocysts surrounds the opening, it is possible that here is foreshadowed the adult mouth.

Serial sections reveal the fact that at the place of origin of a bud considerable activity is shown by the endoderm of the hydranth, which is crowded with finely granular protoplasm and engulfed food particles. These features are carried into the endoderm of the developing bud, the cells of the proximal portions of which contain much secretory and food material. The mesogloea of the bud is somewhat less developed than that of the hydranth, and the ectoderm is remarkable for the regularity and high, narrow palisade-like structure of the cells at both extremities. These contain spherical nuclei similar in size to those of the endoderm (3μ in diameter), and many nematocysts of both types undergoing development. As the bud increases in size some of these approach the surface of the ectoderm and lie in position for functioning, although in none of the buds, attached or free, are cnidocils present. An extremely thin cuticle is excreted by the ectoderm. I have seen no indication of the presence of an external coat of cilia.

The neck joining bud to hydranth is formed of ectoderm, mesogloea and endoderm, and in spite of the narrowness due to increasing constriction there is no sign that rupture is preceded by the disappearance of the endoderm, as in the case of the hydroid of *Gonionemus murbachii* (see p. 561), or of the ectoderm, as in the cases of the basal bulb described below, or of the sporosac buds of species of *Dicoryne.*[1]

Longitudinal fission.

In a single specimen longitudinal fission appears to be in progress (see pl. xxx, fig. 3). From the neighbourhood of the tentacle-zone of a well-developed hydranth with in all eleven tentacles a secondary hypostome branches out as if due to the division of the original hypostome. Both hypostomes are normal in character and the mouth of each is connected in the usual way with the common coelenteron of the hydranth. The endoderm layer between one hypostome and the other is of uniform thinness and regularity, and shows no prolongations or other abnormalities in the neighbourhood of the fission angle. The smaller hypostome has appropriated some of the tentacles of the original hydranth, and new smaller ones are arising at its base. Whether this process is to be reckoned as a normal mode of reproduction or whether the instance described is rather an abnormal case of budding or duplication than an example of true fission, I have no means of deciding.

The Basal Bulb.

Reference has already been made to the significance of a structure which I have termed the basal bulb. To judge by the

[1] *See* Ashworth and Ritchie, 1915.

frequency of its occurrence this structure is of first importance in the propagation of the species, for every hydranth examined (except one) bore one and often two bulbs at varying stages of development. The solitary exception was a young individual with 5 tentacles, in which the proximal extremity ended in a sucker-like disc without perisarc, the equivalent of the basal disc or "Fussplatte" of *Hydra* and other forms. The universal presence of at least one basal bulb or its antecedent on these specimens can be readily understood by the fact that all the hydranths examined were growing upon a seaweed; and that as the basal bulb is the only means of attachment its presence was postulated by the stage of growth of the hydranths discovered. The only hydranths likely to be found lacking the basal bulb are individuals belonging to the unattached, probably planktonic, stage.

Position and General Structure of the Basal Bulb (pl. xxx, figs. 1-3). The basal bulb is situated at the lower free end of the hydranth in the position generally occupied by the hydrorhiza Resemblance to a hydrorhiza is further to be found in the fact that it seems to be the habit of the basal bulb to lie with its long axis parallel to the substratum and at right angles to the hydranth. Basal bulbs have been found in the youngest hydranths as well as in the oldest individuals examined; in the former, the hydranth body springs directly from the bulb, in the latter the bulb terminates the stalk-like proximal prolongation of the hydranth.

At all stages the character of the bulb is obscured by masses of organic debris which adhere to it in a dense coating and spread from it for a short distance upon the hydranth. Within this debris, except in the very earliest stages, lies a more or less globular shell of chitin, thin, delicate, and colourless at first, but later becoming strong, immobile and tinted. During its impressionable stages the chitin of the bulb may be moulded upon the particular substance whereon it lies, and this produces considerable modification in the typical rounded form. The chitinous shell contains and protects a simple cellular sac, which in its more mature stages lies loosely within. This sac, the essential portion of the basal bulb, consists of a single layer of ectoderm and of endoderm. In its advanced stages it is connected with the proximal end of the hydranth by a narrow protoplasmic neck which passes through a small circular opening in the chitinous shell—the only aperture connecting the interior of the shell with the exterior.

Detailed Structure of Mature Basal Bulb (pl. xxxa, fig. 9). Well developed basal bulbs were examined in serial sections. In these specimens the cellular sac did not lie in contact with the chitinous investment; but since the chitin showed many regular growth-lines and could only have been secreted by the sac, the hiatus may be artificial, due to shrinkage in preservation.

No special features mark the single layers of ectoderm and endoderm which form the walls of the sac: the latter contains small round nuclei and here and there groups of excretory products; the former contains many large nematocysts in process of formation, especially in the upper region near the aperture in the chitinous investment. It is curious that nematocysts should develop in an enclosed sac the ectoderm of which has no contact whatever with the exterior, but a similar condition has been observed in the hydrorhizal portion of *Myriothela cocksii* (Hardy, 1891, p. 512), and of *Corymorpha* (Torrey, 1907, p. 279), and I have noticed it in the case of some gonophores. In *Annulella* the history of the layers of the sac (see p. 555) offers a simple explanation. The walls of the sac are thin and leave a moderate space for a central cavity which is in direct communication with the coelenteron of the hydranth.

It is a remarkable fact that in the mature bulb there is no direct connection between the ectoderm of the hydranth and that of the basal bulb. At the constriction or neck uniting the two, the ectodermal layer disappears and the chitinous investment abuts against the mesogloea. This may be a preliminary to the severing of the neck at the time when the hydranth escapes from its hold-fast; as such at any rate it would fall into line with the well-defined process which precedes the release of the free-swimming sporosacs of species of *Dicoryne* (*see* Ashworth and Ritchie, 1915).

An unusual feature distinguishes the mesogloea of the basal bulb. It is continuous with the mesogloea of the hydranth, but just beneath the neck and within the aperture of the chitinous shell, it forms a very much thickened rim deeper than either endoderm or ectoderm. From the proximal margin of this ring the mesogloea suddenly tapers away, and throughout the remainder of the bulb forms a layer of extreme tenuity.

The chitinous investment of the basal bulb is of rudely spherical form, sometimes greatly modified by its contact with the solid substratum. The chitin is of very different densities, but the upper portions are always the more solid and deeply tinted. Round the small but very definite aperture through which the neck of the basal bulb passes there is a thickened ring slightly incurved. While the perisarc is well defined in the distal portions and there exhibit definite growth lines, in the central area of the floor of the bulb it gradually loses its compactness and merges into an amorphous gelatinous mass of much greater thickness (pl. xxxa, fig. 9*a*). In this mass are included, along with other debris, large quantities of diatom skeletons in some of which the protoplasm is so well preserved as to indicate that the algae continued to live after their inclusion. This unconsolidated basal area may add to the efficiency of the basal bulb as a hold-fast, or may provide for the expansion of the perisarc-shell during the growth of the sac within.

The secretion of perisarc is confined to the basal bulb, at the neck of which the chitinous covering ends abruptly. Yet masses of gelatinous material containing much debris not only surround the bulb but are continued for a short distance on the lower exposed portion of the hydranth.

Origin and Development of the Basal Bulb (see pl. xxx, fig. 6). The basal bulb is a modified portion of the hydranth body. This is clearly shown by a hydranth which has recently broken away from a former basal bulb, and is in process of forming a new one. The history of this specimen (fig. 6) may be taken as indicating the general development of a basal bulb, and appears to have been as follows.

The hydranth tapers away at its basal end almost to a point, and here the tissues are ruptured. This narrow portion is the neck of a former basal bulb, and the damaged tissues show where the narrow neck, already prepared by increasing constriction and by the disappearance of the ectoderm layer, has broken asunder, allowing the hydranth to escape from its former anchorage. The final rupture of the neck is no doubt due to mechanical strain brought about by the swaying of the hydranth in the water currents.

So far as one can judge the free stage of the hydranth must be of very limited duration, for even before the traces of rupture at the neck of the old basal bulb have disappeared, a new basal bulb is in process of formation.

Four modifications mark the development of a basal bulb. Its origin is first indicated by a slight constriction in the lower portion of the body of the hydranth. This constriction affects all the cell-layers: the endoderm and mesogloea are simply indented, but even at the early stage figured, there is already a disruption in the ectoderm, which, although not yet severed, is reduced to very thin dimensions at the level of the future neck. A second characteristic regards the differentiation of the ectoderm of the basal bulb. Distal to the constriction, that is on the unaltered hydranth, the ectoderm is of the normal ridged type with rather elongated cells, but proximal to the constriction the cells are smaller, more regular and flattened. In the third place, copious masses of mucus in which debris becomes entangled begin at once to be secreted by the ectoderm of the basal bulb; and lastly the formation of mucus is succeeded by the secretion of a chitinous investment, the perisarc, which at the stage figured had only begun to form in the lower regions. The folding over of the bulb until its long axis lies at right angles to that of the hydranth must be a subsequent development.

While the above mode of development of the basal bulb happens to have come to my notice and is, on account of its uniqueness, described in some detail, it probably represents only one of several methods by which a basal bulb may arise. It can hardly be doubted that the original basal bulb of a hydranth develops directly from the basal-disc or "Fussplatte" of the larva, and development from a lateral bud seems to be hinted at by the following facts.

In many cases there are two basal bulbs at the base of a single hydranth (see pl. xxx, figs. 2 and 3) and in such case they arise not in linear succession but one terminally and one laterally.

The latter may have originated as a bud. Of the two the terminal bulb possesses thicker perisarc and more contracted coenosarc and appears, though not the larger, to be the older individual.

Significance of the Basal Bulb. There is no direct evidence as to the reproductive function of the basal bulb: no young hydranths have been observed springing from the coenosarc of an old bulb, unless it be that where two basal bulbs occur on one hydranth, one represents the original bulb from which the hydranth grew while the other is a development of the hydranth itself.

But the evidence of the structures and development of the bulb seem to point clearly to reproductive function. Thus the disappearance of the ectoderm at the junction of bulb and hydranth seems to be analogous with the similar retrogression in the sporosacs of *Dicoryne* and to indicate a regular preparation for the breaking away of the hydranth. An example of a recently released hydranth has been observed. Again basal bulbs are frequently found alone, and in these the coenosarc is in good preservation. Here we seem to have a parallel to the conseiving power of the stolon as exhibited in the hydrorhiza of *Dicoryne conybearei* (Allman) in which, by the development of partitions of chitin within the lumen of the stolon the coenosarc is preserved unharmed in various sections during unfavourable conditions (see Ashworth and Ritchie, 1915). In *D. conybearei* the conservation of the coenosarc in this way is succeeded so soon as favourable conditions return, by a new development of hydranths produced by the coenosarc; and it seems highly probable that a similar recrudescence of hydranth life arises from the coenosarc of the basal bulb.

THE RELATIONSHPS OF *ANNULELLA GEMMATA*.

I have already drawn attention to the curious combination in *Annulella gemmata* of peculiar characters some of which have been found rarely, and generally one at a time, in other species. The most accurate conception of the significance of these resemblances will be attained by a short comparison of each with its analogues.

UNATTACHED HYDROIDS.

Several genera of Hydroids share with the Pennatulid and a few other types of Alcyonarians, the character of gaining a more or less insecure anchorage by simply embedding their proximal end in the mud of the sea-floor. They are generally characterized by solitary habit and by the weak development or absence of perisarc. Amongst such are to be reckoned the Corynids—*Myriothela* and *Blastothela*; the Pennarid, *Heterostephanus*; and the Tubularids—*Corymorpha, Lampra, Gymnogonos, Monocaulus* and *Branchiocerianthus.* It is probable that with these should also be grouped the lake forms—*Moerisia* and *Caspionema.* Many of these gain firmer anchorage by the development of "rootlets," but the majority or all of them have the power of slight movement, and it is possible that they may be able even to withdraw from the mud and creep along the bottom. In any case, as Hartlaub has pointed out[1] these forms, both in their systematic affinities and the in their habit, present a well-defined half-way house between permanently fixed species and those which are able to leave their attachment and move freely on the substratum or in the sea.

Amongst such temporarily creeping or floating forms we have the freshwater *Hydra*, and its relatives *Protohydra* and *Polypodium*; the Tubularid, *Hypolytus*; and *Halcremita* of uncertain relationship, but closely resembling the larval stage of *Gonionemus*. Here also I am inclined to include the Pennarid, *Trichorhiza*, which, found by Russell (1906) on the tentacles of *Corymorpha nutans*, was apparently caught in the act of moving. General but not universally present characters which link these forms (with the exception of *Trichorhiza*) are the almost total absence of perisarc and the presence of a basal thickening of coenosarc—the pedal disc. I have not included definitely recognized larval forms, but perhaps the floating stage of *Acaulis* ought to be mentioned here, since floating individuals bear well developed medusae buds and may be considered adult.

In a slightly more advanced category of unattached hydroids are to be placed the pelagic forms *Margelopsis* or *Nemopsis*, which represent the detached buds of such forms as *Tiarella* (see Bedot, 1911, p. 211); the unique *Microhydra* and the metagenic form of the Trachymedusan, *Liriope* which "is a true hydra, although its free-swimming mode of life and its superficial aspect caused it to be mistaken formerly for a gonosome" (Perkins, 1903, p. 752).

In none of these groups of unattached Hydroids is to be found an exact parallel to our Indian brackish-water species, the adults of which are at one stage firmly attached, and at another are released from their attached portions in order to lead a temporary free (? pelagic) existence. But, as we shall see in discussing the basal bulb, that structure links *Annulella* with the creeping type, especially common in the family Hydridae.

TENTACLES AND CNIDOBLASTS.

The arrangement of the cnidoblast batteries of the tentacles in well-defined projecting rings is characteristic of very few hydroid stages. It is, however, moderately common in the medusoid generations, being exhibited in such well-known forms as *Thamnostylus dinema*, or in the medusoids of *Corymorpha nutans*, *Stauridium* and *Syncoryne eximia*.[2] In the hydroid stage, so far as I know, it is confined to *Trichorhiza brunnei*, Russell 1906, *Heterostephanus annulicornis* (M. Sars 1859), *Hypolytus peregrinus*, Murbach 1899, *Asyncoryne ryniensis*, Warren 1908, and occurs to a limited extent

[1] Hartlaub, 1902, p. 29.

[2] Compare particularly the representations of the last species as drawn by Allman, 1871, pl. v, figs. 3-4.

in *Tiarella singularis*, Schulze 1876, with its three distal rings of cnidoblast batteries and in *Margelopsis stylostoma*, Hartlaub 1903, which has been shown by Bedot (1911, p. 211) to be the free bud of the preceding species.

It is a striking fact that annulated tentacles should be common in the free-swimming medusoid generation, and should occur in the hydroid generation only in a few species, which, with the exception of *Asyncoryne*, are outstanding on account of their free or partially free habit. The connection of habit and structure appears to be no coincidence, and, on the evidence before me, I would suggest that the arrangement of large cushion-shaped cells in prominent rings is an adaptation to a creeping or free-swimming life. Not only would the greatly enlarged surface area, due to the rings, add to the resistance offered by the organism to the surrounding water, and so check the rate of sinking, even were the organism immobile, but the very large vacuolar spaces, which the nodal cells of *Annulella gemmata* contain (and which in absence of direct evidence I assume to occur in the similar cells of other species), may act directly as buoying agencies. It is possible also that these vacuoles in the tentacular rings may by their enlargement and contraction supply in some degree the means of the daily vertical migrations so characteristic of most hydroid medusae.

In none of the cases mentioned above has the detailed structure of the cnidoblast rings been investigated. But the cushion-shaped cells of which the rings or nodes are made up in *Annulella* closely resemble in general appearance and detailed structure the isolated battery-cells which stud the tentacles of *Hydra*. A comparison of the description and figures of these batteries in *Hydra fusca*, as given by Schneider (1890, p. 332, Tab. xvii, fig. 20), with the description and figures of *Annulella* in this paper throws particular emphasis on this resemblance.

The resemblance to *Hydra* is further emphasized by comparison of the structures of the cnidoblasts themselves (see Schneider, 1890, p. 332 and pl. xvii). In *Annulella* I have recognized only two types of nematocyst, but both occur in almost identical form in *Hydra*. It is true that there are differences in detail; that Schneider describes no connection between the basal prolongation of the cytoplasm of the cnidoblast and a "ganglion mass," that he mentions neither the gubernacula within the macrocnides nor the spiral arrangement of barbules on the ejected filament. But these are negative evidences and in the examination of structures so notoriously difficult as cnidoblasts negative evidences are of even less moment than usual.

The conjunction of capitate and scattered tentacles suggests relationship with the family Corynidae, but the capitation is very slight and might be regarded as a terminal development of the tentacle nodes. Some cases of scattered tentacles (without capitation) occur in the family Hydridae.

The solid multiserial endoderm of the tentacles appears to be paralleled in only one other genus, *Tubularia*. Solid endoderm is, indeed, characteristic of almost all Hydroids, but it consists of a single row of central thick-walled cells. Chun (1897, p. 316) says regarding the occurrence of such uniserial solid endoderm "Was zunächst ihr Vorkommen unter den Hydroiden anbelangt, so fehlen sie lediglich der durch hohle Tentakel ausgezeichneten Gattung *Hydra*. Alle übrigen Hydroidpolypen besitzen solide Tentakel, welche von einer einzigen Reihe derbwandiger centraler zellen gestützt werden." With the exceptional case of *Hydra* must be included that of the since described *Moerisia lyonsi*, Boulenger (1908), and possibly that also of *Caspionema pallasi*, Derzhavin (1912), regarding the endoderm of which the author makes no remark. In *Hydra* and *Moerisia* the endoderm consists of several longitudinal rows of thin-walled cells, penetrated by a fine central lumen. But in species of *Tubularia* which I have examined in detail, the lower or aboral whorl of tentacles contains a solid endoderm composed of many small thin-walled cells. These are not arranged in series but fill in irregularly the centre of the tentacles (see also Warren's account of *Tubularia betheris*, 1908, p. 282). The oral tentacles of *Tubularia* contain the ordinary type of uniserial endoderm.

The solid delicate-walled multiserial endoderm of *Annulella* bears no resemblance to the solid uniserial endoderm of the majority of Hydroids, but closely resembles in structure and arrangement (except that there is no central cavity) the multiserial endoderm in the tentacles of *Hydra* and *Moerisia*, and resembles in a general way the solid endoderm of the aboral tentacles of species of *Tubularia*.

BUDDING.

The phenomena of budding in the Hydroid Zoophytes may be divided into three types: (1) where the bud develops on the parent into a miniature adult and remains attached, thus giving rise to colonial formation; (2) where the bud develops on the parent into a miniature adult which is then set free; (3) where the bud is set free at a simple planula-like stage and develops into a miniature adult away from the parent.

(1) The colonial type of budding is exhibited by the majority of cnidoblastic and gymnoblastic hydroids. (2) The setting free of a miniature adult is much less common but is familiar through the example of *Hydra*, and occurs in a few forms such as *Moerisia* (Boulenger 1908, p. 363) and *Tiarella* (Schulze 1876, p. 411). (3) The escape of a planula-like bud is an exceedingly rare mode of propagation, and since it is the type exhibited by *Annulella* calls for some remark. In its ultimate results it closely resembles the phenomena of those peculiar propagating branches of many Hydroids, the separation of which—"Scissiparité"—has been most recently and ably investigated by Dr. A. Billard (1904). "Scissiparité," however, connotes the adaptation of an old structure, stolon or branch, to a new purpose, and can be reduced to a simple form of transverse fission in a portion of the hydroid

already existing. On the other hand, buds which ultimately become free seem to have evolved to this end alone: they are new structures the one purpose of which is the multiplication and distribution of forms like the parent. They are probably the most primitive of the budding types and the forerunners of the other types mentioned above.

Amongst the rare cases of escaping buds that of *Myriothela cocksii*, Vigurs [British specimens of which have been frequently misnamed *Myriothela phrygia* (Fabricius)] stands somewhat apart. In this species the buds are spherical masses attached to the parent by a thick stalk, and appear to reach a miniature adult stage before they are set free. Hardy, however, assures us that "all connection with the body of the parent is lost at a very early period, almost before the bud has reformed its ectoderm and endoderm and enteric cavity. It remains attached to the perisarc, however, by a sucker-like arrangement at the aboral pole until it is fully formed" (Hardy, 1891, p. 513 and pl. xxxvi, fig. 13). This might almost be regarded as a transitional stage, which although in fact a free bud, retains the aspect of an attached miniature adult.

Moerisia furnishes a more definite example of exparental development. The buds of this peculiar form are oval and are attached by short peduncles to the parent body, usually in the proximal region of the hydranth. As indicated above they "occasionally develop one or two tentacles" before they are set free, and some may therefore be regarded as attached miniature adults, but the majority of the buds "become completely detached from the parent body" before they begin to assume polyp structure (Boulenger, 1908, p. 363). Rare as such cases are, *Moerisia* is by no means a unique example.

Halcremita cumulans, Schaudinn, seems to depend entirely upon liberated planula-like buds for its dissemination and multiplication, for no trace of sex-cells has been discovered (Schaudinn, 1894, p. 227). The buds, which at the time of liberation are much elongated and planula-like, arise sometimes just beneath the tentacle zone and sometimes near the base of the hydranth and up to six may be found on a polyp at one time. After being set free they develop a mouth and creep upon the bottom, simple two-layered sacculae, which retain their simplicity for some 1½ months before the tentacles of the adult make their appearance. Some have been observed to develop buds of their own while yet in the saccula stage.

Much resemblance exists between the general structure and bud-formation of *Halcremita* and that of the larva of *Gonionemus murbachii* Mayer, described by Perkins (1903). The unusual stumpy conical shape of *Halcremita* is duplicated in the *Gonionemus* hydroid, and in both there are four tentacles set crosswise in a single whorl. Both lack sex-cells and both reproduce by planula-like buds. *Haleremita* is unusual amongst hydroids is possessing only one type of nematocyst, but Perkins' description (p. 786) indicates that only one form, long and bean-shaped, is present in *Gonionemus*. It is possible that these resemblances points to the true relationship of the problematical *Haleremita*: that it is the metagenetic hydroid phase of a hydrozoon medusa, a larva which in due course will assume medusa form. It is interesting to find some confirmation of this view in the simplicity of structure (to be expected in a larval form) which has led to the relegation of *Haleremita* to the primitive family Hydridae; yet the bud-formation in the two is by no means identical.

Perkins describes some interesting features in the development of the *Gonionemus* hydroid buds. The buds, which occurred singly on the hydroids, arise about halfway between the base of the polyp and the ring of tentacles. During their early growth the endoderm is solid, and in this condition becomes isolated from the endoderm of the polyp by the gradual constriction of the ectoderm at the junction of the two. Finally the bud comes to be attached simply by a long thin neck of transparent ectodermal protoplasm. The release of the bud, in the only case followed throughout its complete development, was accomplished by the gradual stretching and final rupture of the ectodermal neck. The released bud settled down upon its former *free* or *distal* end, and at the other pole, formerly attached, a mouth and tentacles developed. This bud became attached near the parent polyp, but in most cases an escaped bud was discovered after a few days some distance from the parental form During the interval "it seems probable that it is a creeping unciliated form, although my first conjecture that it was a ciliated planula has not been proved erroneous" (Perkins, 1903, p. 771). A general idea of the developmental period of such planula-like buds can be gathered from Perkins' observations. The development of a bud from its first appearance as a simple knob to the completion of the formation of the coelenteron and the appearance of tentacles, lasted from ten to fourteen days, distributed as follows:—" (*a*) the first period including as far as the detachment of the bud, 5 days; (*b*) motile form, 2 to 5 days; (*c*) from attachment to appearance of tentacles, 3 to 5 days (Perkins *loc. cit.*). Schaudinn found that the development of *Haleremita* buds, up to the point of escape from the parent. varied from 5 hours to 6 days (Schaudinn 1894, p. 230).

In all the cases above mentioned, as well as in that of our Indian form, the buds arose equally from ectodermal and endodermal elements, confirming the observations of Braem (1894) and contrasting with Lang's (1892) description of the purely ectodermal origin of Hydroid buds.

So far as can be determined from my examination of the comparatively few buds available in the Indian species, they agree most closely with those of *Haleremita cumulans*. In both species, in contradistinction to *Gonionemus*, the bud possesses a hollow structure from the beginning, and the internal cavity remains in connection with the coelenteron till the time of escape.

The special resemblances to *Gonionemus* buds are slight and of little account. In both the buds seldom occur more than one at a time on a polyp; and in both it seems that the usual polarity of hydroid buds is reversed, and that the free end becomes the area of attachment, and the attached end the oral and tentacle-bearing area. It may possibly be that this remarkable inversion of the general mode of hydroid development is not a regular habit, but simply emphasises that in hydroid buds there exists an indeterminate polarity ready to be determined by external physical conditions. Such has been shown experimentally to characterise the adult stems of forms like *Tubularia* and *Corymorpha*, or in closer analogy exists in the larva of *Corymorpha*. It is likely that here as in these larvae "external factors such as contact and possibly gravity determine the kind of structure (*e.g.* hydranth or holdfast) which will ultimately appear in connection with the area of differentiation. That is they determine the polarity of the adult." (Torrey, 1907, p. 292).

ANALOGUES OF THE BASAL BULB.

The normal organs of attachment of the vegetative stocks of the Hydrozoa fall into two broad classes: (1) those in which the base of the hydranth is simply modified into a fleshy disc or cylinder, occasionally naked, more often covered by a mucous secretion in which foreign debris becomes embedded, or rarely enclosed in a membranous film of chitin; (2) those in which a more specialized structure is apparent, the attachment being due to well-defined root-like strands of coenosarc, enclosed in a distinct coat of perisarc (the stolon or hydrorhiza) and forming simple threads, or branched "roots," or anastomosed networks, or even thick skeletal layers (as in *Hydractinia*).

It seems to me that these two types of hydroid attachment are homologous, that the simple fleshy attachment was the direct forerunner of the hydrorhiza, and may be regarded as a primitive characteristic in those forms in which it occurs. In known species of Hydroids it is possible to trace the steps by which the simple basal disc became branched and split to form a root-like organ, and by which the final complexity of the hydrorhiza was built up. A process parallel to that suggested by a survey of the attachment organs of adult hydroids seems to be followed in summary during the development of certain individuals. One need only point to the early life-history of the colonial form *Eudendrium ramosum*, after the planula has relinquished its free state and settled down, to illustrate the development of a facsimile of the basal disc into a complex hydrorhiza (see Allman, 1871-72, pl. xiii, figs. 12-16 and 2).

It would be out of place, however, to develop a thesis of the evolution of the hydrorhiza in this paper, and I shall merely indicate the forms which seem to stand most closely related to our Indian species as regards their mode of attachment.

The simplest definite attachment is that of the larval hydroid of *Gonionemus*. Here there simply occurs at that surface of the planula-like bud which comes in contact with the substratum an increase in the thickness of the cells, so that the ectoderm of the base becomes a columnar epithelium. There seems to be no secretion of masses of mucus, but at any rate "it has now secured a firm hold upon the bottom, being so closely applied that it is quite hard to dislodge it" (Perkins, 1903, p. 771). In the above sentence Perkins would seem to hint that the adhesion is physical. In *Halcremita* the attaching area has differentiated a stage further: for while it still consists of a simple layer of special elongated epithelium, there are associated with it many gland-cells which exude the secretion by means of which the polyp is attached to the substratum (Schaudinn, 1894, p. 228).

A clear advance is marked by the condition of *Hydra* and *Protohydra*, for here a first organ of attachment, as distinct from a mere differentiation of ectodermal cells, is apparent. Nevertheless this organ (the foot, pedal disc, disque pedieux, Fusscheibe, Fussplatte) retains the condition of elongate epithelium, with associated secretory cells the mucus of which acts as an accessory holdfast, but it is capable of grasping a firm surface and relinquishing its hold at will.

I regard as closely akin to the foot of *Hydra* in differentiation the "sucker-like" adhesive organs of the miniature adults of *Myriothela cocksii* ("*phrygia*"), mentioned by Hardy (1891, p. 513) as remaining attached to the surface of the parent during development.

Greater structural complexity is shown by the problematical "Basalscheibe" of the miniature-adult buds of *Tiarella singularis* minutely described by Schulze (1876, p. 412 and Taf. xxx, fig. 2). In shape and minute structure this curious organ bears a striking resemblance to the naked basal disc observed in one young individual of *Annulella* with five tentacles, to the basal bulb of young specimens (see plate xxx, fig. 1) or to a section of an adult bulb (pl. xxxa, fig. 9). On account of these resemblances I have no hesitation in discarding Hartlaub's suggestion that it may be "ein fur die pelagische Lebensweise wichtiges Organ" (Hartlaub, 1903, p. 34), and regarding it is an attachment organ developed in preparation for the settling down of the pelagic phase. In exactly the same category may be placed the basal discs of *Margelopsis stylostoma*,[1] *Margelopsis gibbesi*, and *Margelopsis haeckelii* discussed by Hartlaub (1903, p. 34).

Subsequent to fixation the flattened disc-like "Fussplatte" of the adult *Tiarella*, with its coats of both dark and amorphous perisarc, continuous with those of the hydrocaulus, seems to have degenerated from the larval state as regards cellular distinctiveness.

---

[1] A designation which since it indicates simply a young phase of *Tiarella singularis*, must lapse (see Bedot 1911, p. 211).

A much greater advance in the differentiation of the simple basal attachment is exhibited in *Myriothela cocksii* Vigurs, well described by Allman (1876) under the name of *Myriothela phrygia*. In the long-tentacled, free-swimming larval stage, a few days old, an increase in the thickness of the ectoderm at the aboral extremity is noticeable (Allman, 1876, p. 567). This appears to be due to the formation of columnar epithelium (see Allman's figure 15, pl. 56). In any case a sucker-like pad is formed by which the larva attaches itself to the substratum (p. 565). At this early stage of fixation the aboral "sucker" is similar in appearance, structure and function to the pedal discs which have been mentioned above, but new developments soon set in. "The proximal extremity of the animal becomes bent at right angles to the rest of the body so as to form a sort of horizontal, stolon-like foot from which small fleshy processes with sucker-like extremities, and having a considerable resemblance to claspers, are emitted. The function of these processes, however, is very different from that of claspers: they serve to attach the animal permanently to some solid support, to which they fix themselves by their extremities. Along with the stolon-like foot they become clothed with perisarc, and the actinula has thus acquired all the essential characters of the adult trophosome" (Allman, 1876, p. 565).

There is some general resemblance here to the final result in *Annulella*, for although the perisarc-covered "foot" of the adult in *Myriothela cocksii* is a direct development of the larval basal disc, it is almost certain that the original basal bulb of any individual of *Annulella* follows the same course; but the absence of a narrow neck between the stolon-like foot and the hydranth of *Myriothela*, as well as the presence of specialized sucker-like processes, mark it as very distinct from the basal bulb of *Annulella*.

Almost as distinct is the perisarc-covered basal attachment of the Tubularid, *Corymorpha*; for not only does it bear many anchoring processes, but the perisarc is really a portion of that which at one time enveloped the whole hydranth and which by a process of recession became later confined to the lower section of the stem (see Torrey, 1907, p. 279).

As regards the development of its basal bulb directly from the proximal portion of the adult hydranth, and of the special development upon the basal bulb of a highly differentiated perisarc, *Annulella* stands alone. It seems to me that its closest affinity in respect of this organ may be with *Tiarella*, beyond the stage of which, however, it has made considerable advance in specialization. It is well to remember, however, that in its phylogenetic origin the basal bulb is undoubtedly a development of the much simpler naked basal discs characteristic of a primitive group of unattached hydroids.

BASAL TRANSVERSE FISSION.

Transverse fission as a means of multiplication in adult hydroids is not unusual, and varies from the separating of a minute terminal section of branch or stolon, as in several Campanularians and Plumularians (Billard, 1904, p. 41 *et seq.*) to vital processes such as the exaggerated "decapitation" of *Moerisia* or the median division of *Hydra* or *Protohydra*.

So far as I am aware, transverse fission in a determinate region of the base of an adult individual, is a normal mode of multiplication in only one hydroid species other than *Annulella gemmata*. Even that solitary case differs from *Annulella*: for in *Hypolytus murbachii* the fission takes place near the proximal end of a distinct hydrocaulus; it proceeds gradually by means of constriction, but without any disappearance of ectoderm (so far as one can judge) so that there are set free successive small naked planula-like segments which, after more or less limited wandering, settle down and develop directly into new hydranths (Murbach, 1899). In *Hypolytus* a wandering "blastolyte" escapes from a free adult; in *Annulella* a wandering adult escapes from an attached basal section.

In this respect *Annulella* comes very near to the hypothetical form postulated by Murbach as a precursor of *Hypolytus* (Murbach, 1899, p. 353); but to me there appears to be no close relationship between the two forms.

The phenomena of transverse fission in *Annulella* naturally bears a general resemblance to other well-marked cases such as the strobilisation of *Moerisia* (Boulenger, 1908, p. 364) or the division of *Protohydra* (Chun, 1894, p. 217). But the transverse fission of *Annulella* stands by itself as regards the structural changes involved (such as the disappearance of ectoderm at the neck, paralleled only in the sporosacs of *Dicoryne*), and as regards the final results, since here a segment specially modified with a view to fission remains attached, while the hydranth which gave it origin escapes.

SYSTEMATIC POSITION OF *ANNULELLA GEMMATA*.

The majority of the outstanding features of *Annulella gemmata* are primitive in character, a few seem to be adaptive. Among the latter may be reckoned the annular arrangement of large cells upon the tentacles (see p. 558), the great length of the tentacles themselves, and the adoption of basal transverse fission. All of these bear upon the free-living stage, the last as the means of attaining freedom, the former as adjuncts to a (supposed) pelagic existence.

The primitive characters include the normal adoption of various types of vegetative budding; but even these are of simple nature. Thus the setting free of minute, non-tentacled, planula-like buds must probably have preceded in evolutionary development even the liberation of buds at a miniature adult-stage, as occurs in *Hydra* and occasionally in *Moerisia*, both of which are included in the family Hydridae. Further, the naked basal disc observed in one young specimen of *Annulella* appears to be homologous with the similar structure common in adult members of the Hydridae and in young specimens of *Tiarella*, and the basal bulb in its first development phylogenetically and ontogenetically, may be taken as a highly specialized form of the basal disc or " Fussplatte."

Finally, the multiserial endoderm of the tentacles finds a close analogy in the similar (but hollow) endoderm of the Hydridae (*Hydra* and *Mocrisia*).

Perhaps one ought to add that if faith be placed in Haeckel's hypothesis of the origin of a capitate tentacle as the thrusting out on a stalk of a cluster of nematocysts, then the capitation of the tentacles may also be placed amongst the primitive characters.

A survey of the systematic distribution of the distinctive characters of *Annulella* shows that they are confined to four families of the Hydroidae—Hydridae, Corynidae, Pennaridae and Tubulariidae: but that they preponderate towards the more primitive end of the series—the Hydridae and Corynidae. Systematists have long regarded the tentacles as a primary basis of distinction, special stress being laid upon their capitate or filiform condition and their distribution upon the hydranth body. This basis being adopted, the capitate and scattered tentacles of *Annulella* place it definitely in the family Corynidae, but there are clear affinities in the multiserial endoderm of the tentacles, in the simple budding, and in the basal disc and bulb to members of the family Hydridae.

In the Corynidae, where, agreeing with Mme. Motz-Kossowska (1905, p. 45), I would place *Tiarella*, in preference to the position with the Pennaridae assigned to it by Schulze, there is no genus closely comparable to *Annulella*. But it bears some relationship to *Tiarella* from which it differs most markedly in possessing scattered tentacles, and beyond which it has advanced in the specialization of its basal bulb and of the nematocyst rings on its tentacles. In respect of the distribution of tentacles and of the general absence of perisarc except on the basal extremity, *Annulella* approaches *Myriothela*, and, since no more satisfactory alternative presents itself, I rank it with this genus in the sub-family Myriothelinae.

In these days of many tentative classifications misunderstanding may be avoided if I state that I consider the family Corynidae to contain those Gymnoblastic Hydroids in which the tentacles are all capitate and are either scattered or distributed in several whorls; and that in its sub-family Myriothelinae I would place such Corynids as possess scattered tentacles, are solitary, and lack a supporting skeleton of perisarc.

GENERIC AND SPECIFIC DIAGNOSES OF *ANNULELLA GEMMATA.*

Annulella,[1] nov. gen.

GENERIC CHARACTERS.

Trophōsome.—Ployps solitary and naked, with conical proboscis, and long, scattered, capitate tentacles bearing nematocyst batteries arranged in many rings and furnished with solid multiserial endoderm. During their fixed stage the polyps are attached by an adherent base, connected to them by a narrow neck and enclosed in perisarc. Multiplication by vegetative reproduction is the rule.

Gonosome.—? Gonophores producing free medusae.

Annulella gemmatta,[2] nov. sp.

SPECIFIC CHARACTERS.

Minute solitary polyps, 0·15 mm. to 1·0 mm. in height, bearing from 4 to 12 scattered capitate tentacles with nematocyst rings (nodes) along their whole length, and delicate solid endoderm. Tentacles and polyp-body are furnished with two types of nematocysts (macrocnides and microcnides). The polyp is alternately fixed and free, escaping from its basal bulb by rupture of the connecting neck, and again developing a new basal bulb by a modification of its proximal end.

Reproduction is normally asexual, by means of buds set free in a planula-like stage by means of the detached basal bulb, and possibly by means of longitudinal fission. The type of sexual phase is not known with certainty.

Locality.—A brackish pond, Port Canning, Lower Bengal, India.

Type Specimens.—In the collections of the Indian Museum.

LIST OF WORKS REFERRED TO IN TEXT.

Allman, J. G., 1871-72. A Monograph of the Gymnoblastic or Tubularian Hydroids. Ray Society, London.

Allman, J. G., 1876. "On the Structure and Development of *Myriothela.*" *Phil. Trans. Royal Soc. London*, vol. 165, p. 549.

Ashworth, J. H. and Ritchie, J., 1915. "The Morphology and Development of the Free-swimming Sporosacs of the Hydroid Genus *Dicoryne* (including *Heterocordyle*)." *Trans. Roy. Soc. Edinburgh*, vol. 51 (in press).

Bedot, M., 1911. "Notes sur les Hydroïdes de Roscoff." *Arch. Zool. exp. et gén.*, Ser. 5, T. 6, p. 201.

---

[1] Feminine diminutive from Lat. *annulus*, a ring, signifying the ringed tentacles.

[2] Lat. *gemmatus*- budded.

Billard, A., 1904. "Contribution à l'étude des Hydroïdes." *Ann. Sci. Nat. Zool.*, Ser. 8, T. 20, p. 1.

Boulenger, C. L., 1908. "On *Moerisia lyonsi*, a new Hydromedusan from Lake Qurun." *Quart. Journ. Micr. Sci.*, n.s., vol. 52, p. 357.

Braem, F., 1894. "Über die Knospung bei mehrschichtigen Thieren, ins besondere bei Hydroiden." *Biol. Centralbl.*, Bd. 14, p. 140.

Chun, C., 1894. "Coelenterata" in Bronn's *Klassen u. Ordnungen des Thier Reichs*, Bd. 2, Abt. 2, Lief. 9 u. 10.

Chun, C., 1897. *op. cit.*, Lief. 15-17.

Dendy, A., 1902. "On a free-swimming Hydroid, *Pelagohydra mirabilis*, n. gen., n. sp." *Quart. Journ. Micr. Sci.*, n.s., vol. 46, p. 1.

Derzhavin, A., 1912. "*Caspionema pallasi*, eine Meduse des Kaspischen Meeres." *Zool. Anz.*, Bd. 39, p. 390.

Hadži, J., 1911. "Über die Nesselzellverhältnisse bei den Hydromedusen." *Zool. Anz.*, Bd. 37, p. 471.

Hardy, W. B., 1891. "On some points in the Histology and Morphology of *Myriothela phrygia*." *Quart. Journ. Micr. Sci.*, n. s., vol. 32, p. 505.

Hartlaub, Cl., 1903. [Summary remarks on free-swimming Hydroids included in a review of Dendy's 1902 paper cited above]. *Zool. Zentralbl.*, Jahrg. 10, p. 27.

Lang, A., 1892. "Über die Knospung bei Hydra und einigen Hydroidpolypen." *Zeits. f. wiss. Zool.*, Bd. 54, p. 365.

Motz-Kossowska, S., 1905. "Contribution à la connaissance des Hydraires de la Méditerranée occidentale." *Arch. Zool. exp. et gén.*, Ser. 4, T. 3, p. 39.

Murbach, L., 1899. "Hydroids from Wood's Holl, Mass." *Quart. Journ. Micr. Sci.*, n. s., vol. 42, p. 341.

Perkins, H. F., 1903. "The Development of *Gonionema murbachii*." *Proc. Acad. Nat. Sci. Philadelphia*, vol. 54, p. 750.

Ritchie, J., see under Ashworth.

Russell, E. S., 1906. "On *Trichorhiza*, a new Hydroid Genus." *Proc. Zool. Soc. London*, 1906, vol. 1, p. 99.

Schaudinn, F., 1894. "Über *Haleremita cumulans*, n. g., n. sp., einen neuen marinen Hydroidpolypen." *S. B. Ges. Naturfors. Fr. Berlin*, Jahr. 1894, p. 226.

Schneider, K. C., 1890. "Histologie von *Hydra fusca* mit besonderer Berücksichtung des Nervensystems der Hydropolypen." *Arch. f. mikr. Anat.*, Bd. 35, p. 321.

Schulze, F. E., 1876. "*Tiarella singularis*, ein neuer Hydroidpolyp." *Zeits. wiss. Zool.*, Bd. 27, p. 403.

Torrey, H. B., 1907. "Biological Studies on *Corymorpha*, II. The Development of *C. palma* from the Egg." *Univ. California Public., Zool.*, vol. 3, p. 253.

Warren, E., 1908. "On a Collection of Hydroids mostly from the Natal Coast." *Ann. Natal Govern. Mus.*, vol. 1, p. 269.

EXPLANATION OF PLATE XXX.

Annulella gemmata, nov. gen. et sp.

FIG. 1.—Young sessile polyp shortly after fixation, with four tentacles, and thin perisarc covering the basal disc (Fussplatte), × 75. *b.b.* basal bulb; *deb.* coat of mucus and debris.

,, 2.—Adult polyp with twelve tentacles, lateral buds (*bud* 1 & 2, the latter possibly a young gonophore), and two basal bulbs; *b.b.* 1 & 2, the former terminal. In both bulbs the internal coenosarc-sac and external chitinous investment are shown, × 55.

,, 3.—Adult polyp with two hypostomes, *p.* 1 & 2, possibly in process of longitudinal fission, and two basal bulbs; *b.b.* 1 & 2, the former terminal, showing clearly the narrow connecting neck, and the perisarc investment, × 55.

,, 4.—Median longitudinal section of bud attached to wall of polyp, × 230. *cav.* large central cavity in communication with coelenteron; *cut.* cuticle; *mes.* mesogloea; *ect.* ectoderm; *end.* endoderm.

,, 5.—Optical section of bud after its release from polyp, and before it has settled down, × 230. *cav.* central cavity, now much constricted by increase of endoderm cells; *cut.* cuticle; *mes.* mesogloea; *end.* endoderm; *ect.* ectoderm; the pale area at the left pole (the end formerly attached) shows the position of the opening which probably persists as the mouth of the polyp.

,, 6.—Proximal end of an adult polyp, showing the origin of a new basal bulb, × 75. *h.* lower end of hydranth body; *e.h.* ectoderm of hydranth; *mes.* mesogloea; *n.* 2, constriction beginning to mark off neck of new basal bulb; *b.b.* basal bulb in process of differentiation; *e.b.b.* ectoderm of basal bulb; *m.* debrisladen mucus; *n.* 1, ruptured tissues indicating the broken neck of an earlier basal bulb, whence the polyp has escaped.

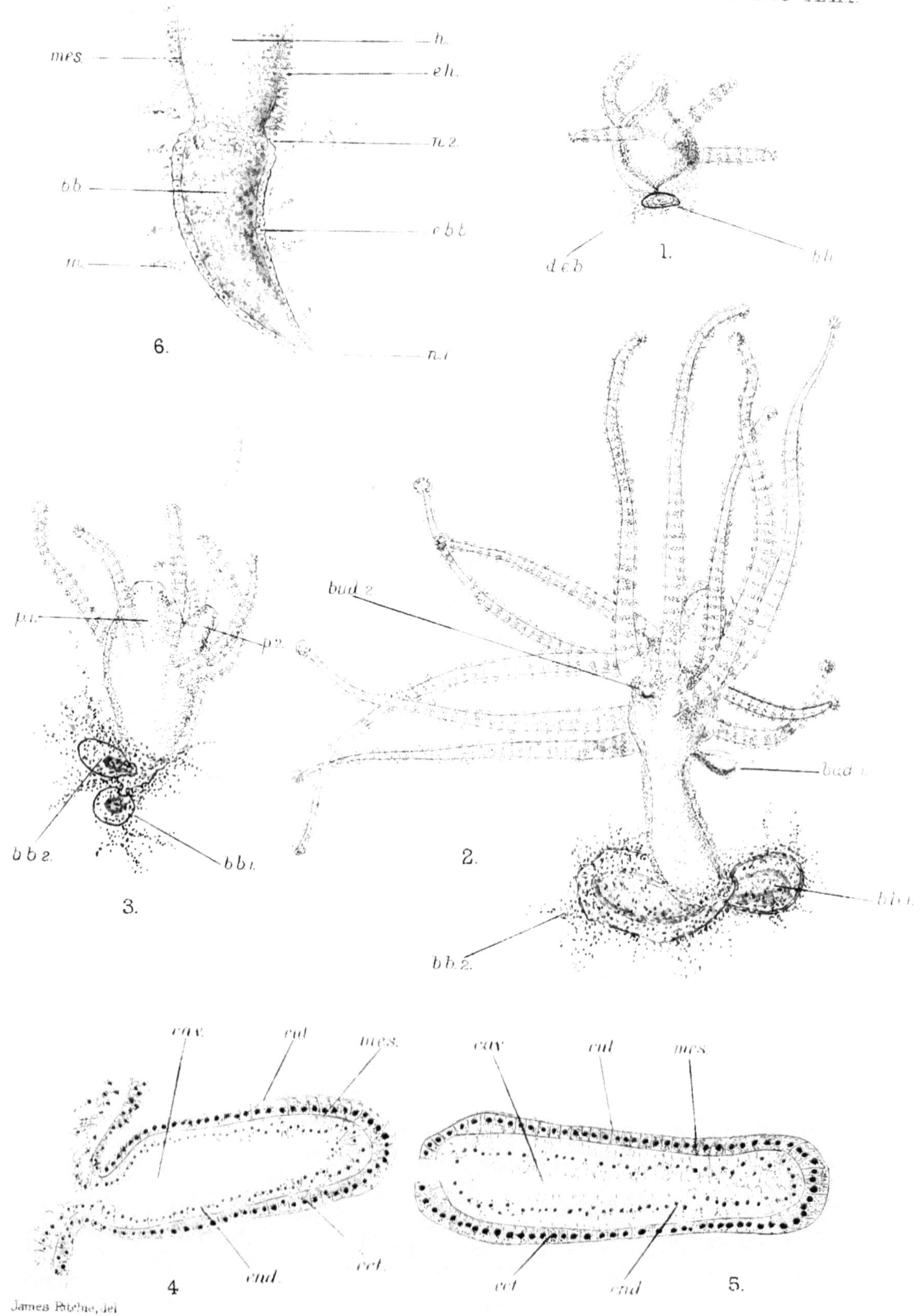

James Ritchie, del.

Bemrose Collo, Derby

EXPLANATION OF PLATE XXXa.

Annulella gemmata, nov. gen. et sp.

FIG. 7.—Portion from near tip of much extended tentacle, showing nodal and internodal areas, and arrangement of nematocysts in nodal cells, × 320.

,, 8.—Median longitudinal section of contracted tentacle, × 770. *n.c.* central nematocyst (macrocnide) of nodal cell; *n p.* peripheral nematocyst (microcnide) of nodal cell; *ect.* ectoderm; *end.* endoderm; *mes.* mesogloea; *cn.* cnidocil.

,, 9.—Median longitudinal section of lower portion of polyp and basal bulb, × 540. *ect.* ectoderm (absent at neck of basal bulb); *mes.* mesogloea (note great thickening within neck of basal bulb); *end.* endoderm; *coel.* coelenteron; *gel.* hyaline, gelatinous secretion (it is difficult to distinguish the relationships of this secretion to the cellular layers, in places it appears to be continuous with the walls of the ectoderm cells, and in one area it seemed to come in contact with the mesogloea); *per.* perisarc of basal bulb; *a.* unconsolidated area of debris-laden secretion into which the firm perisarc merges.

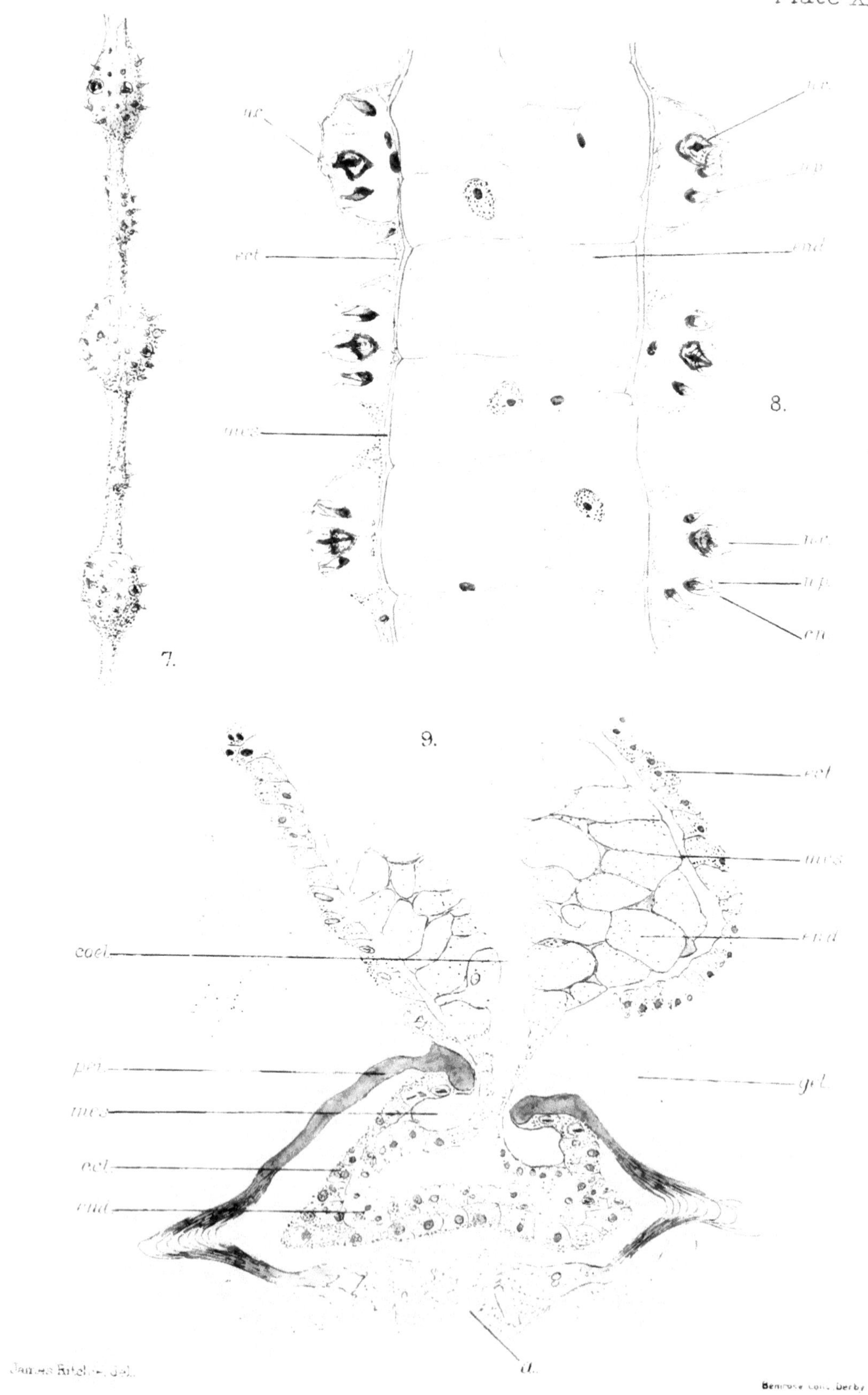

James Ritchie, del.

Bemrose coll. Derby

Lightning Source UK Ltd.
Milton Keynes UK
UKHW031433220520
363714UK00003B/299